Electron Crystallography

NATO ASI Series

Advanced Science Institutes Series

A Series presenting the results of activities sponsored by the NATO Science Committee, which aims at the dissemination of advanced scientific and technological knowledge, with a view to strengthening links between scientific communities.

The Series is published by an international board of publishers in conjunction with the NATO Scientific Affairs Division

A Life Sciences **B Physics**	Plenum Publishing Corporation London and New York
C Mathematical and Physical Sciences **D Behavioural and Social Sciences** **E Applied Sciences**	Kluwer Academic Publishers Dordrecht, Boston and London
F Computer and Systems Sciences **G Ecological Sciences** **H Cell Biology** **I Global Environmental Change**	Springer-Verlag Berlin, Heidelberg, New York, London, Paris and Tokyo

PARTNERSHIP SUB-SERIES

1. Disarmament Technologies	Kluwer Academic Publishers
2. Environment	Springer-Verlag / Kluwer Academic Publishers
3. High Technology	Kluwer Academic Publishers
4. Science and Technology Policy	Kluwer Academic Publishers
5. Computer Networking	Kluwer Academic Publishers

The Partnership Sub-Series incorporates activities undertaken in collaboration with NATO's Cooperation Partners, the countries of the CIS and Central and Eastern Europe, in Priority Areas of concern to those countries.

NATO-PCO-DATA BASE

The electronic index to the NATO ASI Series provides full bibliographical references (with keywords and/or abstracts) to more than 50000 contributions from international scientists published in all sections of the NATO ASI Series.
Access to the NATO-PCO-DATA BASE is possible in two ways:

– via online FILE 128 (NATO-PCO-DATA BASE) hosted by ESRIN, Via Galileo Galilei, I-00044 Frascati, Italy.

– via CD-ROM "NATO-PCO-DATA BASE" with user-friendly retrieval software in English, French and German (© WTV GmbH and DATAWARE Technologies Inc. 1989).

The CD-ROM can be ordered through any member of the Board of Publishers or through NATO-PCO, Overijse, Belgium.

Series E: Applied Sciences - Vol. 347

Electron Crystallography

edited by

Douglas L. Dorset
Houptman-Woodward Institute, Buffalo, NY, U.S.A.

Sven Hovmöller
Stockholm University, Stockholm, Sweden

and

Xiaodong Zou
Stockholm University, Stockholm, Sweden

Springer-Science+Business Media, B.V.

Proceedings of the NATO Advanced Study Institute on
Electron Crystallography
Erice, Sicily
22 May – 2 June, 1997

Library of Congress Cataloging-in-Publication Data

DOI 10.1007/978-94-015-8971-0

Printed on acid-free paper

TABLE OF CONTENTS

4. Structure Refinement

5. Applications

Extended Abstracts

Obituary - BORIS KONSTANTINOVICH VAINSHTEIN (1921-1996)

The crystallographic community was saddened by the death on 28 October 1996 of Boris Konstantinovich Vainshtein, aged 75, one of the great diffraction physicists of the world and a pioneer in the field of electron crystallography. He was to be a keynote speaker at this NATO ASI and this Proceedings is dedicated to his memory.

Born in Moscow, Boris Vainshtein was educated at Moscow State University and at the Moscow Institute of Steel. In 1945, he began his graduate work at the Institute of Crystallography of the USSR Academy of Sciences, the site of much of his life's activity, and here he later became director. This work led to defense of candidate's (1950) and doctoral (1955) dissertations in physics and mathematics.

Fundamental work in the development of electronography, as electron crystallography was originally termed in the Soviet Union, began during his graduate work with Z. G. Pinsker. During his work in this field, Vainshtein developed the geometrical theory for interpreting electron diffraction patterns, including the texture patterns that were the source of many of the crystal structure analyses from his laboratory. He also calculated the first electron scattering factors with J. Ibers and was the first to apply Fourier methods for calculation of potential distributions in the unit cell. The fruit of these labors was expressed by representative structure analyses, primarily to exploit the favorable ratio of light to heavy atom form factors for detection of the former to an accuracy better than then possible in x-ray crystallographic determinations.

A review of his pioneering work was published as **Strukturnaya Elektronografiya** (or Structure Analysis by Electron Diffraction, the title of the English translation by E. Feigl and J. A.

Spink, published by Pergamon Press in 1964), a classic that certainly deserves to be reprinted. Although deviations from kinematical scattering were treated by a two-beam approximation by Vainshtein's group, for reasons justified in a 1967 review by J. M. Cowley, there was considerable criticism of this approximation. In fact, the pessimism engendered by the formulation of a more accurate n-beam dynamical theory led to the impression for many crystallographers that the early structural determinations were, in fact, spurious, and that nothing could be done with such data, particularly for *ab initio* determinations. (Unfortunately, the use of contemporary x-ray structures by his group, to start refinements against electron diffraction data, only intensified this criticism. However, before the appearance of direct phasing methods, no structure determination was easy unless it was obvious from an overlap of point and space group symmetries!) This pessimism was quieted when it was shown entirely objectively by direct phasing methods that the structures found by Vainshtein could be obtained directly from his experimental electron diffraction intensity data. He was correct all the time!

Electron crystallography was not the only area where Vainshtein made his mark. In 1959, he was made head of the Laboratory of Protein Structure in the Institute of Crystallography. In the field of fibrous polymers, his book, **Difraktsiya Rentgenovikh Luchei na Tsepnikh Molekulakh** (Diffraction of X-rays by Chain Molecules in English translation published by Elsevier in 1962) is still a classic work. In addition, he had collaborated on numerous globular protein structure determinations. His work with N. A. Kiselev in electron microscopy also permitted structural results from large proteins and viruses to be obtained, *via* early applications of image analysis and averaging techniques. With I. G. Chistyakov and others, Vainshtein began the Soviet school that studied the symmetry and structure of liquid crystals.

In 1962, he was made director of the Institute of Crystallography, overseeing and participating in a wide diversity of research topics. The extreme breadth of his intellectual view is evidenced by his capabilities as editor-in-chief of the journal **Kristallografiya** (currently available in English as Crystallography Reports, published by Maik Nauka), a crystallographic journal of much broader scope than any rival in the world. With other colleagues he was responsible for writing the four volume treatise, **Sovremennaia Kristallografiya** (English translation as Modern Crystallography, now in its second printing by Springer-Verlag).

Of the many prizes earned by Boris Vainshtein, the one of greatest relevance to crystallographers is the Ewald Prize, given at the Bordeaux IUCr Congress and General Assembly in 1990. He has served as a member of the Executive Committee of the IUCr and was its Vice President. In his own country he received many awards and distinctions. He was also on the editorial board of a number of journals and was a member of many foreign scientific academies and held a number of honorary degrees.

All this betrays the activity of an enthusiastic, keen intellect. To colleagues he was always approachable, and, indeed, always promoted good interactions between his Institute and scientific laboratories world-wide, even when international politics may have been unfavorable. I personally remember Boris Vainshtein playing chess one summer evening with two young Spanish graduate students, during a meeting held near Drottingholm Castle in Ekerö, Sweden. Particularly charming was the quiet smile on his face when he recognized that he had achieved the check-mate. One is told that he also had a great love for cats, perhaps responsible for their being the subject of many symmetry drawings in his Modern Crystallography series. (As one who appreciates felines, I would take this to mean that Boris Vainshtein also had a great respect for individuality and freedom.)

During the ASI, his close colleague and friend, Boris Zvyagin, provided us with many insights into the personal character of Boris Konstantinovich Vainshtein. We have been certainly enriched by his life for pioneering a field in which we now labor with pleasure. We mourn his passing, as will many other crystallographers, because, he still had many new things to tell us.

PREFACE - ELECTRON CRYSTALLOGRAPHY

Electron crystallography is the quantitative use of electron scattering information to determine crystal structures. As a diffraction technique, it was pioneered most prominently in the Soviet Union through the efforts of Boris K. Vainshtein and his co-workers, originally in the laboratory of Z. G. Pinsker. As a microscopic technique it was pioneered e. g. by Aaron Klug and his co-workers, largely for the structural elucidation of biomolecular arrays. An important benchmark for the increased resolution and sophistication in the macromolecular area was given in 1975 by R. Henderson and P. N. T. Unwin via the 7 Å- resolution structure of bacteriorhodopsin in three-dimensions.

Currently, electron crystallography, via image-derived methods for crystallographic phase determination, stands as an important technique for the characterization of protein- and other macromolecular arrays in microcrystals or two-dimensional crystals, a particularly important application being the integral membrane proteins. In a strict electron diffraction approach, electron crystallography has long been a reliable technique for the determination of linear polymer structures, based on single crystal data from chain-folded lamellae, often finding solutions to the phase problem by structural model searches in conformation space. More recently, many new applications of image averaging techniques have been in evidence (e. g. two of us, Hovmöller & Zou), particularly to inorganic materials areas, taking care that the data collected from experimental images or electron diffraction patterns are adequately near the single scattering approximation. The application of direct phase determination, of the kind originated by H. Hauptman and J. Karle in X-ray crystallography, has also flourished, after the initial trial by one of us (Dorset) with experimental electron diffraction data in 1975. Such methods for phase determination have become more and more sophisticated and have involved hybrid studies for resolution enhancement, via phase basis sets enriched with information obtained from the Fourier transform of an average image, a unique possibility unavailable to the x-ray crystallographer. Increasingly, methods for structure refinement are also being developed, sometimes including multiple scattering corrections to the intensity perturbations. Even the existence of dynamical scattering, once the bane of electron crystallographers, has become a boon since it can permit an independent determination of space group symmetry and even crystallographic phases.

In this proceedings of the NATO Advanced Study Institute on Electron Crystallography, held in Erice, Sicily from 22 May to 2 June 1997, one will find many of the current developments in the field, in the form of lectures given to the participants. Also, because of their overall high quality, a number of short papers by participants describing posters given during the institute are included in this volume. As was the case in the 1990 Advanced Research Workshop, Electron Crystallography of Organic Molecules, organized by J. R. Fryer and D. L. Dorset, this Advanced Study Institute was held concurrently with a related ASI on Direct Methods for Solving Macromolecular Structures, organized by S. Fortier. This permittied the use of joint lectures to the combined group of participants, thus fostering thematic and conceptual connections not easily realized in typical scientific conferences.

We are grateful for a number of sponsors for this ASI. First and foremost is the support from NATO (ASI 960401). The European Community (Euroconference and DG XII, Office for Eastern European Countries) gave generous support as did the International Union of Crystallography, the Italian Research Council Committee for Mineralogy, the US National Science Foundation and Astra-Sweden. We are extremely grateful to Prof. Paola Spadon for her constant aid toward making this ASI possible, including her creation of a study book of lecture notes and abstracts from submitted materials. Special thanks are also given to Prof. Lodovico Riva di Sanseverino, not only for his melodious singing voice, but also for his organizational skills, constant attention to all details, valuable advice, and for his competent oversight of finances for the ASI.

We hope that the readers of this book will enjoy their encounter with so many new developments in theory, methods, and applications in the fascinating, reborn field of electron crystallography. Hopefully it will induce some to become participating researchers, exploiting this rapidly developing technique to investigate their own structural problems.

D. L. Dorset, S. Hovmöller, X. D. Zou
Buffalo, Stockholm, August 1997

LIST OF (ATTENDING) CONTRIBUTORS TO THIS VOLUME

D. R. Acosta
Instituto de Fisica
UNAM, A. P. 20-364
01000 Mexico D. F., MEXICO

S. P Ahrenkiel
National Renewable Energy Laboratory
1617 Cole Blvd.
Golden, CO 80401, USA

G. Boyce
Department of Chemistry
University of Glasgow
Glasgow G12 8QQ
Scotland, UK

G. L. Cascarano
Dip. Geomineralogico
Universita di Bari
Via Orabona 4
I-70125 Bari
ITALY

C. Chaillout
Laboratoire de Cristallographie
CNRS, BP 166
F-38042 Grenoble cedex
FRANCE

A. L. Chuvilin
Institute of Catalysis
SB Russian Academy of Sciences
Lavrentieva Ave. 90
Novosibirsk 63005
RUSSIA

D. L. Dorset
Electron Diffraction Department
Hauptman-Woodward Medical Research Institute, Inc.
73 High Street
Buffalo, NY 14203-1196 USA

H. F. Fan
Institute of Physics
Chinese Academy of Sciences
Beijing 100080
P. R. CHINA

J. R. Fryer
Department of Chemistry
University of Glasgow
Glasgow G12 8QQ
Scotland, UK

E. Gautier
Lab. Cristallographie
CNRS, B. P. 166
F-38042 Grenoble cedex
FRANCE

C. Giacovazzo
Dipartimento Geomineralogico
Universita di Bari
Via Orbona 4
I-70125 Bari
ITALY

C. J. Gilmore
Department of Chemistry
University of Glasgow
Glasgow G12 8QQ
Scotland, UK

J. Gjønnes
Department of Physics
University of Oslo
P. O. Box 1048 Blindern
N-0316 Oslo
NORWAY

H. Hauptman
Hauptman-Woodward Medical Researh Institute, Inc.
73 High St.
Buffalo, NY 14203-1196 USA

A. Holzenburg
Dept. Biochemistry and Molecular Biology
Leeds University
Leeds LS2 9JT
England, UK

S. Hovmöller
Structural Chemistry
Stockholm University
S-106 91 Stockholm
SWEDEN

J. Jansen
National Centre for HREM
Laboratory of Materials Science
Delft University of Technology
Rotterdamweg 137
2628 AL Delft
NETHERLANDS

R. R. Keller
National Institute of Standards and Technology
Materials Reliablility Division
325 Broadway
Boulder, CO 80303 USA

R. Kilaas
National Center for Electron Microscopy
Bldg. 72
Lawrence Berkeley Laboratory
Berkeley, CA 94720 USA

U. Kolb
Institut für Physikalische Chemie
Universität Mainz
Welder-Weg 11
D-55099 Mainz
GERMANY

Hans Kothe
Institut für Physikalische Chemie
Universität Mainz
Welder-Weg 11
D-55099 Mainz
GERMANY

E. Landree
Department of Materials Science and Engineering
Northwestern University
Evanston, IL 60208 USA

F. H. Li
Institute of Physics & Center for Condensed Matter Physics
Chinese Academy of Sciences
Beijing 100080
P. R. CHINA

J. Liu
Institute of Physics
Chinese Academy of Sciences
Beijing 100080
P. R. CHINA

X. L. Ma
Dept. Chemical Engineering
Univerity of Dortmund
D-44221 Dormund
GERMANY

A. L. Mackay
Dept. of Crystallography
Birkbeck College, University of London
Malet Street
London WC1E 7HX
England, UK

S. V. Meille
Dipartimento di Chimica
Politecnico di Milano
Via Marncinelli 7
I-20131 Milano
ITALY

M. Mellini
Dip. Scinze della Terra
Universita di Siena
Via della Chechia 3
I-53100 Siena
ITALY

B. Mingler
Institut für Materialphysik
University of Vienna
Boltzmanngaße 5
A-1090 Vienna
AUSTRIA

Z. M. Mo
Beijing Laboratory of Electron Microscopy
P. O. Box 2724
Beijing 100080
P. R. CHINA

R. H. Newman
Protein Structure Lab.
Imperial Cancer Res. Fund.
44 Lincoln
London WC2A 3PX
England, UK

J. Ramirez-Castellanos
Dpto. Q. Inorganica
Fac. Quimicas (UCM)
28040 Madrid
SPAIN

R. Ramlau
Max-Planck-Inst. Festkörperforsch.
Heisnenbergstr. 1
D-70569 Stuttgart
GERMANY

C. Rentenberger
Institut für Materialphysik
Universität Wien
Boltzmanngaße 5
A-1090 Wien
AUSTRIA

M. J. Sayagues
Department of Materials
University of Oxford
Parks Road
Oxford OX1-3PH
England, UK

G. M. Sheldrick
Institut für Anorganische Chemie
Universität Göttingen
Tammannstr. 4
D-37077 Göttingen
GERMANY

M. Tanaka
Research Institute for Scientific Measurements
Tohuku University
Sendai 980-77
JAPAN

F. Vasiliu
National Institute for Material Physics
P. O. Box MG-7
R-76900 Bucharest-Magurele
ROMANIA

I. G. Voigt-Martin
Institut für Physikalische Chemie
Universität Mainz
Welder-Weg 11
D-55099 Mainz
GERMANY

C. Viti
Dip. Scienze della Terra
Universita di Siena
Via delle Cerchia 3
I-53100 Siena
ITALY

T. E. Weirich
Structural Chemistry
Stockholm University
S-106 91 Stockholm
SWEDEN

M. Wolcyrz
Institute of Low Temperature and Structure Research
Polish Academy of Sciences
Ul. Okolna 2
50-950 Wroclaw
POLAND

H. W. Zandbergen
National Centre for HREM
Laboratory of Materials Science
Delft University of Technology
Rotterdamweg 137
2628 AL Delft
NETHERLANDS

X. D. Zou
Structural Chemistry
Stockholm University
S-106 91 Stockholm
SWEDEN

B. B. Zvyagin
Institute of Ore Deposits, Petrology, Mineralogy and Geochemistry
Russian Academy of Sciences
Staromonetny per. 35
Moscow 109017
RUSSIA

THE DEVELOPMENT OF ELECTRON CRYSTALLOGRAPHY
— in memory of **Boris Konstantinovich Vainshtein** (1921-1996).

ALAN L. MACKAY[1],
Dept. of Crystallography,
Birkbeck College, (University of London),
Malet Street, London WC1E 7HX.

Since we meet here in the ancient town of Eryx, full of the monuments of ancient and modern religions, we should remember that Archimedes lived in Sicily[2] and that his work combined science, mathematical theory, technology[3] and computing[4]. Traditional teaching in schools about the ancient Mediterranean classical traditions, dominated by myths and literature, has completely obscured the technological basis of that civilisation[5] which, in fact, produced such devices as the **Anti-Kythera mechanism**[6]. This was an astronomical calculator, made in the first century BC, with some 30 gear wheels and of a complexity comparable to that of a modern mechanical clock. Probably Archimedes knew about the predecessors of such machines[7] and he

---

[1] © The author, while asserting his right to be recognised as the author of this article, renounces his copyright in it, so that the material is thus now in the public domain.
[2] (287-212 BC). An entertaining re-creation of life in ancient Syracuse can be found in Mary Renault's novel, "The Mask of Apollo".
[3] The salt works at Trapani still pumps the brine with Archimedean triple helices.
[4] Archytas (4 cent. BC), renowned for his mechanical devices, had also lived nearby. Archimedes was also in correspondence with Eratosthenes, Librarian of Alexandria, who had performed the supreme feat of the human intellect in measuring the radius of the Earth.
[5] We suffer from the same problem today with 'post-modernism' and the flight from objective reality.
[6] This is the first of the erroneous "**mind-sets**" with which this paper will be concerned. See: "Gears from the Greeks: The AntiKythera Mechanism - a Calendar Computer of 80BC", D. J. de S. Price (1975) and later papers.
[7] Cicero (106-43 BC), Governor of Sicily in 75 BC, who rediscovered and restored the tomb of Archimedes (since again neglected and lost), may even have seen the actual mechanism.

D. L. Dorset et al. (eds.), Electron Crystallography, 1–14.

himself certainly used mechanical models and diagrams[8] to discover mathematical results which he later proved theoretically. He was killed during the Roman invasion of Syracuse, while too absorbed with his thinking machine, his sand-table, the equivalent of the modern PC.

In ancient Sicily it was also known that amber, imported from the Baltic, when rubbed with a cat skin, attracts small pieces of dry vegetable matter. The Greek word for amber is, of course, "electron", but it is still not entirely clear how frictional electrification works, although this year is already the centenary of the characterisation of the electron by J. J. Thompson. Using an instrument presaging the cathode ray oscillograph, J. J. Thompson measured e/m for the electron and regarded an electron as a particle. Later the electron turned out to have wave-like properties too and electrons have proved to be immensely useful in investigating matter. The experiment of producing a diffraction pattern by the interference of electrons passing through two parallel slits still furnishes one of the great paradoxes of physics which show that the world of quantum mechanics is not like our "common sense" everyday world. If the intensity of the beam of electrons is reduced, so that only one electron at a time goes through the system, a diffraction pattern is still obtained. Which slit does it go through?

The development of electron crystallography, and particularly the role of **Boris Vainshtein** in it, provides an interesting case history of the complex interactions of theory, experiment and individual people in our century. All aspects of science are really linked together.

It is not necessary to go over the discovery in 1912 of the diffraction of X-rays by crystals and the subsequent development of crystal structure analysis[9]. Max von Laue, Arnold Sommerfeld (Ewald's supervisor) and Paul Ewald had the idea about 1912 at a café table in Munich. Friedrich and Knipping did the experiment and von Laue got the prize. Paul Ewald provided us with the geometrical tools of the reciprocal lattice and the Ewald sphere which are of the greatest use in visualising what is happening[10].

---

[8] "certain things first became clear to me by a mechanical method, although they had to be demonstrated by geometry afterwards because their investigation by the said method did not furnish an actual demonstration. But of course it is easier, when we have previously acquired, by the method, some knowledge of the questions, to supply the proof that it is to find it without any previous knowledge". The Method. (trans. T. L. Heath).

[9] By far the best single textbook is that of W. L. Bragg, "The Crystalline State", (1933) which, although now 64 years old, is absolutely essential reading.

[10] Everyone concerned to know "wie es eigentlich gewesen ist" should read Ewald's collection: "Fifty years of X-ray diffraction", International Union of Crystallography, 1962. As W. L. Bragg himself said, " you cannot realise how difficult the structure of pyrite was to solve" (having non-intersecting three-fold axes).

X-ray diffraction has been so successful, resulting in the knowledge of the structures of some 200,000 materials, that it has induced a **mind-set** preventing people from considering properly other techniques[11]. W. L. Bragg, the founder of X-ray crystal structure analysis, being involved in the discussions on the wave-particle duality of X-rays and electrons, was well aware of electron diffraction and devoted a chapter to this in his classic book. Following the theoretical predictions of Louis de Broglie made about 1923, it was expected that an electron with a velocity **v** and a mass **m** would have a wavelength $\lambda = \mathbf{h/(mv)}$ associated with it[12]. In 1928 Davisson and Germer demonstrated the diffraction of 65 to 600 volt electrons from the surface of crystal of nickel and at much the same time, G. P. Thompson (the son of J. J. Thompson) showed the photographically recorded diffraction patterns produced by the passage of 60kV electrons through thin films of gold. The techniques of electron diffraction were then rapidly assimilated to those of X-ray diffraction and electron crystallography had begun.

When a new technique appears the usual first strategy is to rush round and look at everything to hand with it[13]. This happened with electron diffraction and the characteristics of the technique rapidly appeared, together with information about various substances examined with it. In the second stage, the new technique is seen in the light of the existing techniques and theories. In particular, since waves are now found to be associated with particles, older techniques using one kind of waves can be transferred to newer techniques with (usually) shorter wavelengths. We see two clear periods in electron crystallography, the first concerned with electron diffraction and the second with electron microscopy. These are perhaps now followed today by a

---

[11] The same **mind-set** was also evident in 1984 in the resistance to the discovery of quasi-crystals. The paradigm of "crystal structure analysis by X-ray diffraction" has now been varied in all three terms; *crystal structure, X-rays,* and *diffraction.* As we will see later there were also other **mind-sets** relating to the state of science in the USSR which were not disrupted until the shock of the Sputnik satellite in 1957.

[12] It is convenient to remember, without confusing them, two similar formulae: (1) The wave length in Ångstroms associated with an electron accelerated by a potential of V volts is given (approximately) by $\lambda = 12.3/\sqrt{V}$. (2) The wavelength in Ångstroms of the X-rays with an energy of V kilovolts is (approximately) $\lambda = 12.3/V$. Thus 100kV electrons have a wavelength of about 0.04 Å. 1.5 A electrons correspond to 70 volts. The characteristic X-rays from a copper anode have a wavelength of 1.54 Å and correspond to a voltage of 8kV. Although 100 kV electron may be stopped by a micron thickness of metal, the most penetrating X-rays which may be associated with these 100kV electrons have a wavelength of 0.12 Å and will require several millimetres of lead for protection. Relativity must be taken into account for more exact calculations of wave-length. Electrons in an electron microscope travel at an appreciable fraction of the speed of light.

[13] "Micrographia" by Robert Hooke (London, 1665) is a classic account of all the discoveries made by one of the first to use the optical microscope.

much more general approach to radiation and structure and the combination of methods.

Electron diffraction

The attractive properties of electron diffraction were at once apparent. To summarise:

1) The wavelength to be expected for an electron beam is extremely short, less than a tenth of the diameter of an atom, and thus there were possibilities of great resolving power.

2) Electrons are very strongly scattered by matter, perhaps 100,000 times as much as X-rays, so that "a minute speck of crystal is sufficient to give diffraction effects" (Bragg). The scattering is so strong that, even for a very thin specimen, electrons are scattered more than once. It is in fact much better here to consider electrons as waves and to picture the wave field in the crystal. For X-ray diffraction it is usually assumed that multiple diffraction (Renninger effect) does not take place. If it does, then a smaller crystal is used and the difficulty is avoided, but just those observations which contain information about the phases are neglected.

With multiple scattering (or standing wave fields) very complex diffraction patterns are produced. This was quickly found by S. Kikuchi (1928), whose name thus became attached to a particular class of diffraction pattern.

Electrons are scattered by atoms and molecules in the gaseous state and this technique has developed steadily from that day to this[14] with increasing refinements, most noticeably by the combination of other techniques like mass spectroscopy.

3) J. J. Thompson had shown that a beam of electrons could be deflected by electrical and magnetic fields.

(a) This meant that, in contrast to X-rays, a beam of electrons could readily be focussed and thus possibilities for making electron-optical instruments, particularly of making a microscope, could be envisaged.

(b) For atoms, electrons are scattered by the electric potential whereas X-rays are scattered by the electron density[15]. The inner potential of a crystal is equivalent to a refractive index. If the X-ray scattering factor for an atom is **f** then the electron scattering factor is proportional to **(Z-f)**, where **Z** is the atomic number (the number of electrons in the atom). The scattering curves for X-rays and electrons thus go differently. For electrons light atoms scatter more compared with heavier atoms than for X-rays and in particular, hydrogen atoms have substantial scattering power.

(c) It followed also that electrons could be selected for their energies (and thus wavelengths) by a combination of electrical and magnetic fields.

---

[14] Hargittai, I., and Hargittai, M., "Stereochemical Applications of Gas-Phase Electron Diffraction", VCH, Weinheim, (1988).

[15] Electron density is not the same as charge density. It means the density of electrons whose charge is neutralised locally by the corresponding positive nuclear charge. The concept of charge density is fraught with difficulties.

4) The notation and concepts developed for X-ray diffraction could be immediately transferred. An electron beam of 100kV energy corresponds to a wavelength of about 0.04 Ångstroms. Bragg's law of diffraction $\lambda=2d \sin\theta$ could be directly applied and indicated that the diffraction angles corresponding to those used for X-ray diffraction were in the range of 1 degree, where $\sin\theta$ is close to θ itself. This meant that when the Ewald sphere representation of the geometry of diffraction was used, the pattern could be seen to be an almost undistorted section through the reciprocal lattice.

We may follow briefly the various schools which developed electron diffraction from 1928.

Crystal structure analysis by electron diffraction

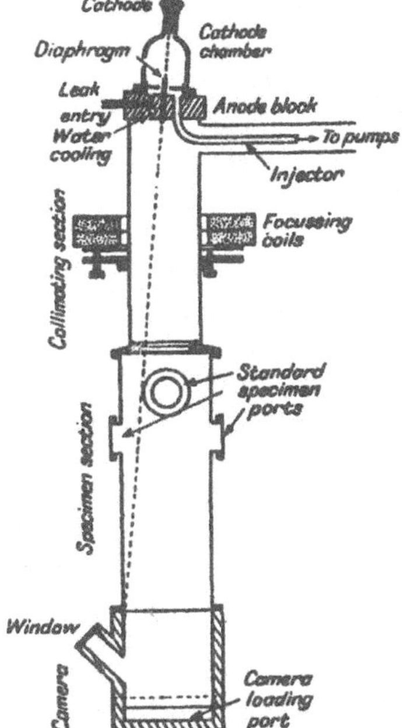

The equipment required for electron diffraction was not complicated and many people began experiments. At Imperial College in London, G. I. Finch, continuing from G. P. Thompson, used reflection diffraction to characterise surfaces, such as the oxide layer on steel piston rings. His apparatus was made locally and required a wine bottle (with the bottom cut off) to use as the main insulator and gun housing. Finch used to explain what a lot of research was necessary to find just the right kind of bottle

Figure 1. The electron diffraction camera used by G. I. Finch at Imperial College, London.

In Japan, S. Kikuchi started a tradition of electron optics which became very important, especially much later when people came back from the Navy after 1945 and founded the Japan Electron Optics Company.

In Russia, where there was a very strong mineralogical and crystallographic tradition, the structure of ammonium chloride was determined using electron diffraction in 1933 by Lashkarev and Usyskin who found the positions of the hydrogen atoms.

Z. G. Pinsker, the head of the electronography laboratory at the Institute of Crystallography in Moscow, said that he had made some significant discoveries[16], but that his most important discovery was that of Boris Vainshtein who joined Pinsker's group after the war and quickly developed his own research style. The Russian

[16] Z. G. Pinsker's book, "Electron diffraction", (Moscow, 1949), appeared in English translation in 1953. B. K. Vainshtein's book, "Structural Electronography" appeared in Russian in 1956 and in English in 1964. B. B. Zvyagin`s book, "Electron Diffraction Analysis of Clay Minerals", appeared in English in 1967 (Plenum).

laboratory built several electron diffraction cameras and began to apply them to crystal structure analysis, building on the pre-war work.

Figure 2. The early electron diffraction camera used at the Institute of Crystallography in Moscow.

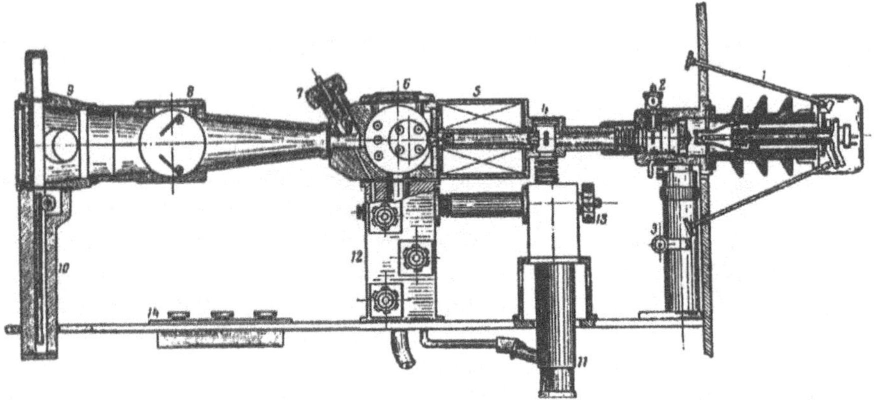

They determined a number of crystal structures, more or less following the procedures for X-ray work. Their techniques had a number of features favouring success in this:

(a) they used very small crystals ordered into textures, that is with strong preferred orientation. Considerable trial and error effort was put into obtaining suitable textures where the crystallites were sufficiently small and well-aligned. The specimen carried could be tilted (to at least 60°) to expose textures where flakes lay parallel to the substrate. The technique was very suitable for clay minerals, where appreciable single crystals did not exist, and Boris Zvyagin developed this topic with great success solving structures which were then impossible by X-ray methods.

(b) the electron beam had a diameter of about 1mm where it struck the specimen (which was usually supported on a collodion film). Thus, many crystallites of various sizes contributed to the diffraction pattern and, with variations in orientation, the intensities of individual crystallites added (rather than the amplitudes). All these factors contributed to the minimisation of the effects of multiple scattering (dynamic scattering). For dynamic scattering a first approximation is that the intensities are proportional to F (the structure factor) rather than to F squared. Usually the exponent of F was refined for the best fit.

(c) at first, at least, the number of parameters to be refined was small compared with the number of reflections available and crystal-chemical considerations reduced the number of possible structures.

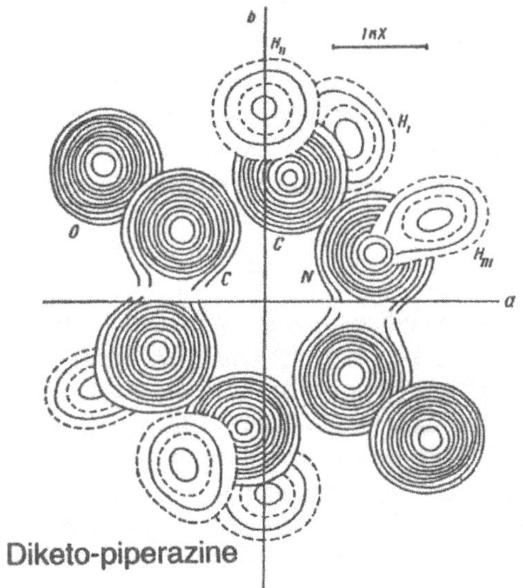

It happens that I first intellectually encountered Boris Vainshtein in translating his early paper "The application of harmonic analysis in electronography" which appeared in the reports of the Soviet Academy of Sciences in 1949[17] in which the determination of the structure of barium chloride monohydrate by electron diffraction was reported. The paper mentioned that hydrogen atoms could be detected and pointed to the difference between the scattering curves of the same atom for electrons and for X-rays. On the basis of this hint that the use of electrons and X-rays for the same material was equivalent to the method of isomorphous replacement[18], I applied for a research studentship at Imperial College. I did not get it and my colleagues, to whom I circulated the translation, took no notice. In 1956 I met Boris Vainshtein at a crystallographic meeting in Madrid and later at the Institute of Crystallography where Bernal was giving lectures. In 1962 I joined Vainshtein's group at the Institute of Crystallography, not very successfully, for a few months. I was at least able to get to know everyone, learn what was going on and see how the equipment worked. The cold war was then in the temporary stage of a modest "thaw". In fact Boris Vainshtein was then, having become Director of the Institute in 1962, moving on to the structure analysis of proteins, which he did also with great success against very considerable odds. To get out a front-line scientific paper in the Soviet Union required several times the effort needed in America. In the period from 1945 to 1960 many structures were determined. These included: $NiCl_2.H_2O$ and related compounds; **diketo-piperazine** (see figure above); Cu/Mn and Cu/Mg alloys; $CuCl_2.3Cu(OH)_2$; poly-γ-methyl-L-glutamate; $PbBi_4Te_7$; thio-urea and urea; they were notable for showing the possibilities of electron diffraction, particularly as regards finding the positions of H atoms and of showing other light atoms in the presence of heavy atoms. From 1962 Boris had to do the strategic scientific planning for all sections of a very large institute (reaching about 1000 people, half in the workshop producing crystals and instruments for sale, these included large single crystal sapphire windows for the US space programme) and he

[17] B. K. Vainshtein and Z. G. Pinsker, Dokl. Akad. Nauk, SSSR, **64** (1), 49-52, (1949).

[18] A. L. Mackay, Suppl. al Nuovo Cimento, **10**, 387-414, (1953). p.395. For some years I abstracted the whole of Kristallografiya and other Russian journals.

contributed important papers on almost all aspects of crystallography. The workshop was necessary because it was easier to have them grind up a lens to order than to get it from outside. It also earned foreign exchange some of which could be used for really essential imports. Life was always rather difficult. Nevertheless Boris continued to supervise work in electron diffraction and in 1993 produced, with Vera Klechkovskaya, a study[19] of Langmuir-Blodgett films (using an electron microscope) for which the technique is well suited.

Electron microscopy

The electron microscope developed from the cathode ray oscillograph with gradual improvements. Since the resolving power of a microscope is roughly $\lambda/2$ and 100kV electrons have a wavelength of 0.04Å the resolving power of the electron microscope is potentially very great but it was (and is) limited by the nature of the magnetic and electric lenses. The spherical aberration of magnetic lenses is always positive and can be minimised but cannot be cancelled out by lenses of negative spherical aberration as is done in the optical case. Useful magnification began to appear about 1939 and immediately after war, the crystallographer R. W. G. Wyckoff was appointed Science Attaché at the American Embassy in London. Since he had just shown the arrangement of protein virus particles in a crystal, he would only take the job if he could bring his electron microscope to the Embassy with him and continue research work. This was really the first direct pictorial demonstration of how identical particles stack up to make a crystal. We will not follow these developments but jump to the date of 1956 when it was realised that the resolving power of the microscope had reached about 10 Ångstroms[20] and that this was comparable with the spacings of the Bragg planes in crystals. The dramatic step was accomplished by Jim Menter with a Siemens electron microscope at Tube Investments near Cambridge, and he showed pictures of the lattice of platinum phthalocyanine (11 A lattice planes). This was also very important because the pictures also showed the presence of dislocations in the lattice thus confirming what had hitherto been theoretical speculations. The chief textbook of the sixties, (Hirsch *et al.*, 1965) was much more concerned with the development of dislocation studies and the complexity of the interaction of electrons with crystals, and structure analysis was hardly mentioned.

With this lattice imaging microscopists slowly began to realise that in the formation of an electron microscope image the relative phases of the beams making up the image are not lost, although they may be distorted, and that microscopy presents possibilities for solving the phase problem which then appeared as the dominating difficulty of X-ray crystal structure analysis.

---

[19] B. K. Vainshtein and V. V. Klechkovskaya, "Electron diffraction by Langmuir-Blodgett films", Proc. R. Soc. Lond. **A 442**, 73-84, (1993).

[20] 1 Ångstrom unit is 10^{-8} cms. Since atoms have radii of 1 to 2 Ångstroms crystallographers prefer to continue with this term rather than to use nanometres (1 nm = 10 Ångstroms) as international obligations require.

Generalised microscopy

At this stage we should look back to the work of Ernst Abbe in Jena who established the wave theory for the resolving power of the optical microscope (1882). He showed that in order to resolve the lines in a diffraction grating the microscope objective must accept, besides the central beam, at least the first order diffracted beam, so that the image is produced as a result of the superposition of these two beams. The equation $\mathbf{d} = \lambda/(2\,\mathbf{n}\,\mathbf{sin}\,\alpha)$[21] is engraved on his memorial in Jena. It was known[22] that tilting the illumination so that the direct beam and one first order diffracted beam pass symmetrically through the system (thus suffering the same aberrations) effectively doubles the resolving power of the microscope. Understanding of this took some time to reach electron microscopists. It was the beginning of the appreciation of the contrast transfer function and the recognition that in the electron microscope the phases of scattered waves are not lost. The contrast transfer function was developed by H. H. Hopkins about 1957 for the design of lenses for television cameras with about 600 lines, where the requirements, aperture rather than resolution, differ from those of, for example, high-resolution surveying cameras. When you buy a high-fi audio amplifier, what you pay for is the frequency response curve. All frequencies should be amplified to the same extent. The same applies to buying an electron microscope but the curve of how phases are changed on transmission through the system is necessary as well as the amplitude response. The ear is rather insensitive to the relative phases of the various waves which add to make up what we hear[23], whereas the eye is very sensitive to phases. The simple symmetrical diagram deriving from Abbe (figure 1) is fundamental for understanding optical and electron microscopy and the applications of image processing.

A small point source of monochromatic radiation is defined by an aperture; waves spread from this and are focussed by a lens on to the plane **D** of the diffraction pattern. If an object, a diffraction grating, is placed at the first lens, it diffracts the incident beam into a series of spectra and these beams also are focussed in the plane D. If we put a screen or film at D the diffraction pattern would be recorded as an

---

[21] With the refractive index n, n sinα is the numerical aperture.

[22] For example the Encyclopaedia Britannica article "Microscope", vol. 18, p. 398, (9th edn. 1910).

[23] That is, a single ear is rather insensitive to the relative phases of the sine waves of different frequencies which are conceived as adding linearly to give a single tone, although it may register such phase differences as differences in "attack". However, the ear is non-linear and produces harmonics and sum and difference tones. The hearing system is, of course, very sensitive to phase differences due to different path lengths of the same pulse as perceived by the two ears together which provides directionality. There is much new information as to how the aural system works and in particular there is evidence of a local oscillator which may supply a reference phase (like a superheterodyne receiver on the brink of oscillation).

intensity distribution. The relative phases of the various beams making the pattern would be lost. However, if there is no screen to stop them, the beams continue diverging from each other until they meet the second lens which converges them so that they overlap and interfere to give the image in the plane S'. If a screen is placed in the diffraction plane we can control which orders of diffraction go to make up the final image. In particular, a doubly periodic object gives a series of point spectra (the reciprocal lattice) whereas non-periodic noise gives diffracted intensity distributed over the spectrum, so that the points of the reciprocal lattice can be selected and the noise can be reduced. As demonstrated by M. J. Buerger, the relative phases of the diffracted beams can also be altered. Thus, image processing can begin. This was developed first by this optical analogy and later by computer. Thus a two-stage optical diffractometer was developed by Klug and de Rosier[24]. The laser appeared at much the same time and greatly facilitated the instrumentation.

Figure 3.
The Optical diffractometer.

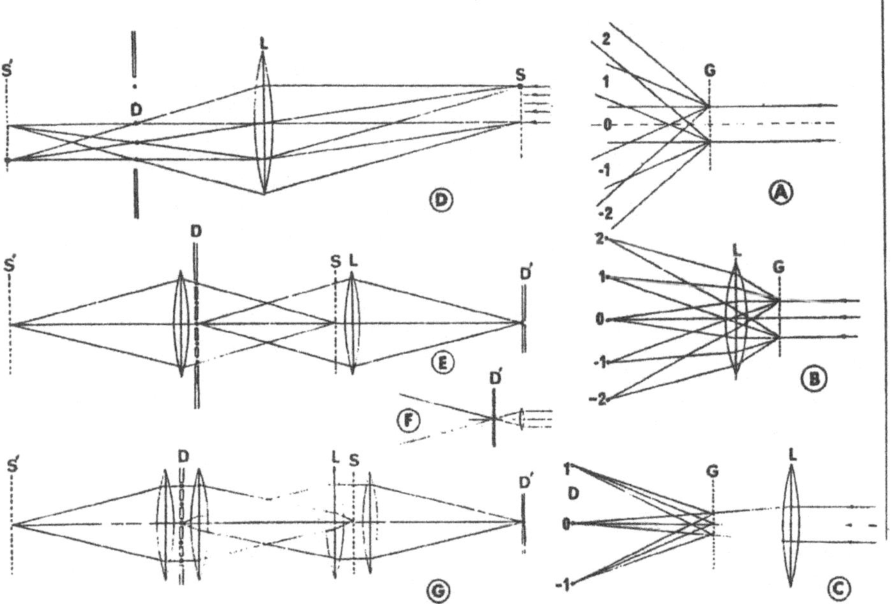

a) Diffraction of a parallel monochromatic beam by a grating of period **a**. Diffracted beams appear at deviations θ given by **λ=na sinθ.**

If the grating **G** has a sinusoidal distribution of scattering power only the first order appears.

b) The parallel diffracted beams can be focussed with a lens **L** to produce points in the

[24] Klug, A. and DeRosier, D. J., Nature, **217**, 130, (1966) and Klug, A., Chemica Scripta, 14, 245-256, (1978-79) [Nobel Symposium "Direct Imaging of Atoms in Crystals and Molecules", 6-10 August, 1979].

diffraction plane.
c) The lens may be either side of the grating.
d) If the diffracted beams continue beyond the diffraction plane they superimpose in the image plane to give an image **S'** of the grating **S**.
e) Monochromatic light from a pinhole **D'** is focussed in the diffraction plane **D**, where each diffracted beam gives an image of the pinhole. Note the symmetry of the system where **S'** is also the image of **S**.
f) A broad parallel beam from a laser may be focussed on to the pinhole to give an intense monochromatic source.
g) With a screen in the plane of the diffraction pattern **D** selected parts of the diffraction pattern can be allowed to proceed through holes in the screen to combine to give a filtered image of S at **S'**.

We may note, in yet another example of the **mind-set** phenomenon, that Frits Zernicke (1888-1966) had great difficulty in finding acceptance of his invention in 1938 of the phase contrast microscope. Zeiss (Jena) resisted it for a decade, it being believed that Abbe had said the last word on the subject. Further, only recently, Carl Zeiss (Oberkochen) found that their key development of energy filtering of the electron microscope image was very slow to be appreciated. This permits selection, on the basis of their characteristic losses, of what atoms are to contribute to the picture.

In 1948 Denis Gabor published his key paper "A new microscopic principle[25]" which described the principles of holography. At much the same time both W. L. Bragg[26] and Martin Buerger (MIT) were coming rather close to the same discovery. Buerger invented the precession camera (for X-ray diffraction) so that he could produce an undistorted picture of a section of the reciprocal lattice on a crystal. This film reversed, black spots becoming white holes, was then to become the source of diffracted light rays in "the two-wavelength microscope". The idea was to perform the first step, diffraction, with X-rays and the second stage, superposition of the diffracted beams, with monochromatic light. A magnification factor of the ratio of the wave-lengths would result. Buerger managed to reconstruct an image of iron sulphide, but only by inserting the **known** phases of each beam with a rather cumbrous optical phase shifting device. Bragg pointed out that the principle of holography was already in use in that the phases of X-ray beams diffracted by a central heavy atom, as Pt in platinum phthalocyanine (J. M. Robertson) or iodine in cholesteryl iodide (C. H. Carlisle and D. Crowfoot) effectively determined the phases, by swamping the contributions of the other atoms. As in everything he touched, W. L. Bragg cut straight through to a simple pictorial understanding of the physical principles involved.

---

[25] Nature, **161**, 777-778, (15 May 1948) and Proc. Roy. Soc., **A 197**, 454, (1950).
[26] "Microscopy by reconstructed wave-fronts", Nature, 166, 399-400, (2 Sept. 1950). Nature, **143**, 678, (1939) and Nature, **149**, 470, (1942).

The first image processing to give an actual picture of a crystal structure was done in 1929 by W. H. Bragg for diopside[27] who had pointed out the extremely important physical principle that a grating with a sinusoidal distribution of scattering power gives only the first order of reflection (all higher order are zero) and correspondingly, in a diffraction pattern each reflection can be attributed to a sinusoidal density wave of appropriate amplitude spatial frequency. Thus, a **picture** of diopside was produced by the **linear superposition** of sine waves on a photographic plate. It is, of course, necessary that these waves should also be combined in the correct phase. Later, (around 1955) it was realised that about 80% of the 'information' in a crystal structure resided in the phases and only about 20% in the amplitudes of the reflections. Thus, exact measurement of the amplitudes is not critical[28] but preservation of the phase relationships of the scattered beams, even in distorted form, is vital. In an **actual** crystal the sinusoidal electron density waves are not independent of each other and add **non-linearly**. The Karle-Hauptman determinant, which is the basis of the direct methods of determining phases[29], derives directly as an expression of this non-linearity.

Image processing

Ideas of image processing developed slowly, step by step, using the concept of the optical diffractometer with diffraction, filtering and then reconstruction. It began with the utilisation of *a priori* knowledge of symmetry to get better information about a single repeated element. Thus Roy Markham, dealing with an object with six-fold rotational symmetry, printed the plate six times to superpose the symmetrically related units. This was essentially what Francis Galton[30] had done to produce a face of an average criminal, the average tubercular face and even the face of Alexander the Great (by superposing the images of a dozen coins). It is important to realise that you get nothing for nothing, but that any foreknowledge can be used to limit the uncertainties remaining. Galton assumed in advance that the average portrait would have two eyes in a particular orientation and he lined up his individual pictures

---

[27] Zeit. f. Krist., 70, 488, (1929).

[28] A. D. Booth proposed to determine the structures of centro-symmetric crystals from the accidentally absent (or very weak) reflections on the grounds that for these the scattering by all the atoms together cancels out and that this is thus very sensitive to their exact positioning (personal communication about 1950 when he was at Birkbeck College).

[29] Jerome and Isabella Karle and Herbert Hauptman, encountering still another **mind-set**, this time among crystallographers, had some difficulty in getting their methods accepted, but now these are universally applied for the solution of small and medium crystal structures. Crystals with very large molecules (now up to MW 700,000) rely on the heavy atom method which resembles holography.

[30] "Francis Galton; the Life and Work of a Victorian Genius", D. W. Forrest, London, (1974). "Composite Portraiture", Appendix (pp. 221-241) to "Inquiries into Human Faculty", (1883) Everyman edn., Dent, London, (1907).

correspondingly without subjective guesswork. F. P. Ottensmeyer, seeking to avoid getting protein molecules lined up in parallel in a crystal according to the very exact requirements of nature, tried the same method of superimposing many electron microscope images of molecules of a particular protein lying in random orientations. He was cruelly slaughtered by J. Dubochet[31] at the Nobel Symposium of 1979 who showed images exactly like those of protein molecules produced by Ottensmeyer but where there had been no protein specimen whatever. The recognition of molecules, still less identification of their orientations had been a subjective delusion. Aaron Klug's discussion of this was less cruel and more analytic but equally devastating. Nevertheless the method begins to work when the number of details to be recognised in the object is large (compared with the 5 parameters - position in the direction of the beam is not needed - necessary for specifying the orientation and position). This is to be seen in the remarkable work on the core protein of the hepatitis B virus very recently published by two groups, where resolutions of 7.4Å and 9.0Å were reached from 6,400 and 600 particles respectively[32]. Identity of the particles used is based on prior physico-chemical methods of selection and represents hard-won information.

Thus direct methods of X-ray crystal structure analysis depend on the *a priori* knowledge that the crystal is composed of atoms and that the electron density is nowhere negative. Similarly the Rayleigh criterion for resolving two point objects, that the first diffraction peak of one should lie on the first diffraction minimum of the other, corresponding to a resolution of 0.5 λ, could be greatly improved on, if we knew exactly the expected diffraction from each object. This applies, of course, also to radar, and permits the distinction between one big aeroplane and two smaller ones if we know the characteristic scattering from each type.

Crystals

It is a test for purity that a material crystallises and it is the definition of a crystal that it is composed entirely of identical units identically situated. This assumption is the price paid for X-ray crystal structure analysis. However, not all crystals answer to this description and they are then characterised as poorly crystallised, disordered, flaky, etc. in disparagement. However, real matter knows other modes of ordering and does not always correspond to the preconceptions of the crystallographers. Quasi-crystals were just one of many surprises and there must be many others to come.

Thus, this workshop is devoted to the art of obtaining by electron microscopy three-dimensional images of various objects, about which we have certain other information. We may know (or believe, perhaps erroneously) something about their

---

[31] J. Dubochet, Chemica Scripta, **14**, 293, (1978-1979).

[32] S. Böttcher, S. A. Wynne and R. A. Crowther, "Determination of the fold of the core protein of hepatitis B virus by electron cryomicroscopy", Nature, **386**, 88-91, (6 March 1997) and J. F. Conway, N. Cheng, A. Zlotnick, P. T. Wingfield, S. J. Stahl and A. C. Steven, "Visualisation of a 4-helix bundle in the hepatitis B virus capsid by cryo-electron microscopy", Nature, 386, 91-94, (6 March 1997).

constitution and about their symmetry and this *a priori* information may be used to separate out from the incoming images what we wish to know from what we already know. We will realise here that seeing is not a simple passive act but that, as in many other fields of life, what we see in a situation depends on the cultural preconceptions which we bring with us.

80 years ago, W. H. and W. L. Bragg had the superb vision that the atomic world, miraculously forseen by Democritos and Lucretius, was just like the real world, only smaller[33]. With quantum mechanics our preconceptions have changed as the wave/particle duality has turned out to be much more complex and counter-intuitive than W. H.'s explanation that physicists use the wave theory on Mondays, Wednesdays and Fridays and the particle theory on Tuesdays, Thursdays and Saturdays. **The electron microscope now serves as a kind of matching transformer** which connects directly our every-day human senses with the world of atoms, just as Galileo's telescope connected him to the cosmos, revealing phenomena, like the moons of Jupiter, which were counter-intuitive to the prevailing world view of that time and place[34].

References

Bragg, W. L., "The Crystalline State", Vol.1, Bell, London, (1933).

Hirsch, P. B., Howie, A., Nicholson, R. B., Pashley, D. W. and Whelan, M. J., "Electron Microscopy of Thin Crystals", Butterworths, London, (1965)

Pinsker, Z. G., "Electron diffraction", (Russian edn., Moscow, 1949), (English edn., Butterworth, London, 1953).

Ruska, E., "The Early development of Electron Lenses and Electron Microscopy", (English trans. by T. Mulvey), Hirzel Verlag, Stuttgart, (1980).

Vainshtein, B. K., "Structural Electronography" (Russian edn. Acad. Sci. USSR, 1956 and English edn. Pergamon, Oxford, 1964).

Vainshtein, B. K., "Fundamentals of Crystals, Symmetry and Methods of Structural Crystallography", Springer, (2nd. edn. 1994).(= "Modern Crystallography", Vol. 1, Nauka, Moscow, 1979)

Vainshtein, B. K., Friedkin, V. M. and Indenbom, V. L., "Structure of Crystals", Springer, (2nd. edn., 1995) (= "Modern Crystallography", Vol. 2, Nauka, Moscow, 1979).

Zvyagin, B. B., "Electron Diffraction Analysis of Clay Minerals", (English edn. 1967), Plenum, New York.

---

[33] W. H. Bragg used Lucretius' title, "On the Nature of Things" for his own series of lectures at the Royal Institution.

[34] For Galileo's connection with atomism and trans-substantiation see P. Redondi, "Galileo — heretic".

SOLID STATE STRUCTURES

J.R.Fryer
Electron Microscope Centre, University of Glasgow, Glasgow G12 8QQ, Scotland ,U.K.

1.Types of Structure

Can consider four classes (excluding metals):
1. *Ionic* - packing of spheres, dominated by the largest ions. Electrostatic, non-directional bonding between ions e.g. NaCl, CsF.
2. *Weakly covalent* - packing of spheres with distortion caused by the sterochemistry of the central atom. Covalent bonding within molecules, but electrostatic bonding between molecules e.g. TiO_2 extending to a balance between the sterochemistry and ionic charge as in ReO_3 structures.
3. *Covalent* - infinite structures determined by the covalent bonds between the atoms or molecules comprising the structure. e.g. Carbon in graphite or diamond, extended silicates.
4. *Molecular crystals*. The packing of molecular shapes with only weak dispersive (Van der Waals) forces between the molecules. e.g. neutral(non - polar) organic molecules, inert gases, S, Se, Te. halogens.

Mixtures of all of these types of structure can occur within one crystal.

2. Organic Crystals

Forces holding the molecules together:

2.1. NON-POLAR MOLECULES

2.1.1 *Attractive. Dispersion forces.*
The interactions between non - polar molecules are described in terms of London dispersion forces caused by the changing distribution of electrons around a molecule or atom creating an oscillating dipole.

$$U_{dis} = C_6r^{-6} + C_8r^{-8} + C_{10}r^{-10}............$$

Where C_n is an attractive coefficient
r is the shortest distance between molecules. i.e. between peripheral atoms.

2.1.2 Repulsive forces.
If these are assumed to be isotropic then the repulsive potential r^{-n} has a value of r^{-12}.
Where forces are anisotropic e.g. lone pairs, d orbitals etc then n may be a different number than 12.
The repulsive forces are generally expressed as Ar^{-12} where A is a repulsive coefficent.
General expression for the interaction energy U between two atoms is:

$$U = C/r^6 - A/r^{12}$$

D. L. Dorset et al. (eds.), Electron Crystallography, 15–28.

2.2 POLAR MOLECULES

In this case there is a permanent dipole present, and for a dipole μ at a distance r the electric field is $2\mu/r^3$. Thus for two molecules aligned head to tail the interaction energy is:

$$U_{pol} = 2\mu_1 \mu_2 / r^3$$

The effective interaction range of polar molecules is much greater than non-polar molecules ($r^{-3} : r^{-6}$). Therefore effects such as hydrogen bonding will often dominate the packing of such molecules.
If a polar molecule(1) is incorporated into a non-polar matrix(2) it causes an induced dipole in the surrounding molecules, so that the effect of this polar impurity is much greater than its molecular size. The induced dipole has an interaction energy:

$$U_{12} = 4\alpha_2\mu_1^2/r^6$$

where α_2 is the molecular polarisability of the non-polar surrounding molecules and μ_1 is the dipole moment of the polar molecule.
The lattice energies of different polymorphs of organic crystals are often very similar, so that small changes can affect the particular crystal phase which is obtained. The illustration above of the effect of a polar impurity, is one factor which can affect the product. If the impurity is present in any significant concentration - e.g. > 0.01% then the dipoles of the impurities may be sufficiently close to interact and the crystal packing may be dominated by permanent and induced dipole interactions.

$$U_{total} = \frac{2\mu_1\mu_2}{r^3} + \sum \frac{4\alpha_2\mu_1^2}{r^6}$$

2.3 PACKING OF ORGANIC MOLECULES

An organic molecule is not spherically symmetrical , like an atom, but may have any shape. Therefore it is necessary to consider how a collection of irregularly shaped objects can be stacked to achieve the optimum close packing - ie. each molecule must be in contact with 6 neighbours.
In two dimensions non-equiaxed molecules can either form arrays which are based on rectangles or parallelograms, but with the close packing requirement there are two forms of packing possible in the two dimensional plane.

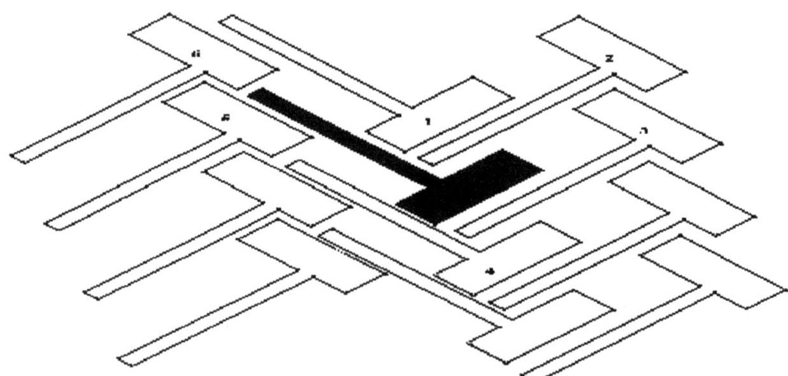

This is herringbone stacking with the molecules lying in alternate directions. The other form is tile stacking:

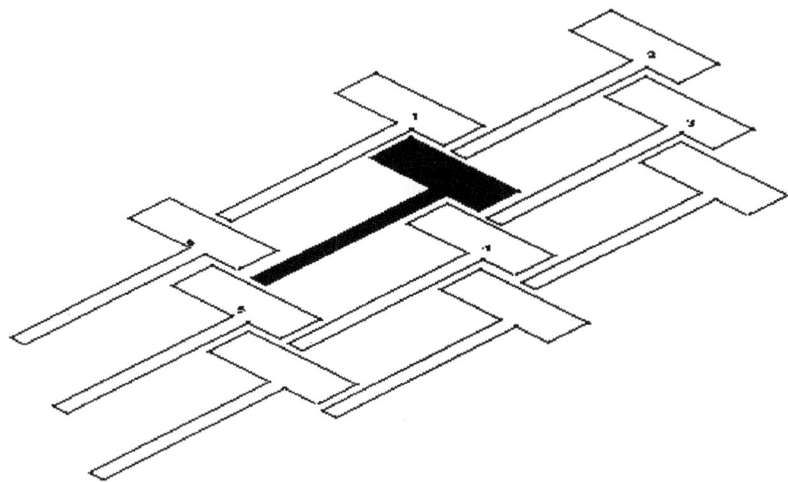

To fulfil these criteria the protrusions of each molecule must fit into the hollows of the adjacent molecule.
The restriction of either a rectangle or a parallelogram and a co-ordination number of 6, in two dimensions, limits the number of space groups which such structures can adopt. Out of the 230 space groups only 13 are possible and of these ~ 10 are likely for irregular shaped molecules. If the molecules have a higher symmetry then other space groups are possible. These 13 space groups are all of low symmetry, and with large molecules the difficulty of achieving the correct crystal geometry during crystallisation, will inhibit and sometimes prevent the formation of crystals. The more irregular the molecular shape, the more difficult is the material to crystallise.

2.4 PACKING COEFFICIENT K

A measure of the ability to form a crystal is the packing coefficient which describes the ratio between the volume occupied by the molecules and the unit cell size.

For example:

Compound	**K**
benzene	0.681
naphthalene	0.702
anthracene	0.722
perylene	0.805
graphite	0.887
ccp spheres in an ionic crystal	0.741

When a molecule in a crystal has a value of K below 0.6 then it is unlikely to crystallise. An example is 2,6,di-n-octyl naphthalene which has a value of K of 0.59 and can only be crystallised with difficulty. Normally such molecules will form a glass or a two dimensional liquid crystal, depending upon the molecular shape.

The importance of the properties and stacking of molecular crystals on experimental procedures is that they determine our ability to obtain crystalline specimens.

2.5 CONSIDERATIONS FOR SPECIMEN PREPARATION

1. Consider the molecular shape (if known) and estimate the degree of difficulty with which it is likely to form a crystal. This is very subjective, but if a molecule has a several large substituent groups then difficulty may be encountered.
2. Consider the nature of the substituent groups. Are they polar? If so then the criteria described above need not apply.
3. For a non-polar molecule, how pure is your sample? Is it contaminated with polar molecules e.g. water, pyridine etc?
4. For a non-polar molecule of very irregular shape then it might be advantageous to co-crystallise it with a polar molecule or add a polar impurity.
5. The more difficult a molecule is to crystallise then the slower must be the nucleation procedure during crystallisation, and the longer it must be held in metastable equilibrium at the crystallisation temperature and concentration, to allow the molecules to orient themselves on a growing crystals face.

Recommended reading

Kitaigorodsky,A.I. (1961) *Organic Chemical Crystallography* Consultants Bureau, New York.
Kitaigorodsky,A.I. (1973) *Molecular Crystals and Molecules* Academic Press New York.
Wright,J.D.(1987) *Molecular Crystals* Cambridge University Press, Cambridge.

3. Specimen Preparation

3.1 TYPES OF SPECIMEN

Since over 3000 different organic compounds have been identified, any attempt at generalisations cannot be sharply defined, but three types of compound can be considered, together with some special cases.

1. Long chain molecules such as paraffins, polymers, and molecules with a long chain component which influences their packing. Normally these compounds have low melting points, and being very anisotropic in shape, will have a tendency to form liquid crystals. Unless these compounds have polar substituent groups they are strongly non-polar.
2. Aromatic azo-compounds, and their salts. An example is shown below with charged groups coordinating to a metal cation so that all types of bonding are present in this molecule from ionic at specific locations to the weak dispersive forces spread over the aromatic rings which account for the majority of the volume of this molecule. These compounds dissociate on heating, and are formed by the very rapid azo-coupling reaction, so that one has very little control over the crystal shape. Some crystal growth can be achieved by refluxing the crystals with organic solvents. The growth habit can be affected by the polarity of the solvent chosen.

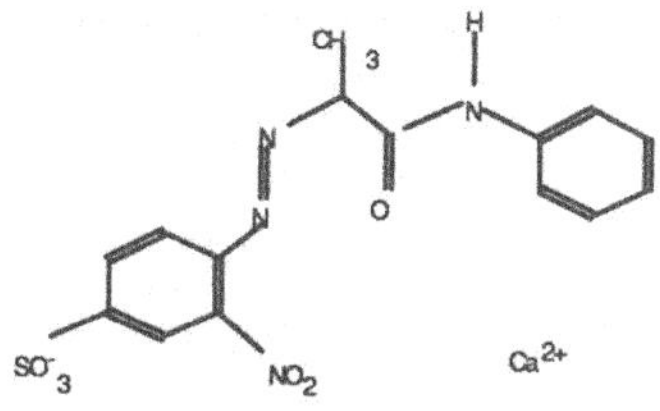

pigment yellow WSC

3. Polynuclear aromatic compounds. Those containing more than four rings have medium to high melting points, and in some cases have melting points over 800°C. The larger molecules are generally insoluble in conventional organic solvents, although they dissolve in concentrated sulphuric acid from which they can be recovered by *careful* dilution with water. If no strongly polar substituent groups are present the compounds are extremely non-polar, and unreactive, although some of the molecules are believed to be carcinogenic.

Compounds containing features of two or three of these types of molecule occur, and it is necessary to make an assessment of which features could be most favourably exploited for specimen preparation.

3.2 THE IDEAL SPECIMEN

This should be an even dispersion of thin (<10nm) crystals. They should be 50-100nm in extent without bends, and oriented so that a unit cell axis is aligned with the electron beam. The distance between crystals should be also approximately 100nm so that single crystals can be used to obtain unambiguous diffraction patterns. This ideal specimen should have a complementary specimen of the same compound prepared in another manner in which the crystals are oriented with another unit cell axis parallel to the beam direction.

3.3 THE NON-IDEAL SPECIMEN

The following bad features are listed in the table below. In some cases they can be overcome by the methods suggested below, otherwise a different specimen preparation technique must be used.

Observed feature	Procedure and consequences	Remedy
Very few crystals on specimen grid	Do not spend time looking at this specimen. It is probable that you will be examining an impurity, or non-typical crystal.	Prepare another specimen at higher concentration. Better support film or less polar solvent if caused by breakage of support film
Specimen too thick	Dynamical scattering effects from even hydrocarbons >10nm	Use lower concentration of material, in preparing another specimen
Specimen charging, variable astigmatism around specimen	Try an find a small crystal which is in good contact with the support film	For a new specimen use a more polar solvent to spread the crystals or overcoat the specimen with carbon.
Crystals too small	Microdiffraction, or high resolution imaging if radiation damage rates permits	Grow crystals in solvent. For vapour grown crystals use a higher substrate temperature, or anneal in vacuum at >1/2mp.
Crystals disappear under electron beam	Too intense a beam has been used resulting in sublimation of crystals which may condense elsewhere on the specimen in a misleading different form.	Initially examine specimen at low magnification(<5k) with a very weak beam so that the typical appearance is known.
Dark bands move over specimen	These are caused by buckling of the specimen under the electron beam. They result in small local specimen tilts which can affect the diffraction pattern, and in long-chain compounds can give rise to misleading symmetry	Use small illumination areas - spot scan imaging. Strengthen specimen with carbon overlayer. For aromatic compounds very short (10ms) exposure times can be advantageous.
Non-crystalline area	The lack of diffraction pattern may be caused by radiation damage, bad orientation, or may be an impurity.	Always obtain data from several areas on the grid, and preferably more than one grid.

3.4.TECHNIQUES OF SPECIMEN PREPARATION

1.Dry, Dusting, adhesive film
2.From a slurry
3. From solution
4.From a melt
5.Vacuum evaporation, epitaxy.
6. Films on a surface, eg water - LB films, protein films.

3.4.1 *Dry Preparation*

In situations where the characteristics and shape of the specimen are important, or the presence of a solvent would destroy, or alter the crystalline phase of a specimen, then it can only be prepared dry. It is necessary to grind the specimen to a fine dust, or rely on examining a thin area of the edge of the specimen, in the assumption that it is typical of the bulk. In most cases it is sufficient to dip a carbon coated grid into the powder, shake off excess, and examine in the microscope. If the specimen charges badly, because of poor contact with the support film, then overcoating with a layer of carbon can be effective.

If the powder does not adhere to the film, then make up a weak(1%) solution in ether or acetone of a commercial adhesive e.g. a simple polymer adhesive such as is used to mount photographic prints. Dip the carbon covered grid into the solution, allow to dry and then drag through the powder.

When the specimen is prepared in this way, the orientation is random and usually it is not a suitable specimen for high resolution or crystallographic examination. It is, however, a method which introduces least artefacts to the specimen, and is quick and simple. It is useful to make such a preparation to check whether more sophisticated specimen preparation methods have introduced artefacts.

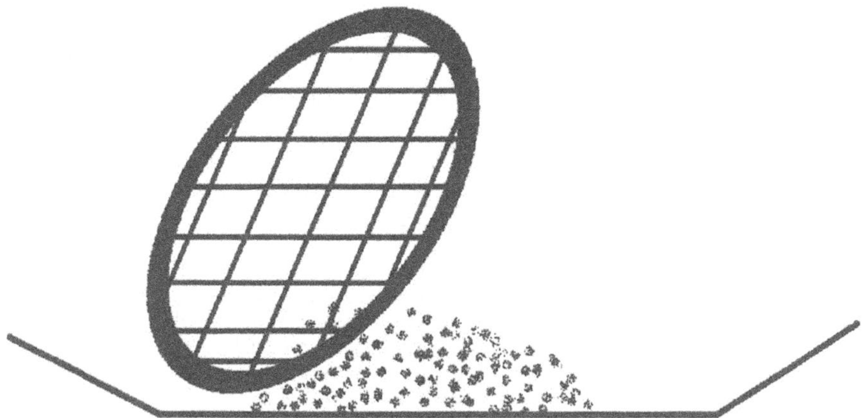

3.4.2 *Preparation from Suspension*

This involves dispersing the solid in a liquid in which it is insoluble. However, if there is no interaction between the liquid and the solid, then the solid will aggregate either at the bottom or top of the liquid in a test tube, and will not form an even dispersion on the specimen. Therefore the liquid should be chosen so that it wets the specimen, but does not dissolve it. An example is in the case of aromatic hydrocarbons which are strongly non-polar. A suitable liquid is isopropanol(propan-2-ol) since this will wet the surface of the crystals, but will not dissolve them. Water would be unsuitable, since the crystals would not disperse.

The dispersion can often be done by shaking, or bubbling air through from a pipette. In more difficult cases ultrasonic dispersion can be used, although this may fragment the crystals, giving a misleading impression if shape is important.

Concentration of the suspension can be done by centrifuging the suspension, removing the supernatent liquid, and redispersing with fresh liquid. This is also a useful method for the removal of soluble impurities.

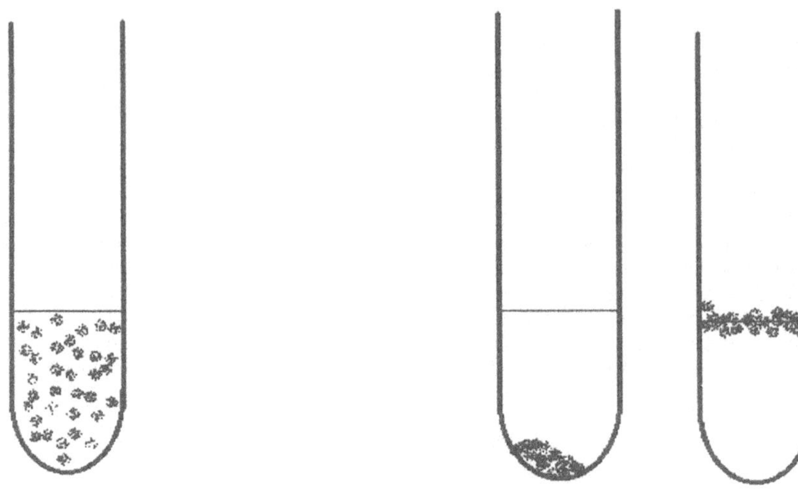

Good dispersion Poor dispersions

Finally a drop of the suspension is put onto a carbon film on the EM grid and allowed to dry. Endeavour to make the drop smaller than the grid area, so that there is no overflow which picks up impurities.

For solids with high surface activity which tend to agglomerate on drying, then spraying the suspension on to the grid with a nasal atomiser can produce good results.

3.4.3. *From solution*

This is very quick, involving drying a drop of solution on to a grid, but it is important to remember that the crystals which form from solution may not be the same polymorph as the original compound, and will almost certainly not have the same crystal form or orientation. Long chain compounds will crystallise with the major chain axis vertical -i.e. normal to the carbon film, and planar molecules tend to form needles with the planar faces across the minor axis of the needle.

Paraffin, chains vertical Needle of planar molecules

3.4.4. *From a melt*

This is a special case of solution growth where the solvent crystallises before the solute and forces the solute molecules into an orientation compatible with the solid solvent. In effect they lie along the void spaces in the host lattice. An example is the crystallisation of paraffin in solution in naphthalene, where the naphthalene crystallises first, forcing the long chain molecule to orient along the (110) direction of the naphthalene lattice. When the naphthalene is removed by sublimation the paraffin molecules form crystals with the chain axes of the molecules oriented parallel to the substrate. This behaviour is followed by all long chain molecules with naphthalene or benzoic acid being favoured 'solvent' materials. Crystallisation in this way will also confer a very uniform orientation of many other compounds, the only criteria are that the solvent has a slightly higher melting point than the solute, and the solute can be easily removed by sublimation or dissolution on a water surface.

The method is to put a drop of the molten solution (approximately 5% concentration) on to a mica sheet and place another mica on top to spread out the solution. Then slowly slide the sheet up and down a metal strip (approximately 0.5m long) which has one end cooled and the other heated. Thus the solution is alternatively melted and frozen several times which allows the molecules time to orient themselves with the host molecules. Remove the mica from the hot surface, allow to cool, and remove the upper mica sheet. Coat the mixture with a carbon layer (or a carbon film could have been put on the mica initially) and float off the carbon with the solution adhering to it which can be picked up in grids. The naphthalene will sublime off under vacuum. The variations of putting the carbon on first or last, and retrieval of the specimen depends on the nature of the compounds used in the experiment.

3.4.5. *Vacuum evaporation.*

Molecules which melt without dissociation can be prepared by heating in a boat under vacuum ($\sim 10^{-5}$torr), and at the melting point the molecules will leave the boat and can be condensed on to a substrate. The vacuum ensures that the molecules travel in straight lines (at this vacuum the mean free path is several metres), so there is no gas phase agglomeration or crystallisation prior to reaching the substrate surface. The choice of substrate and its temperature depend on the molecules concerned, but for aromatic hydrocarbons condensing on an alkali halide substrate(e.g. newly cleaved (100) KCl) a substrate temperature of approximately half of the melting point of the organic compound will generally ensure a well oriented (epitaxial) film. Some compounds are different, and in some cases a 10degree difference in substrate temperature will alter the orientation.

After evaporation the substrate is cooled (although in some cases annealing for an hour or so at high temperature is beneficial), a layer of carbon evaporated on to the substrate, and the carbon film with the compound adhering to floated off on to water, and picked up on a grids. The thickness of the film can be calculated from:

$$T = m/4\pi r^2 \rho$$

m is mass of material of density ρ evaporated on to a surface at a normal distance r

Although this assumes that all the molecules arriving on the substrate do remain there, and do not resublime - i.e. a sticking coefficient of 1, which with heated substrates is not usual. Otherwise the thickness can be measured with a quartz monitor.

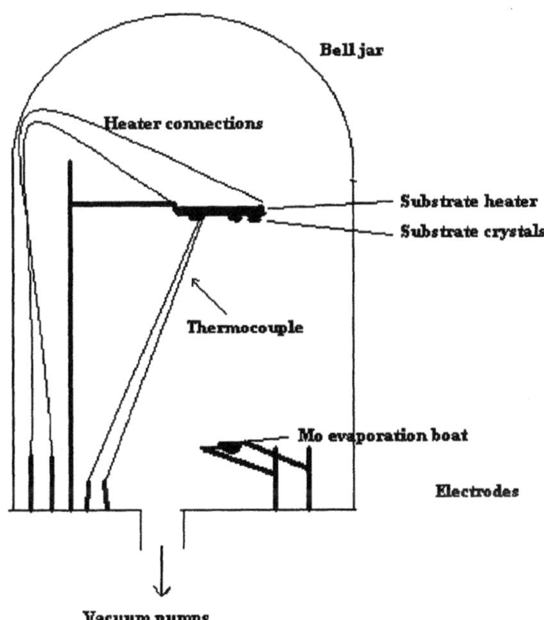

3.4.6.*Films on a liquid surface.*
Instead of drying a drop of solution on to a carbon film, if the drop is put on to an immiscible liquid surface, the drop will spread and dry more evenly and thinner than on a solid surface. This is a good method for polymers and hydrocarbons. A grid supported carbon film is brought up at an angle through the film, and to prevent any adhesion of the film on the underside of the support film, the grid should be mounted on a glass slide which can be removed later.
The procedure can be extended to Langmuir - Blodgett films, and also proteins floating on a water surface with their hydrophilic groups pointing into the water surface. However, to pick up from the underside it is necessary to treat the carbon support films with a plasma discharge to make them also hydrophilic. Alternatively, untreated carbon films can be laid, just touching the water surface so that the hydrophobic ends of the molecules are attracted to the carbon film. The polar groups in a protein dominate its packing so that such molecules normally adopt much higher symmetry space groups than hydrocarbons, and retain an aqueous/lipid envelope which fills in the spaces in the molecule, again making it a more symmetric unit.

4. Microscopy Techniques for Beam Sensitive Materials

4.1. *Calibration.*
In addition to the specimen preparation described previously, it is necessary to take one of the prepared grids and evaporate on to it thin layer of gold. This is to provide a standard calibration for the parameters of the specimen determined by electron diffraction. The gold should provide a ring pattern superimposed on the spot pattern of the specimen.
4.2 *Radiation damage.*
This manifests itself in the diffraction pattern as a broadening of the high index reflections, which spreads inward until all the reflections are extinguished and the structure is totally destroyed. For some compounds the rate of fading can be reduced by putting an overlayer of carbon on to the specimen, and for most specimens the effects can be reduced by a factor of 3 by cooling the specimen to liquid nitrogen temperatures. If it is required to obtain high resolution images of the specimen, the radiation causes buckling and movement of the specimen structure prior to destruction which reduces the resolution. To minimise this a small (<100nm) illuminated area should be used, and preferably very short exposure times.
The fading of the diffraction patterns provides the time and intensity scale under which a specimen can be examined. If it is assumed that one requires 30s to examine and record an image or diffraction pattern, then the intensity should be adjusted so that the fading -i.e. structure loss - is completed in twice this time - 60s. The reason for this is that the exponential fading of the diffraction, also has changes in the relative intensities of specific reflections, so that the errors increase as the structure degradation proceeds. For imaging the same reasoning prevails and the intensity determined by the diffraction pattern should be used, with the magnification reduced to provide the necessary illumination for the recording medium. Remember that high magnification is not necessary for high resolution. Normal em photographic film will record 0.3nm resolution at a magnification of 30,000X. To minimise damage, minimum exposure techniques - i.e. the required area of specimen only exposed to the electron beam during recording - should be used.
The process of specimen degradation is a chemical one with ionisation and free radical formation occurring in $\sim 10^{-14}$s so that this initial interaction with the electron beam is always present. The rate of degradation depends upon the particular compound being studied and also its constituent groups. For aromatic hydrocarbons, bromine substituted molecules are more stable than chlorine substituted, which in turn are more stable than hydrogenated molecules. The OH group is particularly reactive, and hence polysaccarides, and proteins degrade very rapidly. It also is important in inorganic systems, and the radiation sensitivity of many silicates - particularly zeolites - can be much reduced is the specimen is well dried under vacuum prior to examination.

4.3 *Recording media.*

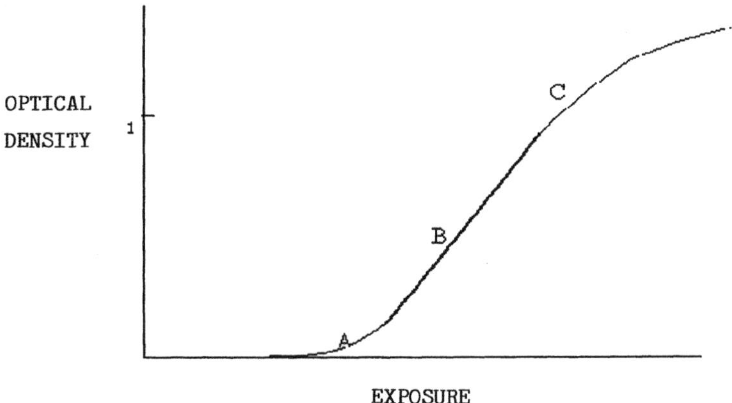

Above is given the response curve for a typical photographic film. The exposure time between 0 and A is an induction period which is shorter for large grain low resolution films such as X-ray. The linear region B is that which is normally used for recording, and has a dynamic range of ~80. That is the range of intensity from the weakest to the strongest. The region C corresponds to saturation and is non linear, eventually becoming horizontal.

All recording systems have similar response curves, with a normal CCD camera having a dynamic range of ~30, a slow scan CCD cameral having a dynamic range of ~3000, and the image plate having a value of several hundred. Their sensitivity varies considerably, but the problem is that the sharp spot diffraction normally recorded on the microscope has a dynamic range of several thousand from the weakest to the zero order reflection. Convenience and cost are also factors to be considered, but it must be recognised that no system is perfect and some compromises have to be made.

For radiation sensitive materials speed of recording is very important, and for diffraction - where the resolution of the film and specimen movement are less important - it is best to use a slow scan CCD camera at very low intensity, or photographic X-ray film. In the case of the latter the diffraction spots should be broadened to reduce the dynamic range of the pattern.

For imaging the dynamic range is less, and the time necessary for use of a slow scan CCD camera, and its lower resolution, make it less important, although it is very convenient. A normal CCD or videcon camera is of value here with a fast fourier transform system so that the power spectrum can be recorded rapidly, and the image retrieved later, although non-periodic detail will not be well resolved because of the lower resolution. It is also possible to use X-ray film at a higher magnification - to compensate for its lower resolution - since the shorter exposure necessary is less affected by specimen movement.

4.4 *Method for obtaining diffraction patterns.*

a. Initially concentrate on a single grid square, and check the microscope alignment with the smallest condenser aperture, and beam spot size, so that when the beam is spread it does not damage crystals adjacent to the area of interest. Correct the voltage centre, beam tilt, and astigmatism of both the image and the diffraction pattern. The object here is to obtain an electron beam optimised for coherent elastic scattering from the specimen, in the same way that such adjustments are done for high resolution imaging.
b. Focus the specimen carefully - Gaussian rather than Scherzer focus - and insert the selected area aperture. The focusing is critical since this brings the back focal plane of the objective lens (the diffraction pattern) coincident with the image plane of the first projector lens. The choice of

aperture size depends upon the crystals under observation. In general a more perfect pattern will be given from a smaller area.

c. Choose a short camera length since it is easier to observe, and when a suitable diffraction pattern is seen then the camera length can be extended appropriately.

d. Move the specimen with the grid square and observe the types and dimensions of the diffraction patterns present. At this stage one is familiarising oneself with the particular specimen, how fast it degrades, and the camera lengths, and exposure times it will be necessary to use. The gold coated specimen is an appropriate specimen to use at this time.

e. Move to another grid square, focus in one corner and then scan for suitable diffraction patterns with the beam stop in position. When a pattern is seen, record so that the minimum intensity spots will be recorded, although the low index spots will be over exposed. Leaving the beam stop in at this stage will not cause a significant loss of data if the spots are far from the centre and will prevent blackening by spill over from the high intensity reflections. Then defocus the diffraction spot size so that the spots become discs, remove the beam stop, reduce the intensity, and record so that the low intensity reflections are not obscured by the central beam.

f. The reason for the multiple exposures at different intensities is that it effectively extends the dynamic range of the film. If a slow scan CCD camera is used, then this is unnecessary. At this stage you should have a collection of reflections corresponding to accurate *d*-values using the gold calibration. In this projection one can take the highest spacings as arbitrary unit cell co-ordinates and index the other reflections. If the specimen is thick the diffraction pattern will not be extensive, since the thinner the specimen the longer are the reciprocal lattice spikes, which correspondingly will intersect the Ewald sphere at greater scattering angles.

g. To explore the full symmetry of the crystals further projections are needed, and hence it is necessary to tilt the specimen, preferably along a major axis so that some reflections are retained which are necessary for indexing and intensity calibration between projections. Preparation of the crystals by another method which gives another orientation is particularly valuable since it enables all of reciprocal space to be explored, instead of the 60degree tilt to which one is normally limited. Remember that tilt involves a vertical displacement of the specimen so that refocusing is necessary when the tilt angle is changed.

5. Treatment of results

5.1 INTENSITIES

The intensities can be obtained from the diffraction patterns by digitisation using a TV (CCD) camera or flat bed scanner. More accurate, but much slower, is a microdensitometer. Intensities from different projections can be scaled and merged, and the true unit cell axes and dimensions determined. It should be possible at this stage to resolve the space group, using the International Crystallography Tables, or at least to confine it to a few possibilities. Programs such as CRISP can assist in this determination. To aid the digitisation procedure it is possible to make prints of the negatives, since the enlargement will aid resolution. However, it is important that the same conditions of exposure and development are carried out to retain the relative intensities.

The intensities can now be converted to E-values, either by hand (very laborious!) or with standard programs.

5.2 PHASES

It is difficult to obtain good high resolution images from beam sensitive materials, but it is only from the image that the phases can be retrieved. Both intensities and phases are necessary for structure determination. Direct methods try to calculate the phases, and are successful if there is a large amount of intensity data, but even two or three relatively low resolution phases can greatly aid the procedure, and strengthen the result.

It is worth considering that for a molecule such a copper phthalocyanine which has 57 atoms per molecule, then despite its fourfold symmetry, it is difficult to get the 171 reflections necessary to give a three dimension resolution of all the atoms. This is a simplified argument which should consider the number of molecules in the unit cell, symmetry related atom positions and reflections, but the overall concept is valid, and it is a fact of electron diffraction that the data is very sparse. In consequence phase information is very valuable, and whilst it is as susceptible to dynamical effects as the intensities, phases of low index reflections should be tried to be obtained by imaging. Higher resolution phases suffer from changes in the contrast transfer function and other lens aberrations, and effects of radiation damage.

6.Recommended Reading

6.1 SPECIMEN PREPARATION AND ELECTRON MICROSCOPY

Ashida,M.,Uyeda,N. and Suito,E. (1966) The orientation overgrowth of metal phthalocyanines on the surface of a single crystal. II Vacuum condensed films of Cu-phthalocyanines on alkali halides. *Bull Chem.Soc.Japan.* **39** 2632-2638.

Ashida,M.,Uyeda,N. and Suito,E. (1966) The orientation overgrowth of metal phthalocyanines on the surface of a single crystal. II Vacuum condensed films of Cu-phthalocyanines on alkali halides. *Bull Chem.Soc.Japan.* **39** 2632-2638.

Dorset,D.L., Massalski,A.K. and Fryer,J.R. (1987) Interpretation of lamellar electron diffraction data from phospholipids. *Z.Naturforsch.* **42a** 381-391.

Buseck,P.R.,Cowley,J.M. and Eyring,L. (1988) *High resolution transmission electron microscopy.* Oxford.U.P. Oxford.

Fryer,J.R. (1989) High resolution imaging of organic crystals. *J.Elec.Microsc.Tech.* 11 310-325.

Fryer,J.R., and.Ewins,C. (1992)Epitaxial growth of thin films of perylene *.Phil Mag.* **A66** 889-898

Fryer,J.R., McConnell,C.H.,Hann,R.A., Eyres,B.L. and Gupta,S.K. (1990) The structure of some Langmuir-Blodgett films.I. Substituted phthalocyanines. *Phil.Mag.* **B 61** 843-852 .

Fryer,J.R., McConnell,C.H., Grant,G.A., Hann,R.A., Eyres,B.L. and Gupta,S.K. (1991) The structure of some Langmuir-Blodgett films.II. Aromatic polar molecules. *Phil.Mag.* **B 63** 1193-1200.

Fryer,J.R. and Kenney,M.E. (1988) Electron microscopy studies of the cofacial phthalocyanine polymers $(AlFPc)_n$ and $(SiOPc)_n$. *Macromol.* **21** 259-262.

Fryer,J.R. and Smith,D.J. (1981) Molecular detail in micrographs of quaterrylene *Nature* **291** 481-482.

Fryer,J.R. and Smith,D.J. (1982) High resolution electron microscopy of molecular crystals.I. Quaterrylene. *Proc.Roy.Soc(Lond.).***A381** 225-240.

Menter, J.W. (1956) The direct study by electron microscopy of crystal lattices and their imperfections. *Proc. Roy.Soc.(Lond.).* **A236** 119-135.

Miller,G., Fryer,J.R., Kunath,W.and Weiss,K. (1990a) The structure of an organo-azo-calcium salt by high resolution electron microscopy and image processing. In *Electron crystallography of organic molecules*(Ed. by J.R.Fryer and D.L.Dorset) pp 343-353 Kluwer,Dordrecht.

Murata,Y.,Fryer,J.R. and Baird,T. (1976) Molecular image of copper phthalocyanine. *J.Microsc.***108** 261-275 .

O'Keefe,M.A. Smith,D.J. and Fryer,J.R. (1983) High resolution electron microscopy of molecular crystals. II. Image simulation. *Acta Cryst.* **A39** 838-847.

Revol,J.F. and Chanzy,H. (1987) High resolution images of linear polymers. *Proc.45th Ann.Meeting EMSA*,Baltimore (Ed. by Bailey,G.W.) 480-483 San Francisco Press, San Francisco.

Smith,D.J., Saxton,W.O., O'Keefe,M.A., Wood,G.J. and Stobbs,W.M. (1983) The importance of beam alignment and crystal tilt in high resolution electron microscopy. *Ultramicrosc.* **11** 263-282.

Williams,R.C. and Fisher, H.W. (1970) Electron microscopy of tobacco mosaic virus under conditions of minimal exposure. *J.Mol.Biol.* **52** 121-123.

Uyeda,N.,Kobayashi,T.,Suito,E.,Harada,Y. and Watanabe,M. (1970) Direct observation of phthalocyanine molecules in epitaxial films. *Proc.7th Int.Cong.EM. Grenoble.***1** (Ed. by P.Favard) pp23-24 Societe Francais de Microscopie Electronique, Paris.

Uyeda,N.,Kobayashi,T.,Suito,E.,Harada,Y. and Watanabe,M. (1972) Molecular resolution in electron microscopy. *J.Appl.Phys.* **43** 5181-5189.

Wittmann,J.C. and Lotz,B.(1982) Crystallization of paraffins and polyethylene from the "vapour phase". A new surface decoration technique for polymer crystals. *Makromol. Chem. Rapid Comm.* **3** 733-738.

Zemlin,F.,Beckmann,E.,Reuber,E.,Zeitler,E. and Dorset,D.L. (1985) Molecular resolution electron micrographs of monolamellar paraffin crystals. *Science* **229** 461-462.

Zuo,J.M. (1996) Electron detection characteristics of slow scan CCD camera. *Ultramicrosc* **66** 21-33.

Zuo,J.M., McCartney,M.R., and Spence,J.C.H. (1996) *Ultramicrosc.* **66** 35-47.

6.2 RADIATION DAMAGE

Clark,W.R.K., Chapman,J.N., McLeod,A.M. and Ferrier,R.P. (1980) Radiation damage in copper phthalocyanine and its chlorinated derivatives. *Ultramicrosc.* **5** 195-208.

Dorset,D.L. and Turner,J.N. (1976) Thermal effects of electron beam damage of organic crystals. *Naturwissenschaften.* **63** 145.

Downing,K.H. and Glaeser,R.M. (1986) Improvement in high resolution image quality of radiation sensitive specimens, achieved with reduced spot size of the electron beam. *Ultramicrosc.***20** 269-278.

Fryer,J.R. (1987) The effect of dose rate on imaging organic crystals. *Ultramicrosc.* **23** 321-328.

Fryer,J.R., Hann,R.A. and Eyres,B.L. (1985) Single organic layer imaging by electron microscopy. *Nature* **313** 382-384.

Fryer,J.R. and Holland,F.M. (1984) High resolution electron microscopy of molecular crystals. III Radiation damage processes at room temperature. *Proc. Roy.Soc.(Lond.)* .**A393** 352-369.

Fryer,J.R.,McConnell,C.H.,Zemlin,F.,and Dorset,D.L.(1992) The effect of temperature on radiation damage to aromatic organic molecules. *Ultramicrosc.* **40** 163-169.

Henderson,R. and Glaeser, R.M. (1985) Quantitative analysis of image contrast in electron micrographs of beam sensitive crystals. *Ultramicrosc.* **4** 201-210.

International Study Group. (1986) Cryoprotection in electron microscopy. *J.Microsc.* **141** 385-391.

Kobayashi,T. and Reimer,L. (1975) Limits in the high resolution electron microscopy of halogen substituted organic molecule single crystals caused by radiation damage. *Bull.Inst.Chem.Res. Kyoto Univ.* **53** 105-116.

Salih,S.M. and Cosslett,V.E. (1984) Reduction in electron irradiation damage to organic compounds by conducting coatings. *Phil.Mag.* **30** 225-228.

Salih,S.M. and Cosslett,V.E. (1985) Radiation damage in electron microscopy of organic materials: effect of low temperatures. *J.Microsc.* **105** 269-276.

Siegle,G. (1972) The influence of very low temperature on the radiation damage of organic crystals irradiated by 100keV electrons. *Naturforschung* **27a** 325-332.

Symonds,M.C.R.(1982) Chemical aspects of electron beam interactions in the solid state. *Ultramicrosc.* **10** 15-24.

Vesely,D. and Downing,K.H. (1990) Radiation beam damage of polyethylene single crystals. In *Electron crystallography of organic molecules.*(ed. by Fryer,J.R. and Dorset,D.L.) 361-364 Kluwer, Dordrecht.

IMAGE FORMATION AND IMAGE CONTRAST IN HREM

F. H. Li
Institute of Physics & Center for Condensed Matter Physics,
Chinese Academy of Sciences, Beijing 100080, P. R. China

1. Introduction

Since the reports about directly observing atoms in crystals by means of high-resolution electron microscope images in the early seventies [1,2], the high-resolution electron microscopy (HREM) has become a powerful technique in determining structures and defects for minute crystals. The image formation and image contrast theories are given in detail in the book "Experimental High-Resolution Electron Microscopy" written by Spence [3]. The image formation process contains two steps —— the interaction of electrons with the object and the formation of image by the lens. There are different ways to describe the electron diffraction theory [4]. The one by means of the physical optics approach [5] is widely accepted in HREM. In the present paper the interaction of substances with electrons and the image formation process by the lens are briefly introduced. Two approximations of image contrast theory —— weak-phase object approximation (WPOA) and projected charge density approximation (PCDA) are given for interpreting the similarity between the image and the structure for very thin and slightly thicker crystals, respectively, and the pseudo-weak-phase object approximation (PWPOA) is introduced for discussing the image contrast change with the crystal thickness.

2. Propagation of electrons

Similar to the propagation of lights in light optics the propagation of electrons can also be described by the Kirchhoff formula which expresses the wave originated from any wave-field and propagated to a certain point [4]. In electron microscopy the electron wavelength is much smaller than the linear dimension of the object and the distance between the object and the observation plane. Under the small angle approximation the propagation of electrons from a plane perpendicular to the optical axis to a parallel observation plane (Fig. 1) becomes the case of Fresnel diffraction, and the Kirchhoff formula degenerates into

$$\psi(\mathbf{r}) = \frac{i\exp(-i\mathrm{k}\mathbf{R})}{\lambda\mathrm{R}}\mathrm{q}(\mathbf{r}) * \exp(\frac{-i\mathrm{kr}^2}{2\mathrm{R}}), \tag{1}$$

here $*$ represents the operation of convolution, q(**r**) the wave function on plane A and

D. L. Dorset et al. (eds.), Electron Crystallography, 29–39.

$\psi(\mathbf{r})$ on plane B, R the distance between plane A and B, **k** the wave vector and $k = 2\pi/\lambda$.

$$p_R(\mathbf{r}) = \frac{i}{\lambda R}\exp(\frac{-i k r^2}{2R}) \tag{2}$$

is named Fresnel propagator.

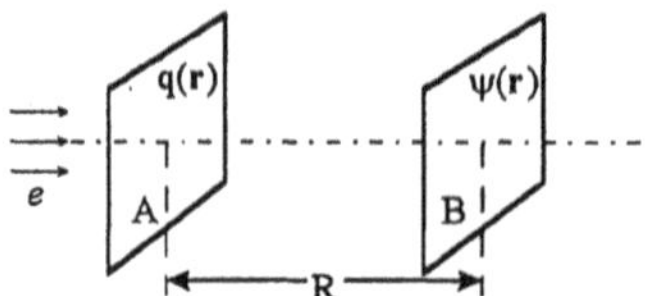

Fig. 1. The propagation of electrons. $q(\mathbf{r})$ and $\psi(\mathbf{r})$ are wave functions on plane A and B, respectively.

Equation (1) indicates that the wave on plane B will be the spherical wave with its amplitude proportional to the convolution of $q(\mathbf{r})$ and $p_R(\mathbf{r})$ and that a phase shift of $\pi/2$ occurs between $\psi(\mathbf{r})$ and $q(\mathbf{r})$. The function $i\exp(-i\mathbf{kR})p_R(\mathbf{r})$ is the wave function on plane B obtained from a point source located at the intersection of plane A with the optical axis. For simplicity, hereafter $\exp(-i\mathbf{kR})$ will be omitted in all wave functions.

3. Interaction of substances and electrons

3. 1. PHASE CONTRAST AND PHASE OBJECT

There are three different image formation mechanisms in transmission electron microscopy, which correspond to three kinds of image contrast, respectively. They are the absorption contrast, diffraction contrast and phase contrast. The phase contrast comes from the phase change of electrons passing through the object, and the amplitude change of electrons is assumed to be negligible. Such object is named the phase object.

3. 2. TRANSMISSION WAVE AND TRANSMISSION FUNCTION

Interacting with the object the electrons are modulated by the electric potential of the object. The modulated electron wave carrying the object structure information is formed in the bottom surface of the object. When the incident electron wave is a plane wave with unit amplitude the modulated electron wave at the bottom surface is named the transmission wave and the corresponding wave function is the transmission function.

3. 2. 1. Transmission function of phase object

The electron wavelength in vacuum is expressed as

$$\lambda = h / \sqrt{2m_0 eE}\,, \tag{3}$$

or

$$\lambda = h / \sqrt{2m_0 eE(1 + \frac{eE}{2m_0 c^2})} \tag{4}$$

when the relativistic correction is considered, where h is the Planck's constant, m_0 the rest mass and e the charge of an electron, E the accelerating voltage and c the light velocity. Apart from the kinetic energy eE included in equation (3), entering the object the electrons gain the potential energy $e\varphi_0(\mathbf{r})$. $\varphi_0(\mathbf{r})$ is the potential distribution function inside the object. Thus the electron wavelength changes into

$$\lambda' = h / \sqrt{2m_0 e(E + \varphi_0(\mathbf{r}))}\,. \tag{5}$$

The phase shift between the incident wave and transmitted wave is

$$\alpha = 2\pi t(\frac{1}{\lambda'} - \frac{1}{\lambda}), \tag{6}$$

t is the thickness of the object. Let $\varphi(\mathbf{r})$ represent the projected potential distribution function (PPDF) of the object in the direction of the incident beam, when t is very small,

$$\varphi(\mathbf{r}) = \int_0^t \varphi_0(\mathbf{r})dz \approx \varphi_0(\mathbf{r})t\,. \tag{7}$$

Because $\varphi_0(\mathbf{r}) << E$ in electron microscopy,

$$\sqrt{E + \varphi_0(\mathbf{r})} \approx \sqrt{E}(1 + \varphi_0(\mathbf{r}) / 2E)\,. \tag{8}$$

To substitute equation (3), (5), (7) and (8) into (6) leads to

$$\alpha = \sigma\, \varphi(\mathbf{r}), \tag{9}$$

where σ ($=\pi/\lambda E$) named interaction constant. The transmission function of a phase object is written as

$$q(\mathbf{r}) = \exp[-i\sigma\varphi(\mathbf{r})]\,. \tag{10}$$

A more complete expression should be

$$q(\mathbf{r}) = \exp[-i\sigma\varphi(\mathbf{r}) - \mu(\mathbf{r})]\,, \tag{11}$$

here $\mu(\mathbf{r})$ represents the change of electron wave amplitude due to the absorption. It can be ignored for a thin object.

3. 2. 2. Weak-phase object

When $\sigma\varphi(\mathbf{r}) << 1$, equation (10) can be approximated to

$$q(\mathbf{r}) \approx 1 - i\sigma\varphi(\mathbf{r})\,. \tag{12}$$

Equation (12) gives the transmission function of the weak-phase object.

3. 3. PHYSICAL OPTICS APPROACH TO THE DYNAMICAL ELECTRON DIFFRACTION [5]

The object is cut into thin slices perpendicular to the incident beam (Fig. 2). Each thin slice can be treated as a two-dimensional phase object.

Let Δz_j, $\varphi_j(\mathbf{r})$ and $q_j(\mathbf{r})$ represent the thickness, the PPDF and the transmission function of the jth slice, respectively. The transmission of electrons through the object

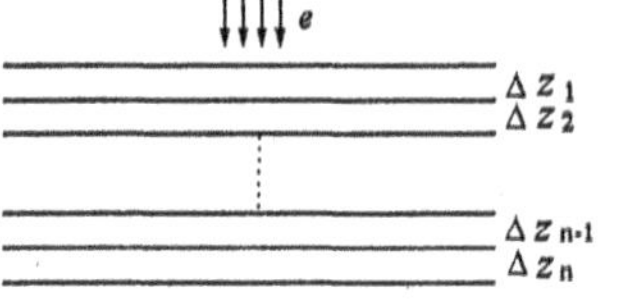

Fig. 2. Schematic diagram of the physical optics approach. The object is cut into n slices. Δz_j denotes the thickness of the jth slice.

can be represented by transmission through each slice and propagation from one slice to the next by Fresnel diffraction. The phase shift of the electron wave passing through each slice is assumed to take place at the top surface. The wave transmitted through the first slice and subsequently propagated to the top surface of the second slice is

$$\psi_1(\mathbf{r}) = q_1(\mathbf{r}) * p_{\Delta z_1}(\mathbf{r})\,. \tag{13}$$

The wave function at the bottom of the second slice, or say, the top surface of the third slice is

$$\psi_2(\mathbf{r}) = q_2(\mathbf{r})[q_1(\mathbf{r}) * p_{\Delta z_1}(\mathbf{r})] * p_2(\mathbf{r}). \tag{14}$$

Finally, the wave function at the bottom surface of the nth slice, namely, the transmission function is

$$q(\mathbf{r}) = \psi_n(\mathbf{r})$$

$$= q_n(\mathbf{r}) \underset{n-1}{[} \; q_{n-1}(\mathbf{r}) \cdots \underset{2}{[} q_2(\mathbf{r}) \underset{1}{[} q_1(\mathbf{r}) * p_1(\mathbf{r}) \underset{1}{]} * p_2(\mathbf{r}) \underset{2}{]} * \cdots * p_{n-1}(\mathbf{r}) \; \underset{n-1}{]} * p_n(\mathbf{r}) \tag{15}$$

When the slice thickness tends to zero and the number of slices tends to infinity, formula (15) can be used to represent rigorously the scattering from any three-dimensional object.

4. Image formation by lens

From the standpoint of the lens the transmission wave of the object is the object wave. Suppose the objective lens is an ideal thin lens. The image formed by the objective lens will be magnified by the intermediate and projective lenses. In the following the process of image formation by the objective lens will be discussed. It can be divided into several steps as follows (Fig. 3).

4. 1. WAVE FUNCTION AT FRONT ENTRANCE OF LENS

Let $\psi_a(\mathbf{r})$ denote the wave function at the front entrance of the objective lens. It is the convolution of the object wave function $q(\mathbf{r})$ and the Frensnel propagation factor $p_a(\mathbf{r})$:

$$\psi_a(\mathbf{r}) = q(\mathbf{r}) * p_a(\mathbf{r}), \qquad (16)$$

here a denotes the distance between the object plane and the lens.

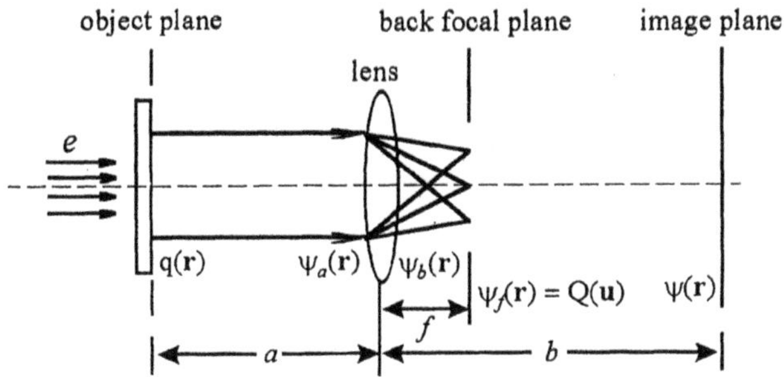

Fig. 3. Image formation process. $q(\mathbf{r})$ and $\psi(\mathbf{r})$ are the object wave and image wave function, respectively. $\psi_a(\mathbf{r})$ and $\psi_b(\mathbf{r})$ the wave function in the front and back entrances of the lens, respectively. $\psi_f(\mathbf{r}) = Q(\mathbf{u})$ the diffracted wave function.

4. 2. LENS ACTION AND WAVE FUNCTION AT BACK ENTRANCE OF LENS

The wave emitted from a point source located at the focus of the lens becomes a plane wave after passing through the lens. Let the Dirac function $\delta(\mathbf{r})$ represents the point source located at the front focus of the lens. After propagating from the front focus to the front entrance of the lens, the wave function becomes

$$\delta(\mathbf{r}) * p_f(\mathbf{r}) = \frac{i}{\lambda f}\exp(-\frac{i\pi}{\lambda f}\mathbf{r}^2), \qquad (17)$$

here f denotes the focus length of the lens. Passing through the lens, the amplitude of wave function turns into unity. Hence, the lens action is a multiplier reversed to equation (17). Therefore, the wave function at the back entrance of the lens is

$$\psi_b(\mathbf{r}) = [q(\mathbf{r}) * p_a(\mathbf{r})][\frac{\lambda f}{i}\exp(\frac{i\pi}{\lambda f}\mathbf{r}^2)]$$

$$= [q(\mathbf{r}) * \exp(\frac{-i\pi}{\lambda a}\mathbf{r}^2)][\frac{f}{a}\exp(\frac{i\pi}{\lambda f}\mathbf{r}^2)]. \qquad (18)$$

4. 3. DIFFRACTED WAVE

The wave function at the back focal plane of the lens is

$$\psi_f(\mathbf{r}) = \psi_b(\mathbf{r}) * \mathrm{p}_f(\mathbf{r})\,. \tag{19}$$

If the object is located at the front focal plane of the lens, namely, $a = f$, and let

$$\mathrm{u} = \mathrm{r} / \lambda f\,, \tag{20}$$

then after substituting (18) into (19) and ignoring the constant, the diffracted wave function is expressed as

$$\mathrm{Q}(\mathbf{u}) = \int \mathrm{q}(\mathbf{r}) \exp(2\pi i \mathbf{u}\mathbf{r}) d\mathbf{r}\,, \tag{21}$$

or

$$\mathrm{Q}(\mathbf{u}) = \mathcal{F}\,[\mathrm{q}(\mathbf{r})]. \tag{22}$$

$\mathcal{F}$ denotes the operation of Fourier Transform (FT). Hence, the diffracted wave function Q(**u**) is the FT of the object wave function.

4. 4. IMAGE FROM AN IDEAL THIN LENS

For an ideal thin lens, the image wave function on the Gaussian image plane is

$$\psi(\mathbf{r}) = \psi_f(\mathbf{r}) * \mathrm{p}_{b-f}(\mathbf{r})\,, \tag{23}$$

here b denotes the distance between the lens and the image plane. From the lens law for conjugate planes $b - f = \mathrm{M}f$, where M is the lateral magnification. Substituting equation (19) and (2) into (23) and for simplicity assuming the magnification equals unity give the image wave function as

$$\psi(\mathbf{r}) = \int \mathrm{Q}(\mathbf{u}) \exp(2\pi i \mathbf{u}\mathbf{r}) d\mathbf{u} = \mathrm{q}(-\mathbf{r})\,. \tag{24}$$

Equation (24) indicates that the image wave reveals the object wave with a rotation angle equal to π. In fact the rotation angle can be arbitrary value other than π. Therefore, it is convenient to write

$$\psi(\mathbf{r}) = \mathrm{q}(\mathbf{r}) = \mathcal{F}^{-1}\,[\mathrm{Q}(\mathbf{u}\,] \tag{25}$$

and to consider the rotation angle separately, here $\mathcal{F}^{-1}$ denotes the operation of inverse FT.

4. 5. CONTRAST TRANSFER FUNCTION

Practically, before propagating to image plane, electrons are modulated by various electron optical parameters such as spherical and chromatic aberration of the lens, the defocus, the crystal and beam tilt, the beam divergence, the astigmatism, etc.. The modulation takes place on the back focal plane so that a multiplier named contrast transfer function (CTF) acts on the diffracted wave.

Let s(**r**) denote the FT of the CTF T(**u**), the image wave function is written as

$$\psi(\mathbf{r}) = \mathcal{F}^{-1}[Q(\mathbf{u})T(\mathbf{u})] = q(\mathbf{r}) * s(\mathbf{r}) . \tag{26}$$

If the object wave is a Dirac Delta function:

$$q(\mathbf{r}) = \delta(\mathbf{r}) , \tag{27}$$

then

$$\psi(\mathbf{r}) = \delta(\mathbf{r}) * s(\mathbf{r}) = s(\mathbf{r}) . \tag{28}$$

Equation (28) means that the image of a point source is no more a point and its figure is described by function s(**r**) which is named the spread function.

The CTF can be written as

$$T(u) = \exp(i\chi_1)\exp(-\chi_2)\frac{2J_1(\chi_3)}{\chi_3} = T_1(\chi_1)T_2(\chi_2)T_3(\chi_3) , \tag{29}$$

$$\chi_1(u) = \frac{\pi}{2}C_s\lambda^3u^4 + \pi\Delta f\lambda u^2 , \tag{30}$$

$$\chi_2(u) = \frac{1}{2}\pi^2\lambda^2D^2u^4 , \tag{31}$$

$$\chi_3(u) = \left|2\pi\theta\ [\Delta f u + \lambda(\lambda C_s - i\pi D^2)u^3]\right| , \tag{32}$$

where J_1 is a Bessal function of the first order, C_s denotes the spherical aberration coefficient, Δf the defocus, D the standard deviation of Gaussian distribution of defocus due to the chromatic aberration [7] and θ the semi-angle of beam divergence. The first part of CTF $T_1(\chi_1)$ is a function oscillating between -1 and +1 (Fig. 4a), while other two parts [$T_2(\chi_2)$and $T_3(\chi_3)$] contribute a damping envelope to the first part (Fig. 4b).

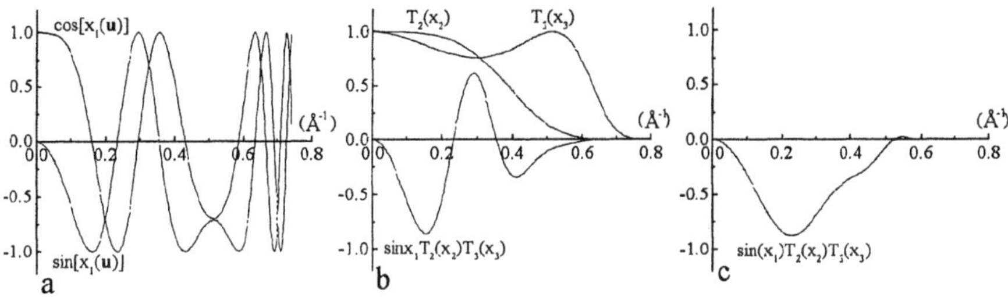

Fig. 4 Curves of (a) sin[χ_1(u)] and cos[χ_1(u)], (b) $T_2(\chi_2)$, $T_3(\chi_3)$ and $\sin\chi_1T_2(\chi_2)T_3(\chi_3)$, (c) $\sin\chi_1T_2(\chi_2)T_3(\chi_3)$ at Scherzer focus.

When $\Delta f = -(\lambda C_s)^{1/2}$, the CTF approximately equals -1 in a wide range of spatial frequencies (Fig. 4c) and most of the diffracted waves will contribute to the image with correct phase. This focus values correspond to the so called Scherzer focus [8].

4. 6. IMAGE WAVE FUNCTION

The image wave function considering various electron-optical parameters is expressed as

$$\psi(\mathbf{r}) = \mathcal{F}^{-1}[Q(\mathbf{u})T(\mathbf{u})A(\mathbf{u})]$$

$$= q(\mathbf{r}) * \mathcal{F}^{-1}[T(\mathbf{u})A(\mathbf{u})], \tag{33}$$

where A(**u**) is the objective aperture function. For a circular aperture

$$A(u) = \begin{cases} 1, & u \leq u_0 \\ 0, & u > u_0 . \end{cases} \tag{34}$$

5. Approximate image contrast theories

5. 1. IMAGE INTENSITY CALCULATION

Equation (33) indicates that the image carries the structure information of the object, but the information is modulated by the CTF. The image intensity depends also on the object thickness. There are different methods to calculate the image intensity [3]. The multi-slice method based on the physical-optics approach to the dynamical electron diffraction theory [5] is widely accepted and the results are in good agreement with the experiments, especially for crystals with a large unit cell. The image intensity equals the product of the image wave given by equation (15) and its conjugate. In this case the influence of the object thickness on the image contrast is caused by the Fresnel propagators contained in equation (15). However, it is far from obvious to understand the image contrast change with crystal thickness from this equation. Some approximate image contrast theories are beneficial to image contrast interpretation.

5. 2. WEAK-PHASE OBJECT APPROXIMATION

The diffracted wave function corresponding to the transmission function given in equation (12) for a weak-phase object is

$$Q(\mathbf{u}) = \delta(\mathbf{u}) - i\sigma F(\mathbf{u}), \tag{35}$$

here F(**u**) is the FT of $\varphi(\mathbf{r})$:

$$F(\mathbf{u}) = \mathcal{F}[\varphi(\mathbf{r})]. \tag{36}$$

For crystalline objects F(**u**) is the structure factor. The modulated diffracted wave is

$$Q(\mathbf{u}) = \delta(\mathbf{u}) - i\sigma F(\mathbf{u})T(\mathbf{u})\,. \tag{37}$$

The image wave function is obtained by inverse Fourier transforming equation (37):

$$\psi(\mathbf{r}) = 1 - i\sigma\varphi(\mathbf{r}) * \mathcal{F}^{-1}[T(\mathbf{u})]\,. \tag{38}$$

If only the first part of CTF is considered, after ignoring the square term of $\sigma\varphi(\mathbf{r})$, the image intensity is expressed as

$$I(\mathbf{r}) = 1 + 2\sigma\varphi(\mathbf{r}) * \mathcal{F}^{-1}[\sin\chi_1(u)]\,. \tag{39}$$

The intensity of structure image taken at the Scherzer focus is

$$I(\mathbf{r}) \approx 1 - 2\sigma\varphi(\mathbf{r})\,. \tag{40}$$

Equation (40) gives a clear interpretation that the experimental image reveal the projected structure intuitively. Such image is named the structure image.

5. 3. PROJECTED CHARGE DENSITY APPROXIMATION

WPOA should be valid only for very thin crystal (of thickness about 10 Å for atoms of medium weight). However, the experimental structure images are obtained from crystals much thicker than a weak phase object. The electron charge projection approximation was developed to explain this. It is based on the phase object approximation and the defocus is taken into consideration while the spherical aberration is ignored. In this case the image wave function is written as

$$\psi(\mathbf{r}) = \mathcal{F}^{-1}[Q(\mathbf{u})\exp(i\pi\Delta f\lambda u^2)] \tag{41}$$

if the envelope function is ignored. When the defocus amount is not very large,

$$\exp(i\pi\Delta f\lambda u^2) \approx 1 + i\pi\Delta f\lambda u^2\,, \tag{42}$$

then

$$\begin{aligned}\psi(\mathbf{r}) &= q(\mathbf{r}) + i\pi\Delta f\lambda\mathcal{F}^{-1}[Q(\mathbf{u})u^2] \\ &= q(\mathbf{r}) - \frac{i\lambda\Delta f}{4\pi}\nabla^2[q(\mathbf{r})]\end{aligned} \tag{43}$$

To substitute equation (10) into (43) and to ignore the second order terms of Δf and σ yield the image intensity

$$I(\mathbf{r}) = 1 - \frac{\lambda\Delta f}{2\pi}\sigma\nabla^2[\varphi(\mathbf{r})]\,. \tag{44}$$

According to the Poisson equation

$$\nabla^2[\varphi(\mathbf{r})] = -4\pi\rho(\mathbf{r})\,, \tag{45}$$

here $\rho(\mathbf{r})$ denotes the projected charge density. Hence,

$$I(\mathbf{r}) = 1 + 2\lambda\Delta f\sigma\rho(\mathbf{r}).\tag{46}$$

According to equation (46), when $\Delta f = 0$, the image is of no contrast; when $\Delta f < 0$ (underfocus), the image intensity is deficient at positions of atoms as under the WPOA; and when $\Delta f > 0$ (overfocus) the image contrast reverses. The PCDA can give an interpretation to structure images of crystals thicker than weak-phase objects, and to the image contrast reverse with the defocus. Although equation (46) was derived without the consideration of spherical aberration, it is still valid qualitatively when the spherical aberration is considered [9]. Detailed calculations of the amplitude and phase of diffracted waves suggest that the phase object approximation is likely to hold to a thickness of about 70 Å for specimens of medium atomic weight. However, the validity of PCDA decreases with the increase of resolution, because the effect of spherical aberration becomes stronger in the high resolution case.

5. 4. PSEUDO WEAK-PHASE OBJECT APPROXIMATION

In fact, even when the resolution is better than 2 Å and the crystal thickness is much larger than that of a weak-phase object, for instance, equal to several tens of Angstroms, the image still can reveal the projected structure of crystals consisting of rather heavy atoms. But the contrast of atoms is not linear to the potential distribution. The contrast of light atoms usually appears higher and that of heavy atoms lower than they should be in WPOA. And heavy atoms may turn to bright at the Scherzer focus when the thickness reaches a certain value. This can neither be explained by WPOA nor by PCDA. Another approximate contrast theory — PWPOA was set up to give an interpretation [10]. It was derived similar to the multi-slice method. The crystal is cut into very thin slices and the transmission of electrons through the crystal is represented by transmission of electrons through each slice and propagation from one slice to the next. In order to see the change of image intensity with the crystal thickness more and less obviously two approximations were made in the derivation of image intensity formula. One is to assume that all slices are the same weak-phase objects. The second one is to ignore the double scattering in the interaction of two adjacent slices. Thus the image intensity at the Scherzer focus condition is expressed as:

$$I(\mathbf{r}) = 1 - 2\sigma[\varphi(\mathbf{r}) + \Delta\varphi(\mathbf{r})],\tag{47}$$

and the increment $\Delta\varphi(\mathbf{r})$ depends on the crystal thickness [10]. When the crystal contains only one slice, the increment in equation (47) is neglible. With the increase of crystal thickness, the image contrast of light atoms increases more rapidly than that of heavy atoms. When the crystal thickness reaches a critical value, the image contrast of heavy atoms vanishes at the centers of atoms. With the further increase of crystal thickness, the image contrast reverses (Fig. 5). The image simulation results by using formula (47) is in agreement with that by using the multi-slice method for crystal thickness below the critical value.

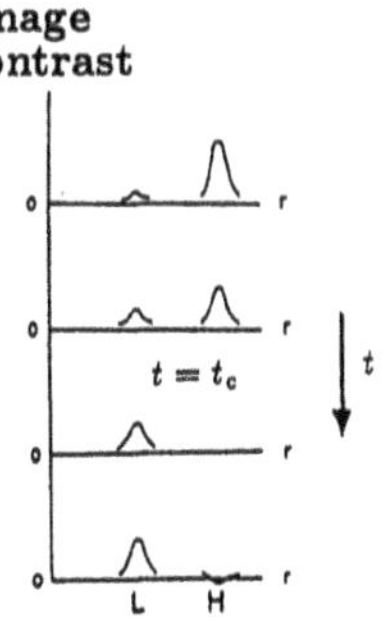

Fig. 5 Schematic diagram showing the image contrast change with object thickness for light (L) and heavy (H) atoms.

6. Validity of WPOA

When one is interested in the position of atoms but not the accurate peak height in the structure image, the WPOA, or say, the modified WPOA is valid if the crystal thickness is below the critical value. The critical crystal thickness depends on the electron wavelength and the weight of the heaviest atoms in the crystal. For the medium accelerating voltage of electrons (200 $\sim$ 400 kV), generally, the critical thickness is about several tens of angstroms. It becomes larger for a smaller electron wavelength and lighter constituent atoms.

References

[1] Cowley, J. M. and Iijima, S. (1972) Electron microscope image contrast for thin crystals, *Z. Naturforsch.* **A27**, 445-451.

[2] Uyeda, N., Kobayashi, T., Suito, E., Harada, Y. and Watanabe, M. (1972) Molecular image resolution in electron microscopy, *J. Appl. Phys.* **43**, 5181-5189.

[3] Spence, J. C. H. (1981) *Experimental high-resolution electron microscopy*, Clarendon Press, Oxford.

[4] Cowley, J. M. (1975) *Diffraction Physics*, North-Holland, Amsterdam.

[5] Cowley, J. M. and Moodie, A. F. (1957) The scattering of electrons by atoms and crystals, *Acta Cryst.* **10**, 609-619.

[6] Hirsch, P. B., Howie, A., Nicholson, R. B., Pashley, D. W. and Whelan, M. J. (1965) *Electron microscopy of thin crystals*. Butterworths, London.

[7] Fejes, P. L. (1977) Approximations for the calculation of high-resolution electron microscope images of thin films, *Acta Cryst.* **A33**, 109-113.

[8] Scherzer, A. (1949) The theoretical resolution limit of electron microscope, *J. Appl. Phys.* **20**, 20-29.

[9] Lynch, D. F., Moodie, A. F. and O'Keefe, M. A.(1975) n-Beam lattice image V. The use of charge density approximation in the interpretation of lattice image, *Acta Cryst.* **A31**, 300-307.

[10] Li, F. H. and Tang, D. (1985) Pseudo-weak-phase-object approximation in high-resolution electron microscopy, *Acta Cryst.* **A41**, 376-382.

ELECTRON MICROSCOPY TECHNIQUES

H.W. ZANDBERGEN
National Centre for HREM, Laboratory of Materials Science, Delft University of Technology, Rotterdamseweg 137, 2628 AL Delft, The Netherlands

1. Introduction

This contribution deals with the broad field of electron microscopy. The main objectives are to give a brief overview of the various techniques, with an emphasis on their application in structure analysis, either by direct imaging or by diffraction. Electron microscopy can be done in transmission or in back-scatter mode. The back-scatter mode is used in scanning electron microscopy (SEM). The transmission mode is used in transmission electron microscopy (TEM).

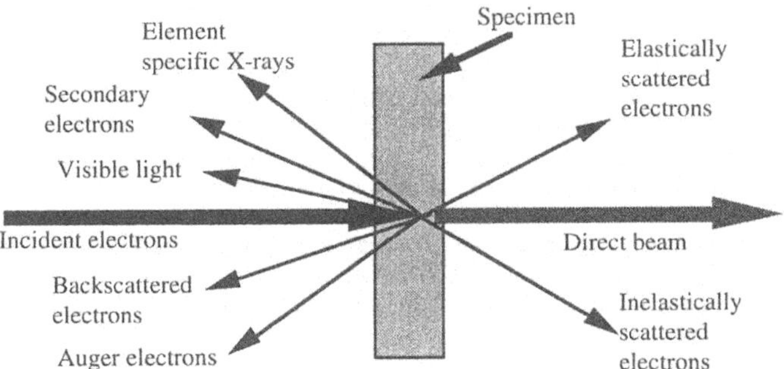

Figure 1. Schematic representation of the most important signals generated by the interaction of the incident electrons with a specimen.

2. Electron-matter interactions

Electrons have, because they are charged, a strong interactions with matter. Also the elastic cross section for the elastic scattering of electron is very high, which implies that the amount of material needed for a diffraction experiment is much less than needed for X ray diffraction For electron diffraction one requires a volume being 10^5 -10^9 times smaller than that of X-rays. Typically the specimen should be thinner than about 50 nm to be sufficiently transparent for the electron beam. High energy electrons as used in electron microscopy are furthermore capable of removing one of the tightly bound inner shell electrons form the attractive field of the nucleus. A major advantage of this ionising nature of the electrons is that this can lead to a range of secondary signals as shown in Figure 1. Some of these signals are frequently used in analytical electron microscopy. The signals most frequently used are the X-rays. This requires the determination of either the wavelength (Wavelength dispersive X-ray analysis(WDX)) or the energy (Energy dispersive X-ray analysis) of the created X-rays. On the other hand the energy loss of the electron can be determined by measuring the energy of the electrons which have passed through the specimen, a technique which is called electron energy loss spectroscopy (EELS). These element analysis techniques will be shortly discussed in section 10.

D. L. Dorset et al. (eds.), Electron Crystallography, 41–54.

3. Types of electron microscopes

Electrons can be focused by magnetic lenses. As such they act similar to glass lenses in an optical microscope. In all electron microscopes it is the quality of the lenses which determines the performance of the microscopes, e.g. the resolution and the smallest probe size.

SEM

In SEM the specimen is scanned by a focused electron beam. The signals created by this incident electron beam and which can escape from the specimen can be detected by placing appropriate analysers above the specimen. Some of these signals are shown in Figure 1. SEM is an excellent technique to determine the morphology of a specimen and by means of EDX the local composition. Little structural information can be obtained, but if one knows the structures of the phases in the specimen it is possible to determine the orientation of grains at the surface. This van be done by Electron back scatter diffraction (EBSD) which uses the structure in the diffraction pattern of the back scattered electrons, which is in first approximation similar to the Kikuchi patterns obtained for rather thick specimens in diffraction patterns recorded in TEM's. The diffraction pattern of the back scattered electrons, which are mainly created near the surface, allows one to determine the orientation of the surface grains and more importantly the orientation relationships of neighbouring grains. This technique works well if the grains are sufficiently large (about 0.5 μm or larger). The Auger electrons, which can only escape from the top 2 nm from the specimen, are the research medium in Scanning Auger Microscopy, which is a typical surface characterisation technique.

TEM

In a TEM the electron beam is passed through the specimen. Similar to the SEM the signals created by the electron/specimen interaction and which leave the top surface of the specimen can be analysed. EDX element analysis is a routine accessory of TEM's. Less frequently detectors for recording of the back-scatter electron are used and then only in combination with a scanning attachment (STEM). This is mainly due to the low yield of back scatter electrons in very thin specimens. Recently an Auger-TEM has been developed which allows the detection of a high fraction of the Auger electrons such that very small areas (about 20 nm^2) can be analysed.
The electrons which have passed through the specimen can be used for analysis of the microstructure (conventional TEM), the atomic arrangement (HREM), electron diffraction (ED) or element analysis (EELS). In the following some important features of a TEM will be briefly described.

STEM

In a STEM one uses a small probe to scan the specimen. The transmitted electrons are subsequently analysed. Probe sizes as small as 0.2 nm have been achieved in dedicated STEM's. In STEM/TEM's the minimum probe size is about 0.5 nm. The advantage of a dedicated STEM over combination instruments is that the whole design of the microscope can be optimised. High resolution microscopy has been shown to be feasible by STEM's with a resolution of approximately 0.2 nm.

4. Electron sources

Electron sources are required to provide either a large total current when one works at a low magnification for which one needs illumination of an area as large as several 100's of μm, or a high intensity probe as small as 0.5 nm for analytical purposes, for high resolution STEM or for phytography. Three types of electron sources are available: the tungsten hairpin filament, the lanthanum hexaboride (LaB_6) single crystal and the field emission gun (FEG).

The source brightness is defined as the current density per unit solid angle and is measured in ($Acm^{-2}sr^{-1}$). This is very different from the intensity of the source, which is a measure of the total number of electrons emitted from the source, which tends to be larger for larger sources. The brightness determines the total current which can be focused on the specimen, which is thus the determining factor for the ease of use of very small probes. The theoretical maximum intensity is

$$I(\text{theoretical}) = B_s(\pi^2\alpha^2 d^2)/4$$

where B_s is the brightness of the source, α the semiangle of the probe-forming lens and d the diameter of the source. The main cause of the energy spread of the electrons is due to the temperature of the tip.

Table 1. Some properties of electron sources

Source	Temperature of tip (K)	Brightness ($Acm^{-2}sr^{-1}$)	Source size (μm)	Energy spread (eV)
W	2800	10^5	50	3.0
LaB_6	2100	10^6	1	1.5
cold FEG	RT	10^9	0.005	0.4
thermal FEG	1800	10^9	0.020	0.7

5. Condenser system

In the present TEM's the condenser system is composed of several lenses, which allow a very flexible illumination, because on the one hand one wants a very broad beam when one works in low magnification, a parallel beam for HREM studies and a very small beam with high intensity for spot analysis. The set-up of the condenser system for a Philips microscope is shown in Figure 2. It consists of two adjustable condenser lenses (C1 and C2), a condenser aperture, a twin lens (T) and the pre-field of the objective lens (Ob).

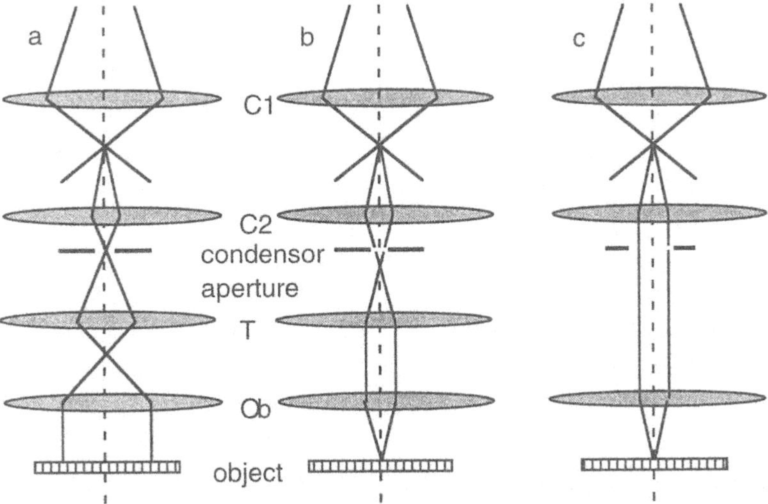

Figure 2. Schematic representation of some electron rays through the condenser system resulting in a) a parallel illumination, b) a small spot size and c) a very small spot size. In the latter configuration the spot size can be smaller because C1 cross-over is imaged on the specimen, which is further away that the C2 cross-over as used in b).

6. Specimen stage

The standard size of the specimens is a thin disk with a diameter of 3 mm. Obviously for a structure analysis one needs the possibility to tilt the specimen in any orientation, which requires two tilts (or a combination of a specimen rotation and a tilt). Apart from this it can be necessary to either cool or heat the specimen. All of these requirements have been realised in commercial specimen holders. The only major improvements to be made are the resolution of the cooling and heating holders and their specimen drift. These improvements are not required for electron diffraction, but are absolutely necessary for obtaining HREM images. In general, the mechanical stability of the specimen stage has to be excellent to obtain images with a high resolution.
In the TEM study of organic materials the cooling of the specimen is often required. Cooling to liquid nitrogen temperature reduces the irradiation damage of the electron beam. By cooling to liquid helium temperature an extra reduction of the beam damage can be obtained.

7. Image forming lenses

The most important lens of the TEM is the objective lens. In particular the spherical aberration of this lens (see section 9.1) determines the resolution of the microscope. The objective lens is followed by several intermediate lenses which allow the imaging of either the object or the diffraction pattern as is schematically shown in Figure 4. The combination of these lenses allow a magnification between about 100x and 1000.000x. The latter magnification is only relevant if the resolution of the microscope is adequate. In some microscopes some extra intermediate lenses are added to correct for the rotation of the image, which is inherent to the use of magnetic lenses for the magnification. An objective aperture can be placed in the backfocal plane which allows one to select specific diffracted beams. If one omits the central beam this application is called dark field imaging. This allows one to determine from which area of the object this diffracted beam originates as is shown in Figure 5.

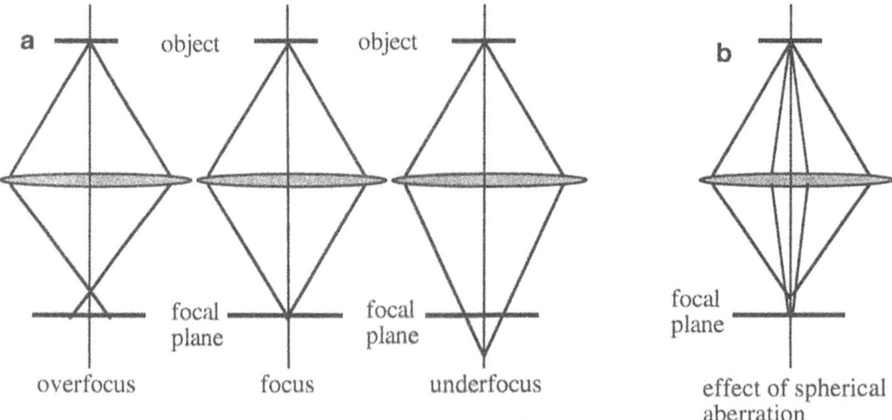

Figure 3. a) definition of the overfocus and underfocus. b) the spherical aberration of the objective lens results in a stronger focusing of the electrons diffracted over a larger angle. This effect can be partly compensated by using an underfocus, as is also evident from the transfer function $\chi(g) = \pi C_S \lambda^3 g^4 + \pi \varepsilon \lambda g^2$*, where* C_S *is the spherical aberration,* λ *the wavelength,* g *a reciprocal lattice vector and* ε *the defocus.*

Figure 4. In the top a schematic representation of the imaging mode and the diffraction mode of a TEM is given. a)-d) show the use of a selected area aperture.

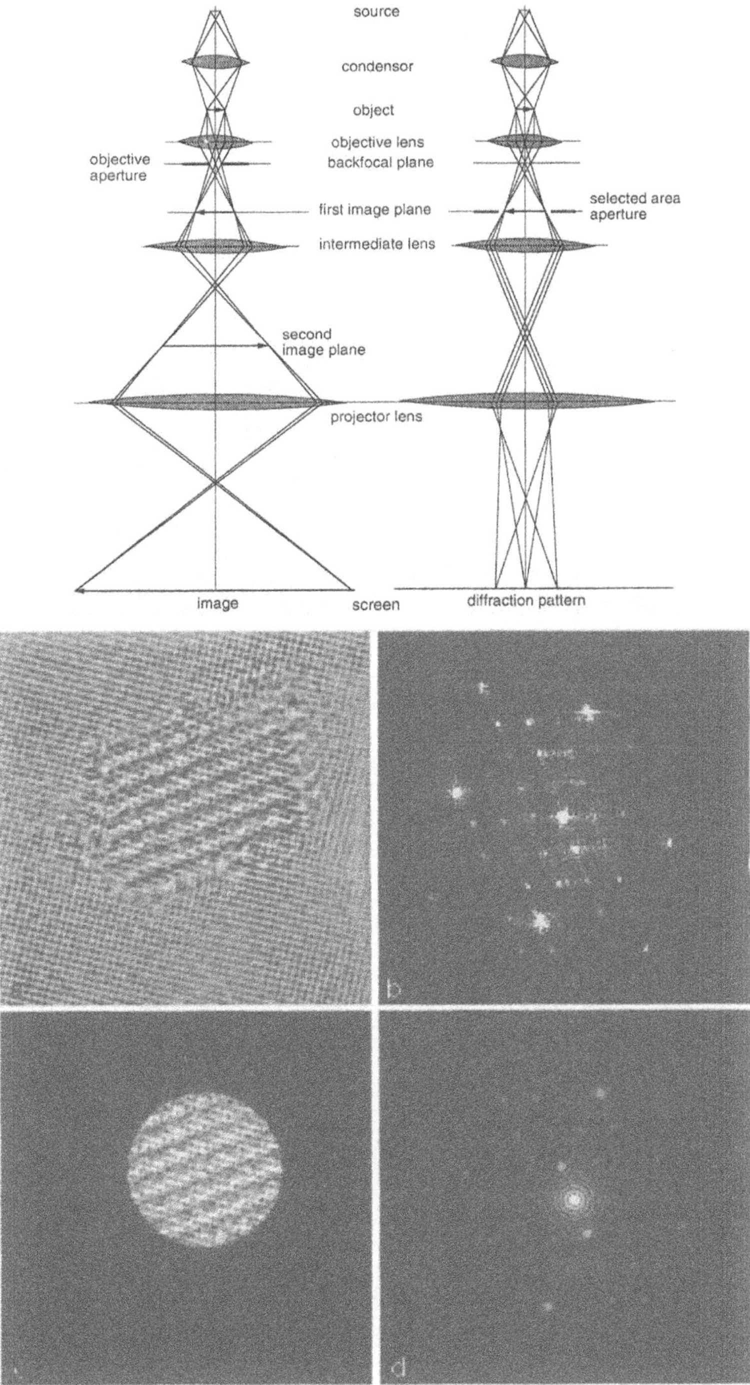

Figure 5. Example of the selection of information in the backfocal plane by an objective aperture. The example given here is done by computer by selection in Fourier space which is analogous.
a) is a high resolution image of a small Mg_5Si_6 precipitate in Al. b) is the Fourier transform of the image in a) which is equivalent to the diffraction pattern in the backfocal plane. c) give the inverse Fourier transform of the information within circle 1. d) gives the inverse Fourier transform of the information within circle 2. Since only reflections of the precipitate occur in this circle, only a lattice image occurs in the precipitate. e) shows the inverse Fourier transform of the information within circle 3 and the central spot. Due to the interference of the information in the circle 3 and that of the central spot, the corresponding interference fringes can be seen. This operation is difficult to perform in the electron microscope because it requires two apertures. f) shows the inverse Fourier transform of the information in circle 4. This selection can be easily done in the electron microscope by placing a small aperture in the appropriate position in the backfocal plane. Since the central beam is omitted no interference fringes are observed.

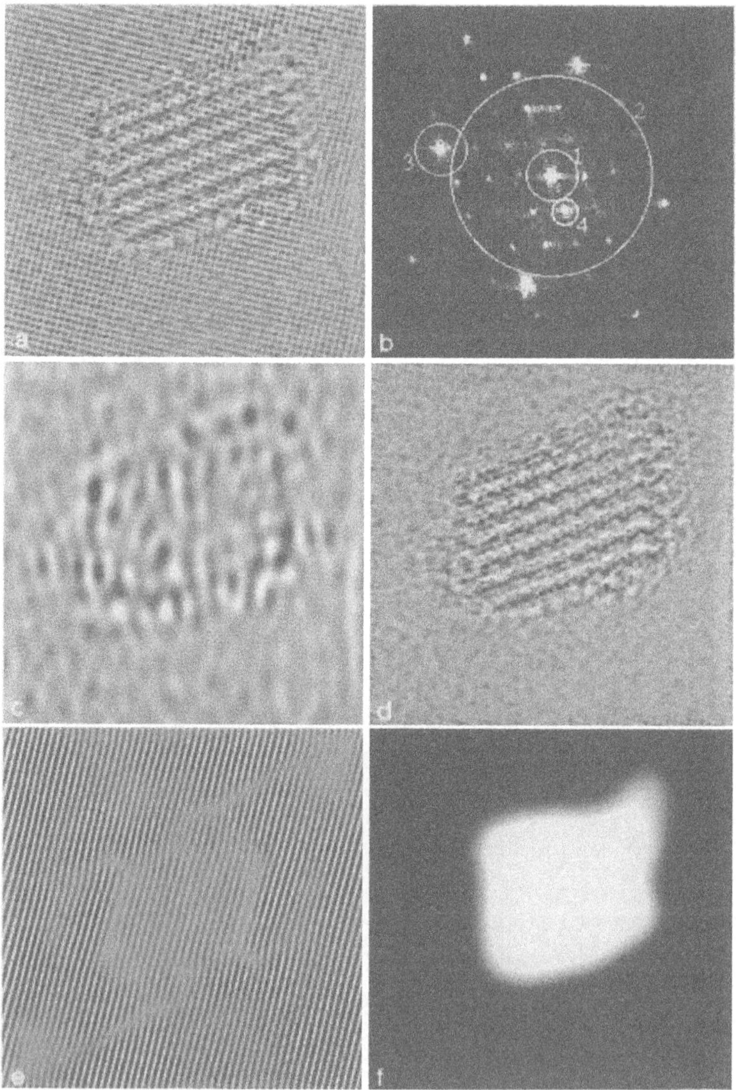

8. Image recording systems

Photographic film
This is still by far the most used medium for recording of transmission electron microscope images. Major advantages are the ease of use, the large field of view combined with a high resolution, the ease of storage. Major disadvantages are the required processing steps (in general it takes at least 1 hour), the very limited linearity of the intensity recording and the need for a digitisation if the images are to be processed in the computer. An additional problem with electrons with energies higher that about 100 keV is that the electrons can cross the photographic plate, so that back-scattered and secondary electrons generated by the supporting holder can lead to an additional background.

Real time cameras
Real time cameras consist of a scintillator to transform electrons into light and either a CCD detector coupled with the scintillator by a fibre-optics device or an optical system (optical lenses) with an image intensifier. The major advantages of these cameras are that one can observe changes such as specimen movement and focusing real time. Major disadvantages are the limited field of view (typically 500 by 500 pixels), the high noise level and the low dynamic range. These cameras are not used to obtain quantitative data, but changes can be followed and recorded easily. There is a continuous improvement in the image quality and the field of view of these cameras. Real time cameras as an absolute must in high resolution electron microscopy.

Slow scan CCD cameras
Slow scan CCD cameras consist of a scintillator to transform electrons into light which is coupled to a high quality CCD detector by a fibre-optics device. The reduce the noise level the CCD camera is mostly cooled to about -20°C. The interaction of the electron with the scintillator results in the creation of a number of photons. This process leads to a broadening of 10-15 mm for 100 keV electrons and 20-30 for 300 keV electrons. Consequently the pixels of the CCD device must be of this size, because for smaller pixel sizes part of the resolution is lost. The pixel numbers of the commercial large pixel CCD devices are 512x512, 1024x1024 and 2048x2048. The major advantages are the linearity in the intensity detection, the dynamic range (2^{12}-2^{16}), the image is immediately available. The major disadvantages are overflow artefacts when electron diffraction patterns are recorded and the price.

Image plates
Image plates consist of a film of chemical molecules which are excited by the electron beam without relaxation to the ground state. Only by irradiating the molecules by light of a certain wavelength the exited state is relaxed under the emission of light. Since the emitted light has another wavelength than the incident light, an exposed image plate can be read out by scanning a laser beam over the image plate and simultaneous detection of the emitted light. The image plate has a very high dynamic range (about 2^{24}), but the detection system for the emitted light has a dynamic range of only 2^{16}. However, the plate can be read out several times such that by using a filter for the light detector, the whole dynamic range of the image plate can in principle be used. The major advantages are the linearity in the intensity detection and the very high dynamic range. The major disadvantages of the image plate are the processing time and the price.

9. Electron microscope modes for structure analysis

9.1 High resolution electron microscopy (HREM)

In HREM two definitions are used to indicate the information content of HREM images. The first is the point-point resolution, which is the minimum distance between two point-like features in the object, which can - with a favourable defocus of the microscope - still be distinguished from each other it can be written as

$$\rho_{\text{p-p}} = 0.65\, C_S^{1/4} \lambda^{3/4}$$

where C_s is the spherical aberration of the objective lens and λ the wavelength of the electrons. Obviously the resolution of the electron microscope can be improved by reducing C_s or λ.. λ is about 4 pm, 2 pm and 1 pm for 100 keV, 300 keV and 1000 keV electrons respectively. For this reason electron microscopes with an acceleration in the range from 1000 to 1500 keV have been built. A decrease in C_s results in a less strong improvement in $\rho_{\text{p-p}}$, because of its 1/4 power. C_s can be reduced by a better choice of materials for the objective lens and a reduction in the space between the two pole pieces of the objective lens. The latter has as disadvantage that the specimen area is reduced and thereby the possibilities for specimen manipulation in particular specimen tilting.

The second useful definition is that of the information limit, ρ_{IL}, which describes the limit below which no significant information is obtained. This limit is mainly determined by the energy spread of the incoming electrons and the current spread of the current in the objective lens, both of which can be described with a defocus spread Δ.

$$\rho_{\text{IL}} = (\pi\lambda\Delta/2)^{1/2} \quad \text{with} \quad \Delta = C_c \left\{ \left(\frac{\Delta E}{E}\right)^2 + \left(\frac{\Delta I}{I}\right)^2 + \left(\frac{\Delta V}{E}\right)^2 \right\}^{1/2}$$

where C_c is the chromatic aberration of the objective lens, ΔI, ΔE, and ΔV the spread of the current I, the energy E of the electrons and the acceleration voltage V respectively. The largest improvements in the information limit can be obtained by reducing the energy spread of the electrons and by reducing the C_c of the objective lens. With the top high resolution electron microscopes the information limit is often determined by the object itself, because of the decrease in the amplitude of the scattering beam with increase of the diffraction angle, the specimen degradation by the electron beam, some charging by the electron beam, and small specimen vibrations. Table 1 gives some values for C_s, C_c and the pole gap for several Philips 300 kV electron microscopes

Table 2. Several data of some Philips 300kV electron microscopes

microscope	C_s (mm)	C_c (mm)	pole gap (mm)	$\rho_{\text{p-p}}$(Å)	ρ_{IL}(Å)
CM300Twin	2.0	2.0	9	2.3	2.0
CM300Supertwin	1.3	1.5	5.4	2.0	1.8
CM300Ultratwin	0.7	1.2	2.5	1.7	1.4
CM300Ultratwin/FEG	0.7	1.2	2.5	1.7	0.9

For lower voltage and intermediate voltage high resolution microscopes the information limit exceeds the point-point resolution. The information between $\rho_{\text{p-p}}$ andρ_{IL} will have a phase shift which differs at least $\pi/2$ from that of the information up to $\rho_{\text{p-p}}$. The

information between $\rho_{p\text{-}p}$ andρ_{IL} is delocalized in the image to such an extent that this information hampers a direct interpretation of the image. For this reason one either has to exclude this information by using an objective aperture or one has to do image calculations or one has to perform exit wave reconstructions which can be performed by either off-axis holography [see section 12.2] or through focus electron holography [see section 12.1]. For high-voltage high-resolution high resolution microscopes the point-point resolution exceeds the information limit. Consequently the delocalisation of information is negligible in this case, provided the proper focus is used.
Other important factors to obtain a high resolution are the mechanical stability of the microscope, the mechanical stability and low specimen drift of the specimen stage and the specimen itself (electron beam damage and contamination).

9.2 Electron diffraction

Selected area diffraction
In this mode of the electron microscope one places an aperture in the first image plane (see Figure 4). In this way an area can be selected of which the diffraction data can be obtained by switching to the diffraction mode. The lower limit of this method is about 1 μm. Selecting smaller areas is not realistic because the accuracy to place the aperture exactly in the first image plane and the further alignment of the microscope do not allow this.

Small beam diffraction
With the present electron microscopes one can focus the electron beam to spots much smaller than 1 μm, which is approximately the limit of selected area diffraction. The focusing of the electron beam will, however, result in a certain convergence. This leads to a broadening of the diffraction spots to diffraction disks. For strongly convergent beams this broadening can be so large that the neighbouring reflections are overlapping. To a certain extend one can use the diffraction lens to focus the disks to spots. On the other hand the diffraction disks do contain a wealth of information, which is the domain of convergent beam electron diffraction. In case of overlapping reflection the phase difference between the two overlapping reflections can be determined. The minimum spot size one can obtain depend on the size of the source and the size of the condenser aperture (virtual source size). Since the source size of the field-emission gun is much smaller than that LaB_6 or W cathodes (see Table 1) and because it has a much higher brightness a field emission gun is much more convenient for quantitative electron diffraction.

10. Element analysis

The element analysis which is most frequently used is Energy Dispersive X-ray Analysis (EDS). This requires the determination of either the wavelength (Wavelength dispersive X-ray analysis(WDX)) or the energy (Energy dispersive X-ray analysis) of the X-rays generated when an electron from a higher energy fall back to the electron hole caused by the removal of an inner shell electron by the primary electron of the electron beam. WDX has a much higher energy resolution than EDX, but it is rather time consuming. This method is most frequently used in electron microprobes (EPMA) which are quite similar to a SEM.

The typical probe size for EPMA's is 1-2 μm. In the EDX or WDX analysis in SEM's and EPMA's one has to take into account that the volume hit by the electron beam is quite large as is shown in Figure 6. X-rays from the deep areas are likely to be absorbed, in particular the low energy X-rays. Consequently a rather complicated correction (including the measurements of standards) is required. EDX is a quick method for obtaining a qualitative estimate of the composition. In TEM's one can obtain with a moderate electron beam

intensity and a spot size of 100 nm a quick fingerprint in only a few seconds. Quantitative EDX is possible with an accuracy better than 1%.
In Electron Energy Loss Spectroscopy (EELS) the energy loss of the electrons is measured. The electrons of different energies are separated by a magnetic prism and are either measured with a photomultiplier behind a slit (serial EELS) or by a CCD camera behind a scintillator and a fibre-optic window (parallel EELS). For quantitative EELS it is essential that the specimen is about as thin as for HREM imaging, which is roughly below 20 nm. The advantage of EELS over EDX is that the light elements (up to F) can be detected more accurately.

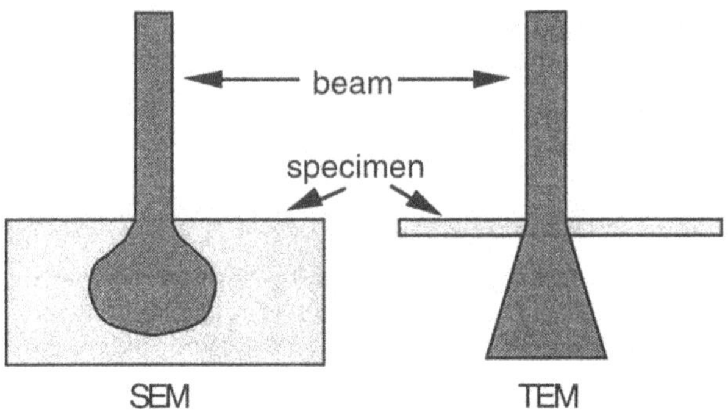

Figure 6. The spread of the electron beam in a SEM and a TEM. Due to the much smaller thickness and the higher acceleration voltage in the TEM analysis the spread of the beam in the specimen is negligible.

11. Specimen preparation

In the TEM study of inorganic materials, four ways of sample preparation are employed: 1) crushing resulting in small sometimes wedge-shaped fragments with a thin enough edge; 2) chemical polishing with a jet-stream of polishing liquid; 3) ion milling in which ions are focused on a specimen under a low angle thus sputtering atoms and conglomerates of atoms from the surface and 4) ultramicrotomy in which very thin slices are cut from the sample. The first technique is mostly applied for some homogeneous bulk materials and can almost never be applied for the study of grain boundaries and interfaces and for materials which change under the pressure of the crushing (metals). With this method the thin area is very small. The second and third technique, presently most widely used in materials science, can provide very thin specimens. However, an inherent problem of these techniques is the presence of a layer (amorphous or crystalline) on the surface which is different from the 'real' material. The fourth method has as major disadvantages the creation of artefacts and the collection of the specimen which is still done on a water surface and can result in changes at the surface of the slices. However, most artefacts created by ultramicrotomy can be identified as due to this technique. Ultramicrotomy is widely used for making cuts of biological materials and is optimised in particular for these types of specimens, but it increasingly used by materials scientists, in particular for polymers and catalysts [1]. Cutting can be performed in a controlled atmosphere as is done routinely in cryo-ultramicrotomy of biological specimens.

In the TEM study of organic materials several methods are used: 1) ultramicrotomy as described above. Also for these materials, artefacts due to the cutting procedure are likely to occur. In particular when one cuts crystals, the resulting crystal slices are likely to be strongly bend, which makes this method less suited for obtaining electron diffraction patterns. 2) growth from a dilute solution after which small crystals are collected on a thin carbon film or direct growth on the carbon film. 3) growth from a vapour phase either directly on a carbon support or on a substrate which can be dissolved (e.g. NaCl), 4) breaking of larger crystals in an ultrasonic bath, 5) melting and subsequent cooling of small pieces of material distributed in some way on the carbon support 6) ion milling was successfully applied by us on several organometallic compounds. In general the sample preparation of organic materials is more time consuming and sometimes even the most important bottleneck.

12. Special techniques to obtain a higher resolution

Figure 7 gives a schematic representation of the distortions of the microscope in the imaging process and in particular where in the imaging process information is lost. Obviously one tries either to reduce the loss of information by improving the optics of the microscope or one tries to restore this information. The last approach is used in two recently developed methods which are shortly described in sections 12.1 and 12.2. Figure 7 also shows the relation between the phase and amplitude in real space and in diffraction space.

In the central column of Figure 7 the course of the evolution of the electron wave is given. It starts with a plane wave which by interaction with the specimen changes into a modified wave, which is called the exit wave. The exit wave can be described in real space (A) and in diffraction space (A"). In both descriptions the exit wave is complex. Both representations can be translated into each other by a mathematical Fourier transform and have the same information content although in another form. The amplitude of the representation in diffraction space contains information from the amplitude as well as the phase of the real space representation. The same holds for the phase of the representation in diffraction space. The transfer function of the microscope (A to A') scrambles the phase and amplitude information. However, by applying the transfer function no information is lost and thus the information content of A, A' and A" is the same. The introduction of the focus spread (A' to B) leads to a loss of information, which can no longer be restored. Thus B cannot be translated into A. The focus spread leads to a reduction of the high frequency information, which can be described easiest in diffraction space by a damping of the amplitude, which has a g^4 dependence of the diffraction vector g. By imaging (image recording) the amplitude is imaged and the phase information is lost, irrespective of whether the imaging is done in diffraction space or real space. The information imaged in real space (C) is very much different from the information imaged in diffraction space (diffraction pattern (D)) and consequently this information cannot be translated into each other. Note that, due to the transfer function, the information in the HREM image contains amplitude as well as phase information of the exit wave, which cannot be unscrambled.

A HREM image provides a good representation of the atomic arrangement up to the point-to-point resolution of the microscope only in the case of a weak phase object (no amplitude in exit wave) and optimum focus of the microscope. If the specimen is not a weak phase object, which is almost always the case, the image is more complicated because it is a mixture (scrambled by the transfer function) of phase and amplitude information. Due to non-linear interaction (which is the interaction between two scattered beams, whereas linear interaction is the interaction between a diffracted beam and the central beam) the information in a HREM image (but not in the exit wave) can reach beyond the information limit of the microscope. It is not possible to unscramble the linear and non-linear information from a single HREM image. This is, however, possible by holographic electron microscopy (see sections 12.1 and 12.2). In the case of exit wave reconstructions by holographic methods one determines the complex exit wave at the image plane and corrects for the transfer function of the microscope. The complex wave obtained in this way does not have the same information as A because of the focus spread. Correction for this spread is not possible because the required information is simply not present in the HREM images.

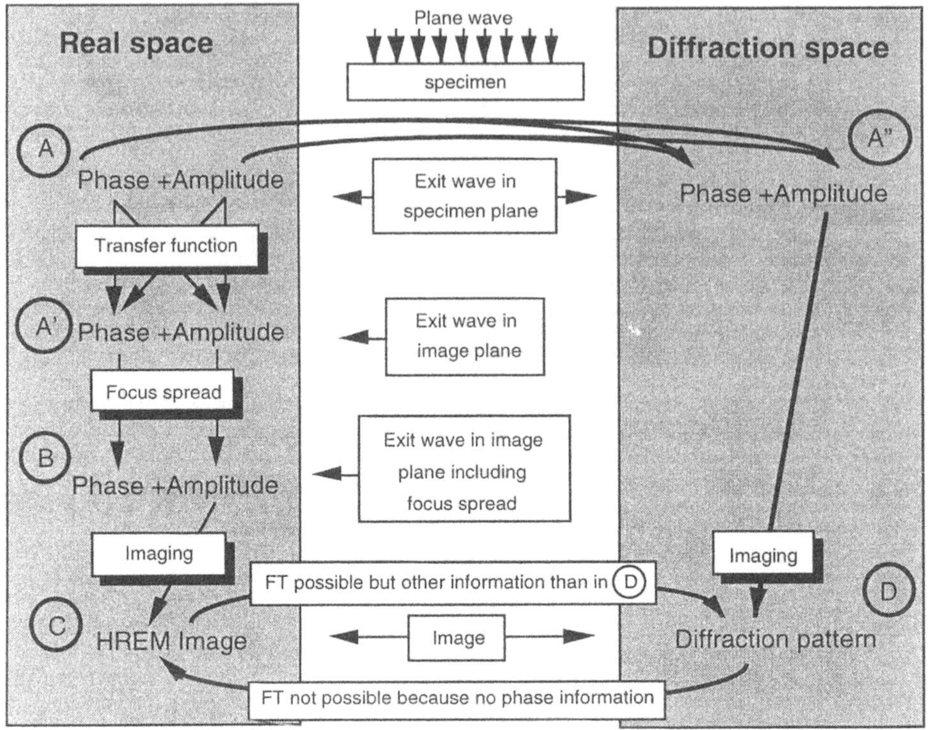

Figure 7. Schematic representation of the relation between the phase and amplitude information in real space and diffraction space and where information is lost in the imaging of the diffraction pattern and of the HREM image. See text for further explanation.

12.1 Through focus electron holography

Through-focus electron holography is an electron microscopy technique [2] which combines the information from a series of high resolution electron microscope images to detailed information which is not distorted by the electron microscope optics. In conventional high resolution electron microscopy one often chooses a defocus setting and objective aperture which provides information up to the point resolution of the electron microscope. This point resolution defines the minimum distance at which, for a very thin specimen, two individual atoms can still be distinguished. Information beyond this point resolution is present in the image, but is scrambled due to the imperfect optics of the microscope. The point resolution is mainly set by the aberrations of the objective lens. The information limit of the microscope, on the other hand, is mainly determined by the energy spread of the electrons and the microscope's vibration. For most electron microscopes the information limit reaches quite far beyond the point resolution.

The through focus exit wave reconstruction uses the focus dependence of the image distortion by the electron microscope to restore the distortion by the microscope optics, as shown schematically in Figure 8. Using algorithms developed by Van Dyck and Coene [2], all the useful information is extracted from a series of high resolution electron microscope images taken at different defocus values with well defined defocus increments. The result is the exit wave function which contains amplitude as well as phase information up to the information limit of the microscope. The exit wave is the electron wave at the exit of the specimen. Since the electron wave passing through the crystal is continuously changed, the exit wave function is still thickness dependent, but not dependent on the distortion introduced by the electron microscope. As such it provides major advantages over conventional high resolution electron microscopy.

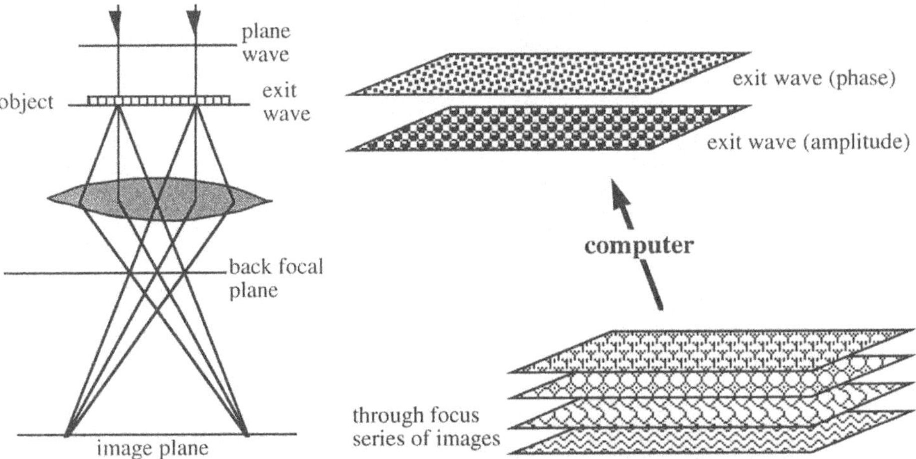

Figure 8. Schematic representation of through focus electron holography. One records a series of HREM images with known focus steps. With these HREM images and the microscope parameters a computer program estimates a first exit wave, which is subjected to the transfer of the microscope allowing a comparison of the experimental HREM images and the calculated ones, which allows a further refinement of the exit wave.

12.2 Off-axis electron holography

In off-axis holography one uses a biprism to create an interference pattern between the part of the beam which has past through a hole with the part which has past through the specimen as is schematically illustrated in Figure 9. The positions of the interference fringes are shifted due to phase changes in the specimen, as is illustrated in Figure 10. This additional information allows one to reconstruct the exit wave [2]. In practice one selects the so-called off-axis diffraction pattern. This off-axis diffraction pattern occurs next to the central (normal) diffraction pattern at a position determined by the spacing and orientation of the interference fringes.

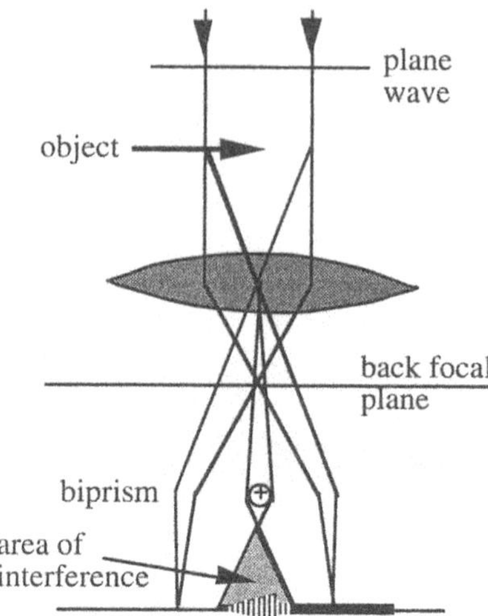

Figure 9 Schematic representation of the experimental set-up for off-axis electron holography. The beam passing through the hole and the beam passing through the object are brought into interference by the biprism.

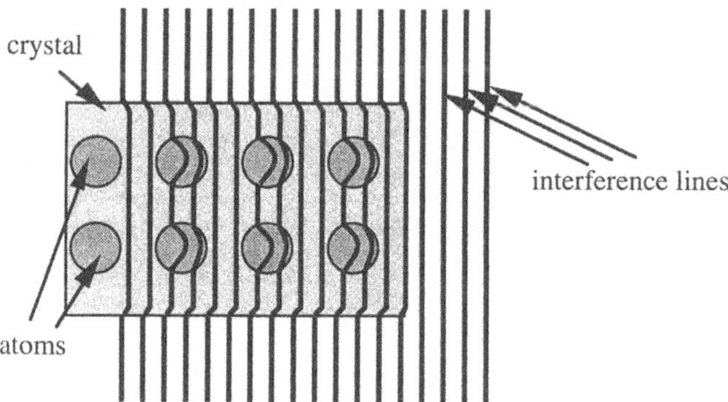

Figure 10. Schematic representation of the interference pattern. Due to the average potential of the crystal the lines of destructive interference are shifted. On atoms these lines are shifted more. The image combines contrast and phase from which the exit wave can be reconstructed.

References

1. T.F. Malis and D. Steele, Workshop on Sample Preparation for TEM of Materials, *Mat. Res. Symp. Proc.* vol 199 (1990) 1
2. Maximum likelihood method for focus-variation image reconstruction in high resolution transmission electron microscopy. W.M.J. Coene, A. Thust, M.Op De Beeck, D.Van Dyck, *Ultramicroscopy* 64, 109 (1996), Sub-Ångström structure characterisation, D. van Dyck, H. Lichte and K.D. van der Mast, *Ultramicriscopy* 64, 1 (1996) and further papers in this volume

DEFINITION, MEASUREMENT AND CALCULATION OF INTENSITIES IN ELECTRON DIFFRACTION

J. GJØNNES
Department of Physics, University of Oslo
PO Box 1048 Blindern
N-0316 OSLO, Norway

1. Introduction

A major problem in electron crystallography has been to obtain intensity data of a quality and extension sufficient for determination and refinement of crystal structure. In this chapter we shall discuss some approaches to this problem, starting from the definition of intensities, in relation to the experimental techniques: *selected area diffraction* (SAD), *micro/nano-diffraction*; *convergent beam diffraction* (CBED), and a *precession technique* - and in relation to calculations. I shall emphasize the need for three-dimensional intensity data, and discuss methods for including or correcting dynamical scattering in connection with structure determination of inorganic materials.

There are two basic principles for recording diffracted intensities from crystals, *viz:*

i) *rocking curves*, *i.e.* intensity profiles through the crystalline reflections G, as function of their excitation errors, s_G, or in two dimensions as function of lateral components of the incident beam wave vector **k**:

$$I_G(s_G) \quad or \quad I_G(k_x,k_y)$$
$$s_G = -(G^2+2k_0G)/2k \qquad (1)$$

ii) *integrated intensities* of the reflections:

$$\int I_G(s_G)ds_G \equiv I_G \qquad (2)$$

i.e. one number is recorded for each reflection, *G*. In X-ray diffraction this number is proportional to the square of the structure factor:

$$\int I_G^X(\boldsymbol{Q}-\boldsymbol{G})d\boldsymbol{Q} = |F_G^X|^2\int|S(\boldsymbol{Q}-\boldsymbol{G})|^2 d\boldsymbol{Q} = const|F_G^X|^2 \qquad (3)$$

Since X-ray scattering is kinematic, the shape factor *S(Q)* is the same for all reflections, and can be integrated out along a path defined by the experiment. The wide range of validity of this simple relation is a chief factor behind the success of X-ray crystallography. X-ray people do not have to worry much about the definition of intensity.

D. L. Dorset et al. (eds.), Electron Crystallography, 55–64.

In electron diffraction the situation is different. The relation between diffracted intensity and structure factor is a complicated one, due to dynamical or multiple scattering effects, which as a rule will include several simultaneously excited diffracted beams. The intensity $I_G(k_x,k_y)$ in a reflection G is then an unknown function of several structure factors, U_G, excitation errors, s_G, and thickness, t:

$$I_G(k_x,k_y)=I_G(U_G;U_H,..;s_G,s_H,...;t) \tag{4}$$

The intensity has to be calculated by numerical procedures, as is explained in other chapters. An unknown intensity function is not readily integrated - but there are situations where integrated intensity is a useful concept also in electron diffraction, we shall return to such cases. Another, third alternative may be to record

iii) a set of diffracted intensities at one particular direction of the incident beam; preferably for conditions where the intensities are not too sensitive to beam direction.

We shall discuss electron diffraction experiments and the specimens we are interested in, in relation to these principles, and establish, as our conclusion: that meaningful electron diffraction intensities can be measured and be useful for structure determination for a wide range of substances, and in some cases be refined to high precision.

2. Electron diffraction experiment

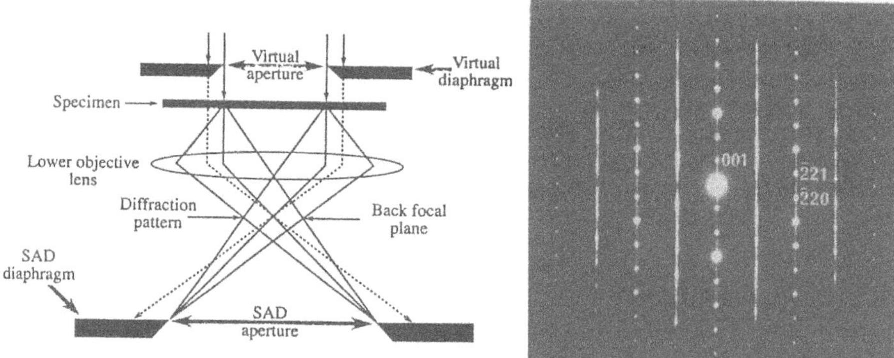

Fig.1 *Principle of selected area diffraction (from Williams and Carter, [1])*

Fig. 2 *Selected area diffraction pattern from a faulted crystal, Al_3Fe*

SAD, the selected area spot pattern taken along a zone axis is the most common type of electron diffraction, Fig1 An aperture in the image plane of the objective lens selects part of the object through its virtual image back on the object. The SAD technique, in combination with movable apertures, workable gonimeter stages and other features of the microscope column established the electron microscope as a crystallographic instrument around 1960. In the classical SAD configuration, with a highly focussed second condensor, small divergence and sharp well-defined diffraction spots are obtained, from crystal areas down to a μm or a little less, as determined by spherical aberration in the objective. SAD is an ideal technique for determination of Bravais lattices, to detect superstructure spots, explore crystallographic relations in polyphase materials *etc*. But the *intensities* of SAD diffraction spots are not so well defined, because:

- thickness and orientation may vary over the μm-size selected area
- with the beam down the zone axis the reflections are off Bragg condition ($s_g<0$)
- dynamical interactions are extensive when many beams are simultaneously excited.

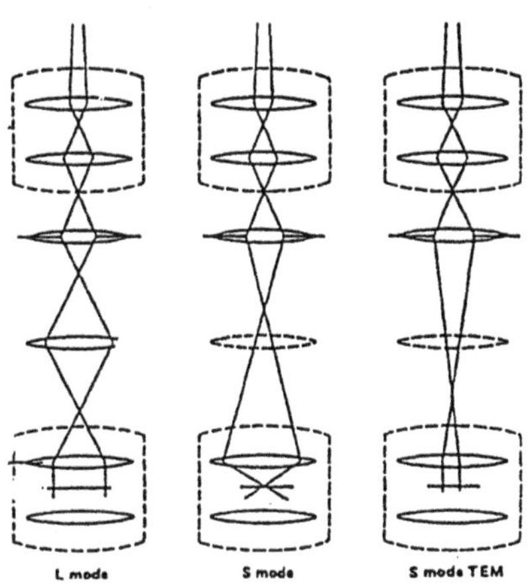

Fig.3 *Diffraction modes in a modern electron microscope. L-mode: SAD; S:CBED; S-mode TEM: microdiffraction (JEM 2000FX)*

Unless the crystal is very thin the SAD-intensities are therefore poorly defined averages over a complicated function. The situation is better in the *micro- and nano-diffraction* modes that became available with modern objective lenses of twin-lens or condensor-objective type, Fig 3. These include a strong fore-field where a parallel beam can be condensed down to about 10nm or less, with high intensity on the crystal, and useful also for analytical techniques. Beam divergence may be somewhat higher than in standard SAD. A slight spread of the intensity over angles (and on the plate) may actually be an advantage; however, it is usually assumed that the recorded intensities refer to one particular incident beam direction. The crystal should therefore be thin, so that the product $s_G t$ of excitation error and thickness is small (<1) for the reflections to be measured, which may call for a thickness of 100Å or less (as for high-resolution imaging). This condition is derived from the geometry of Fig. 3 and simple kinematical or two-beam expressions for the diffracted intensity:

$$I_G(s_G,t)^{kinematical} = |U_G/k|^2\, t\, \frac{\sin^2[\pi t s_G]}{\pi t s_G}$$

$$I_G(s_G,t)^{2\text{-}beam} = \frac{|U_G/k|^2}{s_G^2+|U_G/k|^2}\sin^2[\pi t\sqrt{s_G^2+|U_G/k|^2}] \qquad (5)$$

By comparison of the two equations (5) it is seen that the useful thickness range will be extended by dynamical scattering (and even more in many-beam cases in a dense zone) which tend to widen the central maximum of $I(s_G)$, and of course by bending; both "benefits" will lead to more complicated calculations.

The alternative to spot patterns is *convergent beam electron diffraction* (CBED) which is discussed in detail in later chapters. The standard CBED-pattern is a set of two-dimensional rocking curves, which can be used for accurate determination of structure factors in simple, known structures. In addition to the results obtained in terms of bond charges and ionicities, these studies have demonstrated that the dynamical theory gives a very accurate quantitative description of the scattering - once the crystal shape and size is well specified. But the CBED profile technique does not at present offer a practical way to collect suficient intensity data for determination of an unknown crystal structure with a moderate-size unit cell. The profiles are very sensitive to imperfections; ideally the crystal should be perfect and flat. Calculations tend to be complicated and extensive, and not well suited for unknown structures. For *ab initio* structure determination something different is needed. What are the options?

. integrate CBED-features;
. integrate in the experiment: a precession technique;
. microdiffraction, from thin crystals;
. special points in CBED disc;
. use SAD after all (thin, bent crystals)
. develop CBED-profile as a supplement, check - or refinement?

The variety of approaches correspond to wide variation in structural problems, ranging from

i) *precise refinement of structure factors for small unit-cell crystal, from CBED-profiles,* where a small number, typically half a dozen one or two-dimensional energy filtered profiles are collected from a well defined parallel-plate crystals for refinement of details in a known structure to high precision, based on extensive dynamical scattering calculations [3]; to
ii) *structure determination of biological macromolecules*, notably membrane proteins, by HREM-ED combination whereby 15-20 000 reflections (amplitudes and phases) are collected. [4] Adequate structure solutions is then obtained by kinematical interpretation of electron diffraction from crystals that are very thin with light atoms in large unit cells - as has been confirmed by multislice dynamical scattering calculations [5].

We shall be concerned with an intermediate range: inorganic crystals of irrregular shape and uneven thickness with, say fifty or hundred atoms in the unit cell? We may divide the discussion in two:

3. Patterns from thick crystals (100-1000Å)

The spot patterns intensities collected in dense zones are here of limited value. The CBED-rocking-curves may be extended to somewhat larger cells. We shall deal with this possibility in another chapter. By an iterative procedure signs and amplitudes for structure factors can be extraced from a series of energy filtered profiles along a systematic row of reflections. This is experimentally tedious and calls for extensive computations.

But we can use CBED-patterns to obtain *integrated intensities*. In the outer part of the pattern the reflection profiles are narrow, and therefore separated, because the excitation error, $s_G = -G^2 - 2\mathbf{k}_0\mathbf{G}$, varies rapidly with incident beam direction. By using a wide incident cone, many high-order reflections can therefore be collected simultaneously in one exposure, as illustrated by the systematic case Fig 4. The intensity of each reflection can be represented by just one parameter measured as an integral across the line. Three types of configurations can be used, *viz*: i) a systematic row; ii) parallel rows in an open projection; iii) HOLZ-lines with the incident cone around a zone axis.

The last configuration was used by the Bristol group [6] for intensity measurements and structure determination. With a wide condensor aperture, many HOLZ-line segments appear in the CBED-pattern. They interpreted the intensities of the segments either as kinematical scattering, or as scattering from Bloch waves formed by beams in the zero Laue zone into HOLZ reflections. The number of recorded reflections can be increased by using a range of accelerating voltages; 125 and 43 FOLZ-reflections were measured at two zone axes and refined three position parameters for the unknown structure AuGeAs [6]. HOLZ intensities have also been used to calculate Patterson maps as in generalized projections.

Taftø & Metzger [7] used a systematic row. Fig 4. Intensities for high-order *00l* reflections were collected, and one free z-coordinate was refined in the deuteride V_2D. This technique

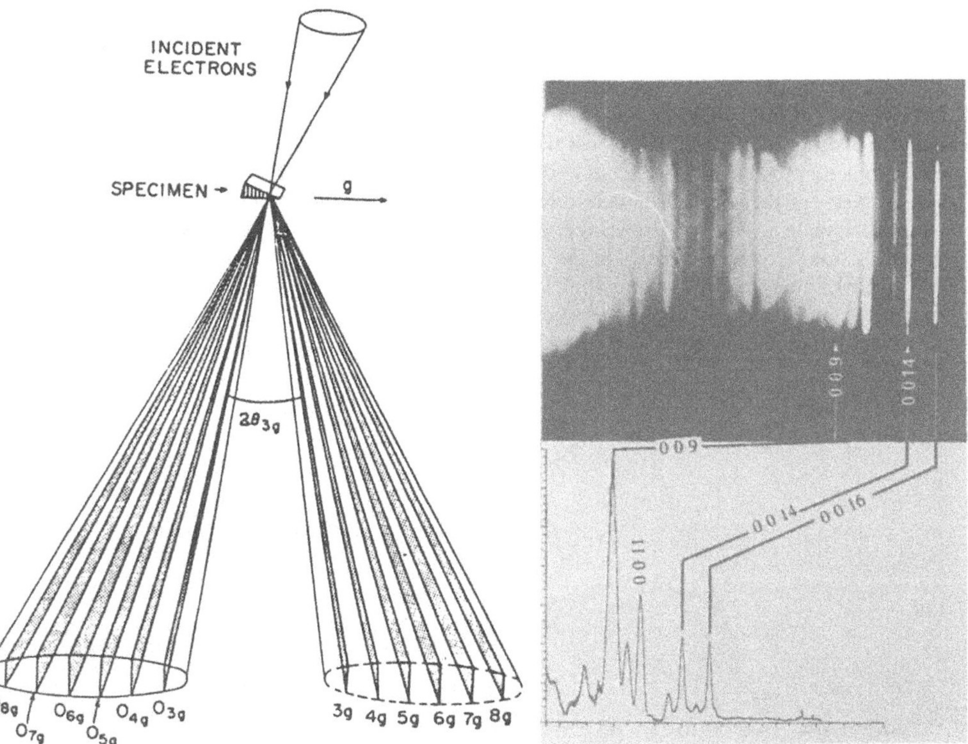

Fig 4. *Principle of wide-aperture CBED from high-orders in a systematic row.*

Fig 5. *Intensity measurement from 00l-reflections in $YBa_2Cu_3O_7$*

was extended to other cases and explored in detail [8] by experiments and calculations on some high-T_C-superconductors, Fig. 5. The validity of various approximations was investigated, in particular the assumption that the profile $I_G(s_G)$ of many reflections exhibit a either kinematic or a two-beam-like form, and hence be represented by one parameter: an effective structure factor U_G^{eff}. Several useful approximations can be found for U_G^{eff}, *e.g.* the so-called Bethe potential, or the gap at the dispersion surface. From the two-beam expression we have the Blackman formula [9] for the integrated intensity

$$\int I_G^{two\text{-}beam}(s_G,t)ds_G \propto U_G\int_0^{U_G t} J_0(x)dx \qquad \textbf{(6)}$$

where an U_G^{eff} can be substituted for U_G. In the high- and low-thickness-limits this becomes

$$\begin{aligned} U_G t \ll 1&: \textit{Quasi kinematic } I_G \propto |U_G^{eff}|^2 t \rightarrow |U_G|^2 t \\ U_G t > 1&, \textit{Quasi two-beam: } I_G \propto |U_G^{eff}| t \end{aligned} \qquad (7)$$

These expressions were tested for the *00l* row in $YBa_2Cu_3O_7$. A set of CBED-patterns were taken, with a wide aperture, tilted off the mid-line, while strong non-systematics were avoided. Intensities were measured either as profiles or as integrals. It was shown that the integrated intensites could be used for refinement of coordinates to quite good values - an order of magnitude less than the best neutron and X-ray values. Structure factors for inner-reflections *e.g.* (001) and (002) could not be measured directly, due to extensive overlap. But values even for these could be determined indirectly from their dynamical scattering effect on reflections in an intermediate range. Tests indicated that such a procedure can include sign determination by phase statistics in the row, and hence be applied to an unknown structure.

The CBED-based procedures outlined above can produce reliable data of fair accuracy, and may be extended. But so far the data collected in this way has been too few for an ab initio structure determination with, say 10-100 atoms in the translation unit.

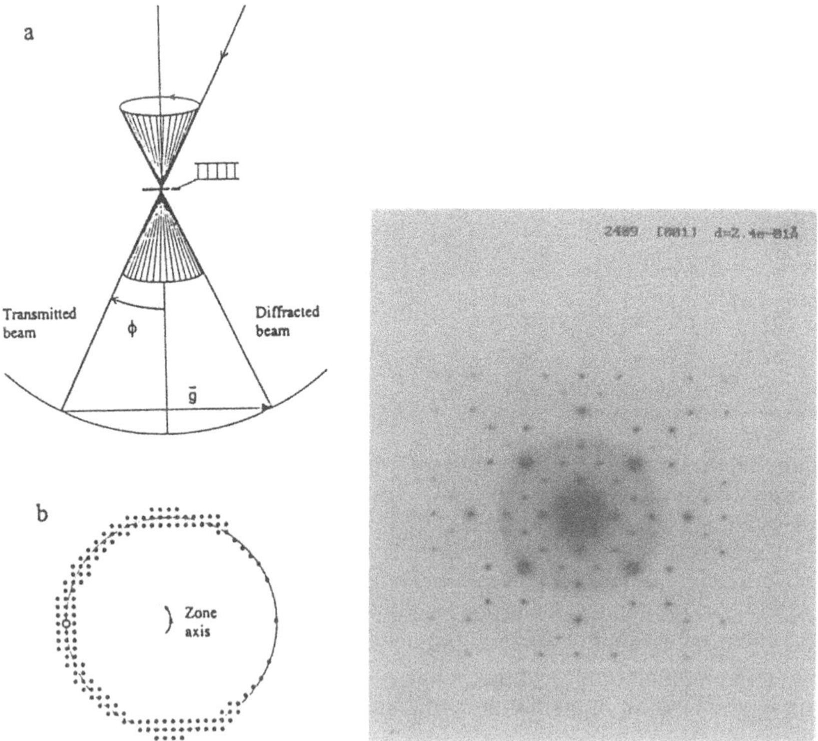

Fig.6. *Vincent-Midgely precession principle and experimental pattern [001]*Al_mFe

Integration in the experiment, as in X-ray diffraction may be better. It is not easy to rotate the crystal mechanically in a controlled way in the electron microscope; but the beam can be rotated, followed by a simultaneous opposite rotation of the pattern. This is done in the precession technique devised by Vincent & Midgley [10] The deflection coils that are standard in modern electron microscopes are used; the incident beam is precessed around a zone axis, typically at an angle of a couple of degrees. The resulting precession patterns look a bit like X-ray precession photographs; intensity variations are usually more marked than in traditional spot patterns from crystals of similar thickness. Dynamical interactions are reduced by the tilt off the zone axis; thickness oscillations are damped by the integration. The pattern extends far out in the zone; reflections from high-order zones are also recorded in the present geometry (it may also be possible to exclude these by a screen as in X-ray precession cameras). Vincent et al. have used the precession patterns especially for collection of HOLZ reflections.

Another application of the precession technique is to collect three-dimensional intensity data from crystals of moderate thickness. Small, *μm*-size particles of a metastable alloy phase Al_mFe, with about 110 atoms in a body-centred tetragonal unit cell (*a*=8.84, *c*=21.6Å, *m*=4.2-4.4) were extracted from an alloy matrix by a special dissolution technique. Structure determination had to be based on three-dimensional electron diffraction intensity data to about *d*=1Å. Selected area or microdiffraction mode were found unsuitable; the precession technique which had just been published was seen as the best possiblity. Eight projections, with two exposures each, were recorded on film with the set-up at Bristol University. The photonegatives were read and digitalized by a simple system consisting of a light-box and a CCD-camera coupled to a standard image analysis system. After subtraction of background and averaging of equivalent reflections in each projection, these were merged to a three-dimensional set. A non-linear least squares procedure was applied to the 28 scaling ratios obtained from common rows of reflection; a linear procedure was found to be inadequate.

Dynamical scattering effects are reduced by the precession technique, due to tilt from the zone axis, and also by the integration, but remains a problem. Even so, the three-dimensional data, about 300 unique reflections out to 1Å, allowed derivation of a structure model by standard crystallographic procedures: Patterson-Fourier, and direct method -assuming kinematical scattering.

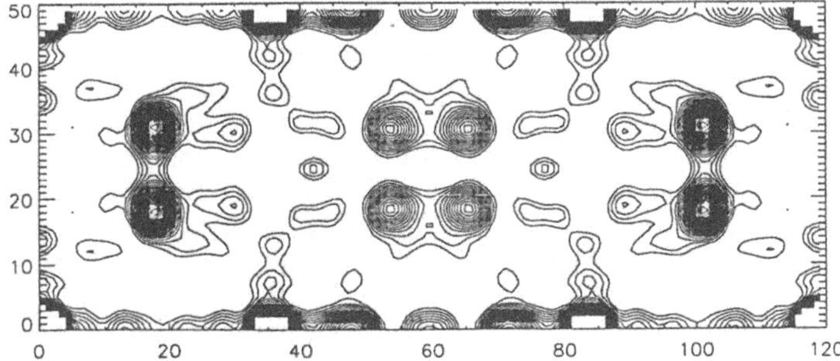

Fig. 7. *Patterson section u,0,w calculated from precession data*

The fit to observed intensities was poor, which raised the question: can intensities collected by these kinds of technique be corrected for dynamical scattering, something like the extinction correction in X-ray crystallography? There may be two possible strategies:

i) Refinement of structure parameters by dynamical calculations which then would have to be carried out three-dimensionally; ii) Correction of structure factors - which can be done in projection.

The second approach has been developpped as an approximate method by K. Gjønnes [11] and was applied to the *hk0*-data. Structure factors for the two systematic rows, *h00* and *hh0* were used as a starting point. The dynamical scattering was assumed to be represented by a two-beam type expression, and the resulting Blackman formula for integrated intensity with an U_G^{eff} given by the Bethe formula:

$$U_G^{eff}=U_G-\sum_H \frac{U_H U_{G-H}}{2ks_H} \tag{8}$$

substituted for UG_G. The main steps were as follows: At first the precsssion intensities for *h00* and *hh0* were fitted to dynamical calculations based on the CBED-values for U_{h00} and U_{hh0}, with thickness as an adjustable parameter. Using the thickness value thus obtained and the Blackman formula (5), the effective potentials $|U_G^{eff}|$ could be derived, as a kind of "primary extinction correction". In order to obtain the U_G from the U_G^{eff}, one must know the signs; these were found by triplet statistics applied to U_G^{eff} by statistics. Four possible solutions for signs and thus for $|U_G|$ were found; the best solution was chosen by a fit of dynamical calculated intensites to the experiment. The resulting U_{hk0} were then used to calculate a Fourier projection -and to calculate an R-value which was reduced by 0.12 to a value below 0.30 for the *hk0* projection. This result was taken as a confirmation of the procedure as well as of the projected structure, and as an indication that correction procedures for dynamical scattering can - and will - be developped.

4. Thin crystals

If the crystals are thin, sensitivity to the excitation eror s_G and to thickenss is reduced. This has been exploited in organic crystals and biological macromoelcules. Patterns from thin crystals are treated in several other chapters, for thin organic crystals by Dorset and by Voigt-Martin; structure refinement by electron diffraction from inorganic crystals is discussed by X.D. Zou and by Zandbergen [12]. Organic specimens are often bent, which effectively will give some integration over directions. For macromolecules data may typically be collected to 3Å or less, whence $s_G t$ <<1. If data has to be recorded to d=1Å, this may indicate a thickness below 100å. Dynamical scattering may extend the useful thickness range considerably in a dense zone - at the expense of the complication associated with dynamical calculations. For inorganic structures that can be solved from projections this may not be so serious. If the atoms are separated in projection the wave function will show the atom positions even by dynamical scattering, as has been pointed out by Van Dyck and Coene in a series of papers [13]. This may be an important factor behind the success of structure determination by electron diffraction in such cases. Electron diffraction patterns are best taken in a microdiffraction mode (in order to reduce variations in thickness and orientation), phases can be taken from HREM micrographs or from direct methods.

Instead of the microdiffraction mode with a concentrated nearly parallel beam on can use CBED mode for a thin crystal, as was first proposed by Goodman [14]. There will then be only smooth intensity varaitions in the CBED-discs; the zone axis position (or other special positions) can be located in the disc, and relative intensites can be measured at corresponding points in the discs. Calculations need only be carried out for one beam orientation. This configuration was used in structure determination for a group of selenides [15] and is currently applied to intermetallic phases [16]

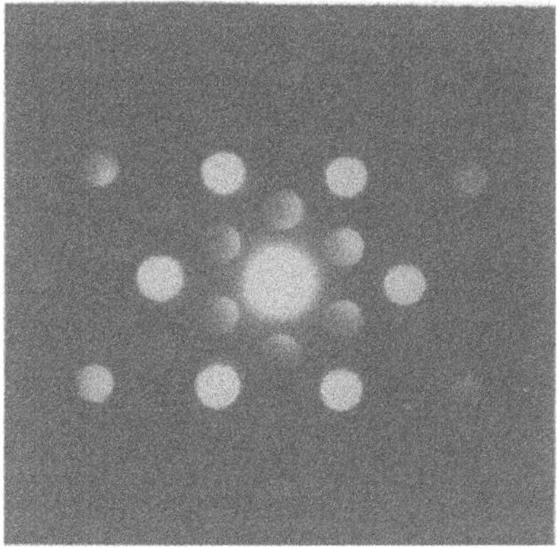
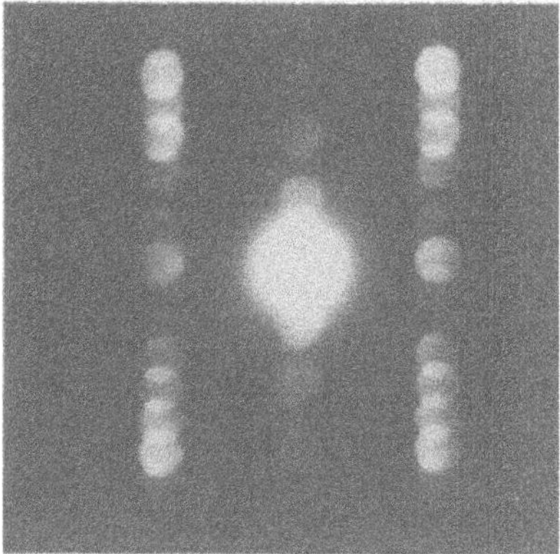

Fig. 8. *Microdiffraction patterns in a CBED-mode from a hexagonal Al-Ni-Si-Zr phase in rapidly solidified Al, [001] and [100] projections.*

5. Recording intensities

Until recently most electron diffraction patterns were recorded on photonegatives, which is the main recording medium in the transmission electron microscope. The film is excellent for reproducing contrast detail, but less suited for intensity measurements, due to a limited dynamic range. A simple system of reading intensity data from photonegatives can be assembled from a light box and a CCD-camera preferably attached to an image analysis system. Recently negative scanners have become available, and ar probaley better and simpler. Correction for over-exposure in the diffraction spot has been developped as part of the commercial ELD program [17] for reading intensites in diffraction patterns.

Considerable improvement over film detection can be obtained with either a slow-scan CCD-camera attached to a YAG crystal [18] or so-called imaging plates [19], both with a wide dynamic range. The small area, presently a few cm^2 is the main drawback of the CCD-system, a very short camera length is needed in order to record an full diffraction pattern; with new microscopse model may have an automtic option for changing camera length when switchin from the fluarsecent screen to the CCD-system. The CCD-camera is commercially available together with energy filter. Imaging plates are made with the same area as the photographic films and can be used together with films in the microscsope. The exposed plates are read with a laser beam. The sensitivity is claimed to be at least a factor two better than film, which may be imprtant for many beam-sensitive materials. Both systems are currently expensive, but will increase the potential of electron diffraction in structure analysis, make extensive dynamical calculations more meaningful, and represent an experimental breakthrough that may be compared with the introduction of automatic diffractiometer for X-rays.

6. Conclusions

Intensities can be measured for thick and thin crystals. For thick crystals one can reduce dynamical interactions in several ways, by moving off major axes and to outer refelctions. Dynamical effects may then have a two-beam character in many reflection, this can be exploited in approximate correction procedures. Integration in CBED-patterns or by a precession system will reduce thickness effects. CBED profiles in one or two dimensions can produce very accurate results for perfect specimens, and be helpful also in determination of unknown strucutres. For thin crystals spot pattern good intensity data can be obtained in dense projections along short axis. Dynamical calculations can be introduced quite early in the calculations.

References

[1] Williams, D.B. and Carter, C.B. (1996) Transmission Electron Microscopy, vol II Diffraction, Plenum, New York.
[2] Reimer, L. (1989) Transmission Electron Microscopy, 2.ed. Springer, Berlin.
[3] Spence, J.C.H. and Zuo., J.M. (1992) Electron Microdiffraction Plenum, New York.
[4] Kühlbrandt, W., Wang, D.N. and Fujiyoshi, Y. (1994) Nature **367**, 614.
[5) Spargo, A.E.C. (1994) Proc. 13. Int. Cong. Electron Microscopy, Paris 1994. Vol 1, pp 959
[6] Vincent,R., Bird, D.M. and Steeds, J.W. (1984) Phil. Mag, **A50**, 765.
[7] Taftø, J. and Metzger T.H. (1985) J. appl. Cryst. **6**, 10.
[8] Gjønnes, K. and Bøe, N (1994) Micron and Microscopy Acta **25**, 29.
[9] Blackman, M. (1939) Proc. Roy. Soc. **A 173**, 68.
[10] Vincent, R. and Midgley, P.A. (1994) Ultramicroscopy **53**, 271.
[11] Gjønnes, K. to appear in Acta Cryst.
[12] In this volume
[13] Van Dyck, D. and Coene, W. (1984) Ultramicroscopy **15**. 29.
[14] Goodman, P. (1976) Acta Cryst. **A32**, 793.
[15] Olsen, A., Goodman, P. and Whitfiled, H. (1985) J. Sol. State. Chem. **60**, 305.
[16] Runde, P. and Olsen, A. to be published.
[17] Zou, X.D., Sukharev, Y. and Hovmöller, S. (1993) Ultramicroscopy **52**, 436.
[18] Krivanek. O.L., Mooney, P.E, Fan. G.Y. Leber, M.L. & Meyer, C.E. (1991) Electron Microscopy and Analysis 1991. Proc Inst Phys Bristol. pp 523.
[19) Mori, N. Oikawa, T & Harada, Y. (1990) J. Electron Microscopy (Japan) **39**, 433-436.

CONVERGENT BEAM ELECTRON DIFFRACTION
Basic principles

J. GJØNNES
Department of Physics, University of Oslo
PO Box 1048 Blindern
N-0371 OSLO, Norway

1. Introduction

Convergent beam electron diffraction (CBED) is fairly recent as a standard technique in commercial electron microscopes. It was invented already in 1939 by Möllenstedt [1]; applications in crystallography were proposed by MacGillavry [2] and Ackerman [3] in the 1940's. In the mid 1960's Goodman and Lehmpfuhl [4] revived CBED as an alternative to SAD spot patterns, which they saw as poor substitutes for integrated intensities as used in X-ray crystallography. They turned to the Kossel-Möllenstedt or CBED patterns as a way of recording one- or two-dimensional rocking curves instead.

The principle is simple: by focussing the beam onto the specimen by a strong condensor lens a cone of incident rays is formed. The standard optics is shown in Figure 1a, other configurations are described in the literature [5], e.g. the LACBED or Tanaka method 1b. Strong inner spots in the diffraction pattern are expanded into circular discs defined by the aperture, often with a detailed fine structure. Reflections further out in the pattern appear more narrow, because their excitation errors s_G, which measure the deviation from the Bragg condition vary more steeply with incident beam direction. An outer reflections will not fill the aperture, but appear as lines defining the Bragg condition -frequently with some perturbations as seen in the examples Fig 2. Many of the details are sensitive to thickness; for thinner crystals less intensity variation are seen inside the disks .

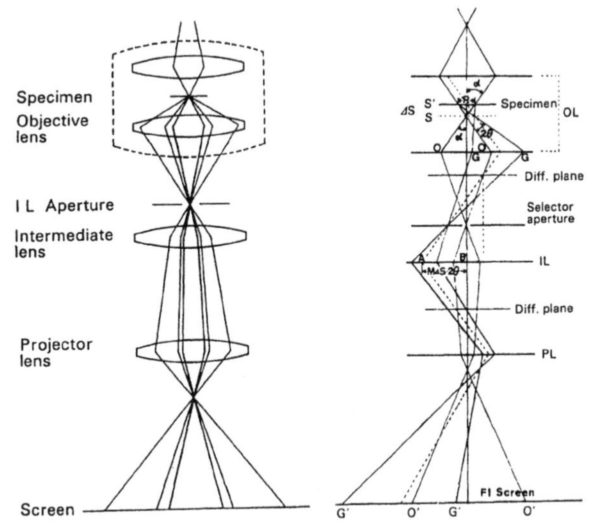

Fig.1a) *standard CBED; b) LACBED (Tanaka & Terauchi)*

The main characteristic of CBED patterns, Fig. 2-5. is the extensive intensity details, which contain information about the object, its crystal structure, and the scattering process. Another advantage of the focussing is that the area contributing to the pattern is quite small, typically a few hundred Å - much less than the μm-size area selected in the conventional SAD technique. Within this small area the thickness and crystal orientation may be regarded as

D. L. Dorset et al. (eds.), Electron Crystallography, 65–75.

constant: *diffracted intensity as a function of incident beam directions is recorded in one exposure for a well defined crystal.* The CBED intensity profiles, in one or two dimensions, were thus found ideally suited for detailed comparison with theoretical scattering calculations.

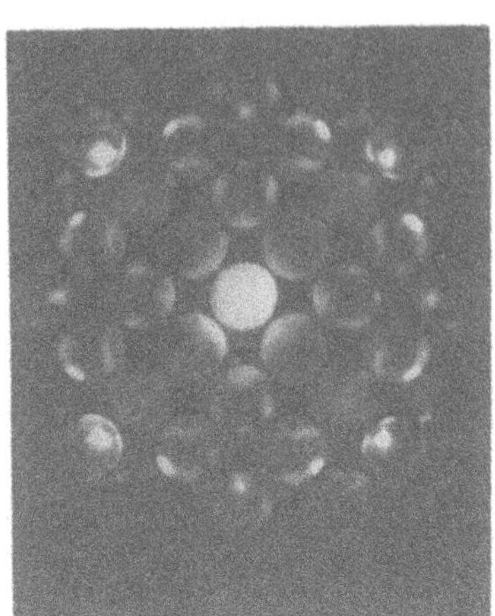

Fig.2 *Al_mFe [001]*

Fig 3.LACBED bright field, Si [111]

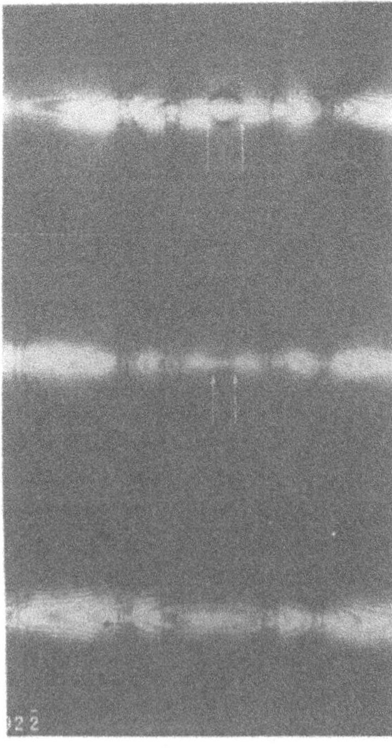

Fig 4. LACBED MgO 220 at three acc. voltages

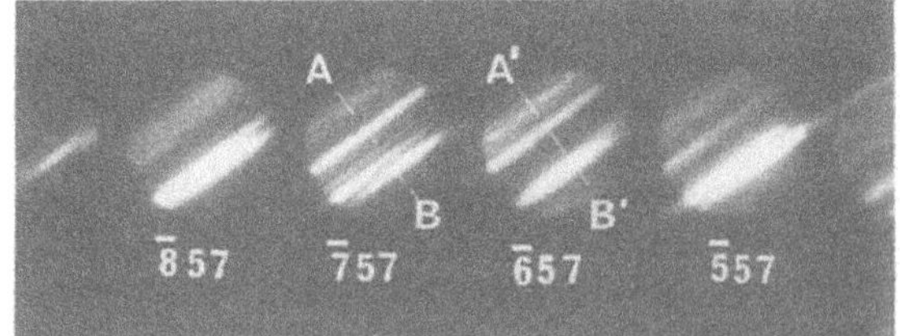

Fig.5. Ordered Cu_3Au -row of reflections

Two important crystallographic applications were soon established through the Goodman and Lehmpfuhl effort [4]: i) *crystal symmetry* can be derived: from special features that are connected with space-group forbidden reflections, and from deviations from Friedels law. These findings were later systematized by Goodman [6]; Buxton *et al* [7] and in particular by M. Tanaka and his coworkers [8]. ii) *Absolute structure factor amplitudes* $|U_g|$ can be determined from intensity profiles for low-order reflections - as had already been proposed by MacGillavry [2]. This is now a very active field, see *e.g.* [9] and other lectures at this school. For some time CBED remained a speciality, cultivated by a handful of dedicated laboratories, in Sendai, Melbourne and Bristol. But from around 1980 onwards, improvements of the commercial TEM has made the technique available, as a standard feature in commercial instruments: modern *objective lenses* include a strong forefield which serve as an effective probe-forming lens to produce the incident cone for CBED; *improved vacuum* condition has reduced contamination by the focussed beam; by *energy filter* systems the inelastic background can be reduced; the new *detection media: slow-scan CCD-camera*, and *imaging plates* offer tremendous improvement over photographic film. This has led to a series of crystallographic applications:

. *determination of crystal symmetry;*
. *precise determination of local lattice constants;*
. *study of defects: faults, dislocations, interfaces;*
. *refinement of structure details, e.g. bond charges;*
. *ab inito determination of structure factor amplitudes and signs;*
. *measurement of integrated intensities*

The richness in detail illustrates many facets of scattering theory. CBED will not replace the SAD parallel beam diffraction; it has limitations for beam-sensitive materials, superstructures, modulated structures, and large unit cells. But CBED can reveal important crystallographic detail with high precision - which will be the theme of subsequent talks. Here we shall concentrate on principles: geometry, intensity features, relation to theory and applications.

2. Geometry of CBED and Kikuchi patterns

Fig. 6. *Kikuchi patterns from Si: high-index direction, ca 80kV.*

Geometry is important in electron crystallography, were so much of the experiment relies on manual operation - and especially in CBED. In order to emphasize the geometrical aspect we may eliminate thickness-dependent detail, and consider the Kikuchi pattern which is related to the CBED pattern by a thickness average. The Kikuchi pattern consists of a number of bright, excess lines and complementary dark, deficient lines that appear in the diffuse, mainly inelastic background scattering. It is best observed when the crystal is tilted off a major zone axis, as in Fig. 6. The lines represent beam directions k_G that satisfy Bragg condition for the various reflections G, and occur in pairs, G and -G, around a center-line which is the trace of the reflecting plane, Fig. 7:

$$k_G G = -G^2/2 \ ; \ \ k_{-G} G = G^2/2 \tag{1}$$

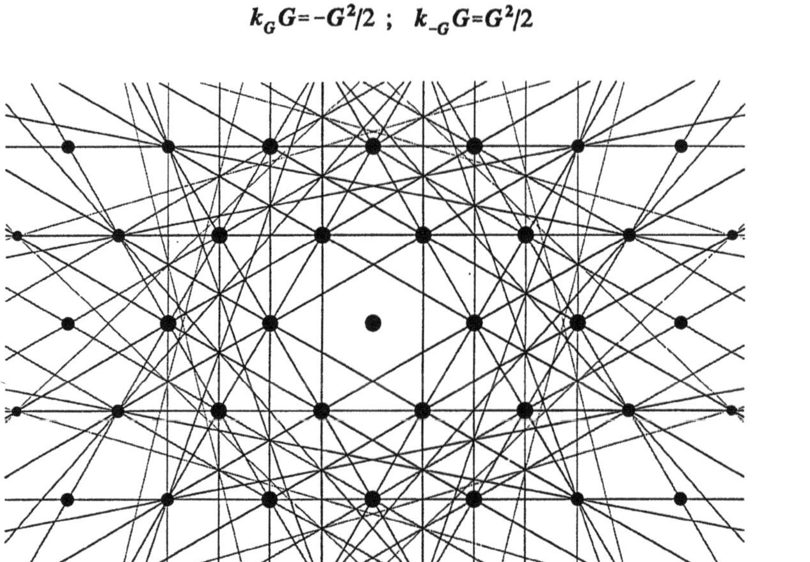

Fig. 7. *Geometrical Kikuchi pattern around a hexagonal/trigonal axis*

Detailed descriptions of how this geometrical pattern of lines can be constructed are found in many textbooks. The Kikuchi pattern is formed by Bragg reflection of rays in the fan of diffuse, mainly inelastic scattering which is peaked about the incident beam. Hence the deficient line is always closer to the central spot with the corresponding excess lines at a reciprocal distance G further out in the pattern. The relation between excess Kikuchi line contrast and CBED profile is shown by two-beam expressions:

$$I_g^{CBED} = \frac{|U_G/k|^2}{s_G^2 + |U_G/k|^2} \sin^2[\pi t \sqrt{s_G^2 + |U_G/k|^2}\ ] \ ; \qquad I_g^{Kikuchi} \propto \frac{|U_G/k|^2}{2(s_G^2 + |U_G/k|^2)} \tag{2}$$

The excitation error $s_g = -g^2 - k_x g$, where k_x is the wave-vector component normal to the plane g. Strong, inner reflections produce broad bands, narrow lines appear for outer reflections.

Kikuchi linelines from different zones (e.g. from higher order Laue zones) will move relative to each other when the accelerating voltage is altered. This is the principle behind determination of lattice constant from Kikuchi lines or CBED [11], Fig 8.

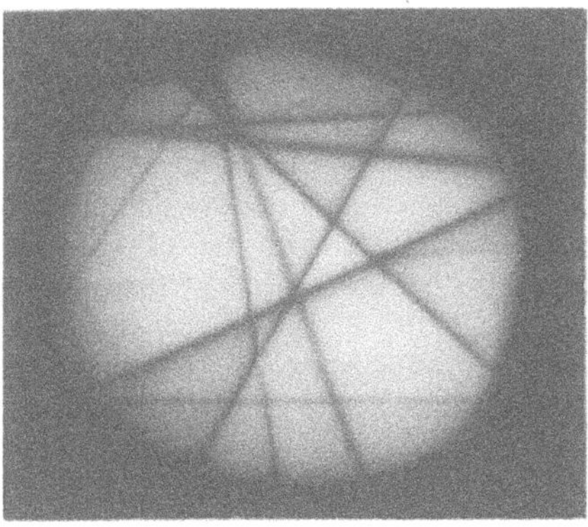

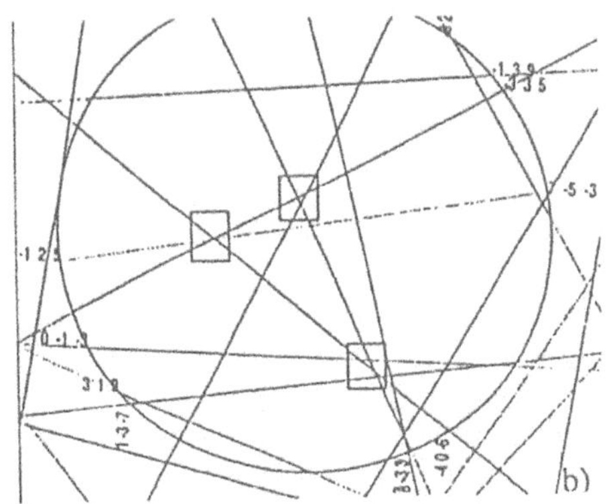

Fig.8. *HOLZ-(high-order Laue zone)lines intersecting in central CBED-disc from an intermetallic phase Al_3Ti, used for determination of lattice constant, accuracy 0.1%. [12]*

Before CBED became generally available contrast features in Kikuchi patterns were studied extensively, and several methods were developped for determination of structure factors. In the critical voltage method [13] vanishing contrast is observed at a particular voltage or diffraction condition. In the intersecting Kikuchi-line method [14] the separation between two segments of a split line at an intersection with a strong band is measured. Kikuchi pattern may still be useful as introduction to many of the effects that are exploited by CBED.

3. CBED intensity features

The detailed contrast in CBED patterns depend in a complicated way on structure factors, thickness and lattice constant. For precise interpretation extensive calculations are needed, usually based on the Bloch wave theory. Here we shall discuss characteristic features in typical configurations of beams, Fig. 9, in a general and qualitative way:

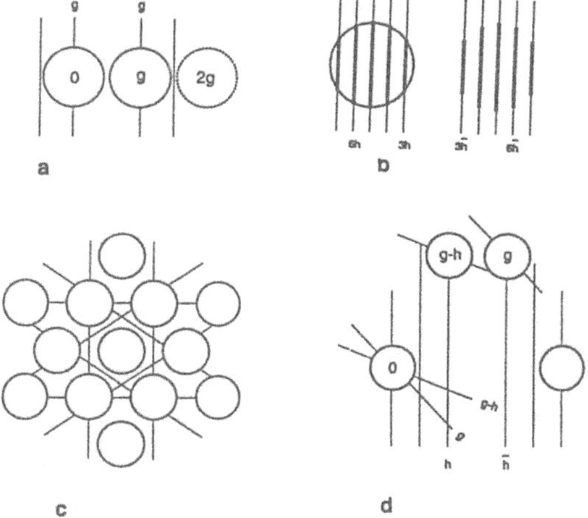

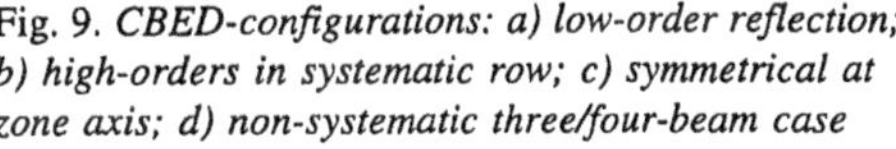

Fig. 9. *CBED-configurations: a) low-order reflection; b) high-orders in systematic row; c) symmetrical at zone axis; d) non-systematic three/four-beam case*

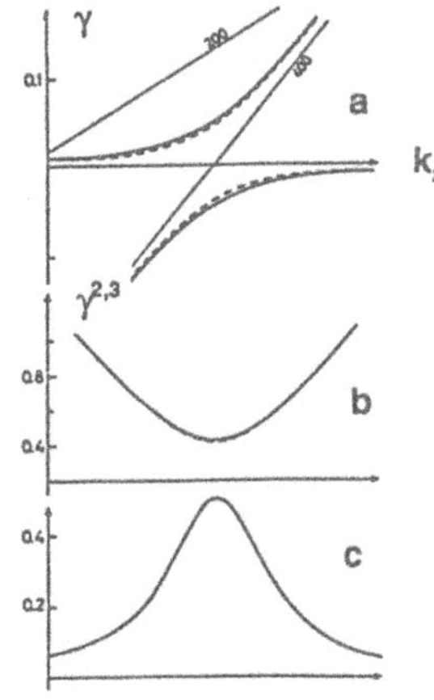

Fig. 10a) *Two-beam-like dispersion surface $\gamma(k_x)$ calculated at $GaAs_{400}$; b) separation $\gamma^{(2)}$-$\gamma^{(3)}$; c) intensity profile, thickness average [15]*

two-beam case: one strong reflection (structure factor, crystal thickness); sucessive beams in a *systematic row of reflections* (dispersion surface, effective structure factor, refinement); *three- and four-beam non-systematic cases:* intersections; *dynamical extinctions*: by symmetry, by accidental interactions; *zone-axis* pattern; *HOLZ*-lines (high-order Laue zones)

We shall relate these intensity features to the Bloch wave description of electron scattering, in particular the so-called *dispersion surface*, and make extensive use of two-beam expressions Eq. 2, Fig. 10, which gives a useful qualitative description of many of the features seen in the pattern.

The two-beam rocking curve. When one strong Bragg reflection is excited the characteristic two-beam-like profile is obtained, Fig. 10. The intensity expression is already given, Eqn. 2 (without absorption, which we ignore in this lecture). In many textbooks this is expression is derived from the amplitude, which is the sum of contributions from two Bloch waves:

$$\begin{aligned}\Psi_G(s_G,t) &= (c_0^{(1)})^* c_G^{(1)} \exp[2\pi i\gamma^{(1)}t] + (c_0^{(1)})^* c_G^{(2)} \exp[2\pi i\gamma^{(2)}t] \\ &= \frac{U_G/k}{\gamma^{(1)}-\gamma^{(2)}} \sin[\pi t(\gamma^{(1)}-\gamma^{(2)})]\exp[\pi i t s_G]\end{aligned} \tag{3}$$

where we may focus on the expression for the difference between the two *eigenvalues* $\gamma^{(1,2)}$ which define the dispersion surface. At the Bragg condition, $s_G=0$ this difference:

$$\gamma^{(1)}-\gamma^{(2)}=\sqrt{s_G^2+|U_G/k|^2} \tag{4}$$

is minimum and equal to U_G/k. In the exact two-beam case this separation or *gap at the dispersion surface* is thus a measure of the structure factor or Fourier potential U_G; in other cases this magnitude may be used as an effective Fourier potential for G. The two-beam curve $I_G(s_G)$, can be used for determination of the structure amplitude U_G and the thickness, t. Approximate values can be found by plotting the minima or maxima of the curve:

$$s_G^2+|U_G/k|^2=\frac{n^2}{t^2} \qquad (\text{max } for\ n=even;\ \text{min } for\ n=odd) \tag{5}$$

We may plot the corresponding s_G^2-values against n^2, and determine $1/t^2$ from the slope, and U_G^2 from the intersection with the axis. (The starting value for n may not be known, but must be found by trial). More precise determination should be based on curve fitting.

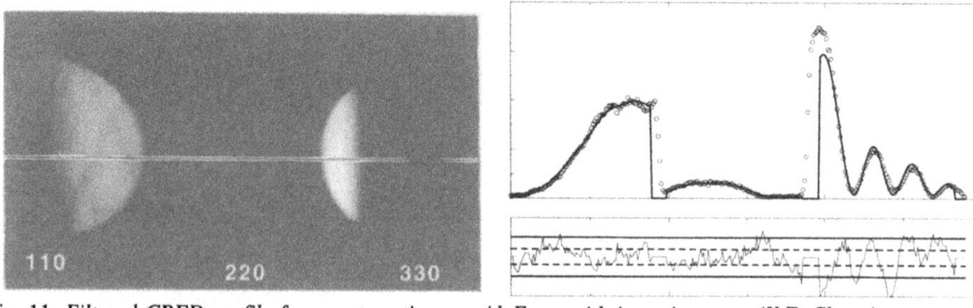

Fig. 11. *Filtered CBED-profile from systematic row: Al_mFe_{330}, with intensity curve (Y.F. Cheng)*

Profiles in a systematic row. Rocking curves collected as energy-filtered CBED-profiles from low-order reflections (Figs. 9a and 11) can be used for refinement of structure factors to high precision accuracy, and hence for determination of the distribution of bond charge in small-unit cell crystal; for a review, see [2]. Relative intensities of high-order lines in a dense systematic row, cf Fig. 9b, can be collected as described in a previous chapter and used for refinement of coordinates and temperature factors [17]. The one-dimensonal intensity profiles along systematic rows are often found to have two-beam like shape, as discussed above, and hence to be described by one parameter: an effective Fourier potential, U_G^{eff}. Comparisons between exact multiple-beam calculations and such a *quasi two-beam* approximation have been made for [200] and [111]-rows in GaAs [15] and [00l]-row in $YBa_2Cu_3O_7$ [17]; as one proceeds towards higher orders, the profiles become more kinematic - provided non-systematic interactions can be avoided. Several approximations have been proposed for calculating U_G^{eff}, the oldest is the Bethe potential:

$$U_G^{Bethe} = U_G - \sum_{H \neq 0,G} \frac{U_H U_{G-H}}{2ks_H} \tag{6}$$

A Bloch wave hybridization approach, proposed by the Bristol group is another possibility.

Ab initio determination of structure factors from CBED-profiles along a systematic row was shown by Cheng *et al* [18] for the intermetallic phase Al_mFe, a tetragonal phase with 110 atoms in the unit cell. Filtered CBED profiles were measured for the *h00* reflections 200 to 1400, and *hh0* from 110 to 11,11,0 with non-systematic reflections avoided as much as possible. At first a preliminary value was obtained for the strongest reflection U_{600} by fitting a two-beam expression to the 600-profile. The next step was to refine the inner reflections U_{200} and U_{400} from their profiles, assuming the preliminary value for U_{600}, and two possible sign combinations. At that stage these two could not be distinguished; but the next step in the refinement, where U_{600} was again varied to give a best fit for the 600-profile gave a uniqe answer. The procedure was then extended successively for further profiles along the row, up to U_{1400}. The same procedure was followed for the *hh0*-row. Altogether 18 structure amplitudes and signs were determined, with uncertainties in the percentage range, Table 1. Non-systematic reflections were suppressed in the experiment and neglected in calculations, so was absorption.

Table 1 *Structure factors in Å^{-2} obtained from filtered CBED-profiles.*

U_{200} = -0.01191 (31)
U_{400} = +0.01592 (44)
U_{600} = +0.01275 (16)
U_{800} = -0.00653 (72)
$U_{10,00}$= +0.00359 (33)
$U_{12,00}$= -0.00135 (18)
$U_{14,00}$= -0.00214 (31)
U_{110} = +0.00290 (45)
U_{220} = -0.00199 (23)
U_{330} = +0.03872 (09)
U_{440} = +0.00750 (34)
U_{550} = +0.00773(107)
U_{660} = +0.00226 (51)
U_{770} = +0.00158 (27)
U_{880} = +0.00514 (40)
U_{990} = +0.00300 (18)
$U_{10,10,0}$=+0.00140 (11)
$U_{11,11,0}$=+0.00210 (08)

Non-systematic interactions: K-line intersections. Characteristic contrast effects appear where Kikuchi-lines or CBED Kossel-lines cross a strong band. The line is split into two segments, and the intensity or contrast of the line is often seen to be different on the two sides of the band edge. These three- and four-beam effects were first studied in Kikuchi patterns, and have been used for determination of low-order structure factors [13, 14]. A qualitative interpretation may again be based on an effective Fourier potential corresponding to the gap at the dispersion surface, as in the three beam case Fig. 10, from SiC [19]

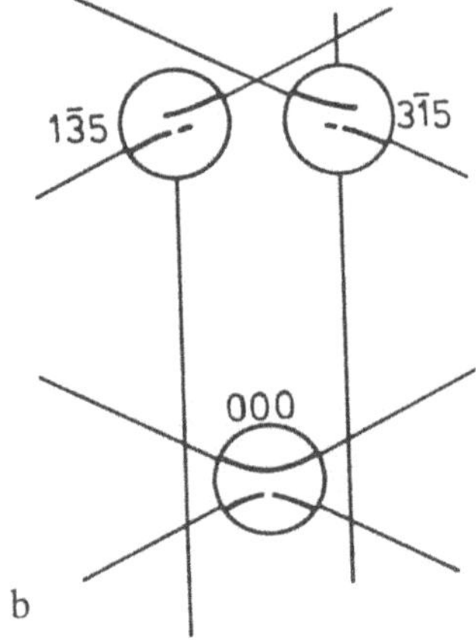

b

Fig. 12a) *Central CBED-disc SiC in the configuration b)*

In the zero CBED disc we see the intersection between the deficient lines 3$\bar{1}$5 and 1$\bar{3}$5 in the middle of the strong 220-band; the simultaneous reflections are coupled by the strong 220. The three-beam Bloch wave solution along the center-line, Fig. 12c include three branches, two symmetrical $\gamma_S^{(1,2)}$ and one antisymmetrical branch, γ_A. The structure factor for 220 can be determined from the split at the intersection, which corresponds to the separation between the degeneracy (γ_A=$\gamma_S^{(2)}$) and the position of the gap between the symmetrical branches 1 and 2. Three- and four-beam cases have been studied by several workers [20] Recent experimental work indicate considerable potential for phase determination.

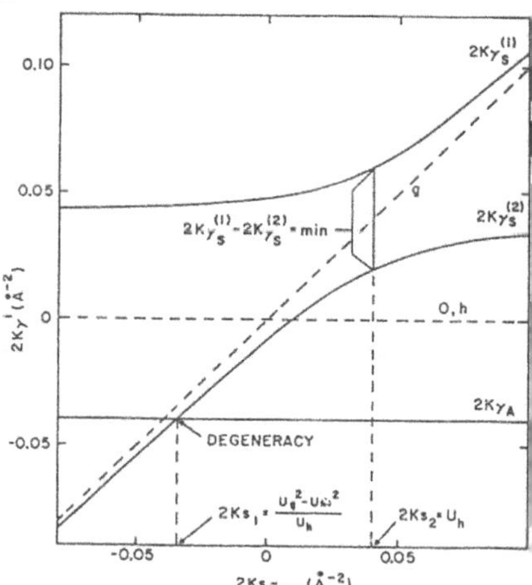

Fig. 12c. *Three-beam dispersion surface*

Extinctions, by symmetry and by accident.
Under certain conditions the contrast of a Kikuchi- or CBED-pattern K-line may vanish, along a line or at a point on the line. This may be due to symmetry - or because the effective potential vanishes due to dynamical scattering. The vanishing contrast along certain lines in CBED-discs for space-group forbidden reflections played a part in the development of the CBED-technique in the 1960s. It was known that these reflections could appear with appreciable intensity due to the Renninger effect, or "umweganregung". The preservation of the extinction of these space-group forbidden reflections along certain lines in CBED-discs Fig. 13 was shown by Goodman & Lehmpfuhl [4] The condition can be derived by consideration of scattering paths [21], or from symmetry properties of Bloch waves [22].

Vanishing contrast can be associated with zero gap at the dispersion surface, *i.e.* a Bloch wave degeneracy. This can occur accidentally at a particular accelerating voltage and is utilzed in the critical voltage method [13]. This is essentially a three-beam effect and can be explained by the Bethe expression for the effective Fourier potential

$$U_{2G}^{eff}=(m/m_0)U^{2G}-(m/m_0)^2\frac{U_G^2}{2ks_G}=0 \quad for \quad U_G=\sqrt{(m_0/m)2ks_G U^{2G}} \tag{7}$$

- here written as condition for vanishing contrast of a second-order line in a row of strong reflections. Similar effects occur in non-systematic cases, *e.g.* the four-beam diamond configuration in the Fig. 14, [23]. In SnO_2 five such cases were measured and used to determine the ionicitiy. An important advantage of these methods is that no scaling is involved, absolute values of few structure factors are obtained. A limitation is that many other structure factors has to be entered as known parameters in the calculations.

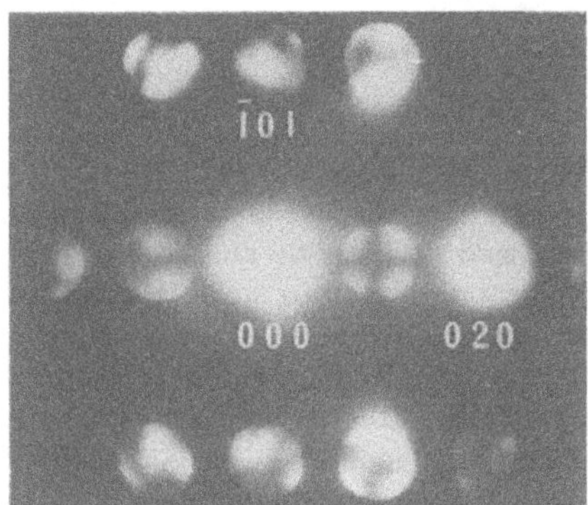

Fig.13 *Extinctions in space-group forbidden 010 in SnO_2*

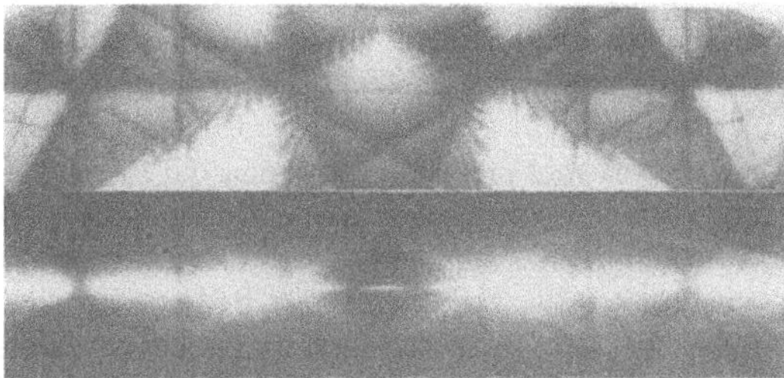

Fig. 14. *Four-beam accidental extinction: 422-line in Si, 100kV. LACBED 422 (a) and 000 (b).*

Zone axis pattern. When the CBED pattern is taken along a major zone axis extensive dynamical scattering effects will be presemt, in calculations a large number of beams must be included. Zone axis patterns can be used for refinement of structure factors. The Bristol group claim this to be the most sensitive configuration for precise refinement - it may also be the one that needs most extensive computations. High symmetry in the projection may be of considerable help in calculations at special points. Projections along short axis are often characterized by well separated atom columns, which facilitates interpretation. For thin crystals zone-axis CBED can be used to record intensities at special points.

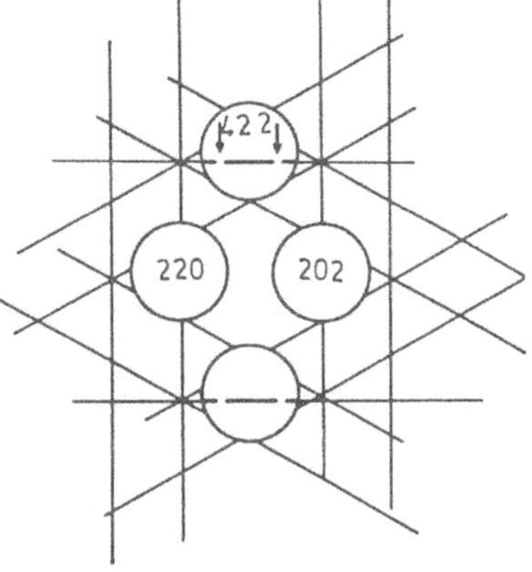

Fig. 15 *Beam configuration for Fig 14.*

HOLZ-lines. An interesting feature in zone axis patterns is the lines associated with high-order Laue zones (HOLZ); these appear as deficient lines in the zero disc, Fig. 8 and as a ring of excess segments in HOLZ rings, Fig. 16. HOLZ-lines contain information about three-dimensional symmetry, and can be used to determine lattice constant - (but beware of the influence of dynamical effects on position). The HOLZ-lines may be split in several segments, which can be associated with the Bloch waves in the "ZOLZ".

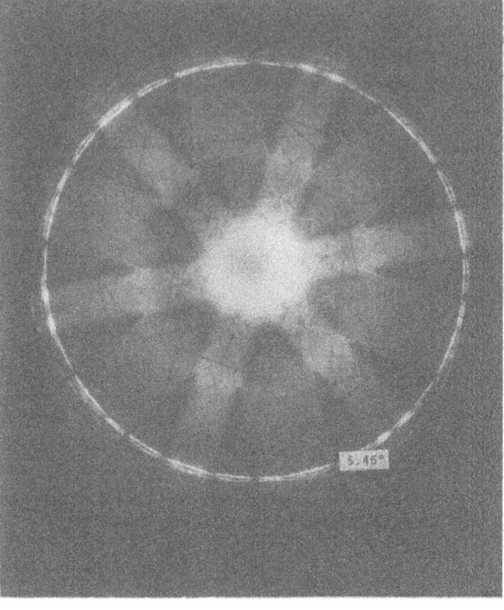

Fig. 16. *CBED-pattern with HOLZ-ring, Si [111]*

4. Conclusions

CBED-patterns recorded from small, perfect crystal regions contain a wealth of detailed and precise information, about crystal structure and lattice constant. Extensive calculations will as a rule be needed in order to extract precise information but the qualitative or semiquantitative considerations of the kind discussed above are useful as guide. Crystallographic applications are increasing in the inorganic field, mostly on structures with small unit cells. The quality of CBED-patterns become poorer if crystals are bent or contain many defects - or is damaged by radiation. The latter is a serious limitation for organic and other beam-sensitive material; to some extent this may be overcome by more sensitive detection, and by using larger area -renouncing finer details. CBED is not suited for weak superstructure reflections or diffuse scattering; but other large application fields remains to be explored, e.g. experimental phase determination from few-beam configurations.

References

[1] Kossel, W. and Möllenstedt, G. (1939) Ann. der Physik **36**, 113.
[2] MacGillavry, C. (1940) Physica, **7**, 329.
[3] Ackermann, I. (1948) Ann.Phys. (Leipzig) **2**, 41.
[4] Goodman, P and Lehmpfuhl, G. (1964) Z. Naturforsch. **19a**, 818; (1967) Acta Cryst. **22**, 14.
[5] Tanaka, M. and Terauchi, M. (1985) Convergent-Beam Electron Diffraction JEOL, Tokyo.
[6] Goodman (1975) Acta Cryst **A31**, 793.
[7] Buxton, B.F., Eades, J.A., Steeds, J.W. and Rackham G.M. (1976) Phil Trans R.Soc. London **281**, 171.
[8] Tanaka ,M. Saito R and Sekii H. (1983) Acta Cryst **A39**, 357; Tanaka M. Sekii, H. and Nagasawa, T. *ibid*, 825.
[9] Spence, J.C.H. and Zuo, J.M. (1992) Electron Microdiffraction Plenum, New York.
[10] Reimer,L. (1989) Transmission Electron Microscopy 2.ed. Springer, Berlin.
[11] Olsen, A. (1992) in A.J.C. Wilson, ed. International Tables of Crystallography Kluwer, Dordrecht vol **C**, pp463.
[12] Gunnæs, A. and Olsen, A. to be published.
[13] Watanabe, D., Uyeda, R. and Fukuhara, A. (1969) Acta Cryst. **A25**, 138.
[14] Gjønnes, J, and Høier, R. (1971) Acta Cryst. **A27**, 313.
[15] Gjønnes, K.,Gjønnes, J, Zuo, J. and Spence, J.C.H.(1988) Acta Cryst **A44**, 810.
[16] Holmestad, R. (1994) Quantitative Electron Diffraction, Dissertation University of Trondheim.

[17] Gjønnes, K. and Bøe, N. (1994) Micron and Microscopy Acta **25**, 29.
[18] Cheng, Y.F., Nüchter, W, Mayer, J., Weickenmeier, A. and Gjønnes, J. (1996) Acta Cryst. **A52**, 923.
[19] Taftø, J. & Gjønnes, J. (1985) Ultramicrscopy **17**, 329.
[20] Moodie, A.F., Etheridge, J. and Humphreys, C.J. (1996) Acta Cryst **A52**, 596.
[21] Gjønnes, J and Moodie,A.F (1965) Acta Cryst. **19**, 65.
[22] Gjønnes, J and Taftø, J. (1993) Ultramicroscopy **52**, 445.
[23] Matsuhata, H. and Gjønnes, J. (1994) Acta Cryst. **A50**, 107; Matsuahta, H., Gjønnes, J. and Taftø, J. *ibid*, 115.

CONVERGENT-BEAM ELECTRON DIFFRACTION

M. TANAKA
Research Institute for Scientific Measurements,
Tohoku University, Sendai 980-77, Japan

Abstract. Convergent-beam electron diffraction (CBED) is a very powerful method for determing crystal symmetries, lattice defects and crystal structures of small specimen areas. We briefly describe the procedures of symmetry determination of crystals and structure refinement, and coherent effects using the CBED techniques.

1. Introduction

CBED obtains diffraction patterns using a conical electron beam with an angle of more than 10^{-3} rad. on a uniform and non-bent specimen area about 10nm in diameter (Fig.1 and 2). Instead of usual diffraction spots, diffraction disks are produced [1].

All the point groups can be determined uniquely by inspecting the symmetries appearing in the disks. A 2_1 screw axis and glide planes are identified by CBED through a conspicuous dynamical diffraction effect. As a result almost all space groups can be identified by the CBED method.

The large-angle CBED (LACBED) method [1] provides intensity variation for parameters $\mathbf{r}$ (position) and θ (orientation), I ($\mathbf{r}$, θ). The ray-path diagram of the LACBED method and a pattern obtained are respectively shown in Fig.3 and Fig.4. Thus, the LACBED method is quite an effective method to identify lattice defects. The shift vector at a stacking fault, the Burgers vector of a dislocation and the angular change at a twin boundary are successfully determined by CBED [2], [3].

Crystal-structure analysis is one of the development goals of the CBED technique [3]. Its greatest advantage exists in making possible structure analysis from a crystal area about one nanometer or less in size, which X-ray and neutron diffraction can not achieve. HOLZ reflections are used for

D. L. Dorset et al. (eds.), Electron Crystallography, 77–113.

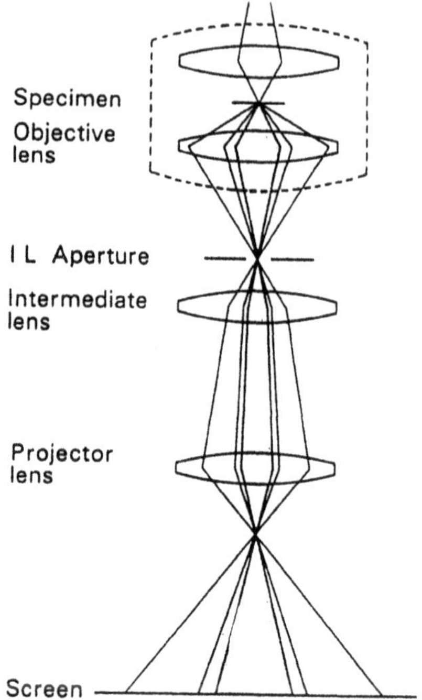

Figure 1. Ray-path diagram of CBED.

structure analysis by CBED, because the intensities of HOLZ reflections with large reciprocal lattice vectors **g** are more sensitive to the displacements of atoms than those of ZOLZ reflections. This can be understood from the fact that the crystal structure factor F_g includes the phase factor $\exp(-2\pi i \mathbf{g} \cdot \mathbf{r})$.

When an electron source with high brightness (*i.e.*, field-emission gun: FEG) is used, interference fringes are produced in the overlapping regions of CBED disks due to a coherent effect [3]. If the fringes in the two disks which are related by mirror symmetry show a shift of a half period, they exhibit to the existence of glide symmetry. The phases of crystal structure factors can be determined from the relative positions of the fringes. Various applications of coherent CBED are open to imperfect crystals.

2. Symmetry Determination

2.1. POINT-GROUP DETERMINATION

The point-group determination method was established by Buxton et al. [4]. They considered a perfect crystalline specimen that is parallel-sided

Figure 2. CBED pattern of [111] Si

and infinite in two dimensions. The symmetry elements of the specimen form "diffraction groups", which are isomorphic to the point groups of diperiodic plane figures and Shubnikov groups of colored plane figures. The diffraction groups of a specimen are determined from the symmetries of CBED patterns taken at various orientations of the specimen. The crystal point group of the specimen is identified by referring to the table that gives the relation between diffraction groups and crystal point groups.

A specimen that is parallel-sided and infinitely extended in the x and y directions has ten symmetry elements. The symmetry elements consist of six two-dimensional symmetry elements and four three-dimensional ones. A vertical mirror plane m and one-, two-, three-, four- and six-fold rotation axes that are parallel to the surface normal z are the two-dimensional symmetry elements. The three-dimensional symmetry elements consist of a horizontal mirror plane m', an inversion center i, a horizontal twofold rotation axis 2' and a fourfold rotary inversion $\bar{4}$ whose axis is parallel to the surface normal as shown in Fig.5. 31 diffraction groups are constructed by combining these symmetry elements.

Because of dynamical diffraction effects, two-dimensional symmetry elements, which belong to a zone axis, exhibit their symmetries in CBED

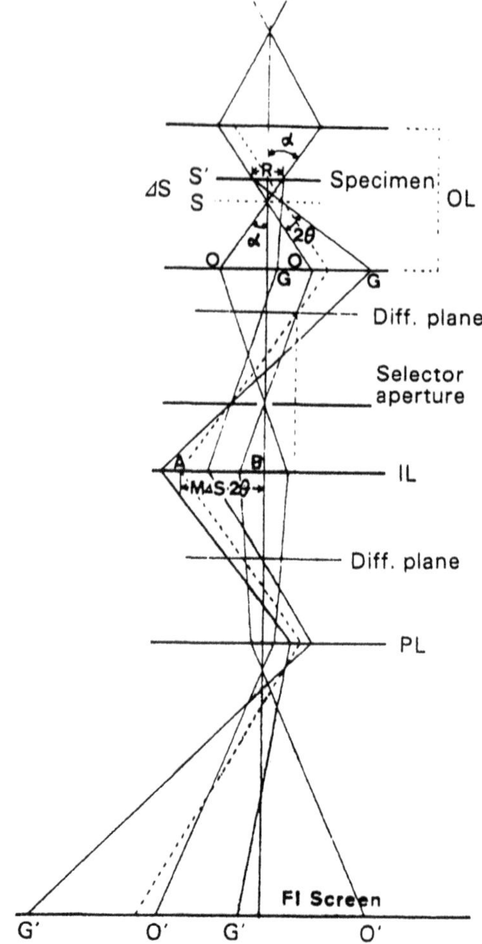

Figure 3. Ray-path diagram of LACBED.

patterns taken at the electron incidence parallel to the zone axis. The patterns are called zone-axis patterns (ZAPs). A ZAP contains a bright-field (BF) pattern and a whole pattern (WP). The BF pattern is the pattern appearing in the BF disk. The WP is composed of the patterns of the BF disk and diffracted disks. It should be noted that since these diffracted disks do not contain exact Bragg positions, the patterns of the disks are not called dark-field (DF) patterns. The two-dimensional symmetry elements m, 1, 2, 3, 4 and 6 yield, respectively, a symmetry m_v and one-, two-, three-, four- and six-fold rotation symmetries in the WP, where the suffix v of m_v is given to distinguish the symmetry from the mirror symmetry m_2 originated from a horizontal twofold axis. All the two-dimensional symmetry

Figure 4. LACBED pattern of [111] Si

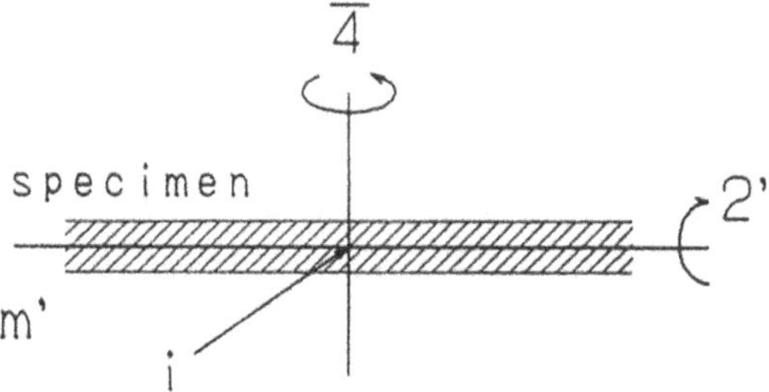

Figure 5. Four three-dimensional symmetry elements of a parallel-sided and infinitely extended specimen.

elements can be identified from WP symmetries.

The four three-dimensional symmetry elements were found to produce different symmetries in CBED patterns as shown in Fig.6. These facts enable us to unambiguously identify these symmetry elements from the symmetries of CBED patterns.

All the symmetry elements of an infinitely extended parallel-sided specimen or all the diffraction groups can be identified by the symmetries of WP and DF patterns. A practical method to determine diffraction groups how-

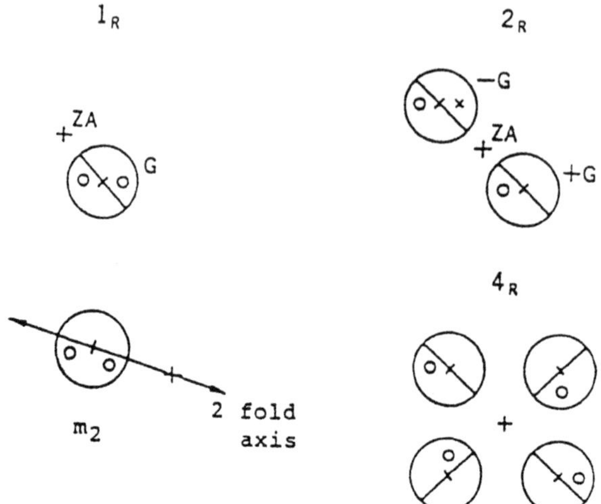

Figure 6. CBED symmetries of the four three-dimensional symmetry elements.

ever, utilizes BF, WP, DF and ±G. The symmetries appearing in BF, WP, DF and ±G are given for the 31 diffraction groups in Table 1. A diffraction group is determined from the four patterns BF, WP, DF and ±G appearing in three photographs, although many groups can be determined from BF and WP, or BF, WP and DF. Possible point groups are selected from the obtained diffraction group by consulting Table 2 (Buxton et al., [4]), in which the relation between the 31 diffraction groups and 32 crystal point groups is given. Each of 11 high-symmetry diffraction groups corresponds to one crystal point groups. When plural point groups are obtained from a diffraction group, another zone axis has to be chosen. Thus, a point group is identified by selecting a common point group existing among the point groups obtained at the different zone axes.

High-symmetry zone axes have to be chosen for point-group determination, because low-symmetry zone axes exhibit only a small number of crystal symmetries in CBED patterns. CBED observes part of the symmetry of a given crystal; that is, CBED cannot observe the crystal symmetries oblique to the incident beam and horizontal three-, four- and sixfold axes. Therefore, even for a crystal with a particular point group, different diffraction groups are expected at different crystal settings.

2.2. SPACE-GROUP DETERMINATION

In the course of point-group determination, orientations of symmetry elements were determined with respect to diffraction patterns. Based on the

TABLE 1. Symmetries of BF(II), WP(III), DF(IV) and ±G(V) patterns for 31 diffraction groups(I).

I	II	III	IV	V	VI
1	1	1	1	1	
1_R	2 (1_R)	1	$2 = 1_R$	1	1_R
2	2	2	1	2	
2_R	1	1	1	2_R	21_R
21_R	2	2	2	21_R	
m_R	m (m_2)	1	1 m_2	1 m_R 1	
m	m_v	m_v	1 m_v	1 m_v 1	$m1_R$
$m1_R$	$2mm$ $[m_v + m_2 + (1_R)]$	m_v	2 $2m_vm_2$	1 m_v1_R 1	
$2m_Rm_R$	$2mm$ $(2 + m_2)$	2	1 m_2	2 $2m_R(m_2)$	
$2mm$	$2m_vm_{v'}$	$2m_vm_{v'}$	1 m_v	2 $2m_{v'}(m_v)$	$2mm1_R$
2_Rmm_R	m_v	m_v	1 m_2 m_v	2_R $2_Rm_{v'}(m_2)$ $2_Rm_R(m_v)$	
$2mm1_R$	$2m_vm_{v'}$	$2m_vm_{v'}$	2 $2m_vm_2$	21_R $21_Rm_{v'}(m_v)$	
4	4	4	1	2	
4_R	4	2	1	2	41_R
41_R	4	4	2	21_R	
$4m_Rm_R$	$4mm$ $(4 + m_2)$	4	1 m_2	2 $2m_R(m_2)$	
$4mm$	$4m_vm_{v'}$	$4m_vm_{v'}$	1 m_v	2 $2m_{v'}(m_v)$	$4mm1_R$
4_Rmm_R	$4mm$ $(2m_vm_{v'} + m_2)$	$2m_vm_{v'}$	1 m_2 m_v	2 $2m_R(m_2)$ $2m_{v'}(m_v)$	
$4mm1_R$	$4m_vm_{v'}$	$4m_vm_{v'}$	2 $2m_vm_2$	21_R $21_Rm_{v'}(m_v)$	
3	3	3	1	1	
31_R	6 $(3 + 1_R)$	3	2	1	31_R
$3m_R$	$3m$ $(3 + m_2)$	3	1 m_2	1 m_R 1	
$3m$	$3m_v$	$3m_v$	1 m_v	1 m_v 1	$3m1_R$
$3m1_R$	$6mm$ $[3m_v + m_2 + (1_R)]$	$3m_v$	2 $2m_vm_2$	1 m_v1_R 1	
6	6	6	1	2	
6_R	3	3	1	2_R	61_R
61_R	6	6	2	21_R	
$6m_Rm_R$	$6mm$ $(6 + m_2)$	6	1 m_2	2 $2m_R(m_2)$	
$6mm$	$6m_vm_{v'}$	$6m_vm_{v'}$	1 m_v	2 $2m_{v'}(m_v)$	$6mm1_R$
6_Rmm_R	$3m_v$	$3m_v$	1 m_2 m_v	2_R $2_Rm_{v'}(m_2)$ $2_Rm_R(m_v)$	
$6mm1_R$	$6m_vm_{v'}$	$6m_vm_{v'}$	2 $2m_vm_2$	21_R $21_Rm_{v'}(m_v)$	

TABLE 2. Relation between diffraction groups and crystal point groups (Buxton et al.[4].)

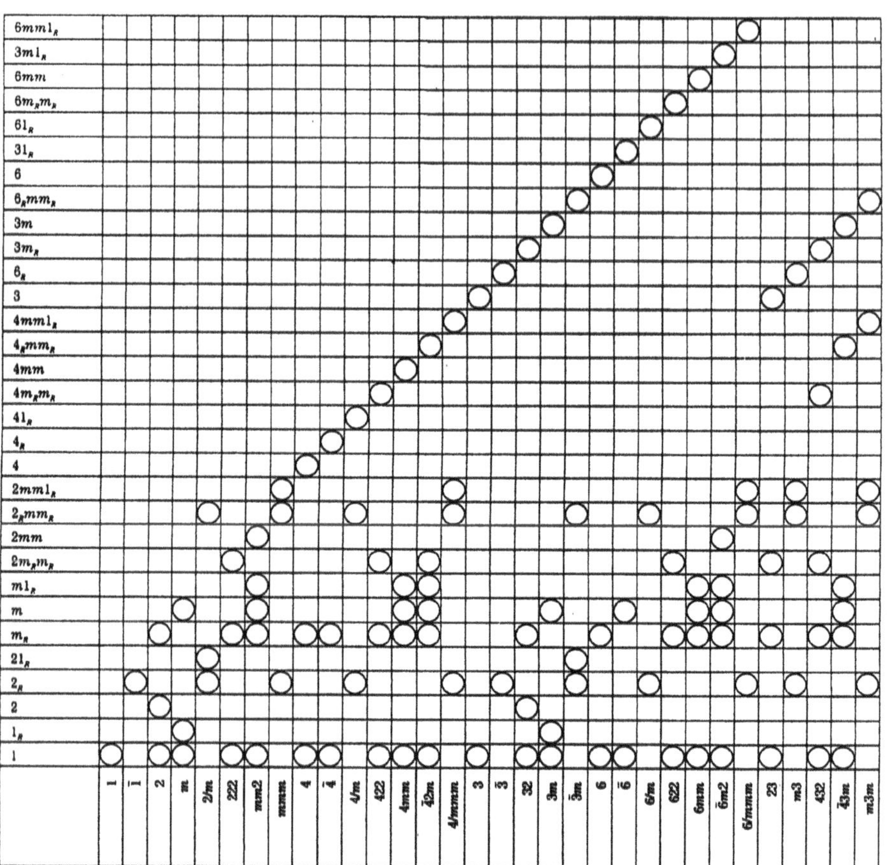

	1	$\bar{1}$	2	m	$2/m$	222	$mm2$	mmm	4	$\bar{4}$	$4/m$	422	$4mm$	$\bar{4}2m$	$4/mmm$	3	$\bar{3}$	32	$3m$	$\bar{3}m$	6	$\bar{6}$	$6/m$	622	$6mm$	$\bar{6}m2$	$6/mmm$	23	$m3$	432	$\bar{4}3m$	$m3m$
$6mm1_R$																											○					
$3m1_R$																										○						
$6mm$																									○							
$6m_Rm_R$																								○								
61_R																							○									
31_R																						○										
6																					○											
6_Rmm_R																				○												○
$3m$																			○												○	
$3m_R$																		○												○		
6_R																	○												○			
3																○												○				
$4mm1_R$															○																	○
4_Rmm_R														○																	○	
$4mm$													○																			
$4m_Rm_R$												○																		○		
41_R											○																					
4_R										○																						
4									○																							
$2mm1_R$								○							○												○		○			○
2_Rmm_R					○			○			○				○					○			○				○		○			○
$2mm$							○																			○						
$2m_Rm_R$						○						○		○										○				○		○		
$m1_R$							○						○	○											○	○					○	
m				○			○						○	○					○			○			○	○					○	
m_R			○			○	○		○	○		○	○	○				○			○			○	○	○		○		○	○	
21_R					○															○												
2_R		○			○			○			○				○		○			○			○				○		○			○
2			○															○														
1_R				○															○													
1	○		○	○		○	○		○	○		○	○	○		○		○	○		○	○		○	○	○		○		○	○	

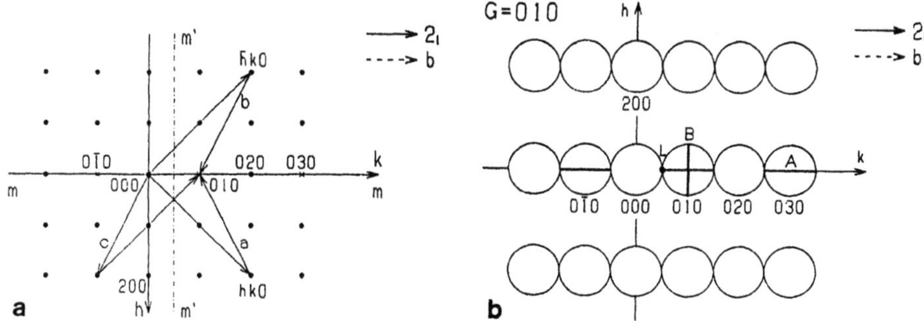

Figure 7. (a) "Umweganregung" paths a, b and c to the 010 kinematically forbidden reflection. (b) *A* and *B* G-M lines in kinematically forbidden reflections.

results, an integral-number index is given to each reflection spot in diffraction patterns. The systematic absence of reflections makes clear the lattice type of a crystal. It should be noted that the reflections forbidden by lattice types are always absent even if dynamical diffraction takes place. By comparing experimentally obtained absences and extinction rules given for lattice types [P, C (A, B), I, F, and R], a lattice type is identified for a crystal examined.

There are three space-group symmetry elements of diperiodic plane figures: 1) a horizontal screw axis $2'_1$, 2) a vertical glide plane g with a horizontal glide vector and 3) a horizontal glide plane g', which are related to the point-group symmetry elements 2', m, and m', respectively. From these elements, together with 10 symmetry elements of the point groups, 80 space groups are produced.

Usual extinction rules for screw axes and glide planes hold only in the approximation of kinematical diffraction. Kinematically forbidden reflections due to these symmetry elements appear due to "Umweganregung" of dynamical diffraction. However, the extinction of intensity still takes place in these reflections for certain crystal settings with respect to the incident beam. This extinction appears as dark lines, called G-M lines after Gjønnes and Moodie [5].

Figure 7(a) illustrates Umweganregung paths to a forbidden reflection. The $0k0$ (k = odd) reflections are kinematically forbidden owing to a b-glide perpendicular to the a-axis and/or a 2_1 screw axis in the b-direction. Let us consider an Umweganregung path "a" in the 0th Laue zone for the

010 forbidden reflection. The path "b" is geometrically equivalent to the path "a" with respect to the glide plane and the 2_1 screw axis. Owing to a translation of one-half of the lattice translation due to the 2_1 screw axis and/or the glide plane, the following relations exist between the crystal structure factors :

$$F(h,k) = F(\bar{h},k) \quad \text{for} \quad k = 2n \tag{1}$$

$$F(h,k) = -F(\bar{h},k) \quad \text{for} \quad k = 2n+1 \tag{2}$$

that is, the structure factor of a reflection $hk0$ located at the lower half of the paper and that of a reflection $\bar{h}k0$ located at the upper half have the same phase for reflections of even order k, but the opposite phases for reflections of odd order k.

Since an Umweganregung path to a kinematically forbidden reflection $0k0$ (k = odd) contains an odd number of reflections with odd k, the following equations hold:

$$F(h_1,k_1)\,F(h_2,k_2) - - - - F(h_n,k_n) \quad \text{for} \quad \text{path } a \tag{3}$$

$$= -F(\bar{h_1},k_1)\,F(\bar{h_2},k_2) - - - - F(\bar{h_n},k_n) \quad \text{for} \quad \text{path } b \tag{4}$$

where

$$\sum_{i=1}^{n} h_i = 0, \quad \sum_{i=1}^{n} k_i = k \quad (k = odd) \tag{5}$$

and functions involving the excitation errors are omitted because we consider the cases where the functions are the same for all these paths. When the projection of the Laue point along the zone axis concerned lies on the axis k, the excitation errors between the paths "a" and "b" are the same. Since the wave passing through the path "a" and that through the path "b" have the same amplitude but opposite signs, these two waves are superposed in the $0k0$ disks (k = odd) and cancel each other, resulting in horizontal dark lines A in the forbidden disks, as shown in Fig.7(b). Line A runs along the direction of the screw axis or the glide translation passing the zone axis of projection. From a similar consideration, it was found that a vertical line B is formed in the exactly excited reflection 010 disk, as shown in Fig.7(b). Line B occurs perpendicularly to line A along the exact Bragg positions. When Umweganregung paths are present only in the 0th Laue zone, the glide plane and screw axis produce the same dynamical extinction lines A and B. We call these lines A_2 and B_2 G-M lines, the subscript 2 indicating two-dimensional interaction. The dynamical extinction rules for space-group symmetry elements are summarized in Table 3. It was found that 177 space groups among 230 can be identified by using G-M lines.

The comments on the symmetry elements observed by CBED are given in a paper [6].

TABLE 3. Dynamical extinction (G-M line) rules for space-group symmetry elements.

Symmetry elements of parallel-sided specimen	Orientation to specimen surface	GM lines	
		Two-dimensional (ZOLZ) interaction	Three-dimensional (HOLZ) interaction
Glide planes	Perpendicular: g	A_2 and B_2	A_3
	Parallel: g'	—	Intersection of A_3 and B_3
Twofold screw axes	Parallel: $2_1'$	A_2 and B_2	B_3

3. Structure Refinement [7], [8]

Firstly, the CBED method enables us to obtain diffraction patterns from a small specimen area about 1nm in diameter. The area illuminated with the incident electron beam is small enough to expect that a crystal is perfect and has a constant thickness and no bending. Consequently, CBED patterns can be directly compared with calculated ones based on the dynamical theory of electron diffraction. The method can be applied not only to the determination of perfect crystal structures but also to that of local crystal structures which change with the specimen position. This is contrasted with the fact that X-ray and neutron diffraction analyses determine the structural parameters averaged over a large specimen volume which consists of mosaic crystals and includes many domains and lattice defects.

Secondly, CBED intensities possess information on the phases of crystal structure factors because of strong dynamical diffraction effects. This fact shows a great difference from X-ray and neutron diffraction analysis, in which the intensities can be explained by the kinematical diffraction theory and information on the phases of crystal structure factors is lost. Therefore, CBED can determine atom positions without encountering the phase problem with which X-ray structure analysis is confronted.

In X-ray analysis, corrections of intensity data especially for extinction and absorption are important while such corrections are not necessary in the case of CBED. Hence, raw intensity data on CBED patterns are of good quality to compare directly with theoretical intensities when the patterns are taken from thin specimens. However, one important thing to be done for accurate structure determination is the subtraction of inelastically scattered electrons.

We should not forget to emphasize that recent technical developments have enabled the structure analysis by the CBED method. The first is new recording tools for electrons - an imaging plate (IP) and a slow scan

CCD camera, which have taken the place of negative films. They have high sensitivity, a wide dynamic range and a linear response for electron doses. The second is energy filtering techniques to eliminate inelastically scattered electrons, using energy filters of the sector-type and the omega-type. A commercial filter of the former type is available from Gatan Inc., and a commercial electron microscope equipped with the latter type filter from LEO Co., Ltd. JEOL recently manufactured a latter type microscope under our research project. Those filters are indispensable for the structure analysis with high precision. The third is high-speed computers for laboratory use or work stations (WS). Many-beam dynamical calculations using more than one hundred beams are necessary for obtaining accurate CBED intensities. The calculations are carried out for a large number of points with different excitation errors, and further repeated by changing structural parameters to obtain their final values, using the nonlinear least square method. Without use of a recent work station, the structure analysis by CBED is impossible.

3.1. ANALYSIS PROCEDURE

In advance of analysis, the space group, the lattice parameters and the tentative positional parameters of atoms of a crystal are assumed to be known, and the accelerating voltage of the incident electrons is separately determined using a HOLZ line pattern of a standard specimen like Si. The analysis procedure is as follows : (i) A CBED pattern is taken from a crystal, using IPs in an electron microscope. A thin area of the specimen has to be chosen to reduce inelastically scattered intensities. If the expected atomic displacements are known in advance, an electron incidence that makes CBED patterns sensitive to the displacements should be selected. (ii) The intensities of the CBED patterns are read out from the IPs in an IP-reader and are transferred to a work station as a digital image. (iii) The distortion of the patterns due to the lens aberrations of the electron microscope is numerically corrected on the work station. The line profiles of HOLZ reflections which run along the radial directions through the centers of the disks are taken from the CBED patterns. Background intensities are subtracted from the line profiles in a certain approximation. (iv) The thickness of the specimen is determined by comparing a ZOLZ pattern with those simulated at different thicknesses. Since the patterns are not so sensitive to the structural details, the dynamical simulations are carried out using the structural parameters initially assumed. (v) The structural parameters are refined by the nonlinear least square method so as to minimize the residual sum of squares S between the experimental HOLZ line profiles and calculated ones. Intensity calculations are carried out by the Bloch

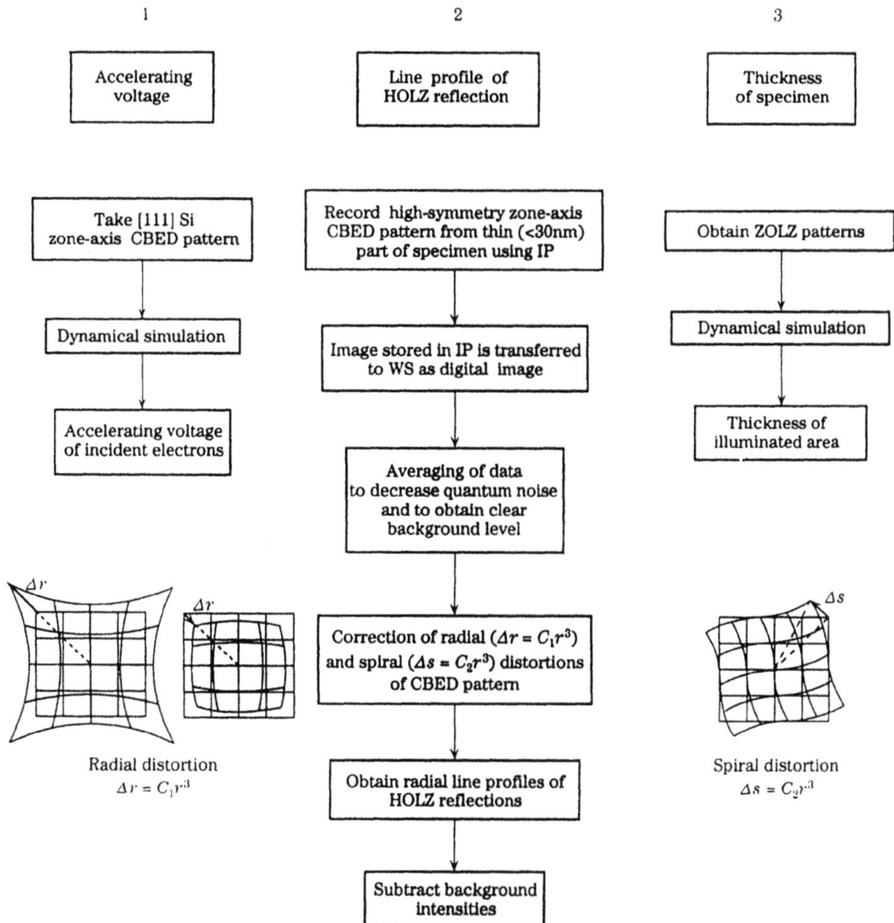

Figure 8. Experimental procedure of the structure refinement by CBED.

wave dynamical theory of electron diffraction with the aid of a generalized Bethe approximation. The analysis procedure is summarized in Fig.8 and 9.

3.2. SRTIO$_3$

We have applied the present method to the low-temperature phase of SrTiO$_3$. SrTiO$_3$ undergoes a second-order phase transformation at 105K from the high-temperature phase of the space-group $Pm3m$ to the low-temperature phase of the space-group $I4/mcm$. The high-temperature phase has a cubic Perovskite structure, in which Sr atoms occupy the lattice cor-

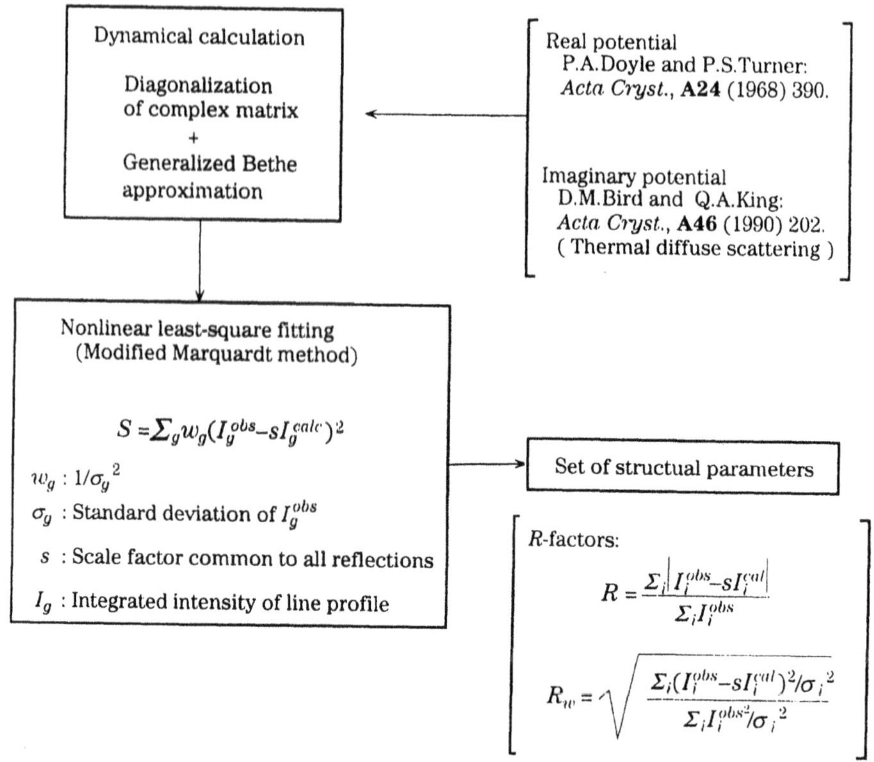

Figure 9. Fitting procedure of the structure parameters by CBED.

ners, oxygen atoms the face centered positions and a Ti atom occupies the body center (Fig.10(a)). In the low-temperature phase, an oxygen octahedron slightly rotates clockwise with respect to the z-axis and the neighbouring oxygen octahedra rotate anticlockwise. This type of atom displacements is described by the condensation of the R_{25}-phonon mode of the high-temperature phase. The positions of Sr and Ti atoms are unchanged. The rotation manner of the octahedra in the successive unit cells is shown in Fig.10(b). The lattice parameters a and c of the low-temperature phase are given by $a \sim \sqrt{2}a_c$ and $c \sim 2a_c$, where a_c is the lattice parameter of the high-temperature phase. Table 4 shows the atomic coordinates of the low-temperature phase. The rotation angle ϕ of the oxygen octahedron is related to the parameter x for the O(2) atoms by the equation

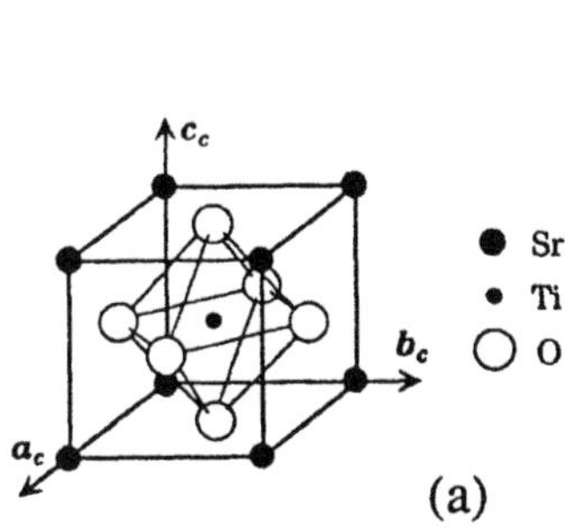

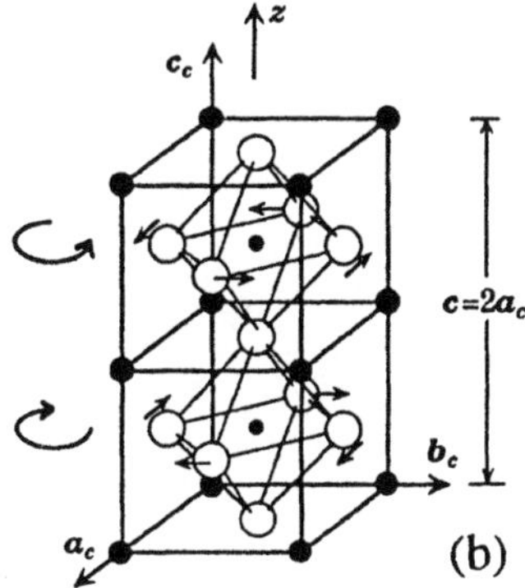

Figure 10. Crystal structures of $SrTiO_3$ (a) high-temperature phase, (b) low-temperature phase.

TABLE 4. Atomic coodinates of the low temperature phase of $SrTiO_3$.

Atom	Wyckoff position	x	y	z
Sr	4b	0	1/2	1/4
Ti	4c	0	0	0
O(1)	4a	0	0	1/4
O(2)	8h	x	1/2+x	0

$x = (1 - tan\phi)/4$. Therefore, the structure analysis of the low-temperature phase implies to determine only one positional parameter, the rotation angle ϕ of the octahedron and the Debye-Waller factor B of the oxygen ions, where B is assumed to be isotropic. $SrTiO_3$ is the most suitable substance for the first application of the present method.

Figure 11(a) shows a CBED pattern of the low-temperature phase of $SrTiO_3$ taken with the [001] incidence at 87K from a 3-nm-diameter area in one domain. The electron microscope used was a JEM-100CX equipped with a field emission gun (FEG). The first-order Laue zone (FOLZ) reflections are superlattice reflections appearing only in the low-temperature phase, where the lattice parameter c is twice that in the high-temperature phase. The intensities of the reflections are governed by the value of the rotation angle of the oxygen octahedron. The second-order Laue zone (SOLZ) reflections were the FOLZ reflections in the high-temperature phase. Only O(2) oxygens contribute to the h k 2 (h, k = even) reflections, but all the atoms to the h k 2 (h, k = odd) reflections. The CBED pattern ex-

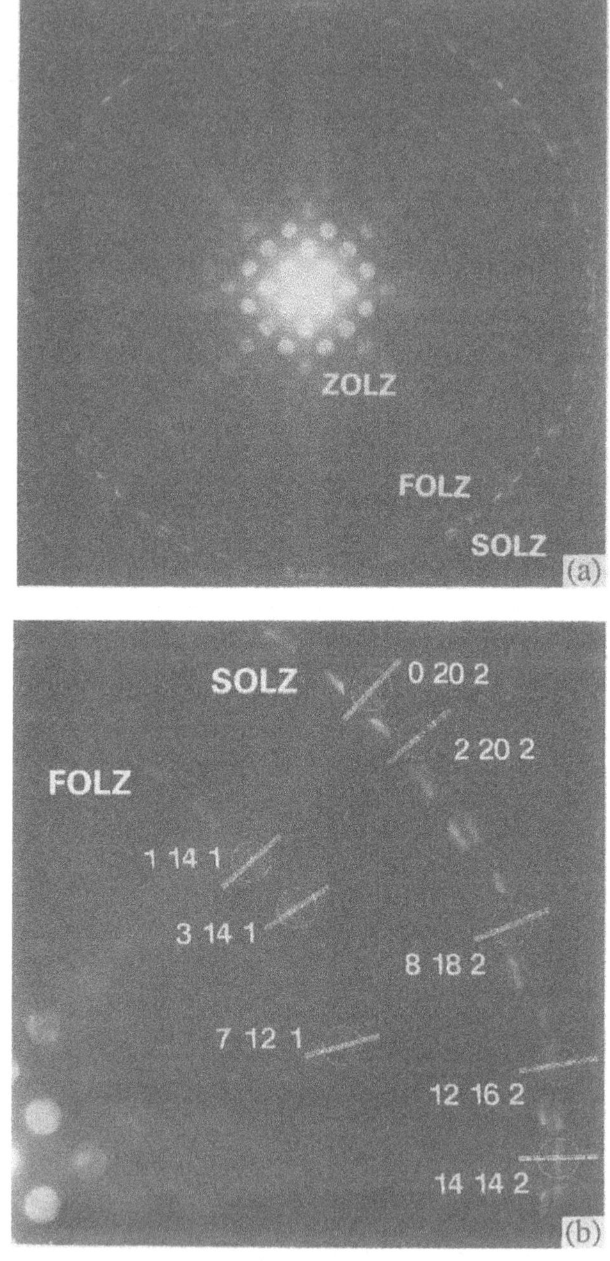

Figure 11. CBED patterns of the low-temperature phase of $SrTiO_3$.

TABLE 5. Final results of the fitting (integrated intensity).

h	k	l	obs.	cal.	res.
7	12	1	3.675	2.480	1.186
1	14	1	4.627	4.563	0.063
3	14	1	5.494	5.734	-0.240
14	14	2	8.518	8.501	0.017
8	18	2	1.675	2.405	-0.731
12	16	2	2.964	3.018	-0.054
0	20	2	4.458	4.429	0.029
2	20	2	1.000	0.353	-0.353

hibited a fourfold rotation and two types of mirror symmetries, the total symmetry being 4mm. Hence, the reflections within one eighth sector are symmetry independent. Figure 11(b) shows a quarter sector of Fig.11(a). After performing distortion correction and averaging of the CBED pattern, three FOLZ reflections and five SOLZ reflections with even values of h and k indicated in Fig.11(b) were used for the analysis, all of these reflections being unaffected by Sr, Ti and O(1) atoms but only by O(2) atoms.

Using the integrated intensities of the HOLZ line profiles, the rotation angle of the oxygen octahedron ϕ and the Debye-Waller factor $B(\mathrm{O}(2))$ were determined to minimize the residual sum of squares $S = \sum_g w_g(I_g^{obs} - sI_g^{calc})^2$ between experimental intensities and calculated ones. The values obtained were $\phi = 1.12 \pm 0.04°$ and $B(\mathrm{O}(2)) = 0.35 \pm 0.06\text{Å}^2$ with R-factor values of $R = 8.2\%$ and $R_w = 5.3\%$. Table 5 gives the final results of the fitting between the experimental and calculated intensities, which show good agreement. Figure 12 shows the contour map of the residual sum of squares $S = \sum_g w_g(I_g^{obs} - sI_g^{calc})^2$ with respect to the parameters ϕ and B. Since no local minimum is seen, it is ascertained that the values of the parameters obtained are the true solution corresponding to the global minimum of S.

In the present analysis, the fitting parameters used were limited to ϕ, $B(\mathrm{O}(2))$ and scale factor s to reduce the calculation time. The method will be extended to add lattice parameters, Debye-Waller factors of the other atoms (Sr, Ti and O(1)), the specimen thickness and low-order structure factors as fitting parameters.

Using the HOLZ line profiles, ϕ and $B(\mathrm{O}(2))$ were determined so as to minimize S. The values refined were $\phi = 1.14 \pm 0.01°$ and $B(\mathrm{O}(2)) = 0.37$ $\pm 0.01°$. This result agrees with that obtained for the integrated intensities,

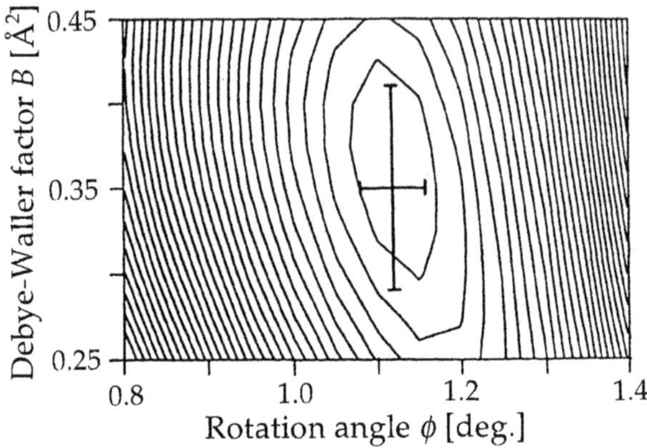

Figure 12. Contour map of the residual sum of squares S.

within error. Calculated intensities are seen to agree well with experimental ones in Fig.13. However, the remaining small discrepancies give not good R-factors, R = 31.6% and R_w = 41.6%. One of the reasons for the large R and R_w factors compared with those for the integrated intensity (case I) is a uniform shift of the line profiles with respect to the calculated ones because of a distortion of the CBED pattern due to the aberration of the image forming lenses.

By electron spin resonance experiments, Unoki and Sakudo [9] and Müller et al.,[10] measured the rotation angle ϕ of the oxygen octahedron as a function of temperature over the temperature range from 4 to 105K. The former obtained $1.4 \pm 0.1°$ as the value of ϕ and the latter 1.25° both at 78K. Shirane and Yamada [11] determined ϕ to be a similar value of $1.37 \pm 0.35°$ with a little low precision at 78K by neutron diffraction. Fujishita et al.,[12] determined ϕ to be $1.6 \pm 0.1°$ at 77K by X-ray diffraction. Figure 14 shows the above results together with the present one. The present result, shown by a black circle, is in good agreement with those of Unoki and Sakudo ($\phi = 1.3°$ at 87K) and Müller *et al.*, ($\phi = 1.1°$ at 87K). The present result also agrees with that of Shirane and Yamada when the temperature dependence obtained by Unoki and Sakudo and Müller et al. is assumed. The value by Fujishita et al. is a little larger than the other results reported and may be a little different from ours even if the temperature difference is taken into account.

4. Coherent CBED [13], [14]

Interference fringes of a lattice spacing $d = 1/|\mathbf{g}|$ can be observed at the overlapping region of the disks due to reflections $\mathbf{g}_1$ and $\mathbf{g}_2$, $\mathbf{g} = \mathbf{g}_1 - \mathbf{g}_2$, when the electron probe size ΔX is smaller than the spacing d. In other

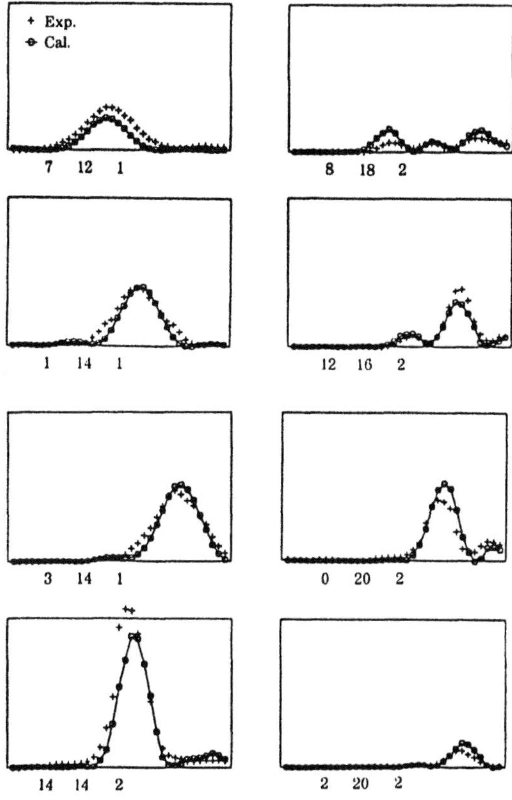

Figure 13. Experimental and calculated line profiles.

words, an intensity distribution which is formed by coherent interference can be seen, if $\Delta X < d$. For $\Delta X > d$, coherent interference is smeared out, which is said to be incoherent. Hence, coherent illumination is not determined by the absolute source size but by the relative size $\Delta X/d < 1$. Using the Bragg equation $2d\theta_B \approx \lambda$, the condition $\Delta X < d$ is written as $\Delta X < \lambda/2\theta_B$ or $\Delta X\theta_B < \lambda 2$. When ΔX is regarded as the source size and θ_B as the illumination angle, the inequality is just a coherent condition in light optics. In ordinary CBED, a specimen is illuminated with a probe about 10nm in diameter. Since the lattice spacings usually observed are less than 1nm, the lattice fringes cannot be observed even if the disks are overlapped. If the probe focused on the specimen becomes smaller than 1nm, the interference effect due to lattice planes appears at the overlapping regions. When the probe is focussed below (above) the specimen with the

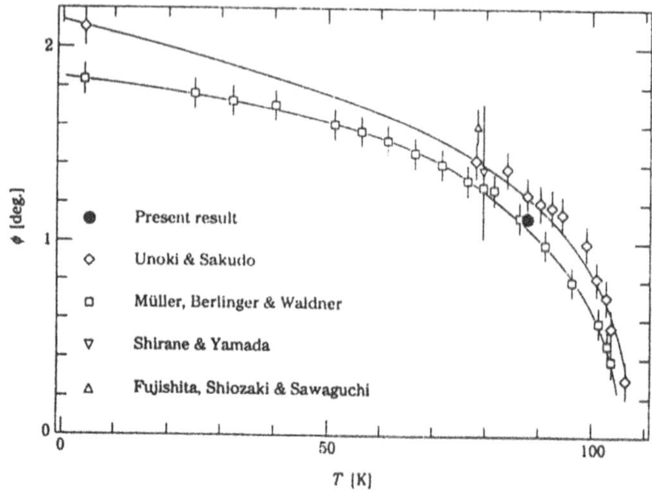

Figure 14. Present result agrees well with other results already reported.

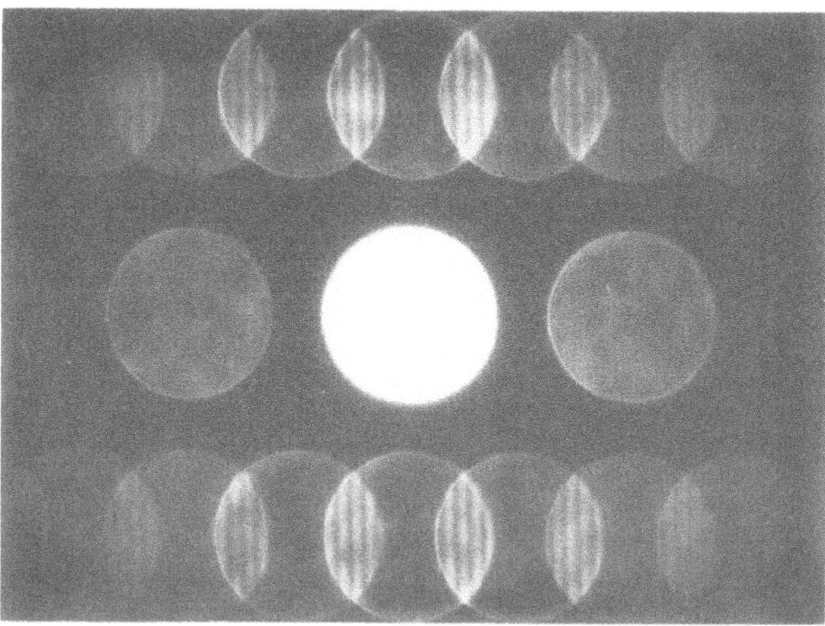

Figure 15. Coherent CBED pattern of [100] FeS_2.

probe size kept unchanged, the interference fringes with the spacing are observed. Figure 15 clearly shows such fringes. That the fringes are formed only at the overlapping regions vividly exhibits that they are produced by the interference of the two diffraction waves.

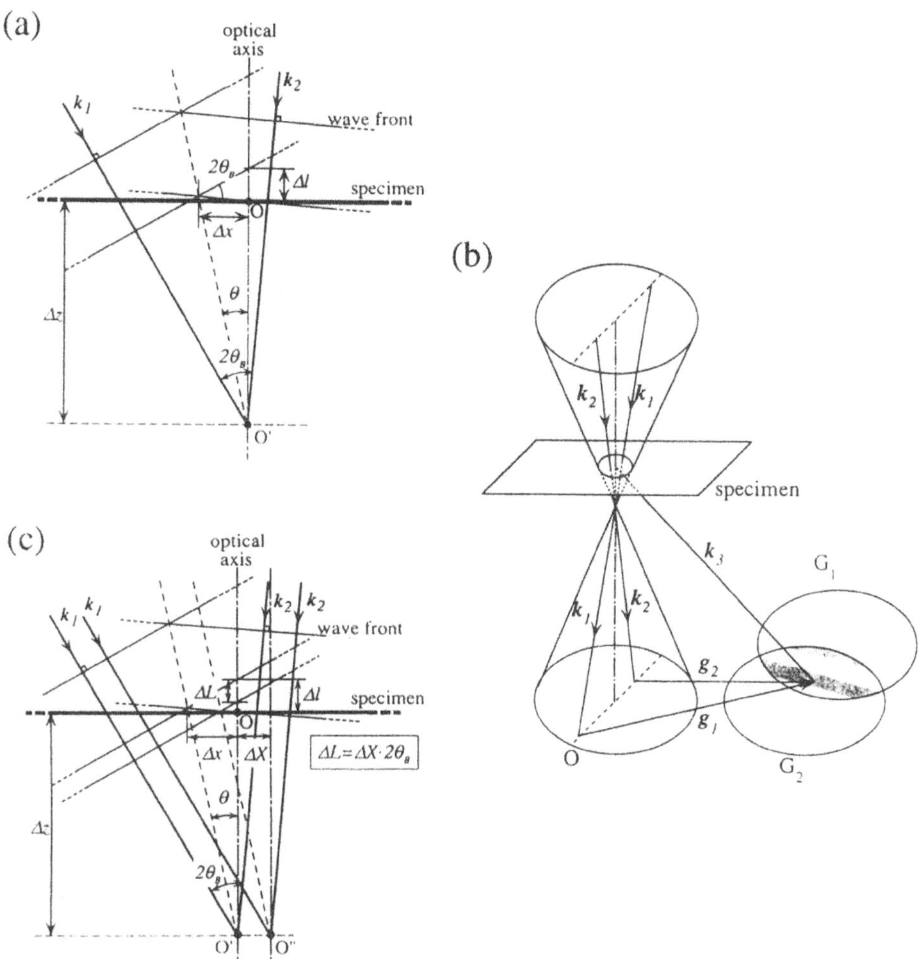

Figure 16. (a) Illustration of the optical path difference between waves $\mathbf{k}_1$ and $\mathbf{k}_2$. (b) Schematic ray-path diagram of coherent CBED. (c) Illustration of beam size effect (see text).

Figure 16(a) illustrates the optical path difference between wave $\mathbf{k}_1$ and $\mathbf{k}_2$. Let the origin O of a crystal unit cell be at the cross point of the specimen and the optical axis of an electron microscope. The two incident waves with wave vectors $\mathbf{k}_1$ and $\mathbf{k}_2$ meet in phase at point O' below the specimen. The two waves are always in phase on the bisector (dotted line) of the two directions. Let us consider the wave $\mathbf{k}_1$ is diffracted in the direction $\mathbf{k}_3 = \mathbf{k}_1 + \mathbf{g}_1$. The wave $\mathbf{k}_2$ is also diffracted in the same direction $\mathbf{k}_3 = \mathbf{k}_2 + \mathbf{g}_2$ when $|\mathbf{g}_1 - \mathbf{g}_2| = |\mathbf{g}| = 2\theta_B/\lambda$, as shown in Fig.16(b), where θ_B is the Bragg angle of reflection $\mathbf{g}$ and λ is the wave length of the incident electron waves.

The phase difference $\Delta\phi$ between the two waves at the origin O is given by $2\pi/\lambda$ times path difference Δl measured along the optical axis, where

$$\Delta l \approx x \cdot \tan 2\theta_B \approx \Delta x \cdot 2\theta_B, \tag{6}$$

Δx being the distance between the origin O and the cross point of the bisector with the specimen. The amount Δx is given by the product of defocus Δz and tilt angle of the bisector θ, $\Delta x = \Delta z\theta$. Then, $\Delta\phi$ is written as

$$\Delta\phi = 2\pi\Delta l/\lambda = 2\pi\Delta x/d = 2\pi\,|\mathbf{g}| \cdot \Delta x, \tag{7}$$

where $\lambda = 2d\sin\theta_B \approx 2d\theta_B$ and d is the lattice spacing for the Bragg reflection $\mathbf{g}$. The resultant diffraction amplitude Ψ in the direction $\mathbf{k}_3$ from the two waves with $\mathbf{k}_1$ and $\mathbf{k}_2$ is proportional to $F_1\exp(-2\pi i\alpha_1) + F_2\exp(-2\pi i\alpha_2 - \Delta\phi)$, where F_i and α_i (i=1,2) are the amplitude and the phase of the crystal structure factor, respectively. The diffraction intensity I in the direction $\mathbf{k}_3$ is expressed as

$$\begin{aligned} I &= |\Psi|^2 && (8)\\ &\propto \left|F_1\exp(-2\pi i\alpha_1) + F_2\exp\{-2\pi i(\alpha_2 + |\mathbf{g}| \cdot \Delta x)\}\right|^2 && (9)\\ &\propto F_1^2 + F_2^2 + 2F_1F_2\cos\left[2\pi i\{|\mathbf{g}| \cdot \Delta x + (\alpha_2 - \alpha_1)\}\right]. && (10) \end{aligned}$$

The third term of the expression implies that the intensity of the interference pattern in overlapping regions of CBED disks depends not only on the phase difference $\Delta\phi = 2\pi\,|\mathbf{g}| \cdot \Delta x$ between the two waves measured at the origin O but also the phases of the crystal structure factors of the $\mathbf{g}_1$ and $\mathbf{g}_2$ reflections.

4.1. DEFOCUS

Let us first assume the beam size at O' to be small enough, compared with the lattice spacing which we want to observe. The intensity sinusoidally changes with Δx. For a fixed amount of defocus Δz, Δx changes with θ. Thus, a sinusoidal intensity change with a period $1/|\mathbf{g}|$ appears in the overlapped regions of the neighbouring disks. As defocus Δz decreases, the phase change due to the change of θ decreases, the number of fringes observed being smaller. When Δz approaches zero, the phase $2\pi\,|\mathbf{g}| \cdot \Delta x$ is independent from θ, and intensities at the overlapping region become uniform. It should be noted that twice the number n of fringes in the overlapping region times the spacing d equals the diameter of the illuminated specimen area, provided that the second neighbouring disks contact each other. Since the interference fringes are produced as a result of the Bragg reflection or originate in the lattice periodicity, the fringes can be called lattice fringes, though no direct correspondence exists between the positions of real lattice planes and those of the fringes.

4.2. BEAM SIZE

We consider another two incident beams with wave vectors $\mathbf{k}_1$' and $\mathbf{k}_2$', which meet in phase at point O" as shown in Fig.16(c). The parallel line to the optical axis through O" crosses the specimen by ΔX right from the origin O. The path difference ΔL between the wave $\mathbf{k}_1$ toward O' and $\mathbf{k}_1$' toward O" is given by $\Delta L \approx \Delta X \tan 2\theta_B \approx \Delta X 2\theta_B$ resulting in a phase difference $\Delta\Phi = 2\pi\Delta L/\lambda = 2\pi\Delta X/d = 2\pi\,|\mathbf{g}|\cdot\Delta X$. We add the phase difference due to the beam position to the expression of the diffraction intensity:

$$I \propto F_1^2 + F_2^2 + 2F_1F_2\cos\left[2\pi i\left\{|\mathbf{g}|\cdot\Delta x + (\alpha_2-\alpha_1) + |\mathbf{g}|\cdot\Delta X\right\}\right]. \quad (11)$$

When the illuminated specimen area is displaced, the interference fringes move through the term $|\mathbf{g}|\cdot\Delta X$. At $\Delta z = 0$, the intensity of the entire overlapping region changes uniformly with a displacement of the specimen or the incident beam. The effect of a finite beam size is taken into account by integrating the equation with respect to ΔX. When the integration is performed over one lattice spacing $\Delta X = 1/|\mathbf{g}| = d$, the third term of the equation vanishes; interference fringes disappear. The beam size must be smaller than the lattice spacing concerned to observe interference fringes or the coherent effect.

4.3. PHASE OF CRYSTAL STRUCTURE FACTOR

The position of the entire set of fringes depends on $\alpha_2-\alpha_1$, giving important information on the phase of the crystal structure factor.

References

1. Tanaka, M. and Terauchi, M. (1985) *Convergent-Beam Electron Diffraction*, JEOL-Maruzen, Tokyo.
2. Tanaka, M., Terauchi, M. and Kaneyama, T. (1988) *Convergent-Beam Electron Diffraction II*, JEOL-Maruzen, Tokyo.
3. Tanaka, M., Terauchi, M. and Tsuda, K. (1994) *Convergent-Beam Electron Diffraction III*, JEOL-Maruzen, Tokyo.
4. Buxton, B. F., Eades, J. A., Steeds, J. W. and Rackman, G. M. (1976) The Symmetry of electron diffraction zone axis pattern, *Phil. Trans. R. Soc. London.* **281**, pp.171–194.
5. Gjønnes, J. and Moodie, A. F. (1965) Extinction conditions in the dynamic theory of electron diffraction, *Acta Cryst.* **19**, pp.65–67.
6. Tanaka, M. (1989) Symmetry analysis, *J. Electron Microsc. Tech.* **13**, pp.27–39.

7. Vincent, R., Bird, D. M. and Steeds, J. W. (1984) Structure of AuGeAs determined by CBED I, *Phil. Mag.* **A50**, pp.745–763, and II pp.765–786.
8. Tsuda, K. and Tanaka, M. (1995) Refinement of crystal structure parameters convergent-beam electron diffraction: the low-temperature phase of $SrTiO_3$, *Acta Cryst.* **A51**, pp.7–20.
9. Unoki, H. and Sakudo, T. (1967) Electron spin resonance of $Fe^{3}+$ in $SrTiO_3$ with special reference to the 110°K phase transition, *J. Phys. Soc. Jpn.* **23**, pp.546–552.
10. Müller, K. A., Berlinger, W. and Waldner, F. (1968) Characteristic structural phase transition in perovskite-type compounds, *Phys. Rev. Lett.* **21**, pp.814–817
11. Shirane, Y. and Yamada, Y. (1969) Lattice-dynamical study of the 110°K phase transition in $SrTiO_3$, *Phys. Rev.* **177**, pp.858–863
12. Fujishita, H., Shiozaki, Y. and Sawaguchi, E. (1979) X-ray crystal structure analysis of low temperature phase of $SrTiO_3$, *J. Phys. Soc. Jpn.* **46**, pp.581–586
13. Vine, W. J., Vincent, R., Spellward, P. and Steeds, J. W. (1992) Observation of phase contrast in convergent-beam electron diffraction patterns, *Ultramicroscopy* **41**, pp.423–428.
14. Tsuda, K. and Tanaka, M. (1996) Interferometry by coherent convergent-beam electron diffraction, *J. Electron Microsc.* **45**, pp.59–63.

5. Exercises

5.1. POINT GROUP DETERMINATION

5.1.1. TiO_2 *(Rutile)*

The space group of rutile is $P4_2/mnm$, its point group being $4/mmm$, Using following three CBED patterns, determine the point group.

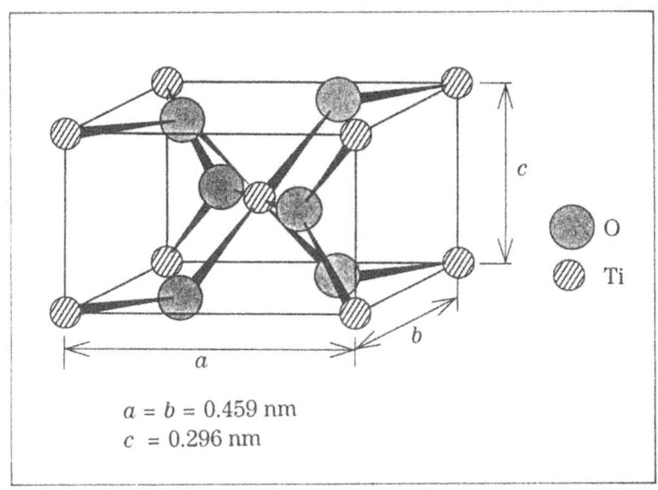

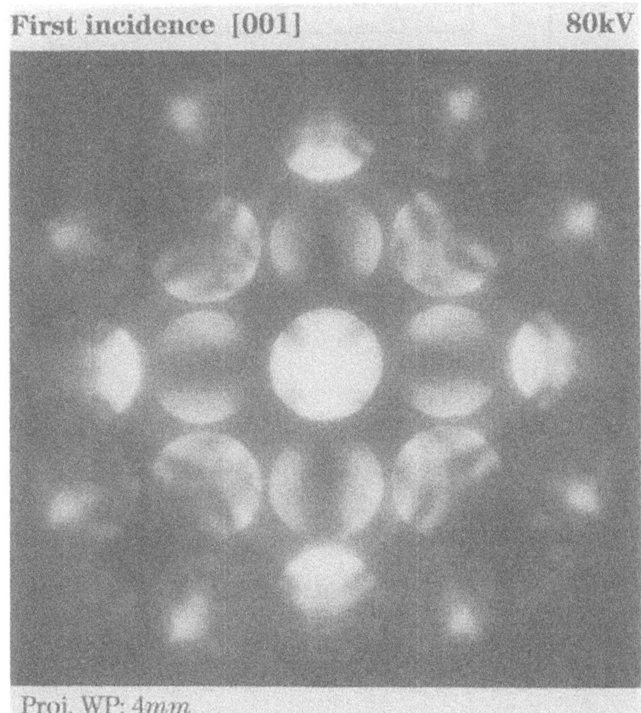

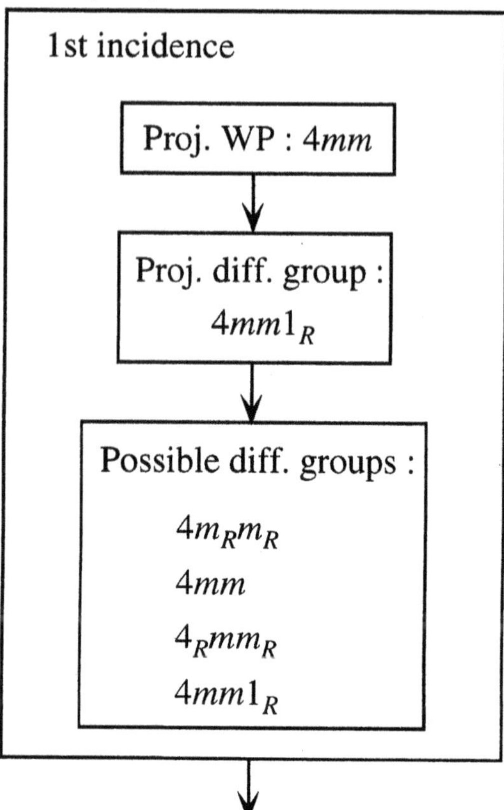

Point group
1st incidence
Proj. WP : $4mm$
Proj. diff. group : $4mm1_R$
Possible diff. groups :
$4m_Rm_R$
$4mm$
4_Rmm_R
$4mm1_R$

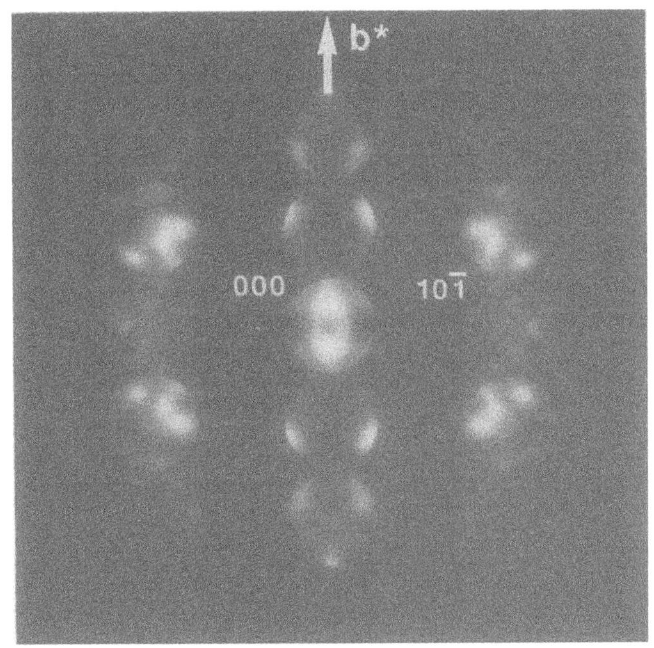
b*
000
10$\bar{1}$

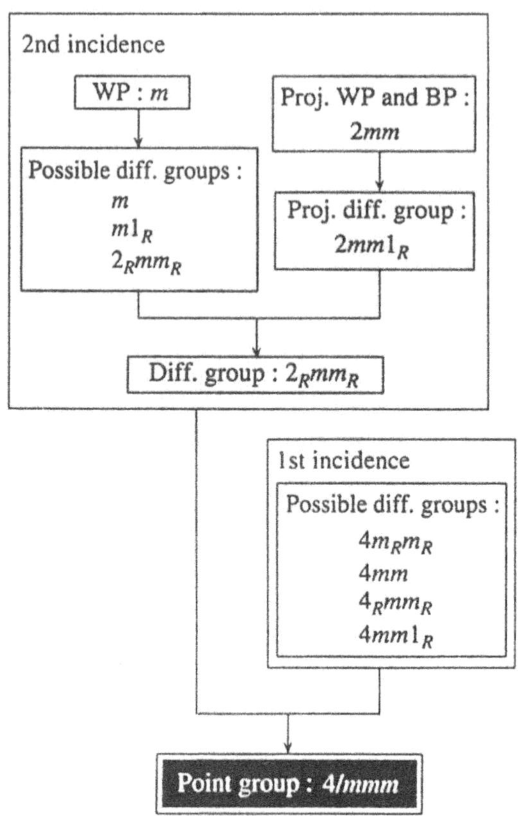

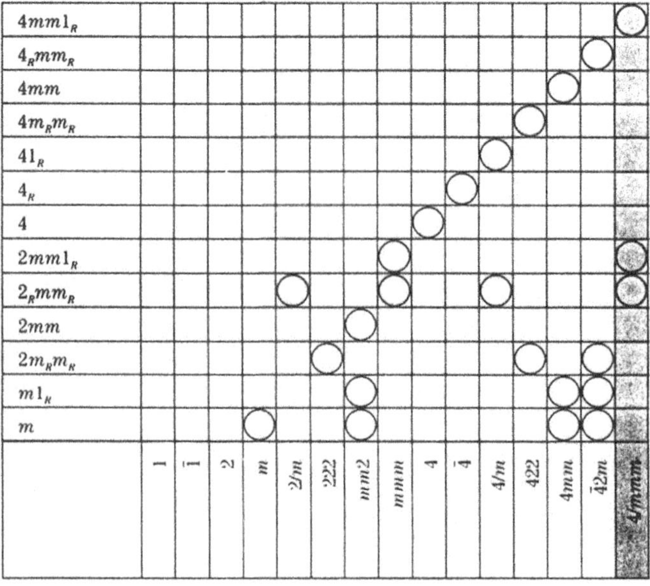

	1	$\bar{1}$	2	m	$2/m$	222	$mm2$	mmm	4	$\bar{4}$	$4/m$	422	$4mm$	$\bar{4}2m$	$4/mmm$
$4mm1_R$															○
4_Rmm_R														○	
$4mm$													○		
$4m_Rm_R$												○			
41_R											○				
4_R										○					
4									○						
$2mm1_R$								○							○
2_Rmm_R					○			○			○				○
$2mm$							○								
$2m_Rm_R$						○						○		○	
$m1_R$							○						○	○	
m				○			○						○	○	

5.1.2. *Sm_3Se_4*

Sm_3Se_4 belongs space group I$\bar{4}$3d and point group 432. Using following four CBED patterns, determine the point group.

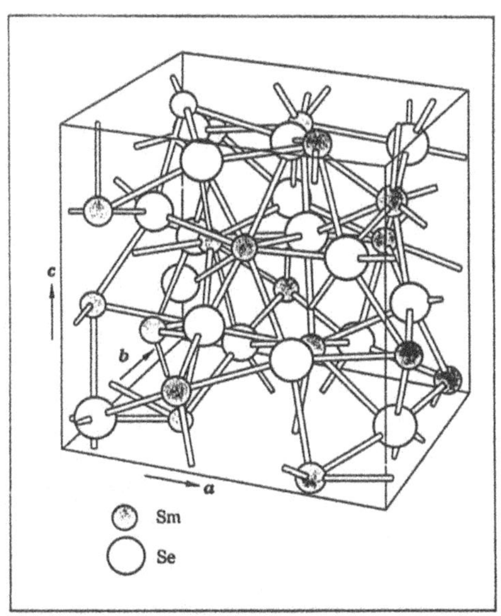

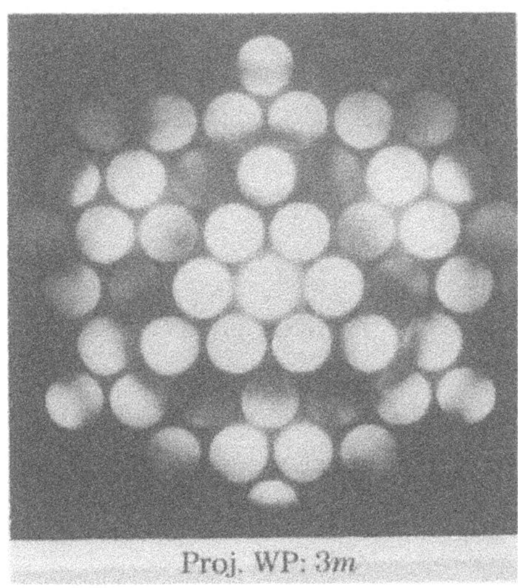

Proj. WP: 3*m*

First incidence [111] T = 100K 80kV

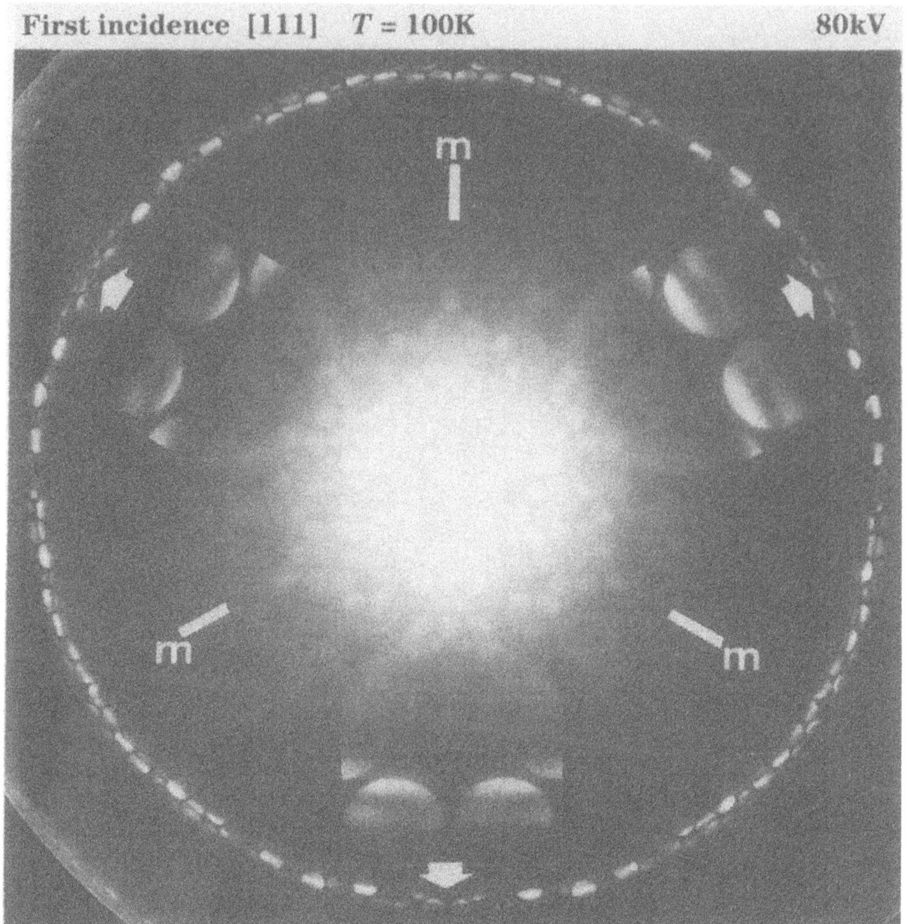

WP: $3m$

Point group
1st incidence
WP : 3m
Proj. WP : 3m
Possible diff. groups :
3m
3m1R
6RmmR
Proj. diff. group :
3m1R
Possible diff. groups :
3m
3m1R
Possible point groups :
3m
4̄3m
6̄m2

m
m
m
m

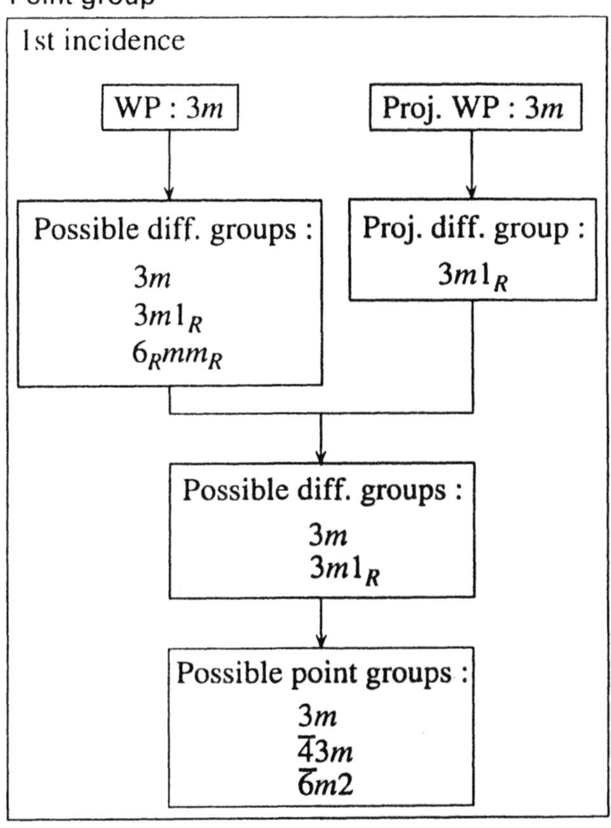

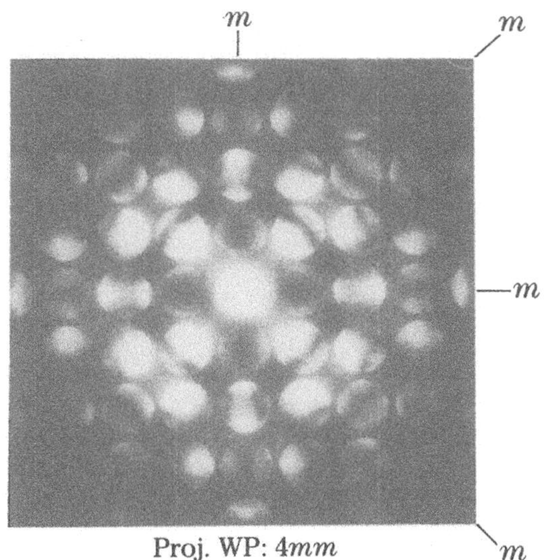

Proj. WP: $4mm$

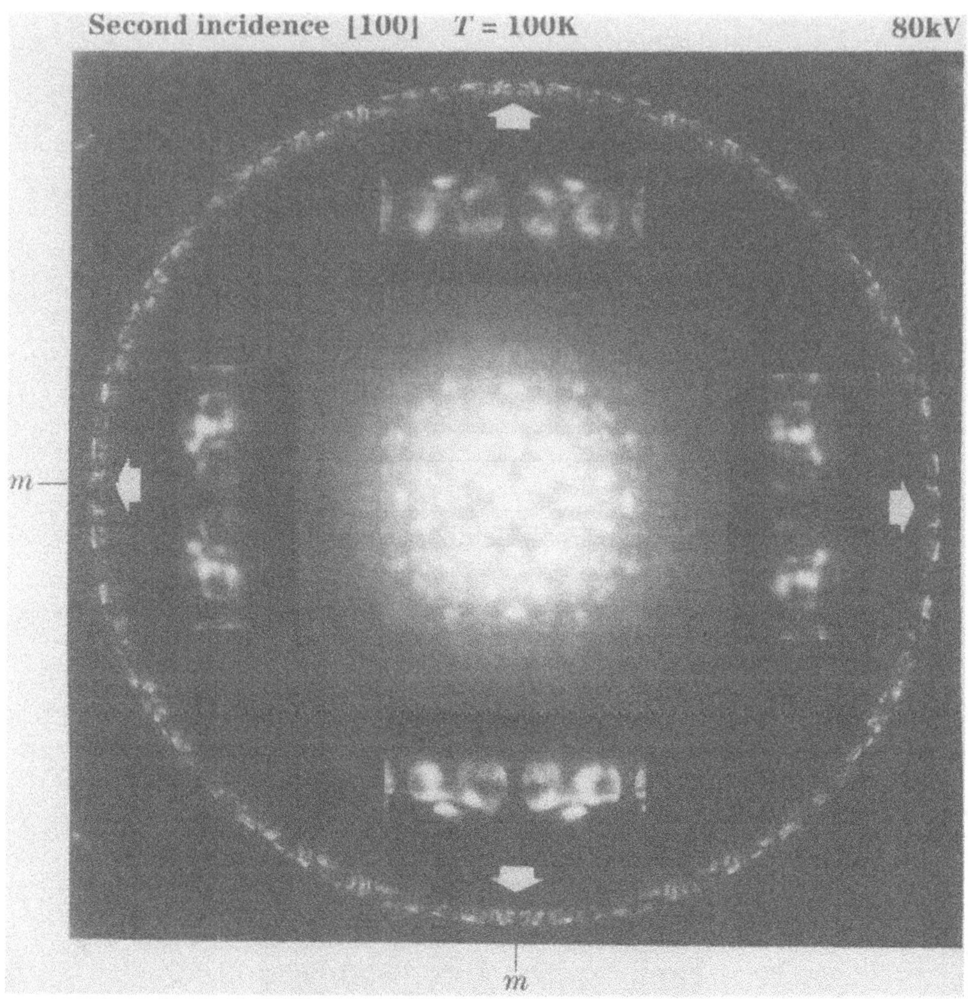

WP: $2mm$

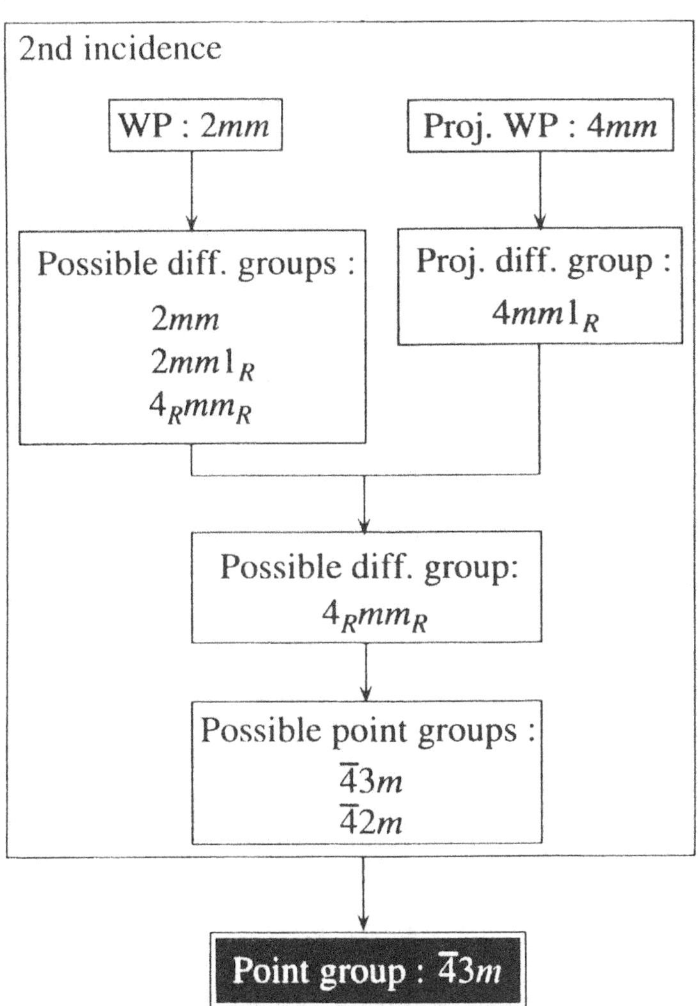
2nd incidence
WP : 2mm
Proj. WP : 4mm
Possible diff. groups :
2mm
2mm1_R
4_Rmm_R
Proj. diff. group :
4mm1_R
Possible diff. group:
4_Rmm_R
Possible point groups :
-43m
-42m
Point group : -43m

5.2. THICKNESS DETERMINATION

We describe here a specimen-thickness determination method which utilizes an equation derived from the two-beam dynamical theory of electron diffraction

$$I = \left(\frac{\pi}{\xi_g}\right)^2 \frac{(\sin \pi t\sqrt{S^2 + \xi_g^{-2}}\,)^2}{\pi^2(S^2 + \xi_g^{-2})},$$

where ξ_g is the extinction distance of an excited reflection g, t is the specimen thickness and S is the deviation parameter from the exact Bragg position. From the equation, the ith subsidiary minimum at S_i from the exact Bragg position is given by

$$t\sqrt{S_i^2 + \xi_g^{-2}} = n_i,$$

where n_i is a whole number. It is rewritten as

$$\left(\frac{S_i}{n_i}\right)^2 = \frac{-1}{\xi_g^2}\left(\frac{1}{n_i}\right)^2 + \frac{1}{t^2}.$$

This shows that the thickness can be determined by measuring the positions of the intensity minima.

The data needed in advance is the lattice spacing of the reflection g and the accelerating voltage of the incident beam. It should be noted that the extinction distance of the reflection is not necessary for the determination, but obtained from the slope of the plot of

$$\left(\frac{S_i}{n_i}\right)^2 \text{ vs. } \left(\frac{1}{n_i}\right)^2.$$

The distance to be measured are L_0 from the center of the diffracted beam profile to the center of the transmitted beam, and L_1, L_2, L_3 and L_4 from the center of the diffracted beam profile to each of the successive minima. These distances are indicated in the photograph, which shows the $2\bar{2}0$ intensity profile of a silicon crystal taken at an approximately two-beam condition.

S_i is driven by the equation.

$$\frac{S_i}{g} = 2\theta_B\left(\frac{L_i}{L_0}\right)$$

$$g = 1/d$$

$$2\theta_B = \lambda/d,$$

where λ is the wavelength of the incident beam and d is the spacing of the reflection planes. Thus, we obtain the equation

$$S_i = \frac{\lambda}{d^2}\left(\frac{L_i}{L_0}\right).$$

The first minimum corresponds not always to $n_1 = 1$ but to the other whole number because of the existence of the term of ξ_g^{-2}. Thus, we set various numbers m for n_1 and make the plot of $(S_i/n_i)^2$ vs. $(1/n_i)^2$, where $n_{i+1} = n_i + 1$ and $n_1 = m$. The plot shows a straight line if the selection of the value of n_1 is collect. The best selection is judged by using a correlation coefficient r. The specimen thickness is given by the intercept of the straight line with the $(S_i/n_i)^2$ axis. The actual procedure is as follows.

By using the values of S_i and by setting $n_i = 1$ and $n_{i+1} = n_i + 1$, the values of $(S_i/n_i)^2$ and $(1/n_i)^2$ are calculated. Next, by setting $n_1 = 2$ and $n_{i+1} = n_i + 1$, these two values are calculated. These calculations are repeated successively by setting $n_1 = 3, 4$.... For a value of n_1, the plot (the regression line) of $(S_i/n_i)^2$ vs. $(1/n_i)^2$ results in a straight line, showing the square of the correlation coefficient r^2 to be a value nearest to unity. We may use a criterior r^4 better to see the regression lines. This method determines the specimen thickness with an accuracy of better than ±2%.

The specimen thickness was determined for the intensity profile of the photograph. The data needed beforehand and the obtained values are given in the tables. The straight line plot of $(S_i/n_i)^2$ vs. $(1/n_i)^2$ was obtained for $n_1 = 2$, and is shown in the graph. The value of the intercept of the line with the vertical axis is 8.67×10^{-7} Å^{-2}, resulting in the thickness of 1070Å. The correlation coefficient is defined by the following equation.

$$r = \frac{S_{xy}}{\sqrt{S_{xx} \cdot S_{yy}}},$$

where

$$S_{xx} = \Sigma x^2 - \frac{(\Sigma x)^2}{n}$$

$$S_{yy} = \Sigma y^2 - \frac{(\Sigma y)^2}{n}$$

$$S_{xy} = \Sigma xy - \frac{\Sigma x \cdot \Sigma y}{n}$$

n : number of data.

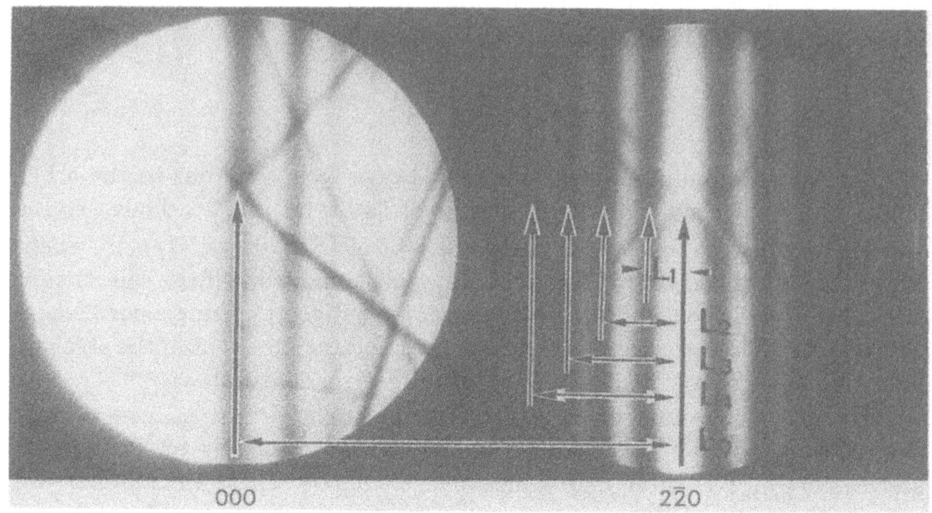

i	(L_i/L_o)	λ	d	S_i
1	7.98×10^{-2}	4.75×10^{-2}(Å)	1.92(Å)	1.03×10^{-3}
2	1.80×10^{-1}			2.32×10^{-3}
3	2.61×10^{-1}			3.36×10^{-3}
4	3.43×10^{-1}			4.42×10^{-3}

		$i=1$	$i=2$	$i=3$	$i=4$
$n_i=1$	$(1/n_i)^2$	1.00			
	$(s_i/n_i)^2$	1.06×10^{-6}			
$n_i=2$	$(1/n_i)^2$	0.250	0.250		
	$(s_i/n_i)^2$	2.65×10^{-7}	1.35×10^{-6}		
$n_i=3$	$(1/n_i)^2$	0.111	0.111	0.111	
	$(s_i/n_i)^2$	1.18×10^{-7}	5.98×10^{-7}	1.25×10^{-6}	
$n_i=4$	$(1/n_i)^2$	6.25×10^{-2}	6.25×10^{-2}	6.25×10^{-2}	6.25×10^{-2}
	$(s_i/n_i)^2$	6.63×10^{-8}	3.36×10^{-7}	7.06×10^{-7}	1.22×10^{-6}
$n_i=5$	$(1/n_i)^2$		4.00×10^{-2}	4.00×10^{-2}	4.00×10^{-2}
	$(s_i/n_i)^2$		2.15×10^{-7}	4.51×10^{-7}	7.81×10^{-7}
$n_i=6$	$(1/n_i)^2$			2.78×10^{-2}	2.78×10^{-2}
	$(s_i/n_i)^2$			3.14×10^{-7}	5.45×10^{-7}
$n_i=7$	$(1/n_i)^2$				2.04×10^{-2}
	$(s_i/n_i)^2$				3.99×10^{-7}

n_1	r^2	t(Å)
1	0.621	8.77×10^{2}
2	0.999	1.07×10^{3}
3	0.991	1.23×10^{3}
4	0.988	1.36×10^{3}

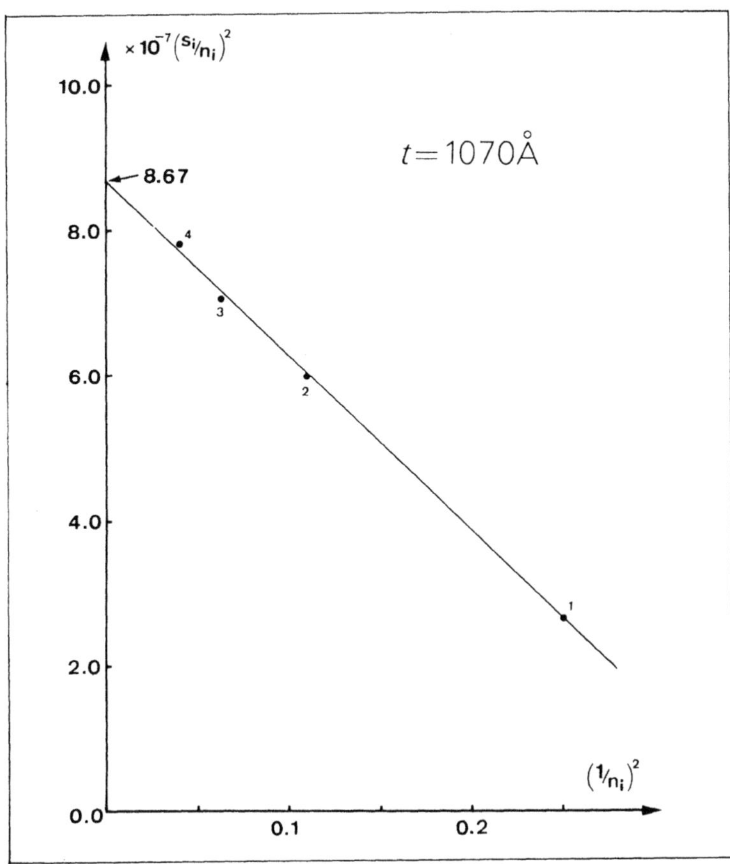

$\times 10^{-7} (s_i/n_i)^2$
10.0
8.0
6.0
4.0
2.0
0.0
8.67
$t = 1070$ Å
4
3
2
1
$(1/n_i)^2$
0.1
0.2

IMAGE SIMULATION IN HIGH RESOLUTION TRANSMISSION ELECTRON MICROSCOPY

ROAR KILAAS
National Center for Electron Microscopy
Lawrence Berkeley National Laboratory,
Berkeley, CA 94720, USA

1. Introduction

The best High Resolution Transmission Electron Microscopes (HRTEM) have a resolution approaching 1 Å which sometimes leads to the erroneous conclusion that using an electron microscope, all atoms in a structure can be resolved. However, it is not the inter-atomic distances that matter, but rather the projected distances between atoms seen from the direction of the incident electron. In order to obtain interpretable results, it is necessary to orient the specimen such that atomic columns are separated by distances that are of the order of the resolution of the microscope or larger. This is a condition that very often is difficult to satisfy and often limits the use of the HRTEM to studies of crystals only in low order zone-axis orientations.

The HRTEM image is a complex function of the interaction between the high energy electrons (typically 200keV - 1MeV) with the electrostatic potential in the specimen and the magnetic fields of the image forming lenses in the microscope. Although images obtained from simple mono-atomic crystals often show white dots separated by spacings that correspond to spacings between atomic columns, these white dots fall on or between atomic columns depending on the thickness of the specimen and the focus setting of the objective lens[1]. Fortunately, in many cases it is only necessary to see the general pattern of image intensities to gain the desired knowledge. However, in general, the image can be best thought of as a complex interference pattern which has the symmetry of the projected atomic configuration, but otherwise has no one-to-one correspondence to atomic positions in the specimen. It is because of this lack of directly interpretable images that the need for image simulation arose. Image simulation grew out of an attempt to explain why electron microscope images of complex oxides sometimes showed black dots in patterns corresponding to the patterns of heavy metal sites in complex oxides, and yet other images sometimes showed white dots in the same patterns[2]. This first application was therefore to characterize the experimental images,

D. L. Dorset et al. (eds.), Electron Crystallography, 115–130.

that is to relate the image character (the patterns of light and dark dots) to known features in the structure.

Most simulations today are carried out for similar reasons, or even as a means of structure determination. Given a number of possible models for the structure under investigation, images are simulated from these models and compared with experimental images obtained on a high-resolution electron microscope. In this way, some of the postulated models can be ruled out until only one remains. If all possible models have been examined, then the remaining model is the correct one for the structure. For this process to produce a correct result, the investigator must ensure that all possible models have been examined, and compared with experimental images over a wide range of crystal thickness and microscope defocus. It is also a good idea to match simulations and experimental images for more than one orientation.

The simulation programs can also be used to study the imaging process itself. By simulating images for imaginary electron microscopes, we can look for ways in which to improve the performance of present-day instruments, or even find that the performance of an existing electron microscope can be improved significantly by minor changes in some instrumental parameter. Alternatively, based on imaging requirements revealed by test simulations, we can adjust the electron microscope to produce suitable images of some particular specimen, or even of some particular feature in a particular specimen.

1.1 DESCRIBING THE TRANSMISSION ELECTRON MICROSCOPE

In order to simulate an electron microscope image, we need firstly to be able to describe the electron microscope in such a way that we can model the manner in which it produces the image. As a first step, we can consider the usual geometrical optics depiction of the transmission electron microscope (TEM).

Figure 1 shows such a diagram of a TEM operated in two distinct modes, set up for microscopy (a), and for diffraction (b). In microscopy mode we see that the TEM consists of an electron source producing a beam of electrons that are focused by a condenser lens onto the specimen; electrons passing through the specimen are focused by the objective lens to form an image called the first intermediate image (I1); this first intermediate image forms the "object" for the next lens, the intermediate lens, which produces a magnified image of it called the second intermediate image (I2); in turn, this second intermediate image becomes the "object" for the projector lens; the projector lens forms the greatly-magnified final image on the viewing screen of the microscope. In microscopy mode, electrons that emerge from the same point on the specimen exit surface are brought together at the same point in the final image.

At the focal plane of the objective lens, we see that electrons are brought together that have left the specimen at different points but at the same angle. The diffraction pattern that is formed at the focal plane of the objective lens can be viewed on the viewing

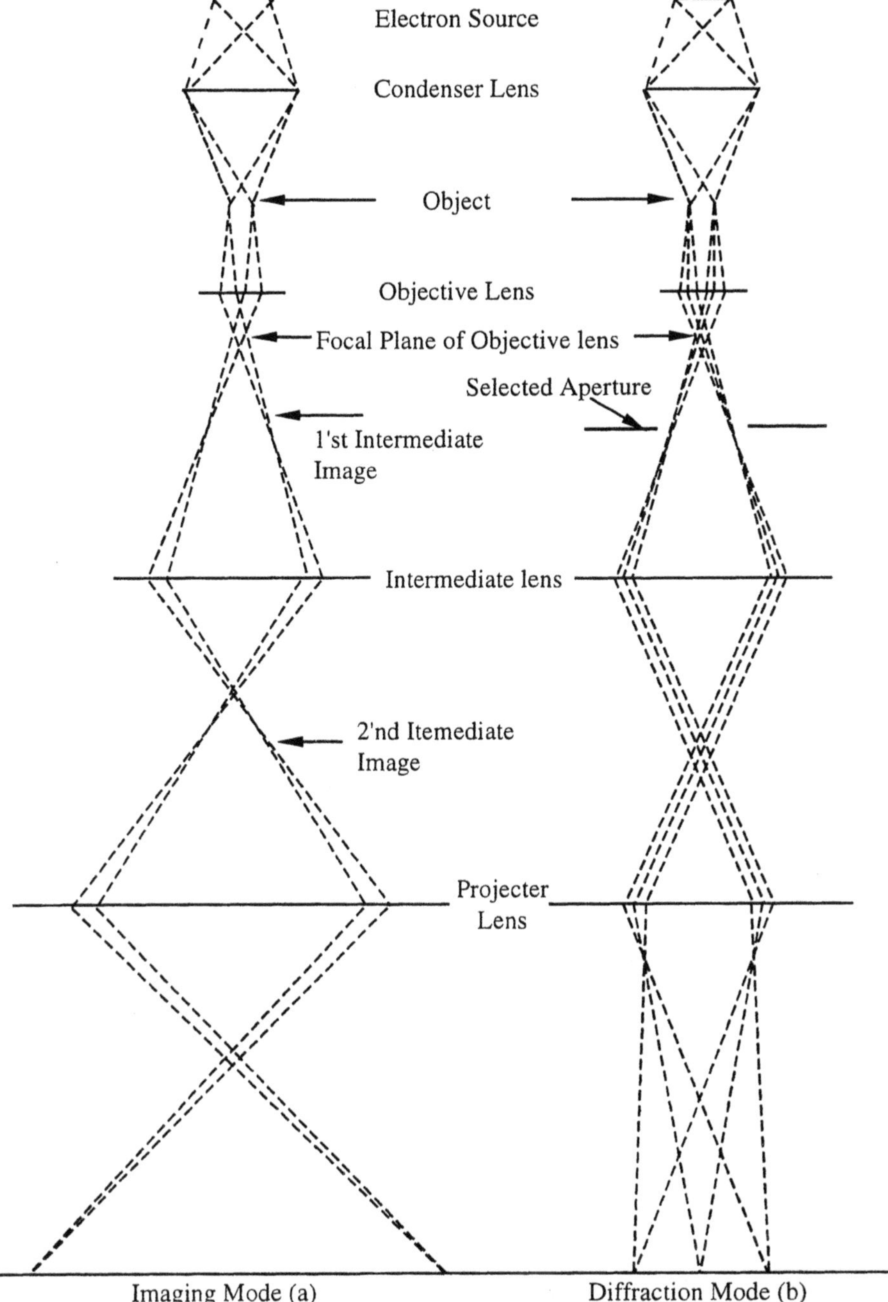

Figure 1. Geometrical optics representation of the TEM in imaging mode (a), and diffraction mode (b).

screen of the TEM by weakening the intermediate lens to place the microscope in diffraction mode (b).

1.2 SIMPLIFYING THE DESCRIPTION OF THE TEM

Consideration of the description of the electron microscope in figure 1 shows that the projector lens and the intermediate lens (or lenses) merely magnify the original image (I1) formed by the objective lens. For the purposes of image simulation we can reduce the TEM to three essential components; (1) an electron beam that passes through (2) a specimen, and then through (3) an objective lens (fig. 2).

Our next step in describing the electron microscope for image simulation is to move from the geometrical optics description of the TEM to a description based on wave optics. In this description of the microscope we examine the amplitude of the electron wavefield on various planes within the TEM, and attempt to determine how the wavefield at the viewing screen comes to contain an image of our specimen.

By treating the electrons as waves, and considering our simplified electron microscope (Figure 2), we see that there are three planes in the TEM at which we need to be able to compute the (complex) amplitude of the electron wavefield.

(1) The image plane:

Working backwards, we start with our desired information, the electron wavefield at the image plane; this wavefield is derived from the wavefield at the focal plane of the objective lens by applying the effects of the objective aperture and the phase changes introduced by the objective lens.

(2) The focal plane of the objective lens:

In turn, the electron wavefield at the focal plane of the lens is derived from the wavefield at the exit surface of the specimen by a simple Fourier transformation.

(3) The specimen exit surface:

In order to know the exit-surface wavefield, we must know with which physical property of the specimen the wave interacts, and describe that physical property for our particular specimen.

1.3 SIMULATING TEM IMAGES

The problem of simulating images thus becomes a problem of computing the electron wavefields at the three microscope planes. Currently, the best way to produce simulated images is to divide the overall calculation into three parts:

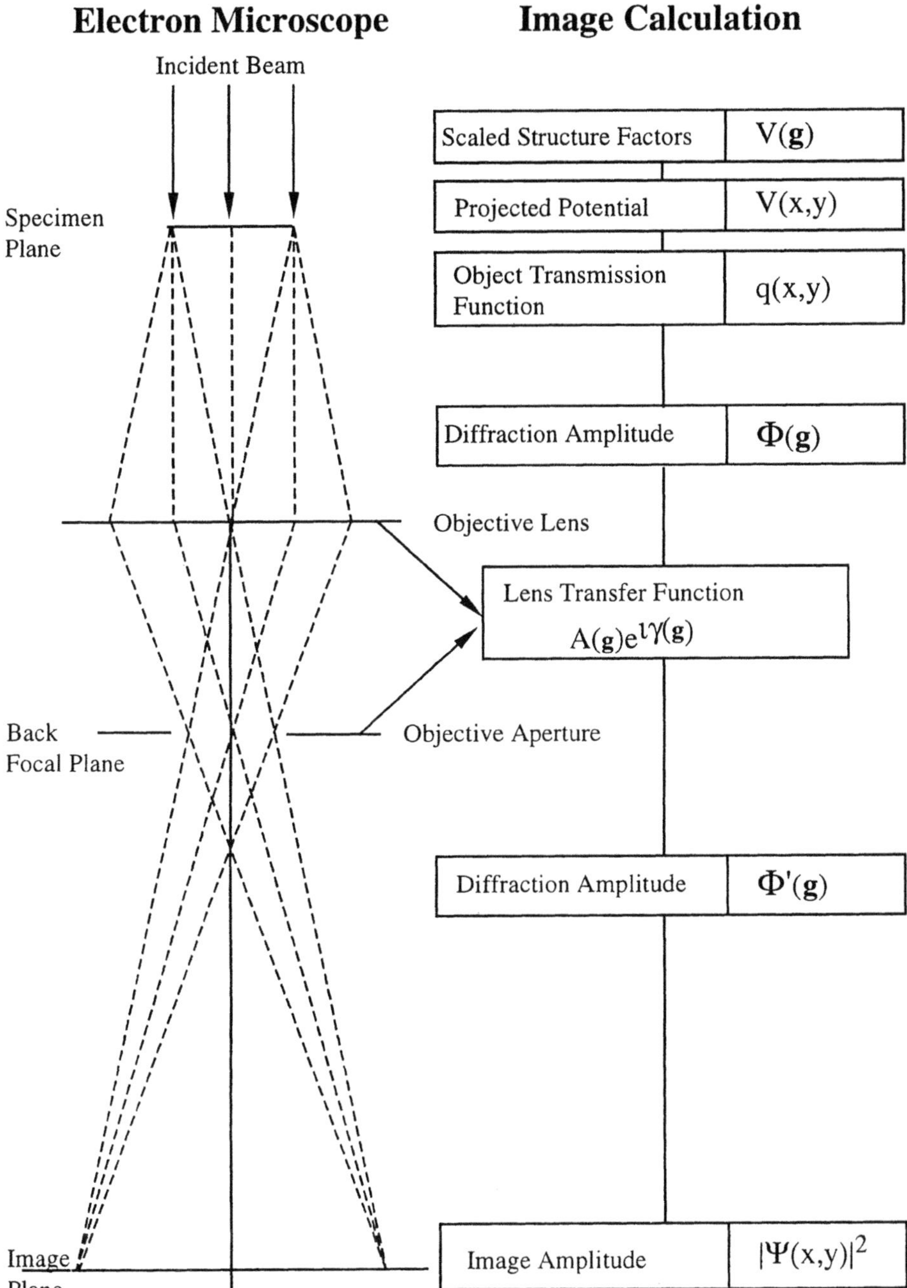

Figure 2. The simplified TEM (left) and the calculations required for an image simulation (right). The three principal planes are marked.

(1) Model the specimen structure to find its potential in the direction of the incident beam.

(2) Produce the exit-surface wavefield by considering the interaction of the incident electron wave on the specimen potential.

(3) Compute the image-plane wavefield by imposing the effects of the objective lens on the specimen exit surface wave.

Each of these steps will be covered in the next sections. However, because of space constraints, it is impossible to cover everything in great depth. For detailed derivation, the reader is encouraged to read the many excellent texts on the subject.

2. Modeling of the specimen

The specimen is a three dimensional objects consisting of a huge number of atoms. From a modeling point of view, it is necessary to reduce the number of parameters to a more manageable number. For crystalline materials described by a repeat of perfect unit cells this is easily accomplished. The unit cell in this case is defined by the lattice parameters A,B and C where A and B are in the plane the specimen perpendicular to the electron beam and C is in the main direction of the incoming electrons. A,B and C are related to the normal lattice vectors a,b, and c depending on the orientation of the specimen. The specimen is thus reduced to M number of unit cells, where M*C is equal to the thickness of the sample, giving in the end a 2D image which covers the area given by A and B.

In the case of a defect structure which no longer can be modeled as a small repeating structure, it is necessary to limit the extent of the calculation by defining a supercell which contains the defect. The resulting image obtained from the calculation will contain artifacts which arise from limiting the structure at arbitrary boundaries and care must be taken to ensure that the image gives a faithful representation of the area of interest.

The entire electrostatic potential of the specimen is now defined by one unit cell with axes **a**,**b**, and **c**, angles *alpha*, *beta* and *gamma*, and N atoms with coordinates x,y,z. For simplicity, we use the nomenclature of the crystallographic unit cell even though we are referring to the transformed unit cell (A,B,C) as described above.

The electrostatic potential in the crystal can be written

$$\phi(\mathbf{r}) = \int d^3\mathbf{r}' \frac{\rho(\mathbf{r}')}{|\mathbf{r} - \mathbf{r}'|} \tag{1}$$

where $\rho(\mathbf{r})$, the charge density is:

$$\rho(\mathbf{r}) = \sum_{\substack{all \\ atoms\ i}} \rho_i(\mathbf{r} - \mathbf{r}_i) \tag{2}$$

with the sum extending over all atoms i at positions $\mathbf{r}_i$, each giving rise to a charge density

$$\rho_i(\mathbf{r}) = Z_i e\delta(\mathbf{r}) - e|\psi_i(\mathbf{r})|^2 \tag{3}$$

where Z_i : atomic number, e: electronic charge, $\psi(\mathbf{r})$: the quantum mechanical many electron wavefunction for the atom.

The potential $\phi(\mathbf{r})$ is described by its Fourier transform $\Phi(\mathbf{u})$ through the relationship

$$\varphi(\mathbf{r}) = \int \Phi(\mathbf{u})e^{-2\pi i\mathbf{u}\cdot\mathbf{r}}d\mathbf{u} = \sum_{\mathbf{H}} \Phi(\mathbf{H})e^{-2\pi i\mathbf{H}\cdot\mathbf{r}} \tag{4}$$

since because of the periodicity of the unit cell, $\Phi(\mathbf{u})$ is non-zero only when $\mathbf{u} = \mathbf{H} = h\mathbf{a}^*+k\mathbf{b}^*+l\mathbf{c}^*$, $\mathbf{H}$ being a reciprocal lattice vector.

The potential $\Phi(\mathbf{H})$ is given as a sum over all atoms in the unit cell

$$\Phi(\mathbf{H}) = \sum_{\substack{all \\ atoms\ i}} f_i^{el}(\mathbf{H})e^{2\pi i\mathbf{u}\cdot\mathbf{r}_i} = \frac{e}{4\pi^2\varepsilon_0} \sum_{\substack{all \\ atoms\ i}} \frac{Z_i - f_i^x(|\mathbf{H}|/2)}{\mathbf{H}^2} e^{2\pi i\mathbf{u}\cdot\mathbf{r}_i} \tag{5}$$

where the electron scattering factors f_i^{el} and the x-ray scattering factors f_i^x have been calculated from relativistic electron wavefunctions and parameterized. They can be found in various tables which are used by image simulation programs[3].

Taking into account any deviation from full occupancy at a particular site and the thermal vibration of the atom, the Fourier coefficients of the crystal potential from one unit cell is calculated as:

$$\Phi(\mathbf{H}) = \sum_{\substack{unit\ cell \\ atoms\ i}} f_i^{el}(\mathbf{H})Occ(\mathbf{r}_i)\exp[-B_i\mathbf{H}^2]e^{2\pi i\mathbf{H}\cdot\mathbf{r}_i} \tag{6}$$

B: Debye Waller factor; $Occ(\mathbf{r}_i)$: The occupancy at position $\mathbf{r}_i$

3. Interaction between the electron and the specimen

The interaction between an electron of energy E and the crystal potential $\phi(\mathbf{r})$ is given by the Schrödinger equation

$$[-\frac{h^2}{8\pi^2 m}\nabla^2 - e\phi(\mathbf{r})]\Psi(\mathbf{r}) = E\Psi(\mathbf{r}) \tag{7}$$

where m is the relativistic electron mass and h is Planck's constant.

Before entering the specimen, the electron is treated as a plane wave with incident wavevector $\mathbf{k}_0$, , $k_0=2\pi/\lambda$, so that the incident electron wave is written

$$\Psi_0(\mathbf{r}) = \exp\{i(\omega t - 2\pi \mathbf{k}_0 \cdot \mathbf{r})\} \tag{8}$$

It is useful to define the quantity V(**r**) which will loosely be referred to as the potential as:

$$V(\mathbf{r}) = \frac{8\pi^2 me}{h^2}\phi(\mathbf{r}) \tag{9}$$

The Schrödinger equation above cannot be solved directly without making various approximations. Depending on how the problem is formulated, one can derive the most common solutions to the electron wavefield at a position T within the specimen.

3.1 THE WEAK PHASE OBJECT APPROXIMATION (WPOA)

In the Phase Object Approximation (POA)[4], the phase of the electron wavefunction after traversing a specimen of thickness T is given as

$$\Psi(x, y, z = T) \approx \Psi(x, y, z = 0)\exp[-i\sigma V_p(x, y)T] \tag{10}$$

with

$$\sigma = 2\pi me\lambda\left[1 + \frac{eE}{mc}\right] \Big/ h^2 \tag{11}$$

where V(x,y) is the average potential per unit length. The specimen is considered thin enough so that electrons only scatter once and are subject only to an average projected potential. In the weak phase object approximation, the exponent is considered much less than one, so that the electron wavefunction emerging from the specimen is:

$$\psi(x, y, z = T) \approx \psi(x, y, z = 0)(1 - i\sigma V_p(x, y)T) \tag{12}$$

The WPOA only applies to very thin specimens of the order of a few tenths of Å, depending on the atomic number of the atoms in the structure[5]. The FT of the wavefunction gives the amplitude and phase of scattered electrons and in the WPOA one has:

$$\Psi(\mathbf{u}) = \delta(\mathbf{u}) - i\sigma \mathrm{V}_p(\mathbf{u})T \tag{13}$$

where **u** is a spatial frequency.

Again, for periodic crystals, $V_p(\mathbf{u})$ are non-zero only for frequencies **u=H** where **H** is a reciprocal lattice vector in the crystal.

We will now use V to mean V_p. Thus for single electron scattering and when the Fourier coefficients V(**H**) are real (true for all centro-symmetric zone axis), the WPOA illustrates clearly that:

i) Upon scattering, the electron undergoes a -90° phase shift.

ii) The amplitude of a scattered electron is proportional to the Fourier coefficient of the crystal potential.

3.2 THE BLOCH WAVE APPROXIMATION (BWA)

In the BWA the electron wavefunction of an electron with wavevector **k** is written as a linear combination of Bloch waves b(**k**,**r**) with coefficients ε[6]. Each Bloch wave is itself expanded into a linear combinations of plane waves which reflect the periodicity of the crystal potential.

$$\psi(\mathbf{r}) = \sum_j \varepsilon^{(j)} b^{(j)}(\mathbf{k},\mathbf{r}) = \sum_j \varepsilon^{(j)} \sum_{\mathbf{g}} c_{\mathbf{g}}^{(j)} \exp[-2\pi i(\mathbf{k}_0^{(j)} + \mathbf{g})\cdot \mathbf{r}] \tag{14}$$

The formulation above gives rise to a set of linear equations expressed as

$$[k_0^2 - (\mathbf{k}^{(j)} + \mathbf{H})^2] c_{\mathbf{H}}^{(j)} + \sum_{\mathbf{H'}} V(\mathbf{H'}) c_{\mathbf{H}-\mathbf{H'}}^{(j)} = 0 \tag{15}$$

which needs to be solved. Detailed derivation of the Bloch wave approximation can be found elsewhere.

Characteristics of the Bloch wave formulation are:
- Requires explicit specification of which reflections **g** are included in the calculation.
- Easy to include reflections outside the zero order Laue zone.
- Very good for perfect crystals, not suited for calculating images from defects.
- The solution is valid for a particular thickness of the specimen.
- Allows rapid calculation of convergent beam electron diffraction patterns.
- Includes dynamic scattering.

3.3 THE MULTISLICE FORMULATION

The multislice formulation[7,8] is by far, the most commonly used method of calculating the electron wavefield emerging from the specimen. Although it does not as easily include scattering outside the zero order Laue zone as the BWA, the multislice formulation is more versatile for use with structures containing any kind of defects, either they be point-defects, stacking faults, interfacial structures, etc. The multislice solution gives the approximate solution to the electron wavefunction at a depth z+dz in the crystal from the wavefunction at z. In the multislice approximation one has:

$$\psi(x,y,z+dz) \approx \exp[-i\sigma dz \nabla_{x,y}^2] \cdot \exp[-i\sigma \int_z^{z+dz} V(x,y,z')dz'] \psi(x,y,z) \tag{16}$$

Thus starting with the wavefunction at z=0, one can iteratively calculate the wavefunction at a thickness n*dz, by applying the multislice solution slice by slice, taking the output of one calculation as the input for the next. Equation 16 is solved in a two step process.

The potential due to the atoms in a slice dz is projected onto the plane t=z, giving rise to a scattered wavefield

$$\psi_1(x,y,z+dz) = \exp[-i\sigma\int_z^{z+dz} V(x,y,z')dz']\psi(x,y,z) \equiv q(x,y)\psi(x,y,z) \qquad (17)$$

The function q(x,y) is referred to as the phasegrating.

Subsequently, the wavefield is propagated in vacuum to the plane t=z+dz, according to

$$\psi(x,y,z+dz) = \exp[-i\sigma dz\nabla^2_{x,y}]\cdot\psi_1(x,y,z) \qquad (18)$$

The last equation represents a convolution in real space and is solved more efficiently in Fourier space[9], where the equation transforms to

$$\Psi(\mathbf{H},z+dz) = \exp[-i\pi\lambda dz\mathbf{H}^2]\cdot\Psi_1(\mathbf{H},z) \equiv p(\mathbf{H},dz)\cdot\Psi_1(\mathbf{H},z) \qquad (19)$$

where $\Psi(\mathbf{H},z)$ are the Fourier coefficients of $\psi(x,y,z)$. $p(\mathbf{H},dz)$ is called the propagator. The multislice formulation is a repeated use of the last two equations and will give the wavefield at any arbitrary thickness T of the specimen. If the slice-thickness is chosen as the repeat distance of the crystal in the direction of the electron beam, only the zero order Laue reflections are included in the calculation as the unit cell content is projected along the direction of the electron beam. Three dimensional information which involves including higher order Laue reflections can be included by reducing the slice thickness[10].

3.4 SAMPLING CRITERIA

Any numeric calculation must be performed for a limited set of data points (x,y) or reciprocal spatial frequencies **u**. Working with periodically repeated structures; if the lateral dimensions of the unit cell is a and b, which we for simplicity make orthogonal so that the axes are associated with an orthogonal x,y coordinate system, then for a given sampling interval dx=dy, we have

$$N = \frac{a}{dx} \quad ; \qquad M = \frac{b}{dy} \qquad (20)$$

defining the calculation to a grid of N*M points. The sampling interval automatically restricts the calculation in reciprocal space as well. The maximum reciprocal lattice vector for orthogonal axes is given as

$$\mathbf{H}_{\max}^2 = \left|h_{\max}\mathbf{a}^* + k_{\max}\mathbf{b}^*\right|^2 = \left(\frac{N}{2a}\right)^2 + \left(\frac{M}{2b}\right)^2 \tag{21}$$

Because most implementations of the multislice formulation makes use of Fourier transforms, the calculation grid N and M is adjusted so that both are powers of 2. This is because Fourier transform algorithms can be performed much faster for powers of 2 rather than arbitrary dimensions. This results in uneven sampling intervals dx,dy when a ≠ b. In order to not impose an arbitrary symmetry on the calculation, a circular aperture is imposed on the propagator. In practice, this aperture is set to 1/2 of the minimum of ($\mathbf{h}_{max}$, $\mathbf{k}_{max}$) as defined above in order to avoid possible aliasing effects associated with digital Fourier transforms. The sampling must be chosen such that the calculation includes all (or sufficiently enough) scattering that takes place in the specimen.

4. The image formation

After the electron wavefield emerge from the specimen, it is subjected to the varies magnetic field of the lenses that form the imaging and magnification part of the microscope. Of these lenses, only the first lens, the objective lens, is considered in the image formation calculation. Since the angle with which the electron forms with the optic axis of the lens varies inversely with the magnification, only the aberrations of the objective lens are important. The remaining lenses serve to just magnify the image formed by the objective lens. The effects of the lens which normally are included in the calculation are spherical aberration, chromatic aberration and lens defocus. Two-fold and three-fold astigmatism, including axial coma, are considered correctable by the operator although they can be included in the equations.

Without any aberrations, no instabilities and with the specimen in the focal plane of the objective lens, the image observed in the electron microscope would be am magnified version of

$$I(x,y) = \left|\psi(x,y,z = exitplane\ \ of\ \ specimen)\right|^2 = \psi_e(x,y)\psi_e^*(x,y) \tag{22}$$

4.1 OBJECTIVE LENS DEFOCUS

Consider an electron traveling from the plane defined by the exit surface of the specimen to the plane given as the plane of focus for the objective lens. This distance is referred to as the objective lens defocus Δf.

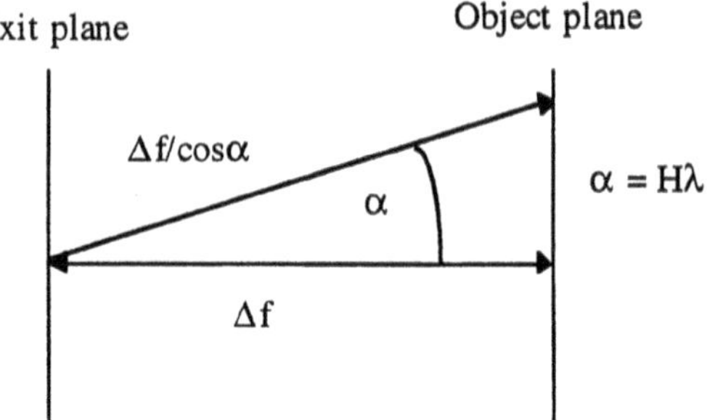

The electron traveling along the optic axis will have a path length of Δf while an electron that has been scattered an angle α=Hλ, will travel a distance Δf /cosα. This can be expressed as a phase difference

$$\frac{2\pi}{\lambda}\left(\frac{\Delta f}{\cos\alpha} - \Delta f\right) \approx \pi\lambda\Delta f\mathbf{H}^2 \tag{23}$$

4.2 SPHERICAL ABERRATION

Electrons crossing the optic axis with an angle a at the focal plane of the objective lens should form parallel paths emerging from the lens.

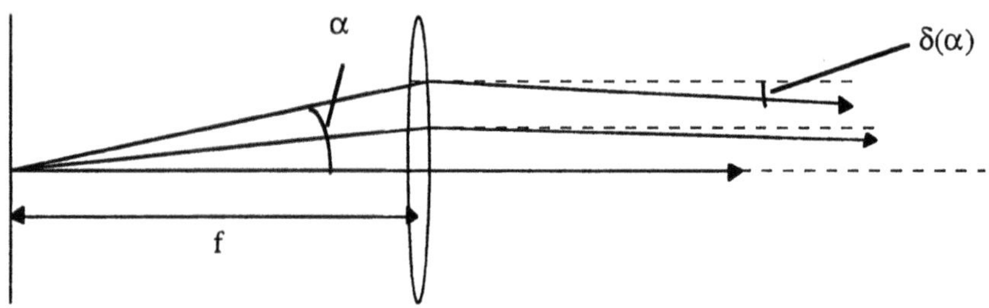

However, the spherical aberration of the lens causes a phase shift relative to the path of the unscattered electron (α=0) which is written as[11]:

$$2\pi/\lambda * 1/4\, C_x\alpha^4 = 1/2\,\pi C_s\lambda^3\mathbf{H}^4 \tag{24}$$

If there were no other effects to consider, the image would be obtained as follows:

- Calculate the wavefield emerging from the specimen according to one of the approximations.
- Fourier transform the wavefield which gives the amplitude and phase of scattered electrons.
- Add the phase shift introduced by the lens defocus and the spherical aberration to the Fourier coefficients.

- Inverse Fourier transform to find the modified wavefunction.
- Calculate the image as the modulus square of the wavefield.

However, there are two more effects that are usually considered. Variations in electron energy and direction.

4.3 CHROMATIC ABERRATION

Electrons do not all have exactly the same energy for various reasons. They emerge from the filament with a spread in energy and the electron microscope accelerating voltage varies over the time of exposure. The chromatic aberration in the objective lens will cause electrons of different energies to focus at different planes. Effectively this can be thought if as rather than having a given defocus f_0, one has a spread in defocus values centered around f_0. The value f_0 is what is normally referred to as Δf as indicating defocus. The images associated with different defocus values add to make the final image. Assuming a Gaussian spread in defocus of the form

$$D(f-f_0) \propto \exp[-\frac{(f-f_0)^2}{\Delta^2}] \tag{25}$$

gives:

$$I = \int |\Psi(f-f_0)|^2 D(f-f_0)df \Rightarrow \Psi(\mathbf{H}) \rightarrow \Psi(\mathbf{H})\exp[-1/2(\pi\lambda\Delta\mathbf{H}^2)^2] \tag{26}$$

This states that each Fourier term (diffracted beam) is damped according to the equation above[11].

4.4 SPREAD IN INCOMING ELECTRON DIRECTION

The electron beam is not an entirely parallel beam of electrons, but form rather a cone of an angle α. This implies that electrons instead of forming a point in the diffraction pattern form a disk with a radius related to the spread in directions. As for a variation in energy, the images formed for different incoming angles are summed up by integrating over the probability function for the incoming direction. It turns out that this also leads to another damping of the diffracted beam[12] so that:

$$I(\mathbf{r}) = \int |\psi(\mathbf{r},\alpha)|^2 D(\alpha)d\alpha \Rightarrow \Psi(\mathbf{H}) \rightarrow \Psi(\mathbf{H})\exp[\pi\alpha\lambda(C_s\mathbf{H}^2\lambda^2 + \Delta f)]^2 \tag{27}$$

4.5 THE "FINAL" IMAGE

Equation 26 and equation 27 are only valid when the intensities of the scattered beams are much smaller than the intensity of the central beam. Thus the image results from scattered beams interfering with the central beam, but not with each other. This is referred to as linear imaging. Although the formulation is slightly more complicated in the general case, the expressions above give sufficient insight into the image formation. Image simulation programs do however include the more general formulation which include non-linear imaging terms[13]. Each Fourier component is damped by the spread in energy and direction and the image is formed by adding this to the recipe in section 4.2

4.6 THE CONTRAST TRANSFER FUNCTION (CTF)

When reading about HRTEM, it is impossible not to encounter the expression "Contrast Transfer Function". Loosely speaking, the CTF of the microscope refers to the degree with which Fourier components of the electron wavefunction (spatial frequencies) are transferred by the microscope and contribute to the Fourier transform of the image. Although the CTF only holds for thin specimen and linear imaging, it is often generalized and wrongly applied to all conditions. However, the CTF does provide insight into the nature of HRTEM images. In order to derive the expression for the CTF, we start by calculating the image intensity as given by the Weak Phase Object approximation. In the WPOA:

$$\Psi(x,y,z=T) \approx 1 - i\sigma V_p(x,y)T \tag{28}$$

and

$$\Psi(\mathbf{H}) = \delta(\mathbf{H}) - i\sigma V_p(\mathbf{H})T \tag{29}$$

Applying the phase shift due to the spherical aberration and the objective lens defocus which we will call $\chi(\mathbf{H})$, we get that the FT of the wavefunction is (for simplicity V = V_p):

$$\Phi(\mathbf{H}) = \delta(\mathbf{H}) - i\sigma V(\mathbf{H})e^{i\chi(\mathbf{H})}A(\mathbf{H}) \tag{30}$$

where A(**H**) is the damping terms arising from partial coherence.

The FT of the intensity is now given as

$$\begin{aligned} I(\mathbf{H}) &= FT(\psi \cdot \psi^*) = \sum_{\mathbf{H}'} \Psi(\mathbf{H}')\Psi^*(\mathbf{H}-\mathbf{H}') \approx \\ &\sum_{\mathbf{H}'} \left(\delta(\mathbf{H}') - i\sigma A(\mathbf{H}')V(\mathbf{H}')e^{i\chi(\mathbf{H}')}\right)\left(\delta(\mathbf{H}-\mathbf{H}') - i\sigma A(\mathbf{H}-\mathbf{H}')V(\mathbf{H}-\mathbf{H}')e^{i\chi(\mathbf{H}-\mathbf{H}')}\right) \approx \\ &\delta(\mathbf{H}) + 2\sigma A(\mathbf{H})V(\mathbf{H})\sin\chi(\mathbf{H}) \end{aligned} \tag{31}$$

The last result is very useful and it leads to the frequently used concept of the Contrast Transfer Function (CTF). The CTF is defined as $A(\mathbf{H})\cdot\sin\chi(\mathbf{H})$ The equation above states that each reflection H contributes to the image intensity spectrum with a weight that is proportional to the CTF. Figure 3. shows a plot of a CTF including $\sin\chi$ and the damping curves. When $\sin\chi$ (**H**) = -1 for a large range of frequencies **H**, which is the condition referred to as Scherzer defocus[11], the image can be thought of as:

$$I(x,y) \approx 1 - 2\sigma U(x,y) \tag{32}$$

where U(x,y) is a potential related to the original crystal potential, but keeping only the Fourier coefficients related to frequencies transferred by the microscope. The equation above shows the often used rule of thumb. For thin specimens, under Scherzer imaging conditions, atoms are black.

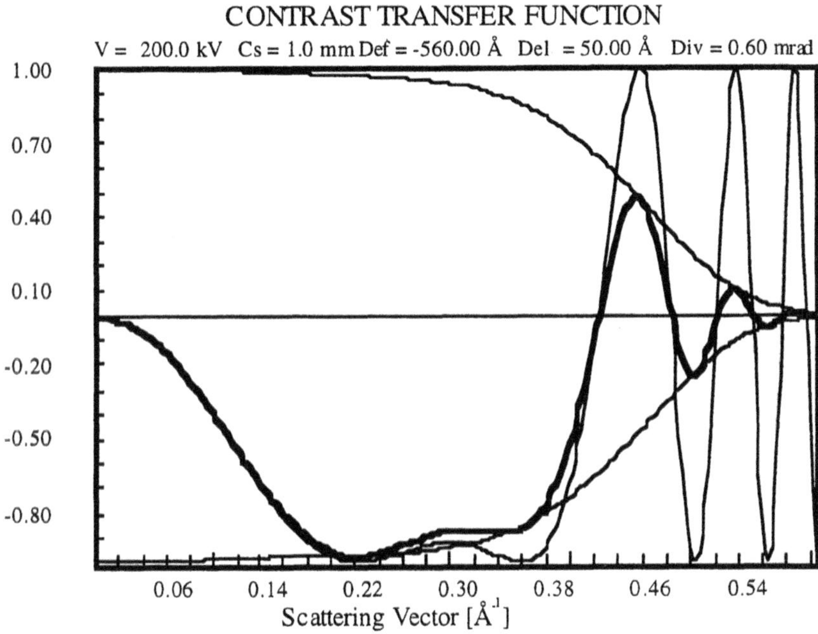

Figure 3. Plot of the Contrast Transfer Function for a 200kV microscope with the parameters indicated.

5. References

1. O'Keefe M.A. et al. (1989) Simulated Image Maps for use in Experimental High-Resolution Electron Microscopy, *Mat. Res. Soc. Symp. Proc.* **159**, 453-458
2. Allpress J.G. et al (1972) n-beam Lattice Images. I. Experimental and Computed Images from $W_4Nb_{26}O_{77}$, *Acta Cryst.* **A 28**, 528-536

3. Doyle P.A. and Turner P.S. (1968) Relativistic Hartree-Fock X-ray and Electron Scattering Factors, *Acta Cryst.*. **A 24**, 390-397
4. Cowley J.M. and Iijima S. (1972) Electron microscope image contrast for thin crystals, *Z. Naturforschung* **27a**, 445-451
5. Gibson J.M. (1994) Breakdown of the weak-phase object approximation in amorphous objects and measurement of high-resolution electron optical parameters, *Ultramicroscopy* **56**, 26-32
6. Howie A. (1963) Inelastic scattering of electrons by crystals, *Proc. Roy. Soc.* **A271**, 268-275
7. Goodman P, Moodie A.F. (1974) Numerical evaluation of N-beam wave functions in electron scattering by the multislice method, *Acta Cryst.* **A30**, 322-324
8. Self P.G.et al. (1983) Practical computation of amplitudes and phases in electron diffraction, *Ultramicroscopy* **11**, 35
8. Van Dyck D. (1983) High-speed computation techniques for the simulation of high resolution electron micrographs, *Journal of Microscopy*, **132**, 31
9. Ishizuka K. and Uyeda N. (1977) A new theoretical and practical approach to the multislice method, *Acta Cryst.* **A 33**, 740
10. Kilaas R. et al. (1987) On the inclusion of upper Laue layers in computational methods in High Resolution Transmission Electron Microscopy, *Ultramicroscopy* **21**, 47-62
11. Scherzer O (1949) The Theoretical Resolution Limit of the Electron Microscope, *Journal of Applied Physics* **20**, 20-29
12. Frank J (1973) The envelope of electron microscope transfer functions for partially coherent illumination, *Optik* **38**, 519-536
13. O'Keefe M.A. (1979) Resolution-damping functions in non-linear images, *Proc. of EMSA* **37**, 556-557

THE PHASE PROBLEM OF X-RAY CRYSTALLOGRAPHY: OVERVIEW

HERBERT A. HAUPTMAN
Hauptman-Woodward Medical Research Institute, Inc., 73 High St., Buffalo, NY 14203 USA

Abstract

The intensities of a sufficient number of X-ray diffraction maxima determine the structure of a crystal, that is, the positions of the atoms in the unit cell of the crystal. The available intensities usually exceed the number of parameters needed to describe the structure. From these intensities a set of numbers $|E_{\mathbf{H}}|$ can be derived, one corresponding to each intensity. However, the elucidation of the crystal structure also requires a knowledge of the complex numbers $E_{\mathbf{H}} = |E_{\mathbf{H}}| exp(i\phi_{\mathbf{H}})$, the normalized structure factors, of which only the magnitudes $|E_{\mathbf{H}}|$ can be determined from experiment. Thus, a "phase" $\phi_{\mathbf{H}}$, unobtainable from the diffraction experiment, must be assigned to each $|E_{\mathbf{H}}|$, and the problem of determining the phases when only the magnitudes $|E_{\mathbf{H}}|$ are known is called the "phase problem". Owing to the known atomicity of crystal structures and the redundancy of observed magnitudes $|E_{\mathbf{H}}|$, the phase problem is solvable in principle.

Probabilistic methods have traditionally played a key role in the solution of this problem. They have led, in particular, to the so-called tangent formula which, in turn, has played the central role in the development of methods for the solution of the phase problem. A number of computer programs, stressing different aspects of the central theme, have proven to be particularly effective.

Finally, the phase problem may be formulated as one in constrained global optimization. A method for avoiding the countless local minima in order to arrive at the constrained global minimum leads to the *Shake-and-Bake* algorithm, a completely automatic solution of the phase problem for structures containing as many as 600 atoms when data are available to atomic resolution.

1. THE PHASE PROBLEM

The electron density function $\rho(\mathbf{r})$ in a crystal is a three-dimensional periodic function of the position vector $\mathbf{r}$ and may therefore be represented by the three-dimensional Fourier series:

$$\rho(\mathbf{r}) = \frac{1}{V}\sum_{\mathbf{H}} F_{\mathbf{H}} \exp(-2\pi i\mathbf{H}\cdot\mathbf{r}) \tag{1}$$

where V is the volume of the fundamental parallelepiped, the so-called unit cell of the crystal, and the three components of the vector $\mathbf{H}$ range over all the integers. The Fourier coefficient $F_{\mathbf{H}}$, said to be the structure factor corresponding to the reciprocal lattice vector $\mathbf{H}$, may then be calculated in the usual way:

$$F_{\mathbf{H}} = \int_V \rho(\mathbf{r})\exp(2\pi i\mathbf{H}\cdot\mathbf{r})dV \tag{2}$$

D. L. Dorset et al. (eds.), Electron Crystallography, 131–138.

where the integration is carried out over the unit cell V. Clearly, each $F_{\mathbf{H}}$ is a complex number that may be written in polar form

$$F_{\mathbf{H}} = |F_{\mathbf{H}}| \exp(i\phi_{\mathbf{H}}) \tag{3}$$

where $\phi_{\mathbf{H}}$ is the phase of the structure factor $F_{\mathbf{H}}$.

When a beam of monochromatic X-rays is incident on a crystal, the radiation is scattered in discrete directions determined by the crystal lattice and labeled by the reciprocal lattice vectors **H**. Both the amplitude and phase of each scattered ray, or reflection, depend on the crystal structure. The amplitude, or intensity, of a reflection leads in a straightforward way to the magnitude $|F_{\mathbf{H}}|$ of the complex structure factor $F_{\mathbf{H}}$. However, the phases $\phi_{\mathbf{H}}$, which are also needed if one is to determine $\rho(\mathrm{r})$ from Eq. 1, are lost in the diffraction experiment. If one uses the known values of the magnitudes $|F_{\mathbf{H}}|$ but arbitrary values for the phases $\phi_{\mathbf{H}}$ in Eq. 1, then density functions $\rho(\mathbf{r})$ consistent with the observed values of the diffraction intensities are obtained. Thus, diffraction intensities alone do not determine a unique density function $\rho(\mathbf{r})$. Even if the known non-negativity of $\rho(\mathbf{r})$ is assumed, thus greatly restricting the values of the phases [1,2] the observed diffraction intensities are, in general, still not sufficient to determine $\rho(\mathbf{r})$ uniquely. It follows that the phase problem, to determine the values of the phases $\phi_{\mathbf{H}}$ of the structure factors $F_{\mathbf{H}}$ when only the magnitudes $|F_{\mathbf{H}}|$ are given, is, in principle, unsolvable when formulated in these terms. It was this argument that led the crystallographic community, prior to 1950, to believe also that crystal structures could not, even in principle, be determined from the diffraction intensities alone. However, by invoking the prior structural knowledge that crystals consist of discrete atoms, one readily refutes this argument, as shown below.

If one replaces the real crystal, with continuous electron density $\rho(\mathbf{r})$, by an idealized one, the unit cell of which consists of N discrete, non-vibrating point atoms, then the structure factor $F_{\mathbf{H}}$ is replaced by the normalized structure factor $E_{\mathbf{H}}$ and Eqs. 3, 2, and 1 are replaced by

$$E_{\mathbf{H}} = |E_{\mathbf{H}}| \exp(i\phi_{\mathbf{H}}) \tag{4}$$

$$E_{\mathbf{H}} = \frac{1}{\sigma_2^{1/2}} \sum_{j=1}^{N} Z_j \exp(2\pi i \mathbf{H} \cdot \mathbf{r}_j) \tag{5}$$

$$\begin{aligned} \langle E_{\mathbf{H}} \exp(-2\pi i \mathbf{H} \cdot \mathbf{r}) \rangle_{\mathbf{H}} &= \frac{1}{\sigma_2^{1/2}} \left\langle \sum_{j=1}^{N} Z_j \exp[2\pi i \mathbf{H} \cdot (\mathbf{r}_j - \mathbf{r})] \right\rangle_{\mathbf{H}} \\ &\left.\begin{aligned} &= \frac{Z_j}{\sigma_2^{1/2}} \text{ if } \mathbf{r} = \mathbf{r}_j \\ &= 0 \text{ if } \mathbf{r} \neq \mathbf{r}_j \end{aligned}\right\} \end{aligned} \tag{6}$$

respectively, where Z_j is the atomic number and $\mathbf{r}_j$ is the position vector of the atom labeled j, and

$$\sigma_n = \sum_{j=1}^{N} Z_j^n, \; n = 1, 2, 3, \ldots \tag{7}$$

In practice, the magnitudes $|E_{\mathbf{H}}|$ of the normalized structure factors $E_{\mathbf{H}}$ are obtainable (at least approximately) from the observed magnitudes $|F_{\mathbf{H}}|$, while the phases $\phi_{\mathbf{H}}$, as defined by Eqs. 4 and 5, cannot be determined experimentally. Since one now requires only the $3N$ components of the N position vectors $\mathbf{r}_j$ rather than the much more complicated electron density function $\rho(\mathbf{r})$, it turns out that, in general, the known magnitudes are more than sufficient. This is most readily seen if we equate the magnitudes of both sides of Eq. 5, thus eliminating the unknown phases $\phi_{\mathbf{H}}$, in order to obtain

$$|E_{\mathbf{H}}| = \frac{1}{\sigma_2^{1/2}} \left| \sum_{j=1}^{N} Z_j \exp(2\pi i \mathbf{H} \cdot \mathbf{r}_j) \right| \tag{8}$$

a system of equations in which the only unknowns are the $3N$ components of the position vectors $\mathbf{r}_j$. Since the number of Eq. 8, equal to the number of reciprocal lattice vectors $\mathbf{H}$ for which the magnitudes $|E_{\mathbf{H}}|$ are observed, usually exceeds the number of unknowns, $3N$, by far, the system in Eq. 8 is redundant. Thus the phase problem is, in principle, solvable when reformulated in terms of fixed point atoms, as reference to Eq. 5 shows.

The system of Eq. 5 implies the existence of relationships among the normalized structure factors $E_{\mathbf{H}}$ since the (relatively few) unknown position vectors $\mathbf{r}_j$ may, at least in principle, be eliminated. In this way we obtain the system of equations among the complex normalized structure factors $E_{\mathbf{H}}$:

$$F(E_{\mathbf{H}}) = F(|E_{\mathbf{H}}|, \phi_{\mathbf{H}}) = 0. \tag{9}$$

Since the magnitudes $|E_{\mathbf{H}}|$ are obtainable from the diffraction experiment, the system (9) leads to the set of identities, dependent on the known magnitudes $|E_{\mathbf{H}}|$, which the phases must of necessity satisfy

$$\mathrm{G}(\phi_{\mathbf{H}} \| E_{\mathbf{H}}|) = 0. \tag{10}$$

By the term "direct methods" is meant that class of methods that exploits relationships among the normalized structure factors in order to go directly from the observed magnitudes $|E|$ to the needed phases ϕ.

An explicit system of identities among the structure factors, valid in the case that the structure consists of N identical atoms in the unit cell and that diffraction data are available at atomic resolution, was first found by Sayre [3]:

$$\mathrm{F}_{\mathbf{H}} = \frac{\mathrm{f}_{\mathbf{H}}}{\mathrm{g}_{\mathbf{H}} \mathrm{V}} \sum_{\mathbf{K}} \mathrm{F}_{\mathbf{K}} \mathrm{F}_{\mathbf{H-K}} \tag{11}$$

where V is the volume of the unit cell and $\mathrm{f}_{\mathbf{H}}$ and $\mathrm{g}_{\mathbf{H}}$ are the scattering factors for the true equal atoms of the structure and for the "squared" structure. When expressed in terms of the normalized structure factors E the Sayre equations retain approximate validity:

$$\mathrm{E}_{\mathbf{H}} = \frac{\mathrm{C}}{\mathrm{g}_{\mathrm{h}}} \sum_{\mathbf{K}} \mathrm{E}_{\mathbf{K}} \mathrm{E}_{\mathbf{H-K}} \tag{12}$$

where C is an overall scale factor.

2. THE STRUCTURE INVARIANTS

Equation 6 implies that the normalized structure factors $E_{\mathbf{H}}$ determine the crystal structure. However, Eq. 5 does not imply that, conversely, the crystal structure determines the values of the normalized structure factors $E_{\mathbf{H}}$ since the position vectors $\mathbf{r}_j$ depend not only on the structure but on the choice of origin as well. It turns out, nevertheless, that the magnitudes $|E_{\mathbf{H}}|$ of the normalized structure factors are in fact uniquely determined by the crystal structure and are independent of the choice of origin, but that the values of the phases $\phi_{\mathbf{H}}$ depend also on the choice of origin. Although the values of the individual phases depend on the structure and the choice of origin, there exist certain linear combinations of the phases, the so-called structure invariants, whose values are determined by the structure alone and are independent of the choice of origin. The most important class of structure invariants, and the only one to be considered here, consists of the three-phase structure invariants (triplets),

$$\phi_{\mathbf{HK}} = \phi_{\mathbf{H}} + \phi_{\mathbf{K}} + \phi_{-\mathbf{H}-\mathbf{K}}, \tag{13}$$

where H and K are arbitrary reciprocal lattice vectors.

3. THE PROBABILISTIC BACKGROUND

The techniques of modern probability theory lead to the joint probability distributions of arbitrary collections of diffraction intensities and their corresponding phases. These distributions constitute the foundation on which direct methods are based. They have provided the unifying thread from the beginning, *ca* 1950, until the present time. They have led, in particular to the tangent formula Eq. (19), the basis of almost all computer programs for the direct determination of phase, and to the (first) minimal principle.

While the tangent based computer programs have steadily become more powerful over the years, they have rarely solved structures having more than some 200 independent non-hydrogen atoms *ab initio*. However, when combined with other techniques, e.g. solvent flattening and histogram matching, they appear likely to be useful for phase refinement and extrapolation even in the macromolecular range.

The minimal principle [5], on the other hand, which has found expression in the *Shake-and-Bake* formalism [6,7], does provide a completely automatic solution to the phase problem, *ab initio*, even for structures having as many as 600 non-H atoms, provided that diffraction data to atomic resolution are available. Its ultimate potential is still unknown.

It is assumed that (i) A crystal structure is specified, (ii) Three non-negative numbers R_1, R_2, R_3 are also specified; and (iii) The reciprocal lattice vectors H and K are the primitive random variables which are assumed to be uniformly and independently distributed in the subset of reciprocal space defined by

$$|E_{\mathbf{H}}| = R_1, \ |E_{\mathbf{K}}| = R_2, \ |E_{\mathbf{H+K}}| = R_3 \tag{14}$$

where the magnitudes $|E|$ are defined by (5). Then the structure invariant ϕ_{HK} (Eq. (13)), as a function of the primitive random variables **H** and **K** (Eq. (5)), is itself a random variable.

4. THE CONDITIONAL PROBABILITY DISTRIBUTION OF ϕ_{HK}, GIVEN $|E_H|$, $|E_K|$, $|E_{H+K}|$

For simplicity it will be assumed henceforth that all N atoms in the unit cell are identical. Under the three assumptions of §3. the conditional probability distribution of the triplet ϕ_{HK} (Eq. (13)), where $|E_H|$, $|E_K|$, and $|E_{H+K}|$ are given by (14), is known to be

$$P(\Phi / R_1, R_2, R_3) = \frac{1}{2\pi I_0(A_{HK})} \exp(A_{HK} \cos \Phi) \tag{15}$$

where Φ represents the triplet ϕ_{HK},

$$A_{HK} = \frac{2}{N^{1/2}} R_1 R_2 R_3 = \frac{2}{N^{1/2}} |E_H E_K E_{H+K}| \tag{16}$$

and I_0 is the Modified Bessel Function. Equation (15) implies that the mode of ϕ_{HK} is zero, and the conditional expectation value (or average) of cos ϕ_{HK}, given A_{HK}, is

$$\mathcal{E}(\cos\phi_{HK} / A_{HK}) = \frac{I_1(A_{HK})}{I_0(A_{HK})} > 0, \tag{17}$$

where I_1 is the Modified Bessel Function. It is also readily confirmed that the larger the value of A_{HK} the smaller is the conditional variance of cosϕ_{HK}, given A_{HK}. It is to be stressed that the conditional expected value of the cosine, Eq. (15), is always positive since $A_{HK}>0$.

5. THE TANGENT FORMULA

It is assumed that a crystal structure consisting of N identical atoms in the unit cell is fixed, but unknown, that the magnitudes $|E|$ of the normalized structure factors E are known, and that a sufficiently large base of phases, corresponding to the largest magnitudes $|E|$, is specified.

The mode of the triplet distribution (Eq. (15)) is zero and the variance of the cosine is small if A_{HK} (Eq. (16)) is large. In this way one obtains the estimate for the triplet ϕ_{HK} (Eq. (13)):

$$\phi_{HK} = \phi_H + \phi_K + \phi_{-H-K} \approx 0 \tag{18}$$

which is particularly good in the favorable case that A_{HK}, (Eq. (16)), is large, i.e. that $|E_H|$, $|E_K|$, and $|E_{H+K}|$ are all large. The estimate given by Eq. (18) is one of the cornerstones of the traditional techniques of direct methods. It is surprising how useful Eq. (18) has proven to be in the applications especially since it yields only the zero estimate of the triplet, and only those estimates are reliable for which $|E_H|$, $|E_K|$, and $|E_{H+K}|$ are all large. Clearly the coefficient $2/N^{1/2}$ in Eq. (16), and therefore A_{HK} as well, both decrease with increasing N, i.e. with increasing structural complexity. Hence the relationship (Eq. (18)) becomes increasingly unreliable for larger structures, and the traditional step-by-step sequential direct methods procedures based on Eq. (18) eventually fail. (However, see §6 for recent advances.)

An important supplement to the relationship (18), and an immediate consequence of the distribution (15), is the so-called tangent formula [4];

$$\tan\phi_H = \frac{\sum_K |E_K E_{H-K}| \sin(\phi_K + \phi_{H-K})}{\sum_K |E_K E_{H-K}| \cos(\phi_K + \phi_{H-K})} \tag{19}$$

in which the summations are taken over the same (arbitrary) set of reciprocal lattice vectors K. The tangent formula, in one form or another, plays a major role in the traditional techniques of direct methods.

6. Some Computer Programs

Probably the most widely used tangent based computer programs are the various versions of MULTAN, SHELXS, SIR, and MITHRIL. Only two of the most popular of these will be briefly described here.

The first is the RANTAN (Random MULTAN) program devised by Yao Jia-xing [8]. In this procedure initial values of all the desired phases are assigned at random, (rather than only a few as had been done in earlier versions of MULTAN) and all relationships (Eq. (18)) which link them are used from the beginning. A suitably weighted version of the tangent formula (Eq. (19)) is then employed to refine the initial values of the individual phases. An important aspect of the program is the selection of the weights which change during the course of the refinement process. Because of the large number of phases involved at the very beginning, the method is relatively insensitive to the failure of one or several relationships (Eq. (18)) which may occur early on in the process, and this aspect of RANTAN undoubtedly contributes to its success in the applications.

Another recent advance is the development of the Sayre Equation Tangent Formula as expressed in the computer program SAYTAN [9]. This variant of the tangent formula tends to develop phases which satisfy the Sayre Equation (Eq. (12)) for both large and small magnitudes $|E|$ and not only exploits the triplet relationships (Eq. (18)) but involves the analogous quartet relationships as well. It has proven to be a particularly effective direct methods technique especially for larger structures [10].

7. The Minimal Principle

Finally, referring to Eqs. (16) and (17) one may formulate the phase problem as one in constrained global minimization [5,6,7]. First, the minimal function $m(\phi)$ is defined by

$$m(\phi) = \frac{1}{\sum_{H,K} A_{HK}} \sum_{H,K} A_{HK} \left\{ \cos\phi_{HK} - \frac{I_1(A_{HK})}{I_0(A_{HK})} \right\}^2 \tag{20}$$

where ϕ_{HK} and A_{HK} are defined by Eqs. (18) and (16), respectively. Then, in view of Eq. (17), one infers that the constrained global minimum of $m(\phi)$, where the phases are constrained to satisfy the identities which they must satisfy, Eq. (10), yields the correct values of all the phases appearing in $m(\phi)$ (the minimal principle). The computer program *Shake-and-Bake*, described in the paper in the Direct Methods Proceedings by Miller and Weeks [11], shows how the constrained global minimum of $m(\phi)$ is reached and identified. *Shake-and-Bake* has solved, *ab initio*, six small protein structures ranging in size from about 200 to 600 non-H atoms. Its major limitation is the requirement that diffraction data to a resolution of 1.1Å be available. Its ultimate potential is still unknown.

8. Progress Since 1950

It seems fair to state that direct methods had their beginning in the Harker-Kasper inequalities [1]. Progress since then has been spectacular. In the decade of the fifties some eight to ten structures, mostly containing perhaps some 30 non-H atoms, were solved by direct methods. Since then the number of structures solved by direct methods has increased dramatically and, of equal importance, the complexity of solved structures has approximately doubled in each decade (Fig. 1). Thus in the present decade structures having as many as 500 non-H atoms have been routinely solved. In view of Figure 1 it seems safe to predict that in the decade beginning with the year 2000 structures having as many as 1000 non-H atoms will be routinely solvable by direct methods.

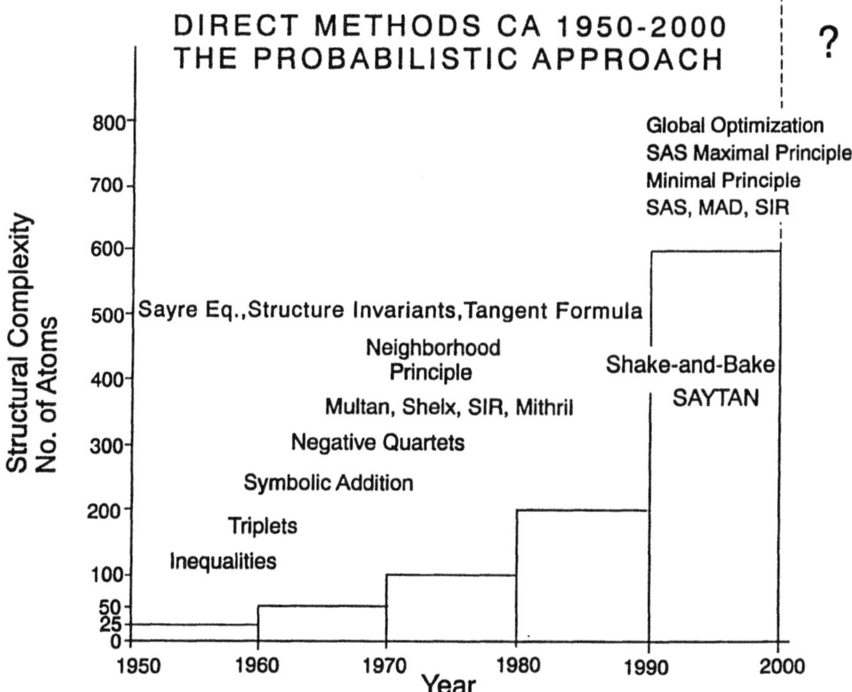

Figure 1. Complexity of structures solved by Direct Methods has doubled in each decade since 1950.

9. Acknowledgments

The *Shake-and-Bake* algorithm and the *SnB* program have been made possible by the financial support of grants GM-46733 from NIH and IRI-9412415 from NSF.

10. References

[1] Harker, D. and Kasper, J.S. (1948) Phases of Fourier coefficients directly from crystal diffraction data, *Acta Cryst.* **1**, 70-75.

[2] Karle, J. and Hauptman, H. (1950) the phases and magnitudes of the structure factors, *Acta Cryst.* **3**, 181-187.

[3] Sayre, D. (1952) The squaring method: a new method for phase determination, *Acta Cryst.* **5**, 60-65.

[4] Karle, J. and Hauptman, H. (1956) A theory of phase determination for the four types of non-centrosymmetric space groups 1P222, 2P22, $3P_12$, $3P_22$, *Acta Cryst.* **9**, 635-651.

[5] Hauptman, H.A. (1991) A Minimal Principle in the Phase Problem, in D. Moras, A.D. Podnarny & J.C. Thierry (eds.), *Crystallographic Computing 5: From Chemistry to Biology*, IUCr Oxford Univ. Press, pp. 324-332.

[6] DeTitta, G.T., Weeks, C.M., Thuman, P., Miller, R. and Hauptman, H.A. (1994) Structure solution by minimal function phase refinement and Fourier filtering: Theoretical basis, *Acta Cryst.* **A50**, 203-210.

[7] Weeks, C.M., DeTitta, G.T., Hauptman, H.A., Thuman, P., and Miller, R., (1994) Structure solution by minimal function phase refinement and Fourier filtering: II. Implementation and applications, *Acta Cryst.* **A50**, 210-220.

[8] Yao, J.-X. (1981) On the application of phase relationships to complex structures. XVIII. RANTAN - random MULTAN, *Acta Cryst.* **A37**, 642-664.

[9] Debaerdemaeker, T., Tate, C. and Woolfson, M.M. (1988) On the application of phase relationships to complex structures. XXVI. Developments of the Sayre-Equation tangent formula, *Acta Cryst.* **A44**, 353-357.

[10] Woolfson, M.M. and Yao Jia-xing (1980) On the application of phase relationships to complex structures. XXX. *Ab initio* solution of a small protein by SAYTAN, *Acta Cryst.* **A46**, 409-413.

[11] Miller, R. and Weeks, C.M. (1997) *Shake-and-Bake*: Applications and advances, in S. Fortier (ed.), *Direct Methods for Solving Macromolecular Structures*, Kluwer Academic Publishers, Dordrecht, The Netherlands, 25th Course, Intl. School of Crystallography, NATO ASI Series, Erice, Sicily, May (1997).

THE EFFECTS OF SYMMETRY IN REAL AND RECIPROCAL SPACE

Sven Hovmöller
Structural Chemistry,
Stockholm University, S-106 91 Stockholm, Sweden

1. Symmetry in real space

Normally, we think of symmetry in ***real space***, for example the mirror symmetry relating our left and right hands, or the 4-fold symmetry of ordinary bath-room tiles. A crystal also has symmetry and its symmetry in real space has effects in reciprocal space, *i.e.* in its diffraction pattern. Conversely, any symmetry present in a diffraction pattern provides information about the symmetry of the crystal itself in real space. In real space a symmetry manifests itself as two or more identical molecules or groups of atoms (a motif) being present within the unit cell. This symmetry may be only local, or it can be global which means all atoms in all unit cells are related by the same symmetry element(s). A global symmetry is called crystallographic symmetry, while the local symmetry is called non-crystallographic. Local symmetries are not unusual in crystals and they can be very important and useful, but they fall outside the scope of this article.

One of the very first steps of a crystal structure determination is to determine the crystallographic symmetry. Once the symmetry is known, we do not need to determine the atomic co-ordinates in the whole unit cell, but can restrict ourselves to those in the ***asymmetric unit,*** which for example in a 4-fold symmetry is just one quarter of the unit cell. The other atoms are given immediately by the symmetry. This greatly facilitates our task of solving a crystal structure. It should be stressed already now, that if we make a mistake in the symmetry determination, we will have to pay for that. If a too high symmetry is assumed, then two or more groups of non-identical atoms will be averaged, resulting in a smearing which can make it impossible to discern individual atoms. If, on the other hand, a too low symmetry is assumed, then groups of atoms that are exactly identical will look slightly different and we will double the noise level and the work involved in solving the structure.

2. The effects of symmetry in reciprocal space

The size and orientation of the ***unit cell in real space*** determines the ***positions of diffraction spots*** in reciprocal space. The contents within each unit cell, *i.e.* the atomic types (atomic number and the charge of ions) and their relative positions within the unit

D. L. Dorset et al. (eds.), Electron Crystallography, 139–150.

cell determine the amplitudes and phases of the diffraction spots in reciprocal space. The symmetry within the unit cell affects the symmetry of amplitudes and phases of the reflections in reciprocal space. It is very unfortunate that the word ***phase*** is used in two quite different meanings within the field of high resolution electron microscopy. This has caused great confusion. In this article, phase is defined as the ***crystallographic structure factor phase***.

Already from the positions of the diffraction spots, we may say something about the symmetry within the crystal. For example, if the distance between reflections in the h and k directions are identical and the angle between these axes are 90° we may suspect a 4-fold symmetry. But it may be just a coincidence that the lengths of two crystal axes are very similar. The proof that the crystal has a certain symmetry lies in the relative intensities of the reflections. The effects of symmetry on amplitudes is very simple: ***All symmetry-related reflections have the same amplitudes***. This holds for all symmetries, both in 2D plane groups and 3D space groups. For crystallographic structure factor phases, the situation is more complicated. When the symmetry elements do not contain a translation (inversion centers and mirrors m and 2-, 3-, 4- and 6-fold rotation axes) the phases of symmetry-related reflections are equal. In the cases where translations are involved (glide planes and 2_1, 3_1, 3_2, 4_1, 4_2, 4_3, 6_1, 6_2, 6_3, 6_4 or 6_5 screw axes) the phases of symmetry-related reflections have a more complex relationship, but their relative phases can always be derived directly from knowledge of the symmetry elements and the Miller indices of the reflections *(hkl)*.

The relations between amplitudes and phases in real space can be derived from the equivalent positions in the unit cell, using the very powerful tool of rotation matrices and translation vectors.

3. Rotation matrices and translation vectors

In crystallography, the symmetry of a crystal is described in the form of ***equivalent positions*** *(x y z)*, *(x', y', z')*...The set of equivalent positions is unique for every one of the 230 3D space groups, and they are listed in the International Tables for Crystallography [1] and in the program Space Group Explorer, which can be downloaded from the World Wide Web [2]. For example, a center of symmetry is seen as the pair of equivalent positions *(x y z)* and *(-x -y -z)*.

A very powerful mathematical tool for expressing the symmetry operations in crystallography are the **rotation matrices and translation vectors**. An introduction to rotation matrices and translation vectors can be found in [3]. Any equivalent position can be derived from the trivial first position *(x y z)* by

$$\begin{pmatrix} -1 & 0 & 0 \\ 0 & -1 & 0 \\ 0 & 0 & -1 \end{pmatrix} \bullet \begin{pmatrix} x \\ y \\ z \end{pmatrix} + \begin{pmatrix} 0 \\ 0 \\ 0 \end{pmatrix} = \begin{pmatrix} -x \\ -y \\ -z \end{pmatrix} \tag{1}$$

The $(x\ y\ z)$ position should be written as a vertical column vector after the rotation matrix. In this case the translation vector is zero. Notice that the order is important; ***first rotating and then translating***. For matrices the commutative law "first a, then b gives the same result as first b, then a" is not valid, as it is for simple mathematics like addition (2+3 = 3+2) and multiplication (5×7 = 7×5). Matrices are more like the real world, where the order of doing things also makes a great difference - think for example of the difference between first putting the tea-cup on the table and then pouring in the tea, and first pouring the tea and then putting the tea-cup on the table!

A more complicated equivalent position, such as that of a 3_1 screw axis (-y, x-y, z+1/3) is derived from its rotation matrix and translation vector as:

$$\begin{pmatrix} 0 & -1 & 0 \\ 1 & -1 & 0 \\ 0 & 0 & 1 \end{pmatrix} \bullet \begin{pmatrix} x \\ y \\ z \end{pmatrix} + \begin{pmatrix} 0 \\ 0 \\ 1/3 \end{pmatrix} = \begin{pmatrix} -y \\ x-y \\ z+1/3 \end{pmatrix} \quad (2)$$

Any symmetry element in real space is reflected in reciprocal space, but in a reciprocal way, such that one reflection $(h\ k\ l)$ generates a symmetry-related reflection $(h'\ k'\ l')$ by multiplying a horizontal row vector by the rotation matrix. If we again apply the 3_1 operation we get:

$$(h\,k\,l) \bullet \begin{pmatrix} 0 & -1 & 0 \\ 1 & -1 & 0 \\ 0 & 0 & 1 \end{pmatrix} = (\,k,\ -h-k,\ l) \quad (3)$$

Notice that this is not the same result as we would get by just replacing x, y and z by h, k and l! The method of rotation matrices and translation vectors is completely general and shows immediately how to derive all symmetry-related reflections in any space group. The amplitudes of symmetry-related reflections are always equal. The phases of symmetry-related reflections are also related in a predictable way, but they are ***not*** necessarily equal.

Recall that phases are positions in real space. If the phase of a cosine wave is changed, it means the position of its maxima and minima are shifted. If the symmetry operation involves a non-zero translation vector $(t_1\ t_2\ t_3)$, such as (0 0 1/3) in the case of 3_1 above, then the reflections related by that symmetry operation have phases φ related in the following way:

$$\varphi\,(h'\,k'\,l') = \varphi\,(h\,k\,l) - 360^\circ(h\cdot t_1 + k\cdot t_2 + l\cdot t_3) \quad (4)$$

The proof for this can be found in [3]. The relative phases of any two symmetry-related reflections can be calculated directly from the $(h\ k\ l)$ index and the translation vector associated with the rotation matrix which relates the two reflections. Note that in one and the same space group, for example $P3_121$ some symmetry-related reflections have the same phase while others can be 180° apart and yet others can differ by for example 120° or 240°. In 2D plane groups and 3D space groups of lower symmetry (up to orthorhombic) symmetry-related reflections are either equal or differ by 180°.

4. Friedel's law

A very special property of all Fourier transforms (remember reciprocal space is the Fourier transform of real space) is the relation between two points (*hkl*) and (*-h -k -l*) on opposite sides of the center of the Fourier transform:

$$F(hkl) = - F(-h\ -k\ -l) \tag{5}$$

This relation is called Friedel's law. It states that the amplitudes |F| are identically the same for *Friedel pairs* of reflections (*hkl*) and (*-h -k -l*), while their phases are related as:

$$\alpha(hkl) = - \alpha(-h\ -k\ -l) \tag{6}$$

Thus, if for example the phase of the reflection (2,3,1) is 25 degrees, then the phase of (-2, -3, -1) is -25 degrees. Friedel's law can be easily understood from Figure 1.

Figure 1. A geometrical explanation of Friedel's law. The Fourier component (1 0) is a cosine wave of density going along positive x (horizontally). The ***phase*** *of this wave is defined as the position of its maximum (black) relative to the origin. The (1 0) wave reaches its maximum after about 1/4 of the unit cell along x. Thus its phase is -90 degrees. Its Friedel pair (-1 0) is the same wave, but going in the opposite direction, i.e. along -x. The maximum of (-1 0) comes* ***before*** *the origin, and thus the phase at the origin is +90 degrees.*

Obviously, the amplitudes of (1 0) and (-1 0) are equal.

Friedel's law has effects on the phases, as will be seen later. Friedel's law is a general property of all Fourier transforms; it is always valid, even if the object is not crystalline.

5. Center of symmetry

The center of symmetry is a very common symmetry element. It means there is a point in which the structure is mirrored onto itself, see Figure 2. If you were to stand on a center of symmetry, you would see exactly the same view if you looked in any arbitrary direction or in exactly the opposite direction. Two molecules related by a center of symmetry are the mirror images of each other; one left-handed and one right-handed molecule, the two different enantiomers. Proteins and most other natural products exist in only one of two possible enantiomers, so there cannot be any centers of symmetry in protein crystals. One could expect that centers of symmetry were of no interest for crystal structure analysis of proteins, but it turns out that even though the crystal does not have this symmetry, a projection may well be centrosymmetric.

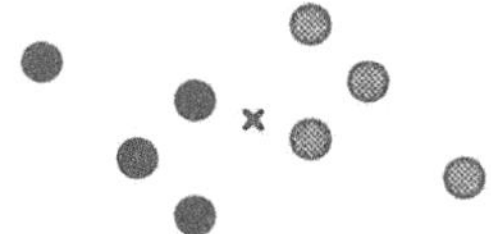

Figure 2. Two molecules, related by a center of symmetry at ×.

Let us look in some detail at the simple 1-dimensional case before going into 2D and 3D.

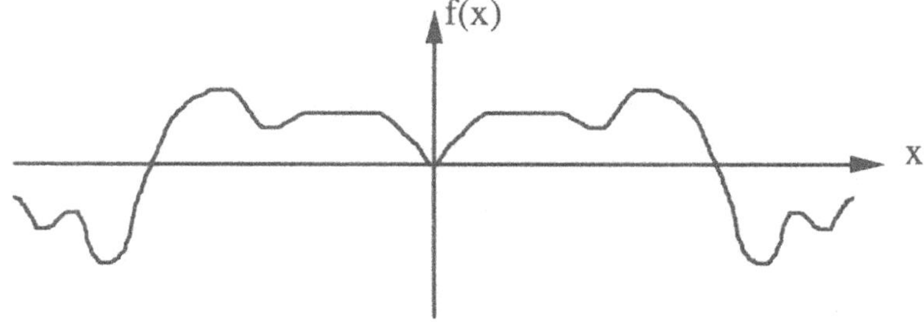

Figure 3. A centrosymmetric function, for which $f(x) = f(-x)$

The function f(x) in Figure 3 is centrosymmetric, since f(x) = f(-x) for every value of x. Such a function is also called an even function. If an even function like this is to be described as a Fourier series, there is no need to include any sine terms, because sin(x) is an odd function, where f(x) = - f(-x). As a consequence, all centrosymmetric functions can be described as a sum of only cosine waves. This leads to the very useful fact that ***all phases must be 0 or 180 degrees for centrosymmetric crystals*** and centrosymmetric projections. This is true only if the origin is chosen on the center of symmetry, as it is in Figure 3, but indeed that is the only reasonable place to position the origin.

6. Mirror symmetry

A mirror symmetry relates a left-handed molecule with its right-handed equivalent. The orientation of a mirror plane is best described by giving the direction of the normal to the mirror plane. A mirror plane perpendicular to the *b* axis has equivalent positions (*x y z*) and (*x -y z*). In reciprocal space this symmetry operation leads to symmetry-related reflections (*h k l*) and (*h -k l*). Since there is no translation involved, the phases of these reflections will be identical. In the electron microscope we always record 2D projections of 3D space. If a mirror plane is projected down along the direction of its normal, then the two mirror-related parts of the unit cell fall exactly on top of each other. If the electron beam goes in the mirror plane (*i.e.* perpendicular to the normal of the mirror plane), then the two mirror-related motifs will be seen as mirror images.

7. Two-fold rotation axis

A 2-fold rotation axis along the *b* axis gives rise to the two equivalent positions (*x y z*) and (-*x y* -*z*) and in reciprocal space relates (*h k l*) with (-*h k* -*l*). When such a structure is viewed in the electron microscope along the 2-fold axis, the projection becomes centrosymmetric, as illustrated in Figure 4.

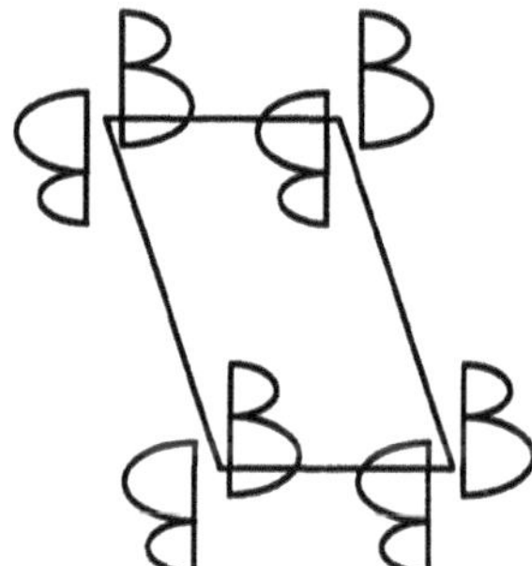

Figure 4. The projection down the 2-fold axis of a non-centrosymmetric crystal is centrosymmetric.

When the projection is perpendicular to the 2-fold axis, on the other hand, the result will be identical to that of projecting down a mirror plane. Thus it is clear that a single projection is not enough to determine the full 3D symmetry. In fact, the 230 space groups can only give rise to 17 different projections, the plane groups. If we want to find out the full 3D symmetry it is necessary to view the crystal along two or more different crystal axes.

8. Symmetry elements including translations; screw axes and glide planes

Symmetry elements having translations, such as screw axes and glide planes, have the same rotation matrices as their corresponding symmetry elements without translation. Since the symmetry-related reflections are generated from the rotation matrices, the symmetry-related reflections are exactly the same for $P2$ and $P2_1$. Similarly, all the space groups $P6$, $P6_1$, $P6_2$, $P6_3$, $P6_4$, $P6_5$ have the same symmetry-related reflections. They do differ in their phase relationships, but since the phases are not visible in diffraction patterns, the diffraction patterns will look the same, except for one thing; the systematically absent reflections along the screw axes. More about those later.

A projection down a screw axis is indistinguishable from a projection along the corresponding rotation axis. This should be evident from the fact that a projection is like collapsing a 3D object into a plane of zero thickness.

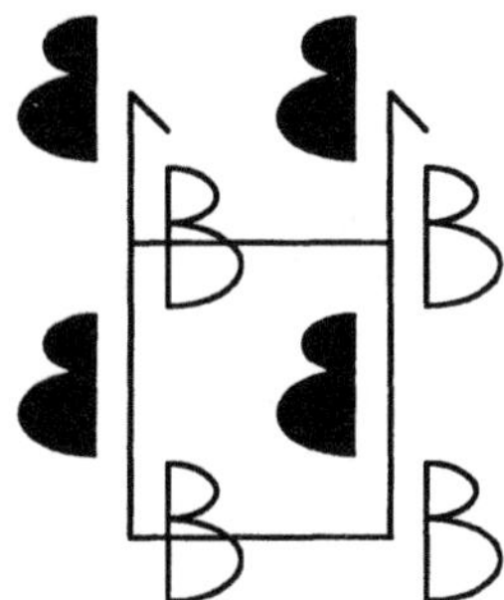

Figure 5. A projection perpendicular to a 2_1 screw axis or parallel to a glide plane give similar results; a zigzag pattern of mirror-related molecules (drawn black and white).

If a 2_1 screw axis is parallel to the plane of the image, it gives rise to a zigzag pattern. Alternating motifs are mirror images of each other and translated half a unit cell along the screw axis. A crystal with a glide plane gives the same projection as a 2_1 screw axis, when viewed along the glide plane. This is another example of two different symmetries which cannot be distinguished from a single projection, for example an EM image.

In reciprocal space, a 2_1 screw axis along y gives rise to the two equivalent positions $(x\ y\ z)$ and $(-x,\ y+1/2,\ -z)$. They leads to a mirror symmetry in the diffraction pattern $|F(h\ k\ l)| = |F(-h\ k\ -l)|$. Because Friedel's law is always present, a single mirror in real space results in a diffraction pattern with two perpendicular mirror planes, denoted *mm* symmetry. The phases φ of two reflections related by a 2_1 screw axis can be found from (4): $\varphi(-h\ k\ -l) = \varphi(h\ k\ l) - 360^{\circ}(k/2)$, *i.e.* symmetry-related pairs of reflections with even k indices have the same phase, while those with odd k indices differ by 180°.

9. Combining Friedel's law with crystallographic symmetry elements

Friedel's law is nearly always valid. Exceptions are due to the small effect of anomalous scattering, which is much studied in X-ray crystallography but not yet in electron crystallography, and the effects of dynamical diffraction of electrons. The following holds for kinematical scattering from a crystal with negligible anomalous dispersion.

When a reflection can be generated in two different ways, namely from the Friedel symmetry and from a rotation matrix, interesting things happen. If the phase of a reflection $(h\ k\ l) = \varphi$ then Friedel's law says the phase of the reflection $(-h\ -k\ -l) = -\varphi$. If $(-h\ -k\ -l)$ is generated also from a rotation matrix operating on $(h\ k\ l)$, we will get another value for the phase. The combined effect of these two demands on the phases may lead to phase restrictions, as shown below.

10. Phase restrictions

If (*-h -k -l*) is generated from a rotation matrix operating on (*h k l*) with the phase φ, we can calculate the phase of (*-h -k -l*) from (4). In the very common case of a centrosymmetric crystal, there is an equivalent position (*-x -y -z*). Since there is no translation element in this symmetry operation, we see directly that the phase shift between (*h k l*) and (*-h -k -l*) is zero, *i.e.* φ(*-h -k -l*) = φ(*h k l*). But we also have Friedel's law, which says that the phase of the reflection φ(*-h -k -l*) = - φ(*h k l*). How can a phase value at the same time be both φ and -φ? That is possible if and only if φ = 0° or 180°, since - 0 = 0 and -180° = 180°. This means that the phases of all reflections in centrosymmetric space groups must be 0° or 180°. We say the reflections have ***phase restrictions*** of 0° (+/- 180°). This is of course a great help in structure determination - we need only to consider two possible phase values for each reflection, rather than anything between 0° and 360°. It may be worth mentioning that even so, it can be difficult to solve also centrosymmetric crystals; since there are two choices of phase for every reflection there will be 2^n possible phase combinations for a crystal with n unique reflections. Already 20 reflections lead to over 1 million possibilities, and that is why it is necessary to find a way of estimating the phase values, either experimentally from EM images or by probabilistic methods, such as direct methods or maximum likelihood.

Also non-centrosymmetric crystals can have individual reflections with phase restrictions. Consider for example a protein in the space group *P2*. The equivalent position (*-x y -z*) causes (*h k l*) and (*-h k -l*) to have the same phase. All reflections of the type (*-h 0 -l*) are generated with the phase φ from (*h 0 l*) but -φ from Friedel's law. This case is the same as that described above for centrosymmetric crystals, so all (*h 0 l*) reflections in crystals with *P2* symmetry (with the 2-fold along *b*) have phase restrictions 0° (+/- 180°).

If the symmetry operation involves a translation, the phase restriction may become different from 0° (+/- 180°). Consider for example the reflection (1 0 2) in the space group $P3_121$. There is an equivalent position (*-x, y-x, 1/3-z*) which makes (*h k l*) become symmetry-related to (*-h-k, k, -l*) with a phase shift of -360°(*l*/3). The reflection (1 0 2) generates its Friedel pair (-1 0 -2) by that operation. If the phase of (1 0 2) is φ, then the phase of (-1 0 -2) is φ - 120° from that symmetry operation, but -φ from Friedel's law. Since a given reflection cannot have two different phase values φ - 120° and -φ must be equal. This happens only for φ = 60° and φ = 240°, so the phase restriction is 60° (as always +/- 180°). Luckily we only have to worry about these strange phase restrictions for some special reflections in non-centrosymmetric 3D space groups. In 2D all phase restrictions are 0° (+/- 180°).

11. Systematic absences

It may happen that the phases derived from a symmetry operator cause a conflict that cannot be resolved. Consider for example the reflection (0 1 0) in $P2_1$. The equivalent position (*-x, ½+y, -z*) generates (0 1 0) from itself with the phase shift -360°(1/2) = -180°. If the phase of (0 1 0) is φ, then its phase is also φ - 180°. But that is not possible, unless the amplitude is exactly zero. Thus (0 1 0) (and all other reflections of the type (0 *k* 0) with *k* odd) are absent in $P2_1$. Such reflections are called systematically extinct or forbidden reflections or systematic absences.

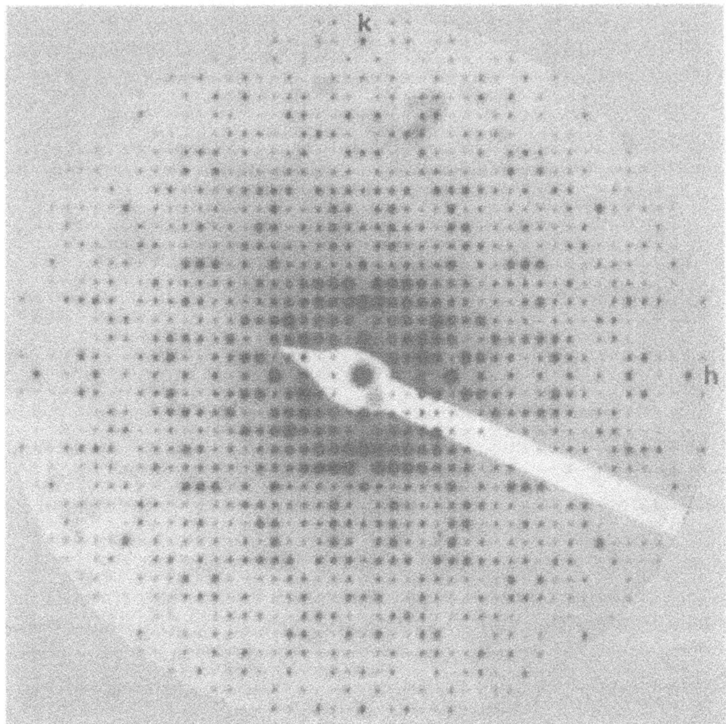

Figure 6. Electron diffraction pattern of Ta_2P. This SAED pattern shows mm-, but not 4-fold symmetry as seen from the intensities of diffraction spots. Notice that all odd reflections along both the h and k axes are absent. This shows there must be 2_1 screw axes or glide planes along both axes. The very faint forbidden reflections (9 0) and (-9 0) are caused by multiple diffraction, but they are so weak that it can be concluded that the crystal must be very thin, and the data thus is quite close to kinematical. Image by Dr. Thomas E. Weirich, see also [4].

Systematic absences are very important in X-ray crystallography, because they are used for space group determination. In electron microscopy these forbidden reflections often have non-zero amplitude, because of dynamical effects. That makes them less useful for symmetry determination, although one must always suspect the possibility of systematic absences if all odd reflections along a crystallographic axis are much weaker than the reflections with even indices. If the crystals are very thin, then the multiple diffraction is small and the crystals are scattering nearly kinematically. In that case, the forbidden reflections are nearly extinct. Thus, it is possible to have an, at least qualitative, estimate of the crystal thickness and deviation from the kinematic case, from inspection of the intensities of systematically forbidden reflections in electron diffraction patterns, see Figure 6.

12. Systematically enhanced reflections

Sometimes a symmetry operator causes a reflection to generate itself, but without any phase shift. Such reflections will on average be stronger than the other, general reflections. This happens mainly along crystal axes. One example is (0 *k* 0) in *P2*. The equivalent position (-*x* *y* -*z*) generates (0 *k* 0) from itself, without a phase shift. Thus all reflections on the *k* axis in *P2* have an expected average intensity twice that of the expected average intensity of all other reflections. This enhancement factor is called ε. It is important to take ε into consideration in direct methods, since it affects the probability of phase relationships.

13. Origin specification

While the amplitudes of reflections may seem very easy to understand as cosine waves of high and low electron density (in the case of X-ray diffraction) or electrostatic potential (in the case of electron diffraction), the phases are often considered as something abstract. One may even get the idea that phases do not have a real physical meaning - that they are mere mathematical constructions. This is unfortunate and unfair to the phases, because these are just as much physical realities as the amplitudes. In a given physical crystal, the atoms are situated at fixed positions within each unit cell. This causes the potential to vary in a systematic way, throughout the crystal. Since a crystal is made up of identical unit cells, the potential varies in a periodic way. All periodic functions can be described in a very efficient way as a Fourier series, *i.e.* as a sum of sine and cosine waves. The positions of the maxima of these waves are given by the positions of the atoms within the unit cell. The crystallographic structure factor phase of a reflection is nothing else than the position of this maximum, relative to a fixed position in the unit cell - the origin.

Unfortunately, there is not a unique point in the unit cell which has to be the origin. In principle any point can be chosen as the origin. Similarly, any directions and lengths of the axes of the co-ordinate system we will use to describe the positions of the atoms in the crystal are possible. However, not much thinking is needed to convince oneself of the usefulness of the practice adopted by the crystallographic community to chose a co-ordinate system with axes along the directions of the unit cell. Even in the common case where this means we have to have different length scales along the *x*-, *y*- and *z*-directions, and sometimes even a non-orthogonal co-ordinate system, it is easier to follow than try to defeat Nature. Similarly, some choices of origin in the unit cell are to be preferred over others simply because they make everything easier. These positions are given for each space group in the International Tables for Crystallography [1].

In a centrosymmetric crystal, the obvious place to put the origin is on a center of symmetry. Only then will we get the very nice pair of equivalent positions (*x* *y* *z*) and (-*x* -*y* -*z*) and can know that all phases must be 0° or 180°. In most (but not all!) tetragonal space groups, the origin is on a 4-fold axis, and so on. Unfortunately, even these specifications are not entirely sufficient for fixing (specifying) the origin. In most cases, there are more than one position in the unit cell which fulfils the specification. Think of a crystal with 4-fold symmetry, for example a wall in a bath-room with square tiles (see Figure 7). There are two distinct points per unit cell which both have a 4-fold surrounding. These are mathematically equivalent, but chemically different positions. The positions of the atoms relative to the origin, and thus the phases of the reflections, will of course be different if one or the other origin is chosen.

It is necessary to specify the origin in an unambiguous way before the phases can be specified. The origin specification can be carried out in real space, as is done in crystallographic image processing where the unit cell is searched for positions with the given symmetry. Origin specification may also be carried out in reciprocal space, by specifying the phase of some reflections (maximum 3 in 3D space and 2 in 2D planes). In direct methods the origin is specified in reciprocal space.

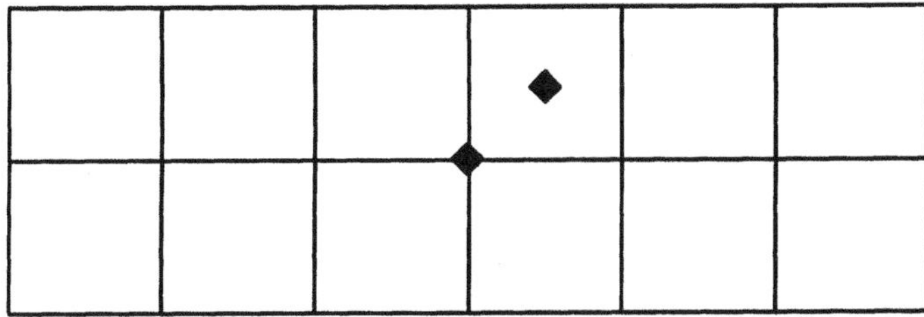

Figure 7. The two chemically distinct but mathematically identical origins with 4-fold symmetry.

14. Symmetry determination

Since the symmetry in real space is reflected in reciprocal space it should be possible to go in the opposite direction, *i.e.* to determine the crystal symmetry from inspecting its diffraction pattern(s). This is indeed possible. In X-ray crystallography, the procedures for symmetry determination are based on finding the symmetry-related reflections and exploiting the information in systematically absent reflections. In electron microscopy the principle is similar, but in practice the procedure has to be quite different. One very elegant way of symmetry determination is from convergent beam electron diffraction, CBED, which is described by Tanaka in this volume [5]. The 3D space group can be determined from the selected area electron diffraction (SAED) patterns, in much the same way as in X-ray crystallography. The difference is that the systematically forbidden reflections, which typically have absolutely zero intensity in the X-ray diffraction pattern, always have at least some intensity in the SAED patterns. Thus one has to take great care in using extremely thin samples, and even so may have to put in a bit of judgement in considering that if all odd axial reflections are much weaker than the even indexed ones, that probably means that they are forbidden by the space group.

An advantage of electron microscopy over X-ray diffraction is that the phases can be determined experimentally from the Fourier transform of the EM image. As mentioned above, a 2_1 screw axis is distinguished from a 2-fold rotation axis not only by the amplitudes (*i.e.* the systematic absences), but also from the phase relations of reflections of the type (*h k l*) and (*-h k -l*). The phases of EM images are more accurately recorded than the amplitudes, and can typically be found within +/- 20°. This makes it quite straight-forward in most cases to distinguish a 2-fold axis (where all symmetry-related reflections have equal phases) from a 2_1 screw axis (where reflections with odd indices along the screw axis should differ by 180°.

15. Summary

Symmetry in real space is always reflected in reciprocal space. Inversely, it is possible to determine the crystal symmetry from the symmetry in electron diffraction patterns (CBED or SAED) and EM images. It is essential in every crystal structure determination to find the correct symmetry. Rotation matrices and translation vectors are powerful tools for describing the symmetry operations in real space and their effects on amplitudes and phases of crystallographic structure factors in reciprocal space.

16. References

1. International Tables for Crystallography Vol A, (1983) T. Hahn, Ed. D. Reidel, Dordrecht, Netherlands.
2. http://www.calidris-em.com/archive.htm
3. Hovmöller, S. Rotation matrices and translation vectors in crystallography. No. 9 in Pamphlet series issued by the International Union of Crystallography, Commission on crystallographic teaching, University College Cardiff Press (1981)
4. Weirich, Th.E., Ramlau, R., Simon, A. and Hovmöller, S. (1997) Exact atom positions by electron microscopy? in D.L. Dorset, S. Hovmöller and X.D. Zou (eds.), *Electron Crystallography* Kluwer Academic Publishers, Dordrecht.
5. Tanaka,M. (1997) Convergent beam electron diffraction, in D.L. Dorset, S. Hovmöller and X.D. Zou (eds.), *Electron Crystallography* Kluwer Academic Publishers, Dordrecht.

OBTAINING PHASES FROM ELECTRON MICROSCOPY FOR SOLVING PROTEIN STRUCTURES

Tribute to Boris Vainshtein (1921-1996)

Sven Hovmöller
Structural Chemistry,
Stockholm University, S-106 91 Stockholm, Sweden

This paper is given in honour of Professor Boris Vainshtein, one of the founders of electron crystallography. He was an invited speaker at this NATO Advanced Study Institute on Electron Crystallography, but regrettably died half a year before it took place. Already in his classical book *Structure Analysis by Electron Diffraction* from 1964 [1] he wrote: "There is no doubt now that electron diffraction may be used for the complete analysis of crystals whose structure is unknown".

1. The phase problem in crystallography

The phase problem is one of the most intriguing phenomena in X-ray crystallography. Among all mathematical problems, the phase problem may be one of the most studied, in terms of man-years. Thousands of crystallographers have tried to solve the phase problem. None has succeeded but many have been successful. Still today there is no general and complete solution to the phase problem in X-ray crystallography, but partial solutions, or solutions of special cases, have been essential parts of at least 8 Nobel prizes in chemistry and physics over the years. The two latest of these were awarded in chemistry in 1982 and 1985. Aaron Klug at the MRC Laboratory of molecular biology in Cambridge received the Prize in 1982 "for his development of crystallographic electron microscopy and his structural elucidation of biologically important nuclei acid-protein complexes"[2]. In 1985 it was awarded to Herbert Hauptman and Jerome Karle "for their outstanding achievements in the development of direct methods for the determination of crystal structures"[2]. These direct methods revolutionised the field of structure determination of organic molecules with up to and even above 100 atoms.

It is easy for X-ray crystallographers to get into the idea that these phases are mere mathematical abstractions, since the crystallographic structure factor phases cannot be directly determined experimentally in an X-ray diffraction pattern. The use of words such as "real" and "imaginary" for phases contribute to this feeling. Actually, the phases are very concrete physical entities, as hopefully will come clear in the following. It is quite the other way with the unit cell - it is a concept which is easily understood although it is

D. L. Dorset et al. (eds.), Electron Crystallography, 151–162.

an abstraction - there are no small boxes in the crystal where the molecules are put inside! One remarkable advantage of electron crystallography is that the enigmatic phases can actually be observed and measured in the electron microscope. This paper will discuss the phase problem and relations between the scattering of X-rays and electrons, real and reciprocal space, images and diffraction. The questions are: Are the phases measured from electron microscopy images the same as those of the X-ray structure factors and can they be used for helping to solve the phase problem in protein X-ray crystallography?

2. Dualism in crystallography

A fascinating aspect of the life of crystallographers is that we tend to spend half our time in real space and the other half in reciprocal space. These dual worlds, also called direct space and Fourier space, are at the same time totally different and closely related. In terms of information content, they may even be considered identical. The phases have a key role in relating these dual spaces.

There are many other important pairs in crystallography, some of which will be the subject of this review, namely:

electrons - X-rays
particles - waves
images - diffraction
wave front phases - structure factor phases
strong interaction - weak interaction
small objects - large objects
low resolution - high resolution
molecular biology - inorganic chemistry
amorphous - crystalline

3. Electrons and X-rays

If we want to determine the positions of atoms, we need a probe which is able to detect distances of the order of 1 Ångström (10^{-10}m). The three kinds of radiation used in crystallography are X-rays, electrons and neutrons. It is possible to tune the wavelength of each of them, such that an optimal wavelength is used for the experiment. By far most crystallographic studies are carried out using X-rays, with neutrons on second place, at least if we talk about determination of atomic co-ordinates. Neutrons are especially useful because they, unlike X-rays, interact relatively strongly with hydrogen, which is an extremely important element in biology. However, here we will concentrate on electrons and X-rays and their interrelations.

X-rays are created by bombarding a metal plate with electrons accelerated to some 40 kV. If the metal is copper, then the main wavelength (λ) of the X-rays is CuK_α at 1.54 Å, incidentally the same as the length of the C-C single bond, the most important bond for life. When X-rays pass through matter, they interact with the electrons surrounding the atomic nuclei, but not with the nuclei themselves. Yet, since most electrons are located very close to the atomic nuclei, X-rays give information mainly about the positions of atoms, rather than about the more diffuse shapes of valence electrons. The interaction is rather weak - the mean free path length is in the order of a millimetre. If the wavelength is shorter (for example MoK_α with $\lambda = 0.71$ Å), the interaction becomes even weaker.

If the accelerated electrons are produced in an electron microscope, they can themselves be used to study the structure of matter. The most common accelerating voltages for electron microscopes today are 100 kV to 400 kV. The electrons reach a velocity of about half the speed of light and have a wavelength of about 0.04 to 0.02 Å. The electrons interact with the electrostatic potential in matter, *i.e.* they sense the positive and negative charges. All atoms are made up of positive nuclei surrounded by negative electrons. Since electrons are themselves negatively charged, they interact very strongly with all these charges when passing through an object. The electrons are attracted by the positive nuclei and repelled by the electrons. The distribution of charges in solids is such that the positive charges are concentrated in the extremely small nuclei, with most of the negatively charged electrons also located close to these nuclei, partially shielding the positive charges. The electrostatic potential is thus a set of rather sharp positive peaks on a low and relatively smooth background. The mean free path length is very short for electrons, in the order of nanometers. Because of the strong interaction of electrons with matter, extremely small samples can be studied. The optimal crystal thickness is less than 100 Å. A crystal structure determination can be carried out on a crystal of less than 100 unit cells, as compared to 10^{12} to 10^{15} needed for single crystal X-ray diffraction [3]. In fact, even single atoms can be studied by electron microscopy, opening up the world of crystal defects and even amorphous materials for study at the atomic level. As with X-rays, if the wavelength is shorter (higher accelerating voltage), the energy becomes higher but the interaction becomes weaker.

In spite of the facts that X-rays and electrons are so different and interact with different features in matter, in both cases it all boils down to sensing the positions of the centers of atoms. We may therefore expect similar results from structure investigations carried out with X-rays and electrons, and a close interaction between the scientific fields of X-ray crystallography and electron microscopy. For various practical and historical reasons, some of which will be discussed in the following, this is not really the case. But there are strong reasons why science in general and crystallography in particular would gain from a stronger interaction between these two fields.

4. Images, diffraction and phases

The mathematical definition of an image involves a one-to-one relation between the points in an object and its image. One fascinating aspect of the Fourier transform (FT) of an object is that it is not at all an image, yet contains the same information as the image. Each point in an object contributes to every point in its FT and *vice versa.* The FT of a periodic object is especially useful, in that the continuous (analogue) function of the object is condensed into a discrete (digital) function; the Fourier coefficients.

Only the first Fourier coefficient is non-zero on average - it gives the average value of the function. In the case of an X-ray diffraction study it is the F(000) term which is the average electron density within the unit cell. For a protein this term is 0.38 electrons/Å^3. All the other Fourier coefficients are cosine waves, only modulating the average electron density, so as to concentrate the density into sharp peaks (where the atomic nuclei are) on a rather smooth background of some 0.01 electrons/Å^3. Each Fourier coefficient is characterised by three numbers; the index (h,k,l) giving the direction and periodicity of the cosine wave, an amplitude and a phase. The phase is the position of the maximum of the wave relative to the origin of the co-ordinate system.

If a function is known, the calculation of its FT is straight-forward, although very time-consuming. The FT of a 2D image consisting of 256 x 256 points (pixels) also consists of 256 x 256 points. Each point in the FT is the sum of every point in the image after having been multiplied by a 2D cosine function. Even at the rate of making one such calculation per second, it would take more than a life-time to calculate a 256 x 256 FT by hand. Thus image processing involving Fourier transforms did not emerge as a science until the computers became fast enough. In the late 1960ies, when Klug and co-workers started image processing, a 256 x 256 FT was an over-night job at the single main-frame computer of the university. Today such FTs can be calculated in a fraction of a second on a personal computer. The inverse FT gives back the original function.

The bright spots seen in a diffraction pattern (often called the reflections) are the squares of the Fourier coefficients. From the positions of the reflections we can immediately get the indices. The intensities are equal to the squares of the amplitudes. However, we can not calculate the inverse Fourier transform, since there is no information about where the maxima of the cosine waves are positioned relative to each other or to the origin of the unit cell. The phase information is lost in a diffraction pattern. If the atomic co-ordinates are known, the FT (with phases) can easily be calculated. The crux is that we do not know the atomic co-ordinates when we start a crystal structure determination.

An image, for example an electron micrograph, is a two-dimensional function. Its FT is easily calculated on modern computers and ***does contain the phases***. That this is really the case should be clear from Figure 1. It is worth emphasising that here the word *phase* is used for crystallographic structure phases. The reason why the crystallographic phases are present in EM images, but not in electron diffraction or X-ray diffraction patterns, is that the different diffracted beams are focused into an image by the lenses in the EM. Each diffracted beam carries information about the relative atomic positions on Bragg planes through the crystal. In the image all the contributing reflections interact, with phases, giving rise to a projection of the crystal structure. Since there are no lenses for X-rays, no X-ray microscope can be constructed and thus we only have X-ray diffraction patterns, where the phase information is lost.

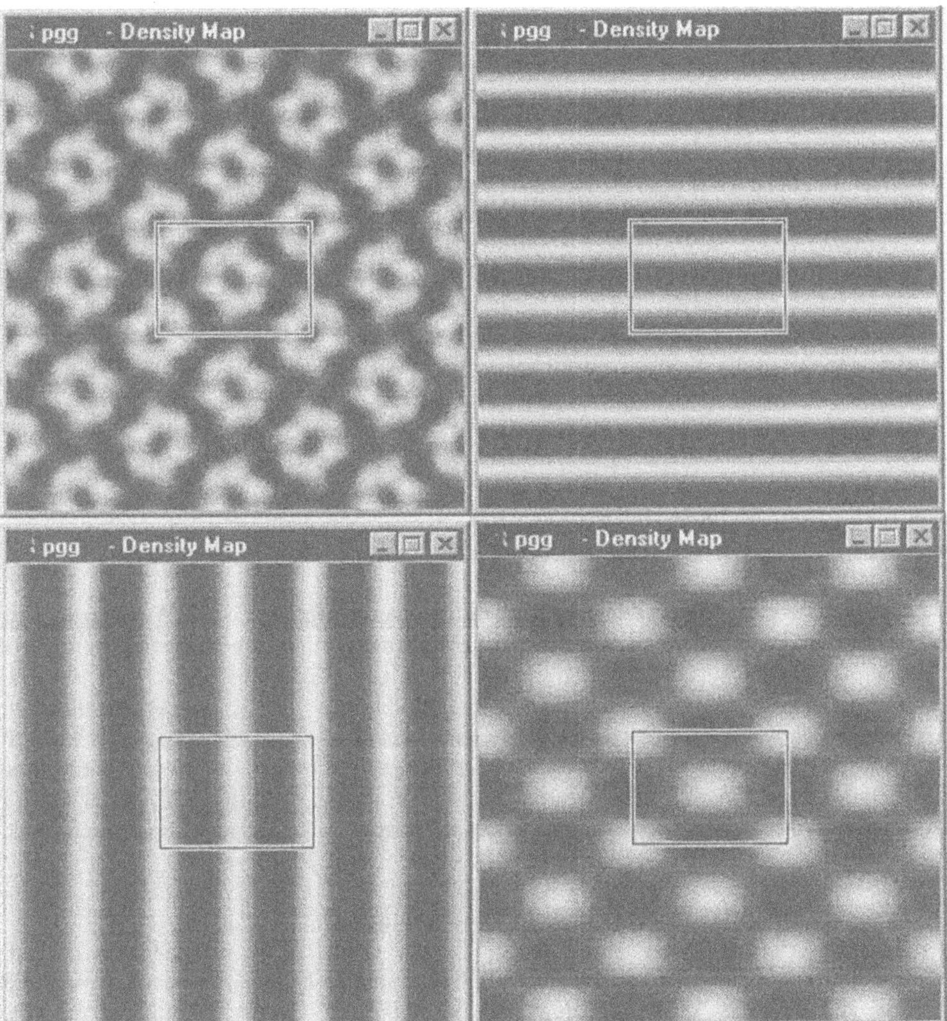

Figure 1. ***The physical meaning of phases***

Top left: An EM image of a negatively stained membrane crystal of cytochrome reductase [17]. One unit cell is indicated; a = 174Å, b = 137Å. Protein is white. At this resolution (25 Å) the protein complex is seen as a white ring. The internal features (α-helices and β-sheets etc.) cannot be seen, since the resolution is too low. The crystal is orthorhombic with a projected centro-symmetric symmetry pgg. There are glide planes g both vertically and horizontally, giving rise to a zigzag arrangement of the molecules. Each ring is a dimer. There are two dimers in each unit cell; one in the centre and one in the corner.

Top right: The (0 2) reflection has a phase of 180° since the electron density of the protein is at its minimum at the origin of the unit cell. (The minimum of the cosine function is -1, obtained for cos180.)

Bottom left: The phase of (2 0) is 0°.

Bottom right: The pair of two symmetry-related reflections (1 1), going \\ and (1 -1) going //, are shown together. Their phases are both 0°, since they have maximum density (white) at the origin of the unit cell.

This electron microscopy image is composed of 26 unique reflections. If their phases were not known from an EM image, each of these 26 reflections could have any of the two phase values allowed for centro-symmetric crystals; 0° or 180°, resulting in as many as $2^{26} = 64{\cdot}10^6$ *possible combinations.*

5. The phase confusion problem

It has frequently been claimed that there should be a phase problem also in electron microscopy. [4,5] It may be more correct to say that there is no phase problem, only a phase confusion problem in electron microscopy. The phase confusion problem has arisen since the word phase is used for (at least) two very different physical entities in the field of electron microscopy. When crystallographers speak about phases, they mean the ***crystallographic structure factor phases***, while physicists usually mean the ***wave front phases***. In order to resolve the phase confusion problem, it is necessary to define clearly which phases are meant. Some of the most prominent differences between the structure factor phases and the wave front phases are shown in the Table 1 below. A full description of the process crystal structure → structure factors → electron wave → lenses → image formation → structure restoration with mathematical and physical formulas has been published by Zou [6].

TABLE 1. A comparison of some central properties of crystallographic structure factor phases and wave front phases.

Entity	Crystallographic structure factor phases	Wave front phases
The waves are	stationary standing waves	moving - propagating
Wave length	≈ 1 - 20 Å	≈ 0.03 Å
Physical entity	electron density or potential	electromagnetic wave
Phases relative to	symmetry elements in crystal	incident electron wave front
Phase information	present in EM images	lost in EM images
Method of retrieval	Fourier transform of EM image	holography, focal series EM
Interrelations	modulates the wave front	records the structure factor

6. Small and large objects at high and low resolution

It is illuminating to compare the sizes of objects studied by electron microscopy and X-ray crystallography during different periods of the 20th century. In spite of the much shorter wavelength of electrons (0.02 - 0.04 Å) compared to that of X-rays (0.5 - 2 Å), electrons are actually best suited for studies of large objects at low resolution, while X-rays are ideal for structure determination of smaller structures at atomic resolution.

X-ray crystallography started in 1912 with the determination of exact atomic co-ordinates for inorganic crystals with small unit cells and then gradually developed to handle larger and larger molecules and unit cells. Electron microscopy, on the other hand, started off 20 years later, with a resolution comparable to that of a light microscope, which is useful for studying large systems such as whole cells, bacteria and organelles. As the electron optical lenses were improved, the resolution of electron microscopes improved. It was only in the early 1970-ies that the EM resolution reached values around 5 Å, allowing the visualisation of details within a unit cell of an inorganic crystal. The historical development of X-ray crystallography and electron microscopy can be considered as running in opposite directions, as shown in Table 2.

TABLE 2. Main new categories of structures studied by X-rays and electrons during different periods of the 20th century

Period	X-ray crystallography	Unit cells (in Å)	Electron microscopy	Resolution (in Å)
1910-1930	Inorganic crystals	3-10	-	-
1930-1950	Organic molecule	5-20	Whole cells and organelles	50
1950-1970	Proteins	30-100	Viruses and ribosomes	15
1970-1990	Viruses	200-400	Proteins, inorganic crystals	5-2
1990-	Ribosomes	600-1000	Atomic resolution	1-2

X-ray crystallography has had a tremendous impact on chemistry and has been the main method providing us with details of how the atoms are arranged in molecules and crystals. The solving of the double helix DNA structure by Watson and Crick and haemoglobin and myoglobin by Perutz and Kendrew in the early 1950-ies marked the birth of molecular biology. At this time, electron microscopy had shown that the inside of a cell was much more complex and ordered than had been thought before. There was not just a nucleus and a soup of cytoplasm, but a large number of highly specialised and structured organelles within the cell, such as the mitochondria, the endoplasmic reticulum, the Golgi apparatus and so on. This marked the birth of another science: cell biology.

7. Membrane protein structures solved by electron microscopy

By 1975, about 200 water soluble proteins had been determined to (near) atomic resolution by X-ray crystallography, but very little was known about the structures of the hydrophobic membrane proteins, because they were so difficult to purify and crystallise. It was then the revolutionary structure determination of the membrane protein bacteriorhodopsin (= purple membrane) was completed by Henderson and Unwin [7]. They took advantage of the natural tendency of membrane proteins to insert themselves into lipid bilayers. If lipids and membrane proteins are mixed in a ratio close to 1:1, the protein molecules sometimes aggregate into a 2D crystal. This happens *in vivo* with the bacteriorhodopsin of *Halobacterium halobium.* Such a membrane crystal is only 40-100 Å thick which is far to thin for single crystal X-ray diffraction, but ideal for electron microscopy. Once Unwin and Henderson had overcome several difficult technical problems, especially how to take EM images of the tremendously radiation sensitive unstained proteins, they obtained a 3D structure at 7Å resolution, showing rods of density perpendicular to the plane of the membrane. This was correctly interpreted as being α-helices composed of 20-25 hydrophobic amino acids. We now know that this is a very general building principle of membrane proteins, and it gave inspiration to the hydrophobicity plots, where amino acid sequences of unknown proteins are searched by computers for stretches of 20-25 consecutive hydrophobic amino acids which, if found, give a good indication that the protein is membrane bound. It took another ten years until Hartmut Michel managed to make 3D crystals of a membrane protein suitable for X-ray crystallography and together with Deisenhofer and Huber solve its 3D structure to atomic resolution. For this work on the photosynthetic reaction centre, they received the Nobel Prize in chemistry 1988.

Two other membrane proteins have been solved in 3D to atomic resolution by electron microscopy; light harvesting complex (Kühlbrandt and Wang, [8]) and porin by Jap *et al.* [9], see below.

8. The large mitochondrial membrane complexes

Until just a couple of years ago, all we knew about the 3D structures of the huge mitochondrial membrane protein complexes, numbered I-V, came from electron microscopy investigations of 2D crystals stained with heavy metal salts {(I) NADH: ubiquinone oxidoreductase, (III) cytochrome reductase, (IV) cytochrome oxidase and (V) the F_1F_0-ATPase}. Unfortunately, severe mistakes were made on the structure interpretations of complex II and cytochrome oxidase.

Crystals that were for a long time claimed to be of the entire complex I [10] turned out only to be from a much smaller sub-complex. This was clear from a calculation of the molecular mass of the protein per asymmetric unit. While the unit cell (a = b = 153 Å, c = about 105Å, giving a total cell volume of 2465 nm^3) could possibly contain two molecules of complex I, the asymmetric unit cell when taking the crystallographic symmetry $p42_12$ into account is only 1/8th of the whole unit cell. Assuming 50% protein in the crystal, the volume of the protein complex is then 2465 000 $Å^3$/16 = 154 000 $Å^3$. The volume per non-hydrogen atom in proteins is 20 $Å^3$, and the average molecular mass per non-H atom is 14.4, resulting in a volume of 20/14.4 = 1.4 $Å^3$/dalton. Thus, the molecular mass of the crystallised protein had to be about 110 kDa; much smaller than that of complex I which was estimated to be at least 600 kDa. The crystals were subsequently identified as made up of only a small subcomplex of complex I [11]. However, Leonard *et al.* later succeeded in crystallising the entire complex I [12].

Cytochrome oxidase was the first membrane protein to be crystallised and studied by electron microscopy, but all interpretations of its 3D structure were wrong since they are based on the hypothesis by Henderson and Capaldi that the crystals formed in Triton were composed of two membranes per unit cell, with the cytochrome oxidase inserted very asymmetrically into the membrane[13], whereas in fact there is only a single membrane in the unit cell passing through the middle of the proteins[14]. Consequently, the crystals made using deoxycholate as the detergent are not monomers[15], but dimers. This interpretation was recently confirmed by the atomic resolution structure of cytochrome oxidase solved by X-ray diffraction [16], where the membrane passes through the centre of the molecule.

The 3D structure of Complex III, cytochrome reductase was solved from electron microscopy images of membrane crystals to 25 Å resolution by Weiss and Leonard [17].

Already in the 1960ies electron microscopy images of the inner mitochondrial membrane had shown protruding "lollipops" which were interpreted as the water soluble part (also called F_1) of the ATPase complex V. The correctness of this interpretation of low-resolution data was confirmed when the complete 3D structure of F_1 to 2.8 Å resolution was solved by John Walker *et al.* [18], using X-ray crystallography.

9. Ribosomes

The ribosomes are among the largest and most complicated of all biological complexes. All we know today about the 3D structures of ribosomes has come from electron microscopy, but in this case mainly from studies of single particles rather than crystals. Randomly oriented ribosomes have been rapidly frozen into ice and studied by cryomicroscopy without adding any stains. Thousands of individual ribosomes have been added after their relative orientations have been found. These procedures make use of advanced multi-variate statistics and has been developed mainly by Joachen Frank and Marin van Heel [19]. It has also been possible to crystallise these huge molecular complexes into 3D crystals which diffract X-rays to high resolution, as shown by Ada Yonath *et al.* [20], but the tremendous task of solving the ribosome structure to atomic resolution will most likely not be finished until next century.

10. Electron crystallographic studies of organic and inorganic compounds

Although the scattering of X-rays and electrons are so different, in effect they both probe positions of atomic nuclei. It is now nearly half a century since Vainshtein and Pinsker in 1950 localised hydrogen atoms in paraffin by electron diffraction data [21]. The Moscow group succeeded in the structure determination of several inorganic compounds from electron diffraction data already in the 1950ies. One advantage of electrons pointed out by Vainshtein was that the relative scattering power of lighter elements, including hydrogen, is higher for electrons than for X-rays. The heavy elements are not so heavy for electrons.

There was an unfortunate and long period of almost no structure determinations by electron crystallography outside the Soviet Union, mainly because of exaggerated fears for the effects of multiple scattering of electrons within the crystals. One of the very few who pursued such investigations was Douglas Dorset whose book [22] is the only modern complete treatise on electron crystallography on organic molecules. Dorset has solved numerous lipids and other organic molecules from electron diffraction data and also shown that it is possible to phase such data by the direct methods used in X-ray crystallography.

The spectacular results on membrane proteins led to a restart of electron crystallography studies also of inorganic compounds, but now using the full power of electron microscopy and image processing rather than only electron diffraction that had been used by the Moscow group. Several metal oxides were solved by electron microscopy and image processing, giving atomic co-ordinates of the metal atoms accurate to within about 0.1 Å [3]. The first refinement of an inorganic structure against selected area electron diffraction data to better than 1 Å resolution, reaching an accuracy of 0.02 Å was recently completed by Weirich *et al.* [23]. After nearly one hundred years, the circle is closed and electrons and X-rays can be used for crystal structure determination in all fields of chemistry, from inorganic solids over organic molecules to proteins, with more or less similar accuracy.

11. Comparison of phases obtained by electron microscopy and X-ray diffraction

As shown above, electron microscopy can give structure factor phases. If these phases are similar to the X-ray structure factor phases, they may be used as a starting set for phase extension, for example by direct methods or maximum entropy. Even if only a few low-resolution reflections can be found, this can be very helpful in determining the packing of proteins in the unit cell. If reliable phases could be determined to intermediate resolution, 5-7 Å or so, then the complete solution of a protein structure should be straight-forward.

We must ask: how similar are the amplitudes and phases obtained from the two techniques X-ray and electron crystallography? Can phases obtained by electron microscopy be used for phasing X-ray diffraction data of proteins? Strangely enough, it turns out that very little work has been carried out trying to answer these questions.

Already in 1975, Unwin and Henderson studied thin crystal plates of the protein catalase by electron microscopy and obtained a projection map to 9Å resolution [24]. Later, several different crystal forms of catalase have been determined by X-ray crystallography, but none of these were from the same crystal form as the one studied by electrons. Only a few other proteins have been studied to high resolution by both electrons and X-rays. Unfortunately, none of these structure determinations were on the same crystal form, as would be necessary if the electron microscopy phases were to be used directly for phasing the X-ray diffraction amplitudes. However, in the case of inorganic structures, it has been shown that indeed the phases obtained from electron microscopy images, are nearly identical to the ones obtained after refinement of X-ray crystallography data on the same crystal, at least to 2.5 Å resolution [25].

The membrane protein porin has been studied by both electron microscopy and X-ray crystallography. The structure is very interesting and beautiful. As had been predicted from optical dichroism measurements, the structure is almost entirely built up from β-pleated sheets. A barrel of 16 anti-parallel β-strands is embedded in the membrane. Unlike the completely hydrophobic α-helices, these β-strands contain many hydrophilic amino acid side chains. In this barrel, alternate side chains point out towards the lipids and in towards the middle of the protein. All the hydrophilic amino acids are on the inside, where they form a hydrophilic channel, allowing transport of specific water-soluble molecules across the membrane. All side chains pointing out towards the lipid bilayer of the membrane are hydrophobic. Again the crystal forms studied by X-rays [26], and electrons [9], are not the same. However, the projected structures of porin obtained independently by the two methods are extremely similar.

A similar case is the structure of the protein streptavidin [27], which also shows very good correspondence between X-ray and electron crystallographic results, but again the molecules crystallised in different space groups, making a direct numerical comparison of phases difficult.

What reasons can there be that so few, if any, proteins crystals have been studied by both X-rays and electrons? One reason may be that only a few laboratories master both of these two techniques. Another may be that the crystal forms ideally suited for electrons are extremely thin plates of one or just a few unit cells, whereas the crystals that can be studied by single crystal X-ray diffraction must be at least 1000 unit cells thick, corresponding to about 10 μm. The chunky crystals that are optimal for X-rays are not suitable for electron diffraction. Early attempts [28], to fix thick 3D protein crystals for example in tannic acid and then cut them with an ultramicrotome seems not to have been pursued.

Cryo-microscopy is today the main technique for studying proteins at atomic resolution by electrons. This has the advantage over the classical method of negatively stained samples (using metal salts) that not only the envelope but also internal features of the protein are visualized. Perhaps small 3D crystals could be deep frozen in ice and then crushed into smaller pieces, thin enough to be transparent to electrons. Considering the amount of work involved in phasing proteins, for example making one or several isomorphous heavy atom derivatives, it should be worth developing this combined approach of electrons and X-rays.

12. References

1. Vainshtein, B.K. (1964) *Structure Analysis by Electron Diffraction.* Pergamon Press, Oxford.
2. http://www.nobel.se/ Home page of the Nobel Foundation, Sweden
3. Hovmöller, S. (1992) Structure Determination of Inorganic Structures by Electron Microscopy and Crystallographic Image Processing. In: Electron Crystallography, Trans. of the American Crystallographic Association. Ed. D.Dorset. **28**, 95-103.
4. Van Dyck, D. and de Beeck, M. (1992) *Scanning Microscopy Supplement* **6**, 115.
5. BRITE-EURAM Project No. BE-3322 (1996) *Ultramicroscopy* **64** Entire volume.
6. Zou, X.D. (1995) *Electron Crystallography of inorganic structures - Theory and practice*
   Chem. Commun. 5. Stockholm University. Doctoral thesis.
7. Henderson, R. and Unwin, P.N.T. (1975) Three-dimensional model of purple membrane obtained by electron microscopy. *Nature* ***257,*** 28-32.
8. Kühlbrandt, W., Wang , D.N., Fujiyoshi, Y. (1994) Atomic model of plant light-harvesting complex by electron crystallography *Nature* **367**, 614-621.
9. Jap, B.K., Walian, P.J. and Gehring, K. (1991) Structure of porin *Nature* **350**, 167-
10. Boekema, E.J., Van Heel, M.G. and Van Bruggen, E.F.J. (1984) *Biochem. Biophys. Acta* **787**, 19-26.
11. Brink,J., Hovmöller, S, Ragan, C.I., Cleeter, M.W.J., Boekema, E.J and van Bruggen,E.F.J. (1988) The Structure of NADH:Ubiquinone Oxidoreductase from Beef Heart Mitochondria. Crystals Containing an Octameric Arrangement of Iron-Sulphur Protein Fragments. *Eur. J. Biochem.* **166**, 287-294.

12. Leonard, K., Haiker, H. and Weiss, H. (1987) Three-dimensional structure of NADH:Ubiquinone reductase (Complex I) *J.Mol.Biol.* **194**, 277-286.
13. Henderson, R., Capaldi, R.A. and Leigh, J.S. (1977) Arrangement of cytochrome oxidase molecules in two-dimensional vesicle crystals. *J.Mol.Biol.* **112**,631-648.
14. Hovmöller, S. (1990) Structure of cytochrome oxidase. An alternative interpretation. Manuscript submitted to *J. Mol. Biol.* but rejected.
15. Fuller, S.D., Capaldi, R.A. and Henderson, R. (1979) Structure of Cytochrome c Oxidase in Deoxycholate-derived Two-dimensional crystals. *J. Mol. Biol.* **134**, 305-327.
16. Tsukihara, T., Aoyama, H., Yamashita, E., Tomizaki, Yamaguchi, H., Shinzawa-Itoh, K., Nakashima, R., Yaono, R. and Yoshikawa, S. (1995) Structures of metal sites of oxidized bovine heart cytochrome c oxidase at 2.8 Å. *Science* **269**, 1069-1074.
17. Leonard, K., Wingfield, P., Arad, T. and Weiss, H. (1981) Three-dimensional structure of Ubiquinol:cytochrome c reductase from Neurospora mitochondria determined by electron microscopy of membrane crystals. J.*Mol.Biol.* **149**, 259-274.
18. Abrahams, J.P., Leslie, A.G.W., Lutter, R. and Walker, J. (1994) Structure at 2.8 Å resolution of F_1- ATPase from bovine heart mitochondria. *Nature* **370**, 621-628.
19. Van Heel, M. and Frank, J. (1981) Use of multivariate statistics in analysing the images of biological macromolecules. *Ultramicroscopy* **6**, 187-194.
20. Evers, U., Franceschi, F., Boeddeker, N., Yonath, A. (1994) Crystallography of Haliphilic Ribosomes: the Isolation of an Internal Ribonucleoprotein Complex, *Bioph. Chem.*, **50**, 3-16.
21. Vainshtein, B.K. and Pinsker, Z.G. (1950) Determination of the hydrogen positions in the crystal structure of a paraffin. *Dokl. Akad. Nauk SSSR* **72**, 53-56.
22. Dorset, D. (1995) *Structural Electron Crystallography,* Plenum Press, New York.
23. Weirich,T., Ramlau,R. Simon,A., Zou,X.D. and Hovmöller,S. (1996) A crystal structure determined to 0.02 Å accuracy by electron microscopy. *Nature* **382**, 144-146.
24. Unwin, P.N.T. and Henderson, R. (1975) Molecular structure determination by electron microscopy of unstained crystalline specimens. *J. Mol. Biol.* **94**, 425-440.
25. Zou, X.D., Sundberg, M., Larine, M. and Hovmöller, S. (1996) Structure projection retrieval by image processing of HREM images taken under non-optimum defocus conditions. *Ultramicroscopy* **62**, 103-121.
26. Weiss, M.S., Abele, U., Weckesser, J. Welte, W. Schiltz, E. and Schulz, G.E. (1991) Molecular architecture and electrostatic properties of a bacterial porin. *Science* **254**, 1627-1630.
27. Avila-Sakar, A.J. and Chiu W. (1996) Visualization of β-sheets and side-chain clusters in two-dimensional periodic arrays of streptavidin on phospholipid monolayers by electron crystallography, *Biophysical Journal* **70**, 57-68.
28. Akey, C.W., Spitsberg, V. and Edelstein, S.J. (1983) Electron microscopy of beef heart F_1-ATPase crystals. *J. Biol. Chem.* **258**, 3222-3229.

CRYSTAL STRUCTURE DETERMINATION BY CRYSTALLOGRAPHIC IMAGE PROCESSING: I. *HREM images, structure factors and projcctcd potential*

X.D. ZOU
Structural Chemistry
Stockholm University
S-106 91 Stockholm
Sweden

1. Introduction

High resolution electron microscopy (HREM) can be used for crystal structure determination. Both the amplitudes and phases of the crystallographic structure factors of a crystal are present in the HREM images and can be used for determining atomic positions in the crystal. Here a brief description is given of why this is possible and how to obtain structure factor amplitudes and phases from HREM images of weak-phase-objects. We will also demonstrate how a projected potential map is reconstructed from the amplitudes and phases by Fourier synthesis. More detailed descriptions and explanations can be found in the book by Zou [1].

2. Atomic positions, potential distribution and structure factors for electrons

Electrons passing through a crystal feel the potential distribution within the crystal. Each atom produces a peak in potential at the center of the atom and scatters electrons. For a crystal containing N atoms in the unit cell, the electrostatic potential distribution $\varphi(\mathbf{r})$ at $\mathbf{r}$ is

$$\varphi(\mathbf{r}) = \sum_{j=1}^{N} \varphi_j(\mathbf{r} - \mathbf{r}_j) \qquad \text{(in Volt)} \qquad (1)$$

where $\varphi_j(\mathbf{r}\text{-}\mathbf{r}_j)$ is the contribution of an atom j at position $\mathbf{r}_j$ to the potential $\varphi(\mathbf{r})$ at position **r.**

The structure factor of a crystal for electrons is defined as

$$F(\mathbf{u}) = \sum_{j=1}^{N} f_j(\mathbf{u}) \exp 2\pi i(\mathbf{u} \cdot \mathbf{r}_j) \qquad \text{(in Å)} \qquad (2)$$

where $f_j(\mathbf{u})$ is the atomic scattering factor of atom j for electrons and $\mathbf{u} = (h\ k\ l) = h\boldsymbol{a}^* + k\boldsymbol{b}^* + l\boldsymbol{c}^*$ is a position vector in reciprocal space ($\boldsymbol{a}^*$, $\boldsymbol{b}^*$ and $\boldsymbol{c}^*$ are the basic vectors of the lattice in reciprocal space). Normally F(**u**) is complex, i.e. $F(\mathbf{u}) = |F(\mathbf{u})|\ e^{i\phi(\mathbf{u})}$, where $|F(\mathbf{u})|$ and $\phi(\mathbf{u})$ are the amplitude and phase of the structure factor, respectively. For centrosymmetric structures, F(**u**) becomes a real number, i.e. the phase $\phi(\mathbf{u})$ can only be either 0° or 180° if the origin is at a center of symmetry.

D. L. Dorset et al. (eds.), Electron Crystallography, 163–172.

The potential distribution $\varphi(\mathbf{r})$ in a crystal is related to the structure factor $F(\mathbf{u})$ by:

$$\varphi(\mathbf{r}) \equiv FT^{-1}[\Phi(\mathbf{u})] = \sum_{\mathbf{u}} \Phi(\mathbf{u})\exp[-2\pi i(\mathbf{u}\cdot\mathbf{r})] = \frac{h^2}{2\pi me\Omega}\sum_{\mathbf{u}} \mathbf{F}(\mathbf{u})\exp[-2\pi i(\mathbf{u}\cdot\mathbf{r})] \quad (3)$$

where $\Phi(\mathbf{u})$ is the Fourier transform of $\varphi(\mathbf{r})$, h the Planck's constant, Ω the volume of the unit cell and m and e the electron relativistic mass and charge[1]. We can see that the structure factor $F(\mathbf{u})$ and the Fourier transform of the crystal potential $\Phi(\mathbf{u})$ are in fact the same thing: $\Phi(\mathbf{u}) = \frac{h^2}{2\pi me\Omega}F(\mathbf{u})$. They just differ by a constant due to the different units for $F(\mathbf{u})$ (Å) and $\Phi(\mathbf{u})$ (Volt). In the following discussion, we will consider the Fourier transform $\Phi(\mathbf{u})$ as the structure factor.

If $\Phi(\mathbf{u})$ for all reflections **u** are known, the potential distribution $\varphi(\mathbf{r})$ can be calculated from (3). When the potential distribution of a crystal is available, atomic positions can be determined from the positions of the peaks in the potential map. Thus to determine crystal structures is equivalent to determining the structure factor $\Phi(\mathbf{u})$ of the crystal potential, which contains the amplitude and phase.

3. HREM image and potential distribution; Fourier transform of the image and structure factors

The relation between an HREM image and the projected crystal potential is usually quite complicated, especially when the crystal is thick. To obtain an image which can be directly interpreted in terms of projected potential, crystals have to be thin enough to be close to weak-phase-objects and the defocus value for the objective lens should be optimal, i.e. at the Scherzer defocus.

For a weak-phase-object, the Fourier transform of the HREM image $I_{im}(\mathbf{u})$ is related to the structure factor $\Phi(\mathbf{u})$ by [1]:

$$I_{im}(\mathbf{u}) = \delta(\mathbf{u}) + 2\sigma N_z T(\mathbf{u})\Phi(\mathbf{u}) \quad (4)$$

where $\sigma = \frac{\pi}{\lambda U} = \frac{2\pi me\lambda}{h^2}$ (in V^{-1} $Å^{-1}$) is the interaction constant and N_z is the number of periods along the incident beam direction. $T(\mathbf{u})$ is called the contrast transfer function (CTF). The effects of the contrast transfer function will be discussed in the next section.

For an image taken at Scherzer defocus, where $T(\mathbf{u}) \approx -1$ over a large range of resolution, the structure factor $\Phi(\mathbf{u})$ can be obtained from the Fourier transform of the HREM image $I_{im}(\mathbf{u})$:

$$\Phi(\mathbf{u}) \approx -\frac{1}{2\sigma N_z}I_{im}(\mathbf{u}) = \frac{1}{2\sigma N_z}\exp(i\pi)I_{im}(\mathbf{u}) \quad (5)$$

The amplitudes of the crystal structure factors are proportional to the amplitudes of the reflections in the Fourier transform $I_{im}(\mathbf{u})$ of the image, while all the phases are shifted by 180° from those of the Fourier transform $I_{im}(\mathbf{u})$ of the image.

The projected potential can be obtained from the Fourier transform $I_{im}(\mathbf{u})$ of the image using (5)

$$\varphi(\mathbf{r}) \approx \frac{-1}{2\sigma N_z} \sum_{\mathbf{u}} I_{im}(\mathbf{u}) \exp[-2\pi i(\mathbf{u} \cdot \mathbf{r})] = -\frac{I_{im}(\mathbf{r})}{2\sigma N_z} \tag{6}$$

The projected potential is proportional to the negative of the image intensity, i.e. black features in HREM positives (low intensity) correspond to atoms (high potential). The corresponding image is called the structure image.

4. Extracting amplitudes and phases from HREM images of weak-phase-objects

Theoretically, HREM images of a weak-phase-object taken at Scherzer defocus represent directly the projected potential. However, in practice this is not always true. Features in different unit cells are slightly different and symmetry-related features in the same unit cell are not exactly identical as they should be, as seen in Fig. 1. Crystallographic image processing is applied to get the average structure over all the unit cells and further impose the crystallographic symmetry in the projection. These two steps are performed in reciprocal space.

An HREM image (Fig.1a) is first digitised and its Fourier transform (FT) is calculated. The image density is a set of real numbers, while the Fourier transform of the image is a set of complex numbers which can be expressed in an amplitude part and a phase part. The amplitude part of the FT is shown in Fig. 1b.

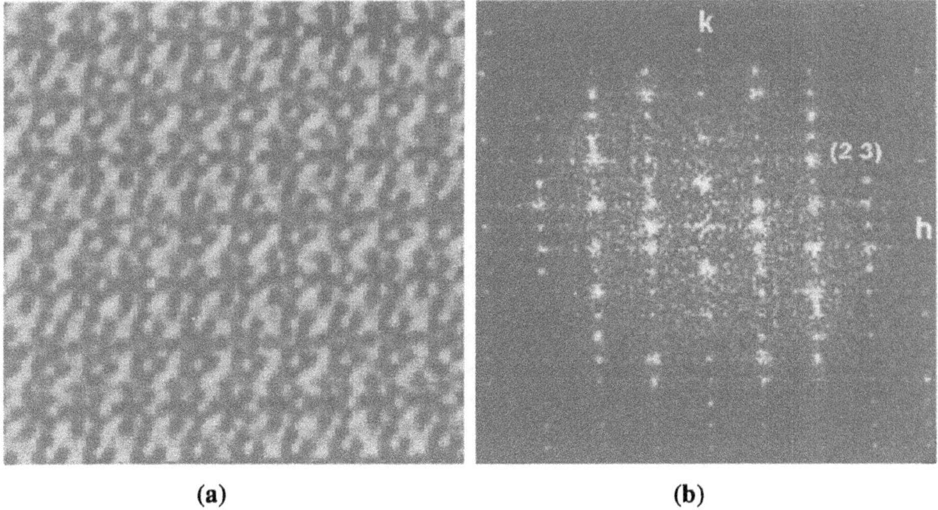

Figure 1. (a) HREM image of $K_{8-x}Nb_{16-x}W_{12+x}O_{80}$ ($x \approx 1$) taken near the Scherzer defocus along the c axis[2]. (b) The Fourier transform $I_{im}(\mathbf{u})$. The image was taken by Margareta Sundberg, Inorganic Chemistry, Stockholm University.

The crystal structure information is concentrated at the diffraction spots in the FT. Fig.2a shows both the amplitude and phase part of the FT in digits around a diffraction spot (2 3) in Fig. 1b. The lattice in the FT is refined using all spots and from the lattice the exact position of each reflection is predicted. Amplitudes and phases for all reflections are extracted from the expected center of the diffraction spots (Fig. 2b).

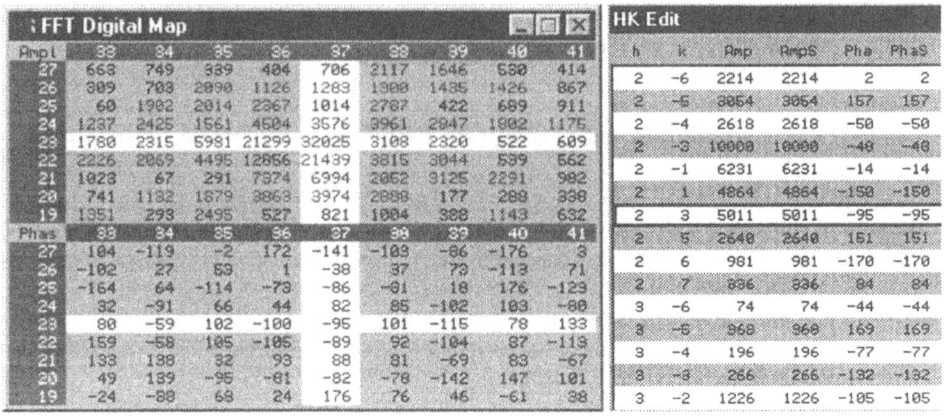

FFT Digital Map

Ampl	33	34	35	36	37	38	39	40	41
27	663	749	339	404	706	2117	1646	530	414
26	309	703	2090	1126	1283	1900	1435	1426	867
25	60	1902	2014	2367	1014	2787	422	689	911
24	1237	2425	1561	4604	3576	3961	2047	1002	1175
23	1780	2315	5981	21299	32025	3108	2320	522	609
22	2226	2069	4495	12856	21439	3815	3044	539	562
21	1023	67	291	7374	6994	2052	3125	2291	982
20	741	1132	1879	3863	3974	2088	177	288	338
19	1351	293	2495	527	821	1004	388	1143	632
Phas	**33**	**34**	**35**	**36**	**37**	**38**	**39**	**40**	**41**
27	104	-119	-2	172	-141	-103	-86	-176	3
26	-102	27	53	1	-38	37	73	-113	71
25	-164	64	-114	-73	-86	-81	18	176	-123
24	32	-91	66	44	82	85	-102	103	-80
23	80	-59	102	-100	-95	101	-115	78	133
22	159	-58	105	-105	-89	92	-104	87	-113
21	133	138	32	93	88	81	-69	83	-67
20	49	139	-95	-81	-82	-78	-142	147	101
19	-24	-88	68	24	176	76	46	-61	38

(a)

HK Edit

h	k	Amp	AmpS	Pha	PhaS
2	-6	2214	2214	2	2
2	-5	3054	3054	157	157
2	-4	2618	2618	-50	-50
2	-3	10000	10000	-40	-40
2	-1	6231	6231	-14	-14
2	1	4864	4864	-150	-150
2	3	5011	5011	-95	-95
2	5	2640	2640	151	151
2	6	981	981	-170	-170
2	7	836	836	84	84
3	-6	74	74	-44	-44
3	-5	960	960	169	169
3	-4	196	196	-77	-77
3	-3	266	266	-132	-132
3	-2	1226	1226	-105	-105

(b)

Figure 2 (a) part of the Fourier transform $I_{im}(\mathbf{u})$ in Fig. 1b in digits around the reflection (2 3). Note that the four pixels with the largest amplitudes have similar phases. The amplitude for reflection (2 3) is extracted by integration of 3x3 pixels around the expected center of reflection (2 3) and then subtracting the averaged background around the spot. The phase is chosen at the pixel nearest to the expected center of the diffraction spot. (b) List of extracted amplitudes and phases. Note that the phase (-95°) for reflection (2 3) is taken from position (37 23) in the FT (Fig. 2a).

The ideal projection of the crystal should have *pmg* symmetry. When the origin is at the intercept of the two-fold axis and the glide plane of the crystal, all phases have to be 0° or 180°. Furthermore, all symmetry-related pairs (h k) and (-h k) should have the same amplitude, while phases are related by $\phi(h\ k) = \phi(-h\ k) + k\cdot180°$[1]. The origin, which was at an arbitrary position in the unit cell, is shifted to the intercept. As a consequence, the phase value of each reflection (h k) is shifted by $360°\cdot(h\cdot x_0 + k\cdot y_0)$, where $(x_0\ y_0)$ is the shift of the origin in fractional coordinates.

At the new origin, phases are still not exactly 0° or 180°, although they are all very close to one of these values. Amplitudes of symmetry-related reflections are not equal, due to crystal tilt and astigmatism. The next step of crystallographic image processing is to impose the crystal symmetry. Phases are changed according to the symmetry and amplitudes of symmetry-related reflections are averaged. Amplitudes and phases before and after imposing the *pmg* symmetry are listed in Table 1. Since the HREM image was taken near the Scherzer focus, the amplitudes and phases after imposing the symmetry should be close to the structure factor amplitudes and phases.

TABLE 1. Amplitudes and phases before (*p1*) and after (*pmg*) imposing the crystal symmetry on the data from the image of $K_{8-x}Nb_{16-x}W_{12+x}O_{80}$ (x≈1) in Figure 1, where a = 8.9 and b = 22.0 Å in the projection along the short c axis (3.9 Å).

Symmetry		*p1*		*pmg*	
h	k	Amp(h k)	ϕ(h k)	Amp_{sym}(h k)	ϕ_{sym}(h k)
0	2	8930	0	8930	0
0	4	2571	-179	2571	180
0	6	1028	5	1028	0
1	0	9848	170	9848	180
1	-1	4353	-5	4825	0
1	1	5297	-179	4825	180
1	-2	3419	-178	2539	180
1	2	1659	177	2539	180
1	-3	2232	8	1663	0
1	3	1093	-166	1663	180
1	-4	112	115	358	180
1	4	604	150	358	180
1	-6	5948	-169	5161	180
1	6	4373	-149	5161	180
2	-1	6205	172	5531	180
2	1	4857	-5	5531	0
2	-3	10000	180	7476	180
2	3	4952	8	7476	0
2	-5	3065	70	2844	0
2	5	2623	-149	2844	180
3	-1	4237	164	3203	180
3	1	2168	-18	3203	0

5. Structure factors to potential projection by Fourier synthesis

A projected potential can be generated from the Fourier transform $I_{im}(hk)$ of an image by an inverse Fourier transformation, as shown in expression (6):

$$\varphi(xy) \approx C\sum_{hk} I_{im}(hk)\exp[-2\pi i(hx+ky)] \quad (7)$$

$I_{im}(hk)$ contains an amplitude and a phase part which can be written as:

$$I_{im}(xy) \approx Amp(hk)\exp i\phi(hk) \quad (8)$$

Thus expression (7) can be rewritten as:

$$\varphi(xy) \approx C\sum_{hk} Amp(hk)\exp i[\phi(hk) - 360°\cdot(hx+ky)] \quad (9)$$

If the summation in (9) is taken over all Friedel pairs (hk) and $(\bar{h}\,\bar{k})$, we get

$$\varphi(xy) \approx C\left\{ Amp(00) + 2\sum_{hk} Amp(hk)\cos\left[\phi(hk) - 360°\cdot(hx + ky)\right]\right\} \tag{10}$$

where Amp(00) is a constant which represents the average projected potential of the crystal. Expression (10) shows that each reflection in the Fourier transform of a HREM image contributes to the projected potential as a cosine wave Amp(hk)·cos[ϕ(hk) - 360·(hx + ky)]. The projected potential φ(xy) can by reconstructed by adding all the cosine waves, i.e. Fourier synthesis.

Figure 3 shows six of the cosine waves which are chosen from the amplitudes and phases listed under the *p1* column in Table 1, directly obtained from the HREM image. The direction and the periodicity of each cosine wave are determined by the indices (h k) and the position of the black/white lines (low/high intensities) in the unit cell (marked in Figure 3) is determined by the phase value ϕ(hk). If phases are near 180°, the contrast at the centre of the unit cell is black (Fig. 3a-c and 3f). When phases are near 0°, the contrast is white (Fig. 3d and 3e).

As mentioned in the previous section, black features in HREM positives correspond to high potential where atoms or atom groups are located. When two cosine waves are added, the contrast at the centre is determined by both the relative amplitudes and the phase values of the two reflections. Assume that the origin (0 0) is at the centre of the marked unit cell. The superposition of the cosine waves from (1 0) and (0 2) indicates that the highest potential is at (0 ¼) (Fig. 3g), (1 1) + (1 -1) at (-¼ ¼) (Fig. 3h) and (1 2) + (1 -2) at (0 0) (Fig. 3i). When (1 0) + (0 2) and (1 1) + (1 -1) are added to each other, the peak with the highest potential moves to (x ¼) (Fig. 3j). Here x is less than -1/8 = [0 + (-¼)]/2 because the amplitudes of (1 0) + (2 0) are larger than those of (1 1) + (1 -1). The amplitude sum of (1 2) and (1 -2) is even smaller so that further addition of (1 2) + (1 -2) only slightly moves the peak position to the left (Fig. 3k).

A step-by-step demonstration of how a projected potential is reconstructed from a HREM image is given in Figure 4. Here amplitudes and phases after symmetrization (the *pmg* columns in Table 1) are used and pairs of symmetry-related reflections are added. The effect of amplitudes on Fourier synthesis is clearly shown in Fig. 4; stronger reflections make more contribution to the final potential map than weaker reflections. When enough strong reflections (9 unique reflections) are included, the map (Fig. 4j) obtained from Fourier synthesis is very close to the final projected potential (Fig. 4l), which is reconstructed from 26 all unique reflections up to 2.77Å resolution.

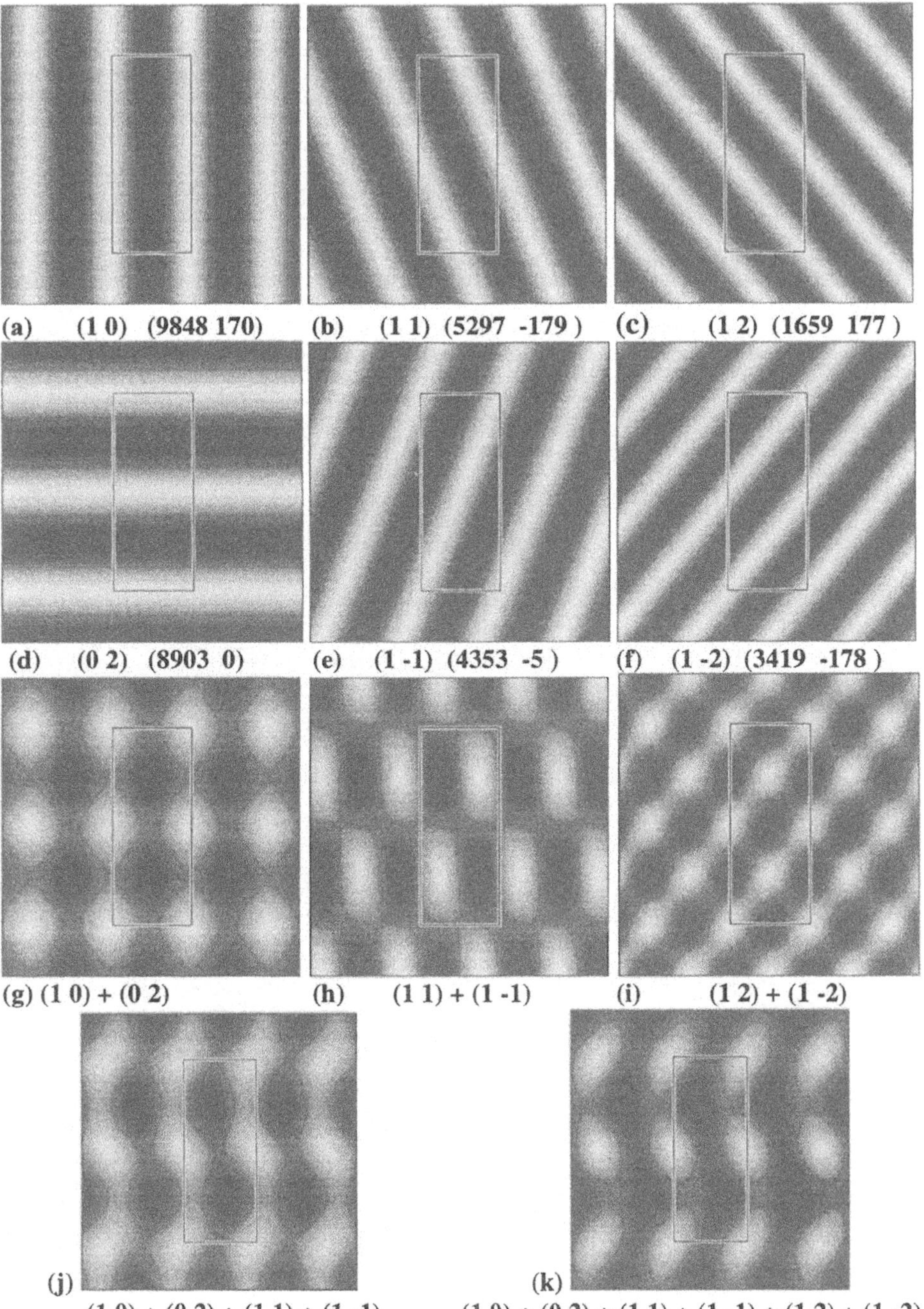

Figure 3. (a) - (f) Cosine waves from six different reflections selected from the *p1*columns in Table 1. The amplitudes and phases are the original ones obtained from the HREM image, as listed under each map. The origin is at the centre of the marked unit cell. (g)-(k) Different steps of the Fourier synthesis.

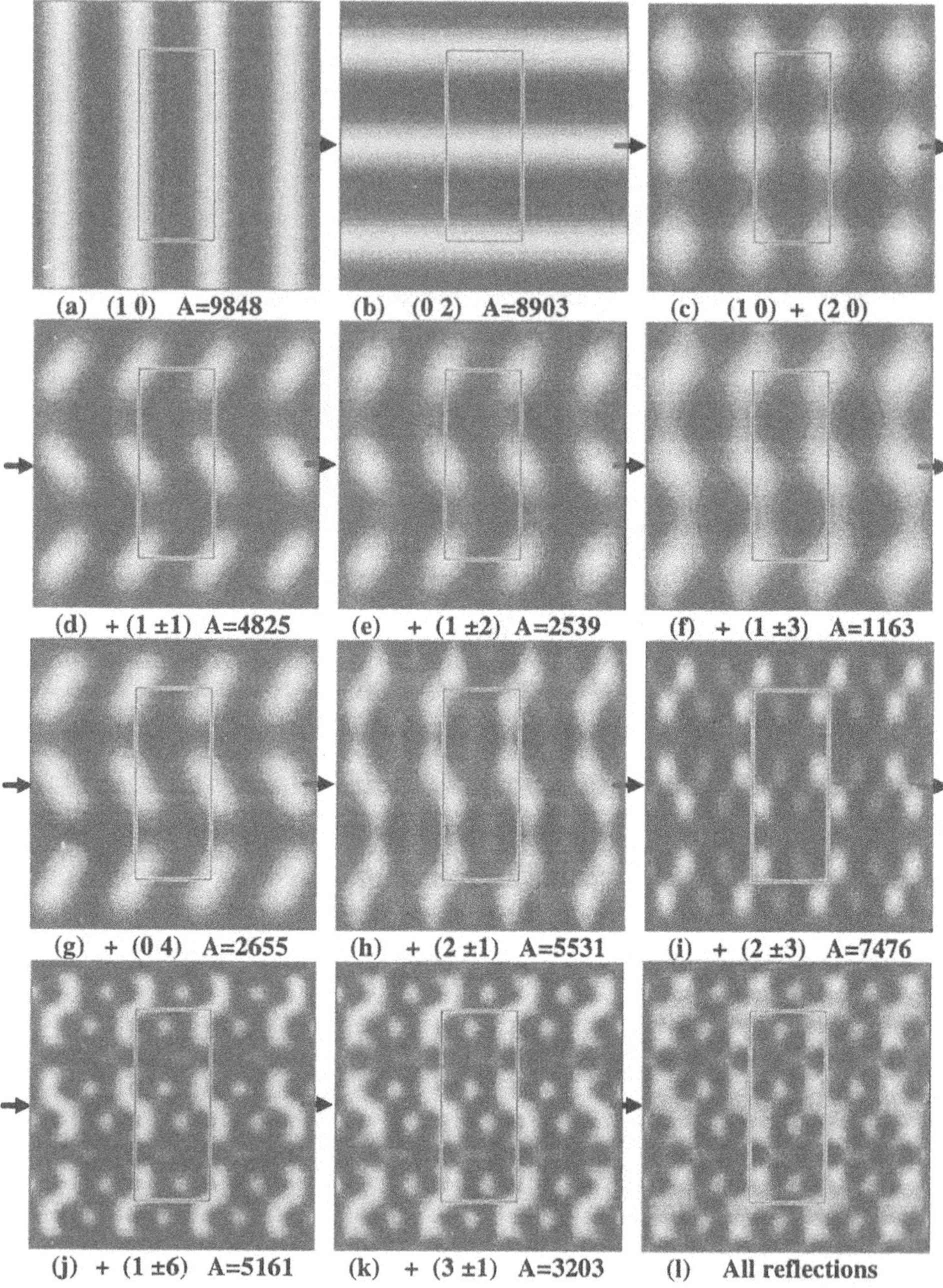

Figure 4 Fourier synthesis step-by-step. A projected potential map calculated from the structure amplitudes and phases obtained from a HREM image. The direction of Fourier synthesis is shown by arrows. From one map to the next one, a pair of symmetry-related reflections is added. The indices and amplitudes (A) are shown under each map. Amplitudes and phases after imposing the symmetry are used. Note that there are significant changes after adding strong pairs while little changes after adding weak reflections.

Amplitudes and phases obtained directly from an HREM image (the *p1* column in Table 1) may not be the same as the crystal structure factors. This can be shown from the reconstructed *p1* map (Fig. 5a). When the symmetry *pmg* is imposed, the reconstructed map (Fig. 5b) is much closer to the projected potential map. This means that amplitudes and phases after imposing the symmetry are closer to the structure factor amplitudes and phases. Atomic positions, which are determined directly from the peaks in the reconstructed map, are within an accuracy of 0.2Å. compared with those determined by single crystal X-ray diffraction [2].

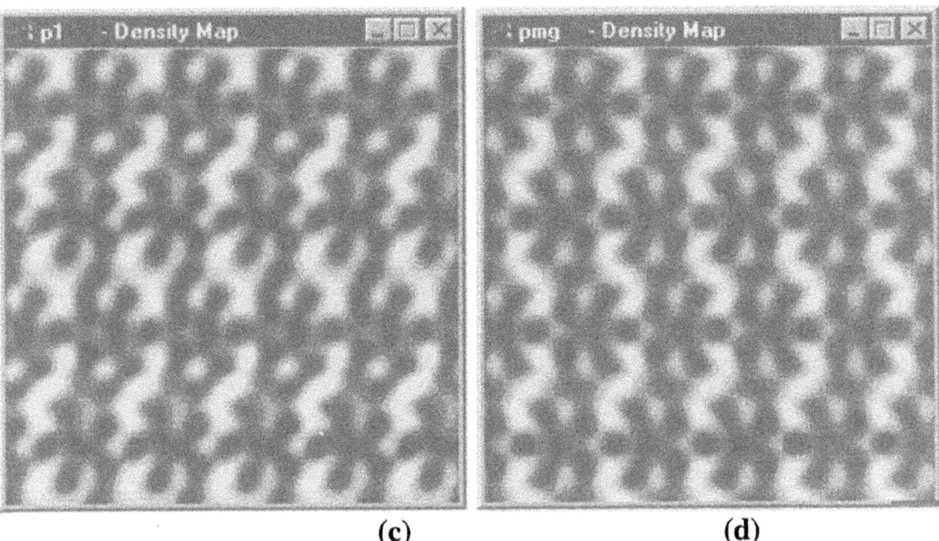

Figure 5 (a) The lattice averaged map with *p1* symmetry. (b) The reconstructed potential map from an HREM image. The crystal symmetry *pmg is imposed on amplitudes and phases.*

It is worth mentioning that phases are more important than amplitudes in determining atomic positions. This is because a phase determines the position of a cosine wave, i.e. the position of the highest potential. To solve a structure, we do not need to have all amplitudes and phases correctly determined. As long as all strong reflections have correct phases, the reconstructed map should be very close to the projected potential map, as shown in Fig. 4.

6. 3D structure reconstruction

We have shown that the crystallographic structure factor amplitudes and phases can be obtained from an HREM image and that the projected potential of the crystal can be reconstructed from them. When atoms do not overlap in the projection and the resolution of a HREM image is high enough, then it is possible to resolve individual atoms from reconstructed potential maps. However, if atoms overlap each other or cannot be resolved from a single projection, we need to construct a 3D potential map to resolve individual atoms.

Images from different projections of the crystal zone axes are needed. Each image is processed individually as described above, whereby the structure factors are determined. The structure factors, initially indexed in 2D, have to be reindexed to their proper 3D indices according to the crystallographic zone axis of the projection. All structure factors obtained from several images are then combined by crystallographic image processing and a 3D inverse Fourier transform is calculated according to expression:

$$\varphi(x\ y\ z) = \frac{\lambda}{\sigma\,\Omega}\sum_{h\,k\,l} F(h\ k\ l)\exp[-2\pi i(hx + ky + lz)] \tag{11}$$

This is the three dimensional potential map of the crystal. If the resolution is about 2 Å or better, atoms are resolved in such a 3D map. This procedure was first used for 2D protein crystals [3] and later for 3D inorganic crystals [4].

Acknowledgement

This work was supported by the Swedish Natural Science Research Council.

References

1. Zou X.D. (1995) Electron Crystallography of inorganic Structures - Theory and Practice, *Chemical Communication* **5**, (Ph.D. thesis).
2. Hovmöller, S., Sjögren, A., Farrants, G., Sundberg, M. and Marinder, B.-O. (1984) Accurate atomic positions from electron microscopy, *Nature (London)* **311**, 238-241.
3. Unwin, P.N.T. and Henderson, R. (1975) Molecular structure determination by electron microscopy of unstained crystalline specimen, *J. Mol. Biol.* **94**, 425-440.
4. Wenk, H.-R., Downing, K.H., Hu, M. and O'Keeefe, M.A. (1992) 3D structure determination from electron-microscope images: electron crystallography of staurolite, *Acta Cryst.* **A48**, 700-716.

CRYSTAL STRUCTURE DETERMINATION BY CRYSTALLOGRAPHIC IMAGE PROCESSING: II. *Compensate for defocus, astigmatism and crystal tilt*

X.D. ZOU
Structural Chemistry
Stockholm University
S-106 91 Stockholm
Sweden

1. Introduction

High resolution electron microscopy (HREM) images are sensitive to focus, astigmatism and crystal orientation. Only images of weak-phase-objects which are taken near the Scherzer focus of a microscope can be directly interpreted in terms of projected potential of the crystal. Furthermore, crystals have to be well aligned. Here we will present how to use crystallographic image processing to compensate for distortions caused by noise, defocus, astigmatism and crystal misalignment. The amplitudes and phases after the CTF and crystal tilt correction can be used for crystal structure determination.

2. Determination of and correction for defocus and astigmatism on HREM images

For a weak-phase-object, the Fourier transform of a HREM image $I_{im}(\mathbf{u})$ is related to the structure factor $\Phi(\mathbf{u})$ by [1]:

$$I_{im}(\mathbf{u}) = \delta(\mathbf{u}) + 2\sigma N_z T(\mathbf{u})\Phi(\mathbf{u}) \tag{1}$$

where $\sigma = \frac{\pi}{\lambda U} = \frac{2\pi me\lambda}{h^2}$ (in V^{-1} Å^{-1}) is the interaction constant and N_z is the number of periods along the incident beam direction. $T(\mathbf{u})$ is called the contrast transfer function (CTF). The Fourier transform $I_{im}(\mathbf{u})$ of a HREM image is proportional to the Fourier transform $\Phi(\mathbf{u})$ of the projected potential multiplied by the contrast transfer function.

The contrast transfer function $T(\mathbf{u})=D(\mathbf{u})\sin\chi(\mathbf{u})$ contains two parts: an envelope part $D(\mathbf{u})$ which dampens the amplitudes of the high resolution components:

$$D(\mathbf{u})=\exp[-\tfrac{1}{2}\pi^2\Delta^2\lambda^2\mathbf{u}^4]\exp[-\pi^2\alpha^2\mathbf{u}^2(\varepsilon+C_s\lambda^2\mathbf{u}^2)^2] \tag{2}$$

and an oscillating part $\sin\chi(\mathbf{u})$ which determines the contrast of the image where

$$\chi(\mathbf{u}) = \pi\varepsilon\lambda\mathbf{u}^2 + \frac{\pi C_s\lambda^3\mathbf{u}^4}{2} \tag{3}$$

ε is the defocus value, C_s the spherical aberration constant, Δ the focus spread and α the electron beam convergence [2]. Two typical contrast transfer functions for an optimum defocus and a non-optimum defocus are shown in Figure 1.

D. L. Dorset et al. (eds.), Electron Crystallography, 173–181.

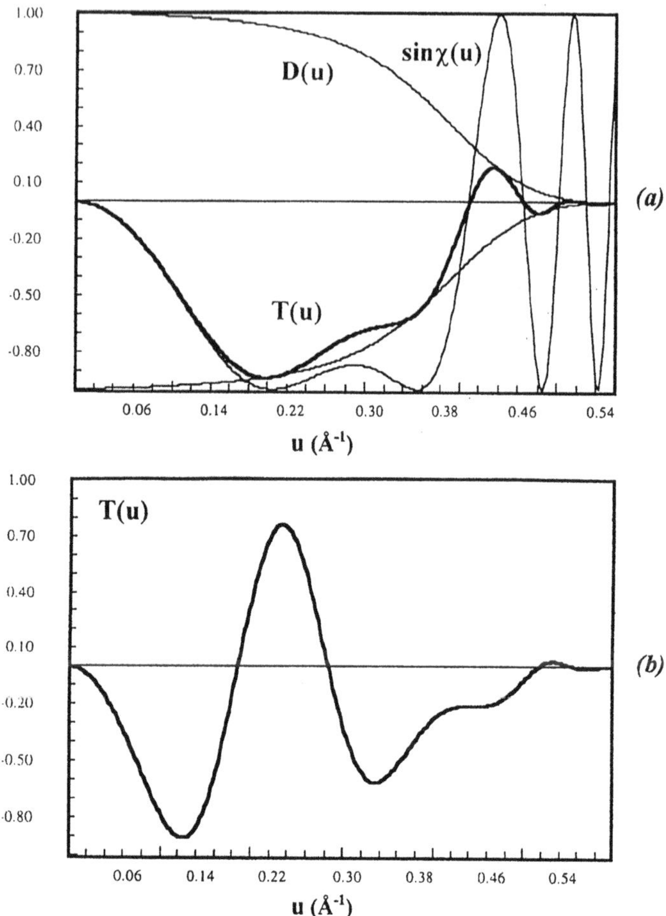

Figure. 1 Contrast transfer function T(**u**) (a) Defocus value $\varepsilon = -633$ Å (Scherzer defocus). The envelope function D(**u**) and sinχ(**u**) are shown in thin lines. (b) Defocus value $\varepsilon = -1300$ Å. The optical parameters are from a JEOL 200CX microscope: U=200 kV, C_s=1.20mm, Δ=00 Å and α= 0.037°.

At a certain defocus, different Fourier components Φ(**u**) of the projected potential, i.e. different structure factors, are transferred by the objective lens into the HREM image in different ways, depending on in which resolution range the Fourier component lies. Fourier components in the range where sinχ(**u**)>0 are transferred to give rise to a contrast proportional to the projected potential. In the range where sinχ(**u**)<0, all Fourier components suffer a phase change of 180°, i.e. the contrast is reversed. An image taken at such a defocus is formed by a complicated mixture of Fourier components; some give correct contrast and some give reversed contrast. Similarly, at different defocus values, the same Fourier component can be transferred in different ways.

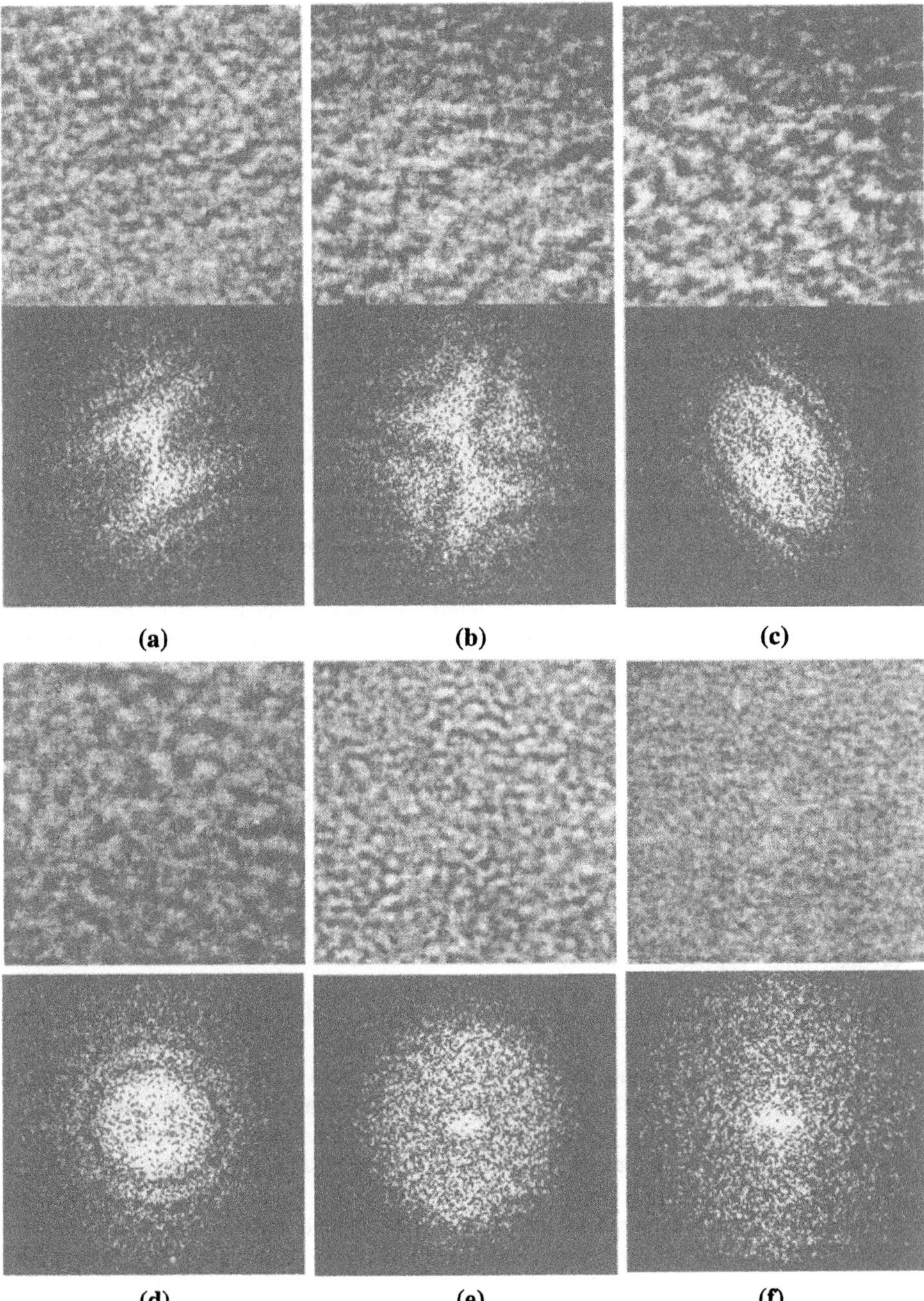

Figure 2 HREM images of amorphous carbon films and the corresponding Fourier transforms (a)-(c) with different astigmatism. (d)-(f) with little astigmatism where (e) is near the Scherzer defocus and (f) near the Gaussian (zero) focus.

Furthermore, the amplitudes of the Fourier components $\Phi(\mathbf{u})$ are attenuated by $|D(\mathbf{u})\sin\chi(\mathbf{u})|$. The Fourier components with $\sin\chi(\mathbf{u}) = \pm 1$ are maximally transferred by the lens, while the Fourier components with $\sin\chi(\mathbf{u}) = 0$ are not transferred at all. In the Fourier transform of a HREM image from amorphous material (Fig. 2), the positions **u** where $\sin\chi(\mathbf{u}) = \pm 1$ will have the highest amplitudes, while the positions **u** where $\sin\chi(\mathbf{u}) = 0$ will have the lowest amplitudes. If the lens is not astigmatic, we will see a set of alternating bright rings corresponding to the $\sin\chi(\mathbf{u}) \approx \pm 1$ and dark rings corresponding to the $\sin\chi(\mathbf{u}) = 0$ in the FT of the image (Fig. 2d). If there is astigmatism, these rings in the FT will become a set of ellipses (Fig. 2c) or in more severe cases hyperbolas (Fig. 2a and 2b).

In general, an image is formed by the combination of Fourier components with both correct and inverted phases with respect to the structure factors. Fortunately, the contrast transfer function $T(\mathbf{u})$ can be determined experimentally from the nodes of the CTF, as seen in the Fourier transform of an HREM image (Fig. 3).

Among those parameters which affect $T(\mathbf{u})$, the defocus value and astigmatism are the main parameters to be determined, since they change from image to image, while the other parameters are either instrument constants, such as C_s, or vary only little from one exposure to the next, as is the case with the focus spread Δ and the electron beam convergence α.

Different methods can be used for determination of defocus [5,6,7]. Here we show how to determine the defocus and astigmatism from the amorphous region of the image (Fig. 3a) [3]. In the Fourier transforms (FT) of an image containing both crystalline and amorphous regions, the sharp diffraction spots come from the periodic features, while the diffuse background corresponds to the FT of the amorphous region (Fig. 3b). The CTF is visible in the diffuse background of the FT.

The positions **u** where $\sin\chi(\mathbf{u}) = 0$ can be read out from the Fourier transform of the images. From (8) the u values at these positions correspond to:

$$\chi(u) = \pi\varepsilon\lambda u^2 + \frac{\pi C_s \lambda^3 u^4}{2} = n\,\pi \qquad (4)$$

where $n = 0, \pm 1, \pm 2, \ldots$ are integers. If both λ and C_s are known, the defocus value ε can be determined [7] by:

$$\varepsilon = \frac{n}{\lambda u^2} - \frac{C_s \lambda^2}{2} u^2 \qquad n = 0, \pm 1, \pm 2 \qquad (5)$$

In expression (5), n is not yet determined. Different n gives different solution for the defocus. n = 0 for the first crossover gives defocus near the optimum defocus; n < 0 gives negative focus value (under-focus) and n > 0 often gives over-focus values. The correct solution of the defocus value is chosen for the best fit of the calculated CTF from the corresponding defocus values and the experimental CTF, which can be estimated from other data, for example the positions of the second and third zero crossovers (if visible in the FT) or by comparing the background noise distribution. A similar scheme was presented by Krivanek [6].

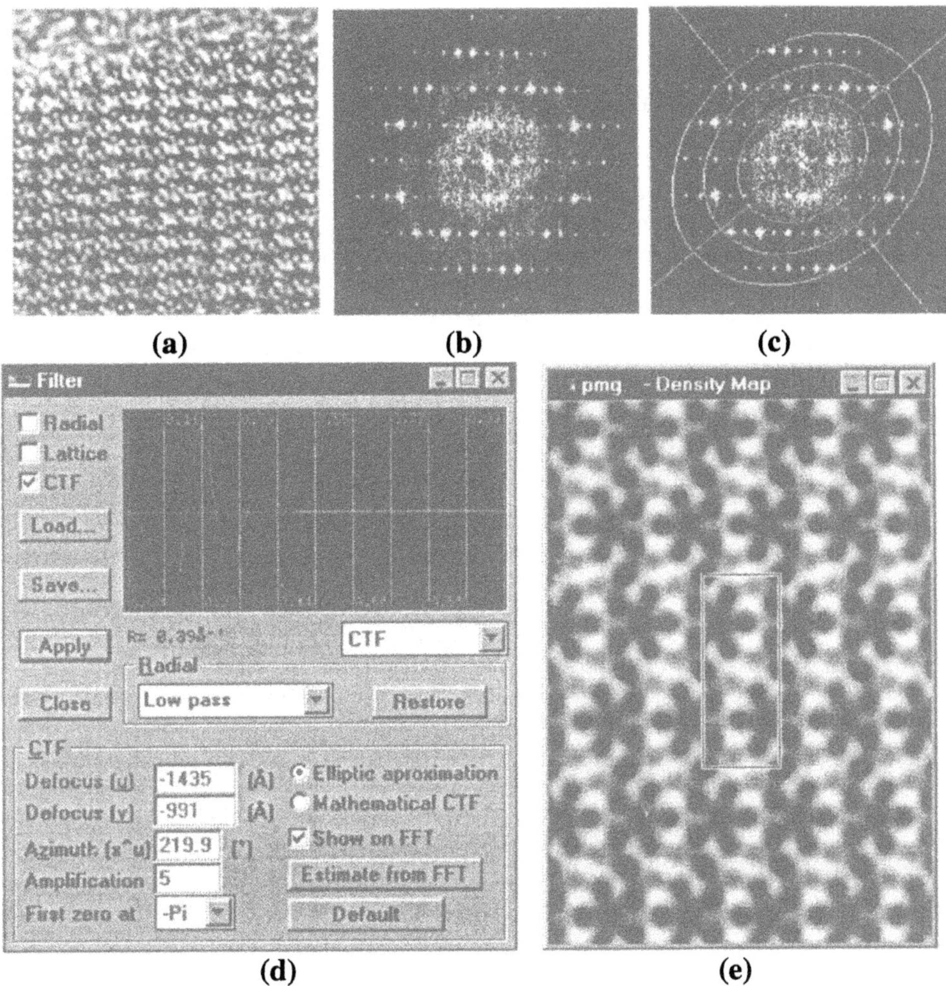

Figure 3. (a) HREM image of $K_{8\text{-}x}Nb_{16\text{-}x}W_{12+x}O_{80}$ ($x \approx 1$) along the c axis taken under a non-Scherzer defocus[3]. (b) The Fourier transform $I_{im}(\mathbf{u})$. An elliptic dark ring can be seen in the background noise of the FT. (c) A set of ellipses are fitted to the dark rings which correspond to the zero crossovers of the CTF. (d) The defocus values along the minor and major axes are estimated from the innermost ellipse to be -1435 and -991Å. The two corresponding CTF curves are shown. (e) The density map after the CTF and astigmatism have been corrected and crystal symmetry imposed. This map is very close to the projected potential map [4]. The HREM image was taken by Margareta Sundberg, Inorganic Chemistry, Stockholm University.

For images with astigmatism, the defocus values are different along different directions in the Fourier transform. The defocus values $\varepsilon_{\mathbf{u}}$ and $\varepsilon_{\mathbf{v}}$ along the major and minor axes of the ellipse (or/and hyperbola) are first calculated according to (5) and the defocus value along each direction in the FT is calculated. The defocus $\varepsilon(\theta)$ along the direction θ degrees away from the x-axis in the FT is

$$\varepsilon(\theta) = \varepsilon_{u}\cos^2(\theta-\theta_0) + \varepsilon_{v}\sin^2(\theta-\theta_0) \tag{6}$$

where θ_0 is the azimuth. When the defocus is estimated, the corresponding contrast transfer function $T(\mathbf{u})=D(\mathbf{u})\sin\chi(\mathbf{u})$ along this direction can be calculated according to (2) and (3).

The contrast transfer function can be compensated for in two ways: a) Using elliptical approximation: a set of ellipses is placed at the crossovers of the CTF in the Fourier transform and the phases for those pixels which lie between the first and the second, the third and the fourth ellipses and so on (which correspond to $\sin\chi(\mathbf{u})>0$ for those image taken at underfocus) are shifted by 180°. All other phases are not changed and no amplitudes are corrected in this case. b) Applying the mathematical CTF calculated from the estimated defocuses. The Fourier transform $\Phi(\mathbf{u})$ of the projected potential is calculated from the Fourier transform $I_{im}(\mathbf{u})$ of the image for all $\mathbf{u}$ except those with $\sin\chi(\mathbf{u}) \approx 0$:

$$\Phi(\mathbf{u}) = \frac{1}{2\sigma N_z} \cdot \frac{I_{im}(\mathbf{u})}{T(\mathbf{u})} \tag{7}$$

Then the projected potential of the crystal can be calculated by inverse Fourier transformation of $\Phi(\mathbf{u})$:

$$\varphi(\mathbf{r}) = \frac{1}{2\sigma N_z} \sum_{\mathbf{u}} \left\{ \frac{I_{im}(\mathbf{u})}{T(\mathbf{u})} \exp[-2\pi i(\mathbf{u} \cdot \mathbf{r})] \right\} \tag{8}$$

In most cases it is possible to retrieve the projected potential map from a single image taken under non-optimum conditions, as shown by Zou et al. [3]. However, some Fourier components which lie within the range of $\mathbf{u}$ where $\sin\chi(\mathbf{u}) \approx 0$ are badly (or not at all) transferred by the objective lens to the image. These lost Fourier components can be obtained from other images taken from the same crystal, but with different defocus conditions, so that the zero crossovers of the CTF are at different positions. By combining components from different images, we can obtain an even more accurate potential projection [3].

3. Effects of crystal tilt and compensation for crystal tilt

Crystal tilt is very common in HREM images. In the microscope, crystal alignment is usually judged from the electron diffraction pattern. Since electron diffraction comes from an area which is typically much (more than 100 times) larger than the area selected for image processing and there is always a risk that the crystal may be bent, the thin crystal area selected for image processing may be more or less tilted.

When the crystal is tilted relative to the incident beam, atoms from different unit cells no longer project exactly on top of each other. The projection of a column of atoms becomes a line rather than a point. The projected potential is smeared out in the direction of that line, giving a lower and more smooth potential distribution than that of a perfectly aligned crystal (Fig. 4). This is equivalent to losing the fine details in the direction perpendicular to the tilt axis. As a consequence of this, in most cases the symmetry of the crystal is lost in the images (Fig. 4c); the amplitudes of reflections away from the tilt axis were attenuated (Fig.5c). Even for the smallest tilts and thinnest crystals, the effect on amplitudes is significant. Pairs of symmetry-related reflections no longer have the same amplitudes if one of the reflections is close to the tilt axis and the other further away. One positive effect of a slight crystal tilt is that the weak-phase-object approximation will be valid for even thicker crystals [8].

The effects of crystal tilt on phases is quite different. The phases are practically unaffected for small tilts and thin crystals [1, 9].

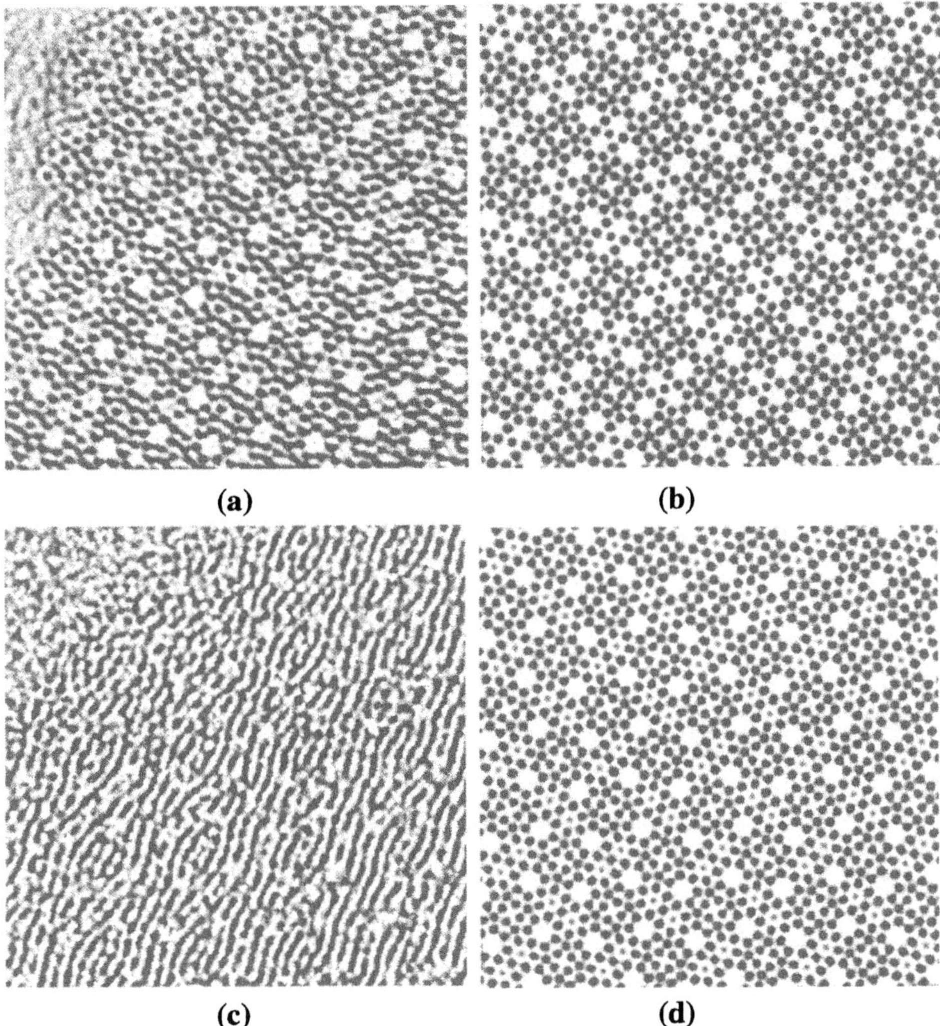

Figure 4. HREM images of $K_2Nb_{14}O_{36}$ along the c-axis [10] from (a) well-aligned crystal and (c) crystal which is tilted 5°. (b) The reconstructed projected potential map from (a). (d) The reconstructed projected potential map from (c).

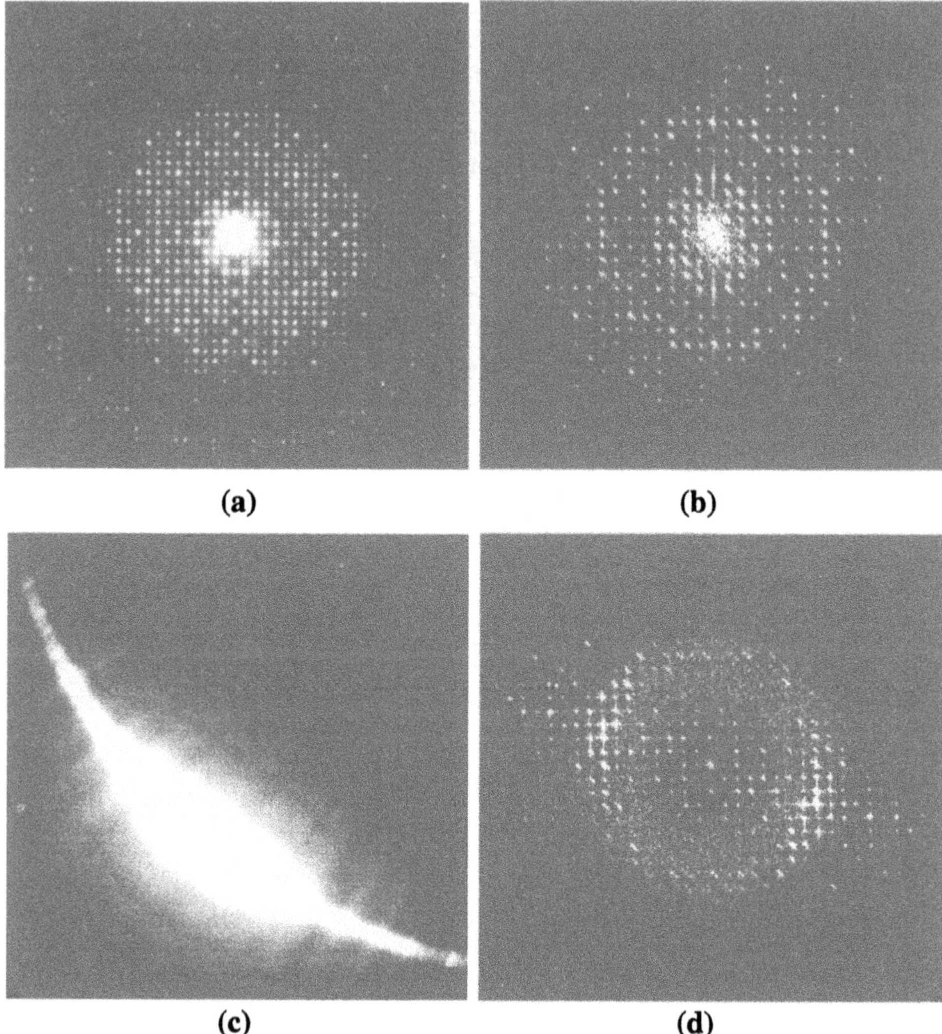

Figure 5. Electron diffraction pattern of $K_2Nb_{14}O_{36}$ along the c-axis from (a) well-aligned crystal and (c) a crystal which is tilted 5°. (b) The Fourier transforms of image (a) in Fig. 6. (d) The Fourier transform of image (c) in Fig. 6.

The effect of crystal tilt depends on the crystal thickness; the thicker the crystal, the more rapidly $I_{im}(\mathbf{u})$ is attenuated. The overall effect of crystal tilt on the image is determined by the product of the crystal thickness t and the tilt angle γ, $t\sin\gamma$ [9].

Furthermore, the effect of crystal tilt on a specific reflection depends on the distance of that reflection to the tilt axis. If the reflection lies on the tilt axis, it will not be affected by crystal tilt. The further away the reflection is from the tilt axis, the more attenuated is $I_{im}(\mathbf{u})$.

One common effect of crystal tilt is that the crystal symmetry no longer is evident in the image. The amplitudes of two symmetry-related reflections are no longer identical (Fig. 5). However, as long as the product $t\sin\gamma$ is small, the phases are unchanged. Both phase relations and phase restrictions are still valid. It is still possible to determine the symmetry from the image of a tilted crystal, using the phases.

For most thin crystals, the distortion of the image due to crystal tilt can be compensated by imposing the crystal symmetry on the image (Fig. 4). This method is especially powerful for crystals with high symmetries (3-, 4- and 6-fold). Also crystals with *mm*-symmetry of the diffraction pattern (i.e. all other symmetries except *p1* and *p2*) can be corrected except for cases where the tilt axis coincides with one of the crystal axes. This is because we impose *mm*-symmetry, which is normally lost by tilting, but it is not lost if the tilting happened to be along a crystal axis. However, the amplitudes of reflections far away from the tilt axis are attenuated by the tilt, although the *mm*-symmetry is not lost. In these cases, if crystal tilt is so small that the phases are not reversed, it is possible to compensate for the tilt by replacing the amplitudes from the image by those from the corresponding ED pattern of a perfectly aligned crystal.

Acknowledgement

This work was supported by the Swedish Natural Science Research Council.

References

1. Zou X.D. (1995) Electron Crystallography of inorganic Structures - Theory and Practice, *Chemical Communication* **5**, (Ph.D. thesis).
2. O'Keefe, M.A. (1992) "Resolution" in high-resolution electron microscopy, *Ultramicroscopy* **47**, 282-297.
3. Zou X.D., Sundberg M., Larine M. and Hovmöller S. (1996) Structure projection retrieval by image processing of HREM images taken under non-optimal defocus conditions, *Ultramicroscopy* **62**, 103-121
4. Hovmöller, S., Sjögren, A., Farrants, G., Sundberg, M. and Marinder, B.-O. (1984) Accurate atomic positions from electron microscopy, *Nature (London)* **311**, 238-241.
5. Han, F.S., Fan, H.F. and Li, F.H. (1986) Image processing in high-resolution electron microscopy using the direct method. II. Image deconvolution, *Acta Cryst.* **A42**, 353-356.
6. Hu, J.J. and Li, H.F. (1991) Maximum entropy image deconvolution in high resolution electron microscopy, *Ultramicroscopy* **35**, 339-350.
7. Krivanek, O.L. (1976) A method for determining the coefficient of spherical aberration from a single electron micrograph, *Optik* **45**, 96-101.
8 O'Keefe, M.A. and Radmilovic, V. (1994) Specimen thickness is wrong in simulated HRTEM images, *Proceedings of the 13th ICEM (Paris)* Vol. **2B**, 361-262.
9. Zou X.D., Ferrow E.A. and Hovmöller S. (1995), Correcting for crystal tilt in HRTEM images of minerals: the case of orthopyroxene, *Physics and Chemistry of Minerals* **22**, 517-523.
10 Hu, J.J., Li, F.H. and Fan, H.F. (1992) Crystal structure determination of $K_2O{\cdot}7Nb_2O_5$ by combining high-resolution electron microscopy and electron diffraction, *Ultramicroscopy* **41**, 387-397.

AN INTRODUCTION TO MAXIMUM ENTROPY IN ACTION

C.J.Gilmore
Department of Chemistry
University of Glasgow
Glasgow G12 8QQ, Scotland, UK.

1. Introduction

1.1 SOME NECESSARY THEORY AND NOTATION

Let us start with a set of unitary structure factors $|U_{\mathbf{h}}|^{obs}$, which are derived from the intensity data by standard normalisation procedures. (For N point atoms of unit weight in space group P1 $|U_{\mathbf{h}}|^{obs} = 1/\sqrt{N}|E_{\mathbf{h}}|^{obs}$) The notation U will be used to define an arbitrary vector of nU's, $\mathbf{U} = (U_{\mathbf{h}1}, U_{\mathbf{h}2}, U_{\mathbf{h}3}, \ldots\ldots U_{\mathbf{h}n})$. We also have a set of phase angles φ_{h} most of which are either unknown (the *ab initio* case) or only approximately determined. For *ab initio* structure solution, because of rules of origin and enantiomorph definition, some phases can usually, but not always (it depends on the space group) be assigned subject to certain well known rules, whereas in the case of macromolecular crystallography, there will be a large set of such phases, albeit rather inaccurately known at times, derived from isomorphous replacement methods. In electron diffraction, the Fourier transform of the corresponding image may give useful phase information.

Those reflections which are phased comprise the basis set {H}; the disjoint set, the non-basis set, of unphased amplitudes is {K}; the phase problem is one of phasing $|U_{\mathbf{h}\in K}|^{obs}$ from $U_{\mathbf{h}\in H}$. In electron diffraction the unmeasured set {U} may also be important. This is shown diagramatically in Figure 1.

At first sight, solving a crystal structure seems straightforward: a set of reflections is chosen to form the basis set. These reflections are used as constraints in an entropy maximisation to generate a maximum entropy map $q^{ME}(x)$. Because it is a maximum entropy map $q^{ME}(\underline{x})$ satisfies the following conditions:

(1) It is optimally unbiased.

(2) Its Fourier transform reproduces the constraints to within experimental error.

D. L. Dorset et al. (eds.), Electron Crystallography, 183–192.

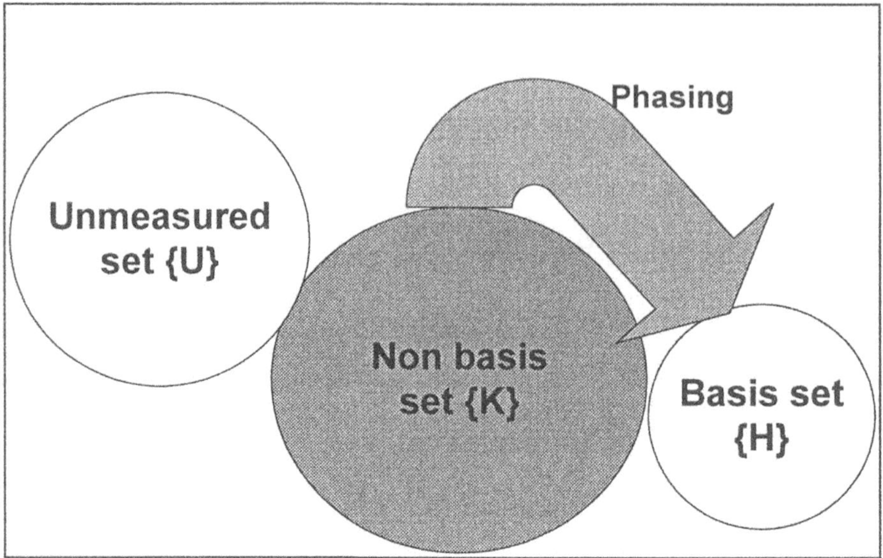

Figure 1. Partitioning the intensity data.

(3) The Fourier transform of $q^{ME}(\underline{x})$ generates estimates of amplitudes and phases for reflections in {K}. This process is called *extrapolation.* This is shown diagramatically in Figure 2 below.

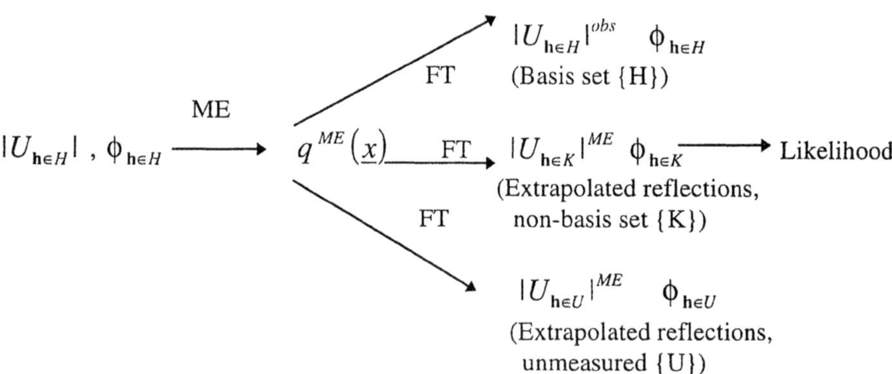

Figure 2. Phase extrapolation

At this stage, for a small basis set, the extrapolation is rather weak, and the structure is certainly not visible in any map one cares to generate. So what can be done? The most obvious answer is to examine the most strongly extrapolated reflections, and to use the phase information from them coupled with the observed U magnitudes, and to pass these into the basis set {H}. This usually leads to disaster: the ME solution gets trapped in a local optimum point in phase space, whole subsets of reflections become wrongly phased, and the structure remains unsolved. This is a manifestation of the *branching problem*: phases are being selected without exploring the relevant U-space in sufficient detail, so that what appears to be an unambiguous choice of new phase is, in reality, no such thing. So what can be done? Extra unphased reflections must be added to the starting set with permuted phases giving rise to a multisolution environment just as in conventional direct methods, and these reflections should be those which are very weakly extrapolated, so that their inclusion in a new basis set offers maximum surprise to the calculations. How then do you select the correct permutation set? Can you choose those which give maps of maximum entropy? The answer to this is *'no'*, and this is discussed in a later chapter, but the Bricogne formalism [1-7] has a way forward.

2. The Bricogne Formalism

Bricogne has outlined a methodology of solving crystal structures based the ME (or saddlepoint) method coupled with likelihood evaluation, that directly addresses the branching problem [1-6].

What have we done when we have defined a basis set and performed the entropy maximisation? We have explored *P(U)* around the basis set phase sub-region in ***U*** space. Why do we need maximum entropy? We only require *P(U)* to reproduce those U-magnitudes which are members of the basis set. It makes no direct reference to non-basis set reflections, and so it is highly indeterminate. The ME method is an obvious way of overcoming this: for such a phase choice maximises the entropy of the distribution subject to the constraint that it must reproduce the basis set amplitudes and phases to generate a non-uniform distribution of atoms $q^{ME}(\underline{x})$. *P(U)* is thus multimodal, and this is the source of the branching problem.

There is a remarkable, and unexpected benefit of using the ME method here: the approximations to *P(U)* are now accurate for large U magnitudes, and we have re-centred the usual Edgeworth series expansion for *P(U)* away from $\boldsymbol{U} = 0$. In real space this is equivalent to leaving behind the Wilson uniform distribution of atoms to a non-uniform one - this is one of the long-term goals of direct methods.

So far all ME phasing methods perform in this way whether we realise it or not, but what can we do next? How do we stop ourselves locking into an incorrect solution? Let us explore new *sub-regions P(U)* where we test various new phase combinations to overcome the multimodality or branching problem. In practice we take the current basis set and add to it a few reflections with large associated U magnitudes which optimally enlarge the second neighbourhood of the basis set. The second neighbourhood is defined by reflections $\mathbf{h}_1 \pm {}^t\underline{\mathbf{R}}_g.\mathbf{h}_2$ for $\mathbf{h}_1,\mathbf{h}_2 \in \mathrm{H}$, where ${}^t\underline{\mathbf{R}}_g$ is the transpose of a rotation matrix obtained from the crystal space group. For acentric reflections phase choices can be $\pm\pi/4$, $\pm 3\pi/4$ i.e. quadrant permutation; for centrics both possible values are used e.g. 0, π or $\pm\pi/2$. We now have several possible basis sets and we now explore *P(U)* around each of these phase sub-regions by performing a constrained entropy maximisation for each phase permutation, and so derive associated phase and amplitude extrapolation. However, we have made no judgement as to which sub-region is the more likely to be the correct one. In the Bricogne formalism, this decision is made using likelihood. For every sub-region, *P(U)* is a multivariate Gaussian centred around the vector $\mathbf{U}^{ME}$ corresponding to $q^{ME}(\mathbf{x})$. It is possible to express this in an analytical form. Several approximations exist, but we will concentrate here on the simplest: the diagonal form in which the extrapolates in ***U*** are decoupled and treated as independent i.e. the covariances between them are zero. This gives rise to readily tractable likelihood functions.

For each acentric extrapolated, non-basis set reflection **k** the likelihood measure, in its diagonal approximation, can be written [1,7,8]

$$\Lambda_{\mathbf{k}} = \frac{|U_{\mathbf{k}}|^{obs}}{\varepsilon_{\mathbf{k}}\Sigma+\sigma_{\mathbf{k}}^{2}} \exp\left\{-\frac{1}{2}\frac{\left(|U_{\mathbf{k}}|^{obs}\right)^{2}+|U_{\mathbf{k}}^{ME}|^{2}}{\varepsilon_{\mathbf{k}}\Sigma+\sigma_{\mathbf{k}}^{2}}\right\} I_{0}\left(\frac{|U_{\mathbf{k}}|^{obs}|U_{\mathbf{k}}^{ME}|}{\varepsilon_{\mathbf{k}}\Sigma+\sigma_{\mathbf{k}}^{2}}\right) \quad (1)$$

where $\varepsilon_{\mathbf{k}}$ is the statistical weight of reflection **k**, $\sigma_{\mathbf{k}}^{2}$ the variance of $|U_{\mathbf{k}}|^{obs}$ and Σ a refinable measure of unit cell contents $\Sigma \approx 1/(2N)$ for N point atoms in the unit cell. The distribution (1) is a Rice distribution comprising a Gaussian (the exponential term) with an offset represented by the Bessel function term (I_0). Note also that this expression is a measure of agreement between $|U_{\mathbf{k}}|^{obs}$ and $|U_{\mathbf{k}}^{ME}|$, indeed it has a maximum where $|U_{\mathbf{k}}|^{obs} = |U_{\mathbf{k}}^{ME}|$. For the centric case the Bessel function is replaced by a cosh term:

$$\Lambda_{\mathbf{k}} = \frac{2|U_{\mathbf{k}}|^{obs}}{\pi\left(2\varepsilon_{\mathbf{k}}\Sigma+\sigma_{\mathbf{k}}^{2}\right)} \exp\left\{-\frac{1}{2}\frac{\left(|U_{\mathbf{k}}|^{obs}\right)^{2}+|U_{\mathbf{k}}^{ME}|^{2}}{2\varepsilon_{\mathbf{k}}\Sigma+\sigma_{\mathbf{k}}^{2}}\right\} \cosh\left(\frac{|U_{\mathbf{k}}|^{obs}|U_{\mathbf{k}}^{ME}|}{2\varepsilon_{\mathbf{k}}\Sigma+\sigma_{\mathbf{k}}^{2}}\right) \quad (2)$$

In the spirit of traditional likelihood analysis, we define a corresponding null hypothesis for the situation of null extrapolation, $|U_{\mathbf{k}}^{ME}| = 0$, which gives the Gaussian distribution of Wilson statistics. For acentric reflections:

$$\Lambda_{\mathbf{k}}^{0} = \frac{|U_{\mathbf{k}}|^{obs}}{\varepsilon_{\mathbf{k}}\Sigma+\sigma_{\mathbf{k}}^{2}} \exp\left\{-\frac{1}{2}\frac{\left(|U_{\mathbf{k}}|^{obs}\right)^{2}}{\varepsilon_{\mathbf{k}}\Sigma+\sigma_{\mathbf{k}}^{2}}\right\} \quad (3)$$

The extension to centric reflections is obvious. Define:

$$L_{\mathbf{k}} = \log\frac{\Lambda_{\mathbf{k}}}{\Lambda_{\mathbf{k}}^{0}} \quad (4)$$

Then the global log-likelihood gain (LLG) is:

$$LLG = \sum_{\mathbf{k}} L_{\mathbf{k}} \tag{5}$$

The LLG will be largest when the phase assumptions for the basis set lead to predictions of deviations from Wilson statistics for the unphased reflections, and in this context it is used as a powerful figure of merit. However, rather than just choose those phase sets with high associated LLG, which is a somewhat subjective process, the Student t-test is used [9]. The LLGs are analysed for phase indications using the t-test. The simplest example involves the detection of the main effect associated with the sign of a single centric phase. The LLG average, μ^+ , and its associated variance V^+ is computed for those sets in which the sign of this permuted phase under test is +. The calculation is then repeated for those sets in which the same sign is - to give the corresponding μ^-, and variance V^-. The t-statistic is then:

$$t = \frac{\left|\mu^+ - \mu^-\right|}{\sqrt{V^+ + V^-}} \tag{6}$$

The use of the t-test enables a sign choice to be derived with an associated significance level. This calculation is repeated for all the single phase indications, and is then extended to combinations of two and three phases. An extension to acentric phases is straightforward by employing two signs to define the phase quadrant both in permutation and in the subsequent analysis. In general, only relationships with associated significance levels <2% are used, but this is sometimes relaxed. This sort of calculation is repeated for all the single phase indications, and is then extended to combinations of two and three phases. Each of the *m* phase relationships, *i*, so generated is given an associated weight w_i

$$w_i = \left(1 - \frac{I_1(s_i)}{I_0(s_i)}\right) \tag{7}$$

where I_1 and I_0 are the appropriate Bessel functions and s_i is the significance level of the *i-th* relationship from the t-test. This weighting function reflects the need for a scheme in which the absolute values of the significance levels are not given undue emphasis since they are themselves subject to errors arising from the nature of the likelihood function used and the lack of error estimates for the LLGs themselves. Each node n is now given a score, s_n :

$$s_n = LLG_n \sum_{j=1}^{m} w_j \tag{8}$$

where the summation spans only those phase relationships where there is agreement between the basis set phases and the t-test derived phase relationships [10]. The scores are sorted and only the top 8-16 nodes are kept; the rest are discarded. New reflections are then permuted and a corresponding set of ME solutions is generated. In this way we build a *phasing tree* in which each phase choice is represented as a *node*, and has a score, or figure of merit, based on its log likelihood gain (LLG). The root node of the tree is defined by the origin defining reflections. The first set of phase permutations defines the second level. Those which do not pass the analysis of likelihood are discarded, then further phase permutations are used to generate the third level, and this continues until a recognisable structure or structural fragment appears. Figure 3 shows a simple outline of a three-level phasing tree.

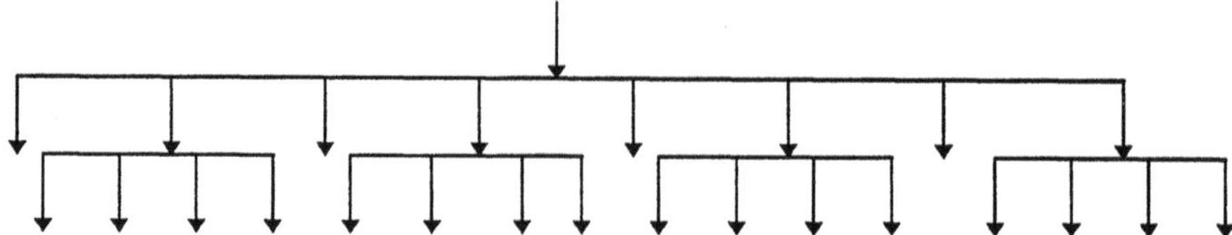

Figure 2. A Three level, 25 node phasing tree.

There is an interesting example of the ME method in action using idealised, one-dimensional data by Carter [11] Note that likelihood is not used in this approach in the Bayesian way of multiplying the prior, instead it becomes a very powerful figure of merit which controls the branching problem by selecting the correct nodes. However, there is a Bayesian connection in the may in which the prior, $q^{ME}(x)$, is continually updated from level to level in the phasing tree by adding new phase information. We shall see later that the power of likelihood is such that it can be used to make other decisions concerning envelopes for macromolecules, parameters for non-crystallographic symmetry and to measure the effective number of atoms in the unit cell at low resolution.

Now $q^{ME}(\underline{x})$ is a probability distribution, and not a map in the traditional sense (although its peaks do correspond to atom positions), and thus it needs conversion to a more conventional one. The trial electron density maps generated in this approach are called *centroid maps*, and can be visualised as Sim filtered maximum entropy maps. For **k** acentric the Fourier coefficients are:

$$|U_{\mathbf{k}}|^{obs}\left[I_1(X_{\mathbf{k}})/I_0(X_{\mathbf{k}})\right]\exp\left(i\varphi_{\mathbf{k}}^{ME}\right) \tag{9}$$

where:

$$X_{\mathbf{k}} = (2N/\varepsilon_{\mathbf{k}})|U_{\mathbf{k}}|^{obs}\left|U_{\mathbf{k}}^{ME}\right| \tag{10}$$

For **k** centric, these coefficients become:

$$|U_{\mathbf{k}}|^{obs}\tanh(X_{\mathbf{k}})\exp\left(i\varphi_{\mathbf{k}}^{ME}\right) \tag{11}$$

with:

$$X_{\mathbf{k}} = (N/\varepsilon_{\mathbf{k}})|U_{\mathbf{k}}|^{obs}\left|U_{\mathbf{k}}^{ME}\right| \tag{12}$$

The practical differences between centroid and maximum entropy maps can be seen with reference to membrane data in Gilmore, Shankland & Fryer [12].

One final practical point needs to made here concerning the tightness with which one fits the constraints in the entropy maximisation which can be very important in a phasing environment. If the fit between $|U_{\mathbf{h}}|^{obs}$ and $\left|U_{\mathbf{h}}^{ME}\right|$ is very slack the ME extrapolation is weaker than it can be, and the phasing power and the discriminating power of likelihood is correspondingly reduced. If however there is overfitting, spurious details (often looking like small stones) appear in $q^{ME}(x)$ which give false phase indications. The latter situation can often be detected by the use of likelihood: if the LLG is monitored through the iterative cycles of entropy maximisation, a maximum is often reached and this can be used as a place to stop. Alternatively, the χ^2 statistic can be used with a default choice of unity as a place to stop.

$$\chi^2 = \frac{1}{n}\sum_{h \in H}\frac{1}{\sigma_h^2}(|U_{\mathbf{h}}|^{obs} - |U_{\mathbf{h}}^{ME}|)^2 \tag{13}$$

where n is the total number of degrees of freedom (2 for each acentric reflection in the basis set, and one for each centric). In practice, for small basis sets either method works well, but for large basis sets maximum LLG is preferred.

The MICE computer program is a practical implementation of a part of the ME formalism [7,8] as is BUSTER [6]. Surveys of the method and its practical applications can be found in [13,14].

Other developments of great importance which will be described elsewhere include:

- The use of error correcting codes to generate amazingly efficient phase permutations [6].
- The use of envelopes [15] including the resolution of envelope ambiguities using likelihood [16].
- Non-crystallographic symmetry [17].

The ME method can incorporate these concepts in a natural way.

3. Where has the method been used in structure solution?

The method has been used successfully to solve:

- Small molecule structures (but it cannot compete with modern direct methods) [8]
- Powder data sets [18-20].
- Membrane proteins using two-dimensional electron diffraction data [12,21].
- Problems with isomorphism and envelope definition, and as an alternative to regular envelope flattening in protein crystallography [15,16].
- Heavy atom positions in protein crystallography [17,22].
- Surface structures [10].
- Electron diffraction data sets on a number of small molecules [14,23,24].

References

1. Bricogne, G. (1984) Maximum entropy and the foundations of direct methods, *Acta Cryst.* **A40**, 410-445.

2. Bricogne, G. (1988) A Bayesian statistical theory of the phase problem. I. A multichannel maximum entropy formalism for constructing generalized joint probability distributions of structure factors, *Acta Cryst.* **A44** (1988). 517-545.
3. Bricogne, G. (1988) Maximum entropy methods in the crystallographic phase problem, in N.W. Isaacs and M.R. Taylor (eds.), *Crystallographic Computing 4,* Clarendon Press, Oxford, pp. 60-79.
4. Bricogne, G, (1991) The X-ray crystallographic phase problem, in B.Buck and V.A.Macaulay (eds.), *Maximum entropy in action*, Oxford University Press, Oxford pp. 187-216.
5. Bricogne, G. (1991) Maximum entropy as a common statistical basis for all phase determination methods, in D.Moras, A.D. Podjarny and J.C. Thierry (eds.), *Crystallographic Computing 5: From chemistry to biology* Oxford University Press, Oxford, pp. 257-297.
6. Bricogne, G. (1993) Direct phase determination by entropy maximisation and likelihood ranking: status report and perspectives. *Acta Cryst.* **D49**, 37-60.
7. Bricogne, G. and Gilmore, C.J. (1990) A multisolution method of phase determination by combined maximisation of entropy and likelihood. I. Theory, algorithms and strategy, *Acta Cryst.* **A46,** 284-297.
8. Gilmore, C.J., Bricogne, G. and Bannister, C. (1990) A multisolution method of phase determination by combined maximisation of entropy and likelihood. II Application to small molecules, *Acta Cryst.* **A46,** 297-308.
9. Shankland, K., Gilmore, C.J., Bricogne G. and Hashizume, H. (1993) A multisolution method of phase determination by combined maximisation of entropy and likelihood. V Automatic Likelihood analysis *via* the student t-test, with an application to the powder structure of magnesium boron nitride, Mg_3BN_3. *Acta Crystallogr.* **A49,** 493-501.
10. Gilmore, C.J., Marks, L.D., Grozea, D., Collazo, C., Landree, E. and Twesten R.D, Direct Solutions of the Si (111) 7x7 Structure, *Surface Science*, in press.
11. Carter, C.W.Jr. (1994) Entropy maximisation permutation, and likelihood scoring methods for improving macromolecular electron density maps, in W.Wolf, E.J.Dodson and S.Glover (eds.), *From first Map to Final Model, Proceedings of the Study Weekend held at Daresbury* Daresbury Laboratory, Warrington, pp. 41-58.
12. Gilmore, C.J., Shankland, K. and Fryer, J.R. (1993), Phase extension in electron crystallography using the maximum entropy method and its application to two-dimensional purple membrane data from *Halobacterium halobium. Ultramicroscopy,* **49**, 147-178.
13. Gilmore, C.J. (1996) Maximum entropy and Bayesian statistics in crystallography: a review of practical applications. *Acta Cryst.* **A52** 561-589.
14. Gilmore, C.J., Shankland, K. and Bricogne, G. (1993) Applications of the maximum entropy method to powder diffraction and electron crystallography, *Proc.Roy.Soc.Ser. A.* **442,** 97-111.
15. Xiang, S., Carter, C.W.C.Jr, Bricogne,G., and Gilmore, C.J., Entropy maximisation constrained by solvent flatness: a new method for macromolecular phase extension and map improvement, *Acta Cryst.* (1993), **D49,** 193-212.

16. Doublié,S., Xiang, S., Gilmore,C.J., Bricogne,G. and Carter C.W.C. Jnr. Overcoming Non-Isomorphism by Phase Permutation an Likelihood Scoring: Solution of the TrpRS Crystal Structure, *Acta Cryst.* (1994), **A50**, 164-182.
17. Gilmore, C.J. & Bricogne, G. (1997) The MICE Computer Program in C.W.C. Carter Jnr *Methods in Enzymology* Academic Press, New York In press.
18. Gilmore, C.J., Henderson, K. and Bricogne G. (1991) A multisolution method of phase determination by combined maximisation of entropy and likelihood. IV The *ab initio* solution of crystal structures from their X-ray powder data, *Acta Cryst.* **A47,** 830-841.
19. Tremayne, M., Lightfoot, P., Mehta, M.A., Bruce, P.G., Harris, K.D.M., Shankland, K., Gilmore, C.J. and Bricogne G. (1992) An *ab initio* determination of $LiCF_3SO_3$ from X-ray powder diffraction data using entropy maximisation and likelihood ranking. *J. Solid State Chem.* **100,** 191-196.
20. Tremayne, M., Lightfoot, Glidewell, C., Harris K.D.M., Shankland, K., Gilmore, C.J., Bricogne G. and Bruce, P.G. (1992) Application of the combined entropy and likelihood method to the *ab initio* determination of an organic crystal structure from X-ray powder diffraction data. *J. Mater. Chem.* **2,** 1301-1302.
21. Gilmore, C.J., Nicholson, W.V. and Dorset, D.L. (1996) Direct methods in protein electron crystallography: the ab initio structure determination of two membrane protein structures in projection using maximum entropy and likelihood, *Acta Cryst.* **A52,** 937-946.
22. Gilmore, C.J., Henderson, A.N. and Bricogne G. (1991) A multisolution method of phase determination by combined maximisation of entropy and likelihood. V The use of likelihood as a discriminator of phase sets produced by the SAYTAN program for a small protein. *Acta Cryst.* **A47,** 832-846.
23. Voigt-Martin, I.G.,Han, D.H., Gilmore,C.J., Shankland, K. and Bricogne, G (1994) The Use of Maximum Entropy and Likelihood Ranking to Determine the Crystal Structure of 4-(4'-(N,N-dimethyl)aminobenzylidene)-pyrazolidine-3,5-dione at 1.4Å Resolution from Electron Diffraction and High Resolution Electron Microscopy Image Data, *Ultramicroscopy,* **56**, 271-288.
24. I.G.Voigt Martin, D.H.Yan, C.J.Gilmore and G.Bricogne (1995) Structure Determination by Electron Crystallography Using both Maximum Entropy and Simulation Approaches, *Acta Cryst,* **A51,** 849-868.

MULTI-DIMENSIONAL DIRECT METHODS

FAN HAI-FU
Institute of Physics, Chinese Academy of Sciences
Beijing 100080, P.R. China

1. Introduction

In single crystal structure analysis, it is usually assumed that crystals are ideal 3-dimensional periodic objects. However real crystals are never perfect. What we obtained under this assumption is not the real structure but just an averaged structure over a large number of unit cells. Unfortunately a knowledge on the averaged structure is often not enough for understanding the properties of many solid state materials. Therefore an important task for methods of solving crystal structures is to extend from ideal periodic crystals to real crystals which contain various kinds of defects. Modulated crystal structures belong to a kind of crystal structures containing periodic defects, i.e. the atoms in which suffer from certain occupational and/or positional fluctuation. If the period of fluctuation is commensurate with that of the three-dimensional unit cell then a superstructure results, otherwise an incommensurate modulated structure is obtained. Incommensurate modulated phases can be found in many important solid state materials. In many cases, the transition to the incommensurate modulated structure corresponds to a change of certain physical properties. Hence it is important to know the structure of incommensurate modulated phases in order to understand the mechanism of the transition and properties in the modulated state. Up to the present many incommensurate modulated structures were solved by using some kind of trial-and-error methods. With these methods it is necessary to make assumption on the property of modulation before we can solve the structure. This often causes difficulties and leads easily to errors. In view of diffraction analysis, it is possible to phase the reflections directly and solve the structure objectively without relying on any assumption about the modulation wave. Multi-dimensional direct methods have been developed for this purpose. The theoretical background and practical applications will be discussed in detail.

2. Incommensurate modulated structures

A modulated structure can be regarded as the result of applying a periodic modulation to a regular structure. *Figure* 1 shows two examples. The modulation wave in *figure* 1a

D. L. Dorset et al. (eds.), Electron Crystallography, 193–202.

represents the fluctuation of atomic occupancy. When it is applied to the background regular structure, the 'heights' of the atoms are modified. A commensurate modulated structure (superstructure) will result (*figure* 1b), if the period **T** of the modulation function is commensurate with the period **t** of the structure, i.e. **T/t** = n, where n is an integer. The resulting superstructure now has a true period **T** and a pseudo period **t**, respectively corresponding to a true unit cell and a pseudo unit cell. On the other hand, if **T** is incommensurate with **t** (*figure* 1c), i.e. **T/t** = r, where r is not an integer, we obtain an incommensurate modulated structure, in which no exact periodicity can be found although **t** remains a pseudo period. A modulation function can also represent the fluctuation of atomic positions and the positional modulation can also be either commensurate or incommensurate. In practice a modulated structure can simultaneously include different kinds of occupational and/or positional modulations.

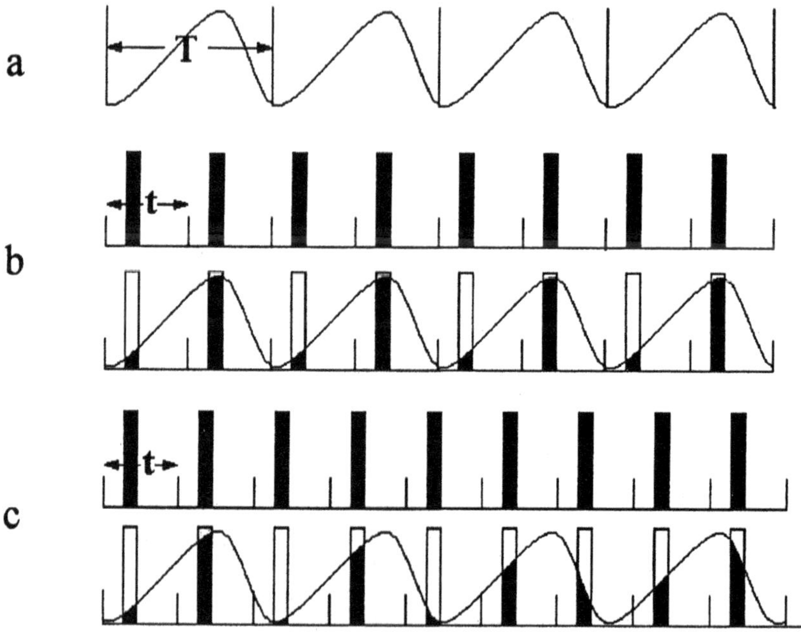

Figure 1. Occupational modulation of a one-dimensional structure
(a) modulation wave with a period equal to T; (*b*) upper row: one-dimensional regular structure with atoms shown as thick vertical lines and with a period equal to t; lower row: the resulting commensurate modulated structure; (*c*) upper row: one-dimensional regular structure; lower row: the resulting incommensurate modulated structure

In the reciprocal space an incommensurate modulated structure produces a 3-dimensional diffraction pattern, which contains satellites round the main reflections. An example of a section of such a 3-dimensional diffraction pattern is shown schematically in *figure* 2.

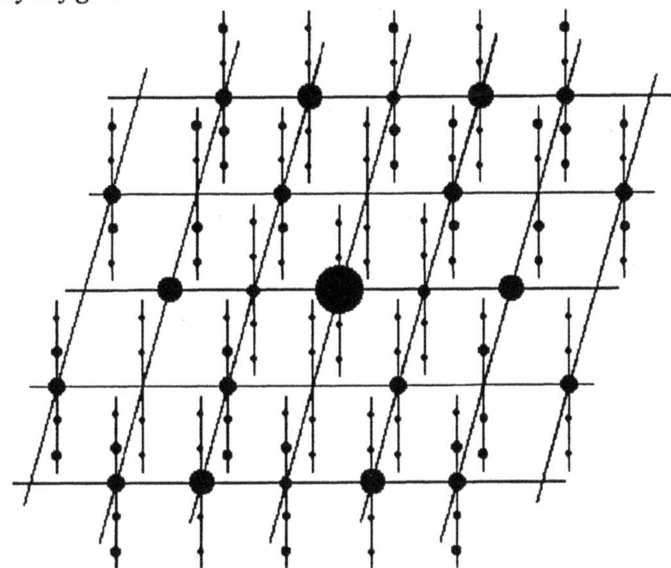

Figure 2. Schematic diffraction pattern of an incommensurate modulated structure. The vertical line segments indicate projected lattice lines parallel to the fourth dimension

The main reflections are consistent with a regular 3-dimensional reciprocal lattice although the satellites do not fit the same lattice. On the other hand, while the satellites are not commensurate with the main reflections, they have their own periodicity. Hence, it can be imagined that the 3-dimensional diffraction pattern is a projection of a 4-dimensional reciprocal lattice, in which the main and the satellite reflections are all regularly situated at the lattice nodes. From the properties of the Fourier transform the incommensurate modulated structure here considered can be regarded as a 3-dimensional "section" of a 4-dimensional periodic structure. This representation was first proposed by de Wolff [1] in 1974 to simplify the structure analysis of the incommensurate modulated structure of γ-Na_2CO_3.

The above example corresponds to a one-dimensional modulation. For an n-dimensional (n=1,2, ...) modulation, it needs a (3+n)-dimensional description. A (3+n)-dimensional reciprocal vector is expressed as

$$\mathbf{h} = h_1\mathbf{b}_1 + h_2\mathbf{b}_2 + h_3\mathbf{b}_3 + \dots + h_{3+n}\mathbf{b}_{3+n} \qquad (n = 1,2,\dots) \qquad (1)$$

where $\mathbf{b}_i$ is the ith translation vector defining the reciprocal unit cell. The structure factor formula is written as

$$F(\mathbf{h}) = \sum_{j=1}^{N} f_j(\mathbf{h}) \exp[i2\pi\,(h_1 \bar{x}_{j1} + h_2 \bar{x}_{j2} + h_3 \bar{x}_{j3})] \tag{2}$$

where

$$f_j(\mathbf{h}) = f_j(h) \int_0^1 d\bar{x}_4 \cdots \int_0^1 d\bar{x}_{3+n} P_j(\bar{x}_4, \cdots, \bar{x}_{3+n}) \times$$
$$\exp\{i2\pi\,[(h_1 U_{j1} + h_2 U_{j2} + h_3 U_{j3}) + (h_4 x_{j4} + \cdots + h_{3+n} x_{j(3+n)})]\} \tag{3}$$

The $f_j(h)$ on the right-hand side of (3) is the ordinary atomic scattering factor, P_j is the occupational modulation function and U_j describes the deviation of the jth atom from its average position $(\bar{x}_{j1}, \bar{x}_{j2}, \bar{x}_{j3})$. For more details on (2) and (3) the reader is referred to the papers by de Wolff [1], Yamamoto [2] and Hao, Liu & Fan [3]. What should be emphasised here is that, according to (2) a modulated structure can be regarded as a set of 'modulated atoms' situated at their average positions in 3-dimensional space. The 'modulated atom' in turn is defined by a 'modulated atomic scattering factor' expressed as (3).

There is a special kind of incommensurate modulated structures called composite structures. The characteristic of which is the coexistence of two or more mutually incommensurate 3-dimensional lattices. Owing to the interaction of coexisting lattices, composite structures are also incommensurate modulated structures. Unlike ordinary incommensurate modulated structures, composite structures do not have a 3-dimensional average (basic) structure. The basic structure of a composite structure corresponds to a 4- or higher-dimensional periodic structure. For a detailed description of composite structures the reader is referred to the paper by van Smaalen [4].

Obviously crystal-structure analysis of incommensurate modulated structures would better be implemented in multi-dimensional space. For this purpose we need firstly a theory on multi-dimensional symmetry and secondly a method to solve directly the multi-dimensional phase problem. The first problem has been solved by Janner and co-workers [5 - 8]. Our work on multi-dimensional direct methods is aiming at the second problem.

3. Modified Sayre equations in multi-dimensional space

It has been proved by Hao, Liu and Fan [3] that the Sayre equation [9] can easily be extended into multi-dimensional space. We have

$$F(\mathbf{h}) = \frac{\theta}{V} \sum_{\mathbf{h}'} F(\mathbf{h}') F(\mathbf{h} - \mathbf{h}') \tag{4}$$

here **h** is a multi-dimensional reciprocal vector defined as (1). The right-hand side of

(4) can be split into three parts, i.e.

$$F(\mathbf{h}) = \frac{\theta}{V}\left\{\sum_{\mathbf{h}'} F_{\mathrm{m}}(\mathbf{h}')F_{\mathrm{m}}(\mathbf{h}-\mathbf{h}') + 2\sum_{\mathbf{h}'} F_{\mathrm{m}}(\mathbf{h}')F_{\mathrm{s}}(\mathbf{h}-\mathbf{h}') + \sum_{\mathbf{h}'} F_{\mathrm{s}}(\mathbf{h}')F_{\mathrm{s}}(\mathbf{h}-\mathbf{h}')\right\} \tag{5}$$

Where subscript m stands for main reflections while subscript s stands for satellites. Since the intensities of satellites are on average much weaker than those of main reflections, the last summation on the right-hand side of (5) is negligible in comparison with the second, while the last two summations on the right-hand side of (5) are negligible in comparison with the first. Letting $F(\mathbf{h})$ on the left-hand side of (5) represents only the structure factor of main reflections we have to first approximation

$$F_{\mathrm{m}}(\mathbf{h}) \approx \frac{\theta}{V}\sum_{\mathbf{h}'} F_{\mathrm{m}}(\mathbf{h}')F_{\mathrm{m}}(\mathbf{h}-\mathbf{h}') \tag{6}$$

On the other hand, if $F(\mathbf{h})$ on the left-hand side of (5) corresponds only to satellites, it follows that

$$F_{\mathrm{s}}(\mathbf{h}) \approx \frac{\theta}{V}\sum_{\mathbf{h}'} F_{\mathrm{m}}(\mathbf{h}')F_{\mathrm{m}}(\mathbf{h}-\mathbf{h}') + \frac{2\theta}{V}\sum_{\mathbf{h}'} F_{\mathrm{m}}(\mathbf{h}')F_{\mathrm{s}}(\mathbf{h}-\mathbf{h}') \tag{7}$$

For ordinary incommensurate modulated structures the first summation on the right-hand side of (7) has vanished, because any three-dimensional reciprocal lattice vector corresponding to a main reflection will have zero components in the extra dimensions so that the sum of two such lattice vectors could never give rise to a lattice vector corresponding to a satellite. We then have

$$F_{\mathrm{s}}(\mathbf{h}) \approx \frac{2\theta}{V}\sum_{\mathbf{h}'} F_{\mathrm{m}}(\mathbf{h}')F_{\mathrm{s}}(\mathbf{h}-\mathbf{h}') \tag{8}$$

For composite structures on the other hand, since the average structure itself is a 4- or higher-dimensional periodic structure, the first summation on the right-hand side of (7) does not vanish. We have instead of (8) the following equation:

$$F_{\mathrm{s}}(\mathbf{h}) \approx \frac{\theta}{V}\sum_{\mathbf{h}'} F_{\mathrm{m}}(\mathbf{h}')F_{\mathrm{m}}(\mathbf{h}-\mathbf{h}') \tag{9}$$

Equation (6) indicates that the phases of main reflections can be derived by a conventional direct method neglecting the satellites. Equation (8) or (9) can be used to extend phases from the main reflections to the satellites respectively for ordinary incommensurate modulated structures or composite structures. This provides a way to determine the modulation functions objectively. The procedure will be in the following stages:

i) derive the phases of main reflections using Equation (6);
ii) derive the phases of satellite reflections using Equation (8) or (9);
iii) calculate a multi-dimensional Fourier map using the observed structure factor magnitudes and the phases from i) and ii);
iv) cut the resulting Fourier map with a 3-dimensional 'hyperplane' to obtain an 'image' of the incommensurate modulated structure in the 3-dimensional physical space;
v) parameters of the modulation functions are measured directly on the multi-dimensional Fourier map resulting from iii).

A program package *DIMS* (Direct methods for Incommensurate Modulated Structures) has been written in Fortran for the implementation of steps i) and ii) [10, 11].

4. Examples

4.1. THE MODULATED STRUCTURE OF γ-Na_2CO_3

This is a one-dimensional displacive modulated structure with $a = 8.904$ Å, $b = 5.239$ Å, $c = 6.042$ Å, $\alpha = \gamma = 90°$, $\beta = 101.35°$ and the modulation wave vector $\mathbf{q} = 0.182\, \boldsymbol{a}^* + 0.318\, \boldsymbol{c}^*$. The superspace group is $P^{C\,2/m}_{-1\;\;s}$. The modulated structure was originally solved by trial-and-error method [12] and was used to verify the multi-dimensional direct methods [3]. The 300 largest $|F(h_1h_2h_3 0)|$, 250 largest $|F(h_1h_2h_3 1)|$ and 150 largest $|F(h_1h_2h_3 2)|$ from the experimental data of γ-Na_2CO_3 were used in the test. Firstly according to equation (6) the phases of $F(h_1h_2h_30)$ were derived by an ordinary direct method. A default run of the program *SAPI* 85 [13] using only the main reflections led automatically to the correct average structure. The signs of 300 strongest main reflections were then calculated, 90% of which were correct. Based on this, phases of the first-order and second-order satellites were derived by using equation (8). The results were sorted in descending order of the structure-factor magnitude and cumulated in Table 1. Reasonably good agreement between the result

TABLE 1. Phase derivation for satellite reflections of γ-Na_2CO_3

Reflection group	Number of reflections	Percentage of reflections with their phase (sign) correctly determined	
		$F(h_1h_2h_31)$	$F(h_1h_2h_32)$
1	50	100	100
2	100	92	90
3	150	81	82
4	200	73	
5	250	70	

of direct methods and that of the original authors can be seen. A 4-dimensional electron density map was calculated using all the above phased reflections. Without bias by any assumed model the modulation parameters were measured directly on the map. 2-dimensional sections of which revealing the modulation of the *Na* and *O*(1,3) atoms are shown in *figure* 3.

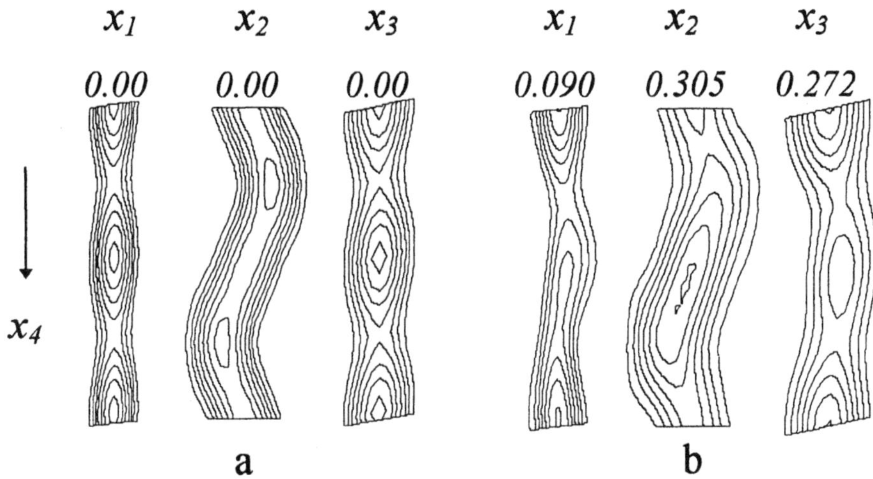

Figure 3. Sections of the 4-dimensional electron density map of γ-Na_2CO_3 showing the modulation of the *Na*(1) and *O*(1,3) atoms. (*a*) Sections through the average position of the *Na* atom. From left to right: $\rho(x_1, 0, 0, x_4)$, $\rho(0, x_2, 0, x_4)$ and $\rho(0, 0, x_3\ x_4)$; (*b*) Sections through the *O*(1,3) atom. From left to right: $\rho(x_1, 0.305, 0.272, x_4)$, $\rho(0.090, x_2, 0.272, x_4)$ and $\rho(0.090, 0.305, x_3\ x_4)$.

4.2. THE MODULATION IN THE 2:2:1:2 Bi-Sr-Ca-Cu-O SUPERCONDUCTOR

This is a one-dimensional incommensurate modulated structure with $a = 5.422$ Å, $b = 5.437$ Å, $c = 30.537$ Å, $\alpha = \beta = \gamma = 90°$ and the modulation wave vector $\mathbf{q} = 0.22\ \boldsymbol{b}^* + \boldsymbol{c}^*$. The superspace group is $N^{B\ b\ m\ b}_{\ \ 1\ -1\ 1}$. The structure has been extensively studied by different authors [14 - 18]. Among the published results there exist some discrepancies, especially on the modulation of oxygen atoms of the Bi-O layer. Based on the known average structure, a default run of the program *DIMS* with 543 main reflections, 867 first-order and 469 second-order satellite reflections resulted in a 4-dimensional Fourier map [19]. The map is then cut by the 3-dimensional physical space to give the modulated structure, the projection of which along the ***a*** axis is shown in *figure* 4. A 2-dimensional section through the 4-dimensional Fourier map shows the sawtooth-like displacive modulation of the O(4) atom (*figure* 5), which is in consistent with the result by Coppens and co-workers [15, 18].

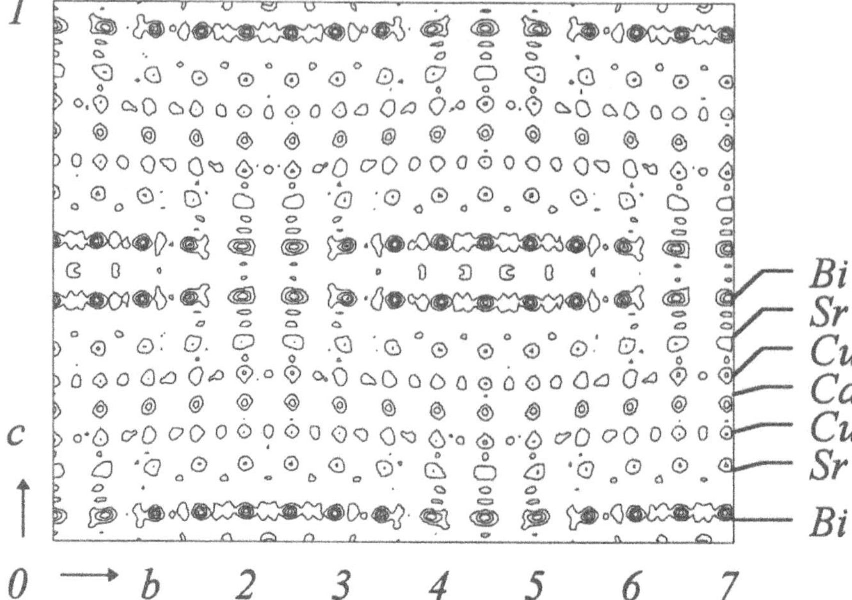

Figure 4. [100] projection of the 3-dimensional electron density map of the modulated structure 2:2:1:2 Bi-Sr-Ca-Cu-O: a region of 7×1 unit cells of the average structure were plotted along the ***b*** and ***c*** axes respectively.

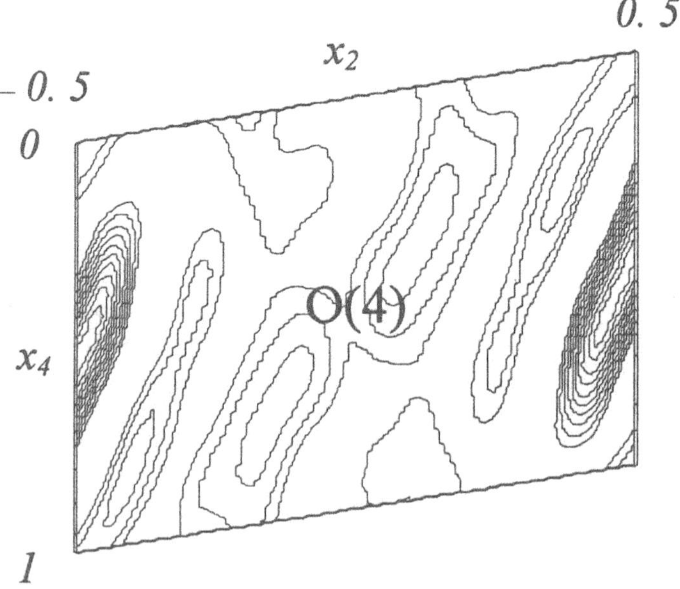

Figure 5. X_2 - X_4 section on the 4-dimensional electron density map of 2:2:1:2 Bi-Sr-Ca-Cu-O through the average position of the O(4) atom (x_1 = 0.145, x_3 = 0.0584) showing the sawtooth-like positional modulation.

4.3. THE MODULATED STRUCTURE OF (Perylene)Co(mnt)$_2$(CH$_2$Cl$_2$)$_{0.5}$

The modulated structure (Perylene)Co(mnt)$_2$(CH$_2$Cl$_2$)$_{0.5}$ was solved straightforwardly by the multi-dimensional direct method [20]. The lattice parameters are $a = 6.544$ Å, $b = 11.717$ Å, $c = 16.425$ Å, $\alpha = 90.09°$, $\beta = 95.34°$, $\gamma = 94.67°$ and the modulation wave vector $\mathbf{q} = 0.211\mathbf{a}^* - 0.137\,\mathbf{b}^* - 0.368\mathbf{c}^*$. The superspace group is *P*[*P-1*]*-1*. Based on the known average structure, two default runs of the program *DIMS* were performed, one assuming the superspace group *P*[*P-1*]*-1* while the other using *P*[*P1*]*1*. The two resulting phase sets are nearly the same. This confirmed the superspace group *P*[*P-1*]*-1* and strengthened the reliability of the resultant phases. The 4-dimensional Fourier map so obtained reveals the correct modulated structure, which is characterised by the sawtooth-like displacive modulation of the *Co* and *S* atoms.

4.4. THE COMPOSITE STRUCTURE OF (PbS)$_{1.18}$TiS$_2$

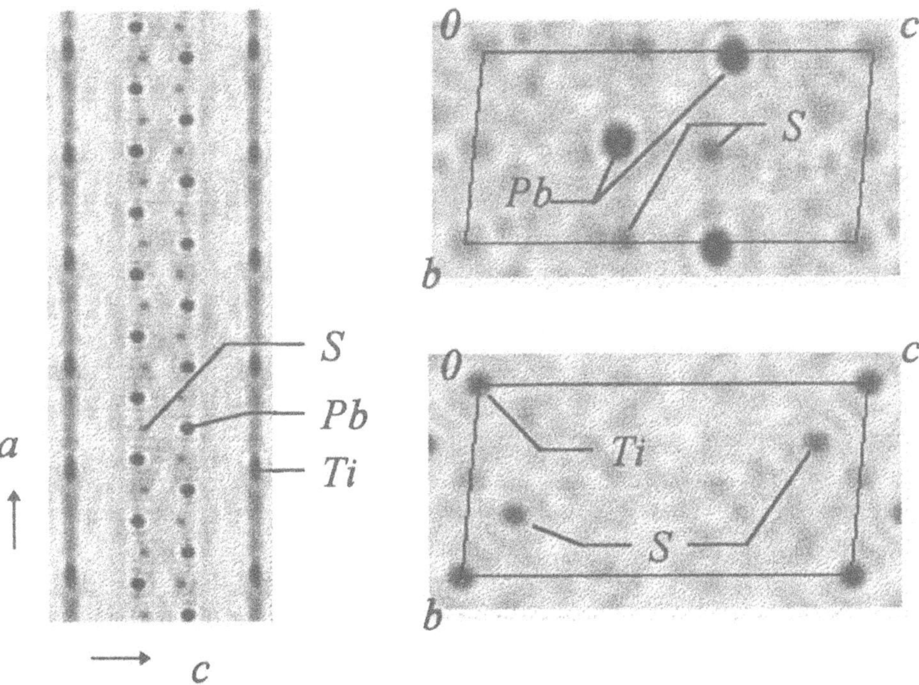

Figure 6. Direct-method phased Fourier sections of (PbS)$_{1.18}$TiS$_2$: space group C2/m (α, *0*, *0*) s-1; sub-system *PbS*: a=3.409, b=5.881, c=11.760 α=95.28°; sub-system *TiS$_2$*: a=5.800, b=5.880, c=11.759 α=95.28°. Left: $\rho(x_1, 0, x_3, 0)$ showing the 'chimney-ladder' structure; Top-right: $\rho(0.25, x_2, x_3, 0.25)$ showing the *PbS* layer; Bottom-right: $\rho(0, x_2, x_3, 0)$ showing the *TiS$_2$* layer.

(PbS)$_{1.18}$TiS$_2$ is a composite structure [21], of which the basic structure is also a 4-dimensional one. A default run of *DIMS* in a test [22] using Equation (6) was able to reveal directly the basic structure (*figure* 6). Determination of the modulation by multi-dimensional direct methods was also successful [23].

References

1. De Wolff, P. M. (1974) The pseudo-symmetry of modulated crystal structures, *Acta Cryst.* **A30**, 777-785.
2. Yamamoto, A. (1982) Structure factor of modulated crystal structures, *Acta Cryst.* **A38**, 87-92.
3. Hao, Q., Liu, Y. W. and Fan, H. F. (1987) Direct methods in superspace I. Preliminary theory and test on the determination of incommensurate modulated structures, *Acta Cryst.* **A43**, 820-824.
4. Van Smaalen, S. (1992) Superspace description of incommensurate intergrowth compounds and the application to inorganic misfit layer compounds, *Mater. Sci. Forum* **100** & **101**, 173-222.
5. Janner, A. and Janssen, T. (1977) Symmetry of periodically distorted crystals, *Phys. Rev. B* **15**, 643-658
6. De Wolff, P. M. Janssen, T. and Janner, A. (1981) The superspace groups for incommensurate crystal structures with a one-dimensional modulation, *Acta Cryst.* **A37**, 625-636.
7. Janner, A. Janssen, T. and De Wolff, P. M. (1983) Bravais classes for incommensurate crystal phases, *Acta Cryst.* **A39**, 658-666.
8. Jansen, T., Janner, A., Looijenga-Vos, A. and De Wolff, P. M. (1992) Incommensurate and commensurate modulated structures, in Wilson, A. J. C. (ed.), *International Tables for Crystallography*, Kluwer Academic Publishers, Dordrecht, pp. 797-835; 843-844.
9. Sayre, D. (1952) The squaring method: a new method for phase determination, *Acta Cryst.* **5**, 60-65.
10. Fu, Z. Q. and Fan, H. F. (1994) *DIMS* — a direct method program for incommensurate modulated structures, *J. Appl. Cryst.* **27**, 124-127.
11. Fu, Z. Q. and Fan, H. F. (1997) A computer program to derive (3+1)-dimensional symmetry operations from two-line symbols, *J. Appl. Cryst.* **30**, 73-78.
12. Van Aalst, W., Den Hollander, J., Peterse, W. J. A. M. and De Wolff, P. M. (1976) The modulated structure of γ-Na_2CO_3 in a harmonic approximation *Acta Cryst.* **B32**, 47-58.
13. Yao, J. X., Zheng, C. D., Qian, J. Z., Han F. S., Gu, Y. X. and Fan, H. F. (1985) A computer program for automatic solution of crystal structures from X-ray diffraction data, Institute of Physics, Chinese Academy of Sciences, Beijing 100080, P.R. China.
14. .Gao, Y, Lee, P., Coppens, P., Subramanian, M. A. and Sleight, A. W. (1988) The incommensurate modulation of the 2212 Bi-Sr-Ca-Cu-O superconductor, *Science* **241**, 954-956.
15. Petricek, V., Gao, Y., Lee, P. and Coppens, P. (1990) X-ray analysis of the incommensurate modulation in the 2:2:1:2 Bi-Sr-Ca-Cu-O superconductor including the oxygen atoms, *Phys. Rev. B* **42**, 387-392.
16. Yamamoto, A., Onoda, M., Takayama-Muromachi, E., Izumi, F., Ishigaki, T. and Asano, H. (1990) Rietveld analysis of the modulated structure in the superconducting oxide $Bi_2(Sr,Ca)_3Cu_2O_{8+x}$, *Phys. Rev. B* **42**, 4228-4239.
17. Kan, X. B. and Moss, S. C. (1992) Four-dimensional crystallographic analysis of the incommensurate modulation in a $Bi_2Sr_2CaCu_2O_8$ single crystal, *Acta Cryst.* **B48**, 122-134.
18. Gao, Y., Coppens, P., Cox, D. E. and Moodenbaugh, A. R. (1993) Combined X-ray single-crystal and neutron powder refinement of modulated structures and application to the incommensurately modulated structure of $Bi_2Sr_2CaCu_2O_{8+y}$, *Acta Cryst.* **A49**, 141-148.
19. Fu, Z. Q., Li, Y., Cheng, T. Z., Zhang, Y. H., Gu, B. L. and Fan, H. F. (1995) Incommensurate modulation in Bi-2212 high-Tc superconductor revealed by single-crystal X-ray analysis using direct methods, *Science in China A* **38**, 210-216.
20. Lam, E. J. W., Beurskens, P. T., Smits, J. M. M., Van Smaalen, S. De Boer, J. L. and Fan, H. F. (1995) Determination of the incommensurately modulated structure of (Perylene)Co(mnt)$_2$(CH_2Cl_2)$_{0.5}$ by direct methods, *Acta Cryst.* **B51**, 779-789.
21. Van Smaalen, S. Meetsma, A., Wiegers, G. A. and De Boer, J. L. (1991) Determination of the modulated structure of the inorganic misfit layer compound $(PbS)_{1.18}TiS_2$, *Acta Cryst.* **B47**, 314-325.
22. Mo, Y. D., Fu, Z. Q., Fan, H. F., Van Smaalen, S., Lam, E. J. W. and Beurskens, P.T. (1996) Direct methods for incommensurate intergrowth compounds. III. Solving the average structure in multidimensional space, *Acta Cryst.* **A52** 640-644.
23. Fan, H. F., Van Smaalen, S., Lam, E. J. W. and Beurskens, P. T. (1993) Direct methods for incommensurate intergrowth compounds. I. Determination of the modulation, *Acta Cryst.* **A49** 704-708.

CRYSTAL STRUCTURE DETERMINATION BY TWO-STAGE IMAGE PROCESSING

F. H. Li, D. X. Huang, B. Lu, W. Liu, Y. X. Gu, Z. H. Wan, H. F. Fan
Institute of Physics & Center for Condensed Matter Physics,
Chinese Academy of Sciences, Beijing 100080, P. R. China

1. Introduction

It was reported that the two-stage image processing technique based on the combination of high resolution electron microscopy and electron diffraction is useful for the *ab initio* crystal structure determination [1]. In the first stage the image taken at an arbitrary defocus condition is transformed into the structure image by image deconvolution [2, 3]. The resolution of such obtained structure image is limited by the resolution of the electron microscope so that not all atoms but usually only heavy atoms can be seen. In the second stage the image resolution is improved by phase extension using the corresponding electron diffraction data [4].

Although electron diffraction patterns (EDPs) yield the amplitudes of diffracted waves only, the number of diffracted wave amplitudes recorded in an EDP is much larger than that in an image. In case of electron diffraction, to combine the information from images affords some initial phases of diffracted waves with low spatial frequencies and makes the solution of phase problem simpler. In HREM, to combine the information of EDPs yields some additional amplitudes of diffracted waves in the high spatial frequency region, which offers the possibility of improving the image resolution. In addition, some equations and methods of diffraction analysis, for instance, the Sayre equation [5], tangent formula [6], heavy atom method, Fourier synthesis, direct methods [7], Wilson statistic [8], etc. are very beneficial to high resolution electron microscope image processing.

Fig. 1 shows the schematic diagram of crystal structure determination by image deconvolution and phase extension. The top right is the calculated projected potential distribution map of chlorinated copper phthalocyanine containing the structure details up to 1 Å with all atoms resolved. The bottom left is the high resolution electron microscope image of resolution 2 Å calculated by referring to the experimental conditions reported in Ref. [9] i.e. with accelerating voltage 500 kV, spherical aberration coefficient 1 mm, standard deviation of the Gaussian distribution of defocus due to the chromatic aberration 150 Å and defocus amount -1000 Å. Because the defocus amount is far from the Scherzer focus [10], the image does not reflect the crystal structure. After image deconvolution it turns into the structure image shown in

D. L. Dorset et al. (eds.), Electron Crystallography, 203–211.

the bottom right. The resolution of this image is limited by the resolution of the electron microscope so that only copper and chlorine atoms can be seen clearly, while carbon and nitrogen atoms are not resolved. After the direct method phase extension the resolution of the final image (top right) is about 1 Å and almost all atoms can be distinguished individually.

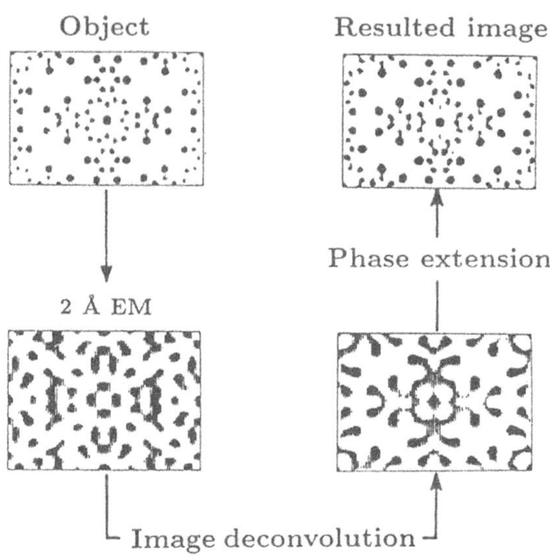

Fig. 1 Schematic diagram of the two-stage image processing.

Because one cannot expect that the structure amplitudes can equal the square roots of corresponding diffraction intensities, an empirical method was developed for correcting the diffraction intensities so that the structure amplitude can be equal to the square root of the corrected diffraction intensity with a certain accuracy [11].

In the present paper the method of electron diffraction intensity correction is introduced briefly and an example of image determination by use of the two stage image processing technique is given.

2. Electron diffraction intensity correction

An empirical method was proposed for correcting the distortion of electron diffraction intensity caused by Ewald sphere's curvature, crystal thickness, crystal bend, dynamical scattering, etc.. The key point of the method is to correct the electron diffraction intensity by means of the structural information afforded from a high resolution electron microscope image [11]. If the reciprocal space is divided into many circular zones, the average square of structure factor modulus inside the ith zone is expressed as $<|F(\mathbf{u})|>_{u_i \pm \Delta u_i}$, here F($\mathbf{u}$) denotes the structure factor, u_i the average of $\mathbf{u}$ in the ith zone, Δu_i the width of the ith zone and < > the symbol of averaging. Let $F_p(\mathbf{u})$ denotes the partial structure factor including only the contribution of atoms seen in the deconvoluted image. By referring to the Wilson statistic [8], it is assumed that

$$I_c(\mathbf{u})\Big|_{u_i \pm \Delta u_i} = k_{u_i} I_0(\mathbf{u}) \tag{1}$$

and

$$k_{u_i} = \frac{<\left|F_p(\mathbf{u})\right|^2>_{u_i \pm \Delta_{u_i}}}{<I_0(\mathbf{u})>_{u_i \pm \Delta_{u_i}}}, \tag{2}$$

where $I_c(\mathbf{u})\Big|_{u_i \pm \Delta u_i}$ denotes the corrected electron diffraction intensity for reflections located in the ith zone and $I_0(\mathbf{u})$ observed intensity.

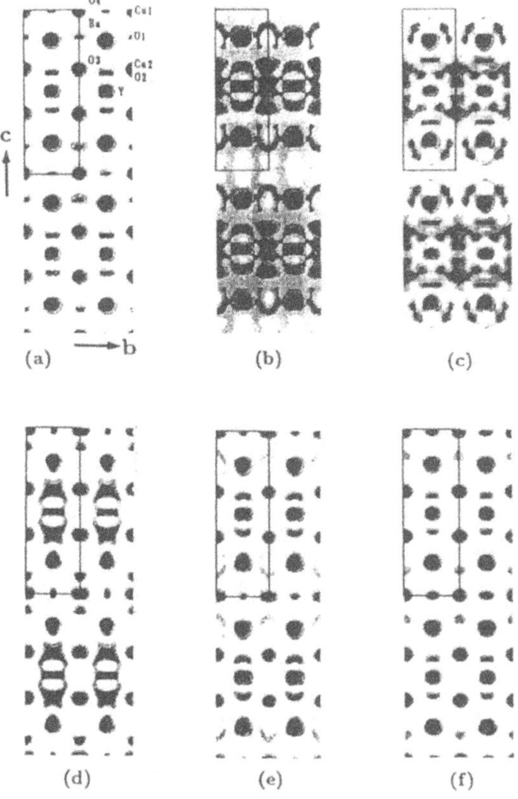

Fig. 2 (a) Projected potential distribution map of $YBa_2Cu_3O_{7-x}$, (b) Fourier transform obtained from square roots of diffraction intensities and phases of calculated structure factors, (c) to (f) projected potential maps obtained by the image deconvolution and then phase extension using diffraction intensity experimental, after the first, second and third correction, respectively.

After the diffraction intensity for all zones have been corrected, the phase extension is carried out by combining the deconvoluted image and diffraction intensity. Some additional atoms would be seen in the image obtained after the phase extension with the resolution about 1 Å. Then a series of structure factors $F_p'(\mathbf{u})$ with an approximation higher than $F_p(\mathbf{u})$ can be calculated by considering all atoms seen in this image. Thus the diffraction intensities can be corrected further by using equations (1) and (2) again, but $F_p(\mathbf{u})$ will be replaced by $F_p'(\mathbf{u})$. The correction procedure can be run continuously until the result of phase extension becomes convergent. The crystal of high temperature superconducting oxide $YBa_2Cu_3O_{7-x}$ was employed to test the method. The result shown in Fig. 1 indicates that the method is effective.

3. Crystal structure determination for $Bi_4(Sr_{0.75}La_{0.25})_8Cu_5O_y$

$Bi_4Sr_8Cu_5O_y$ is noted as a non-superconducting phases because of its three-dimensional Cu-O network [12]. It belongs to the orthorhombic system with the lattice parameters a = 5.373(2), b = 23.966(4) and c = 3.3907(6) Å. The space group is F_{mmm}. When Sr atoms are partially replaced by La the crystal structure changes into the C-centered orthorhombic with the space group C_{cca}. Nevertheless, the lattice parameters do not change much, they are a = 5.39, b = 23.5 and c = 34.0 Å [13]. It would be interesting to study the structure variation after the replacement of Sr atoms by La. Since the bulk sample of $Bi_4(Sr_{0.75}La_{0.25})_8Cu_5O_y$ consists of crystalline grains insufficient large for X-ray diffraction analysis, the crystal structure was studied by electron crystallography.

3.1. EXPERIMENTAL AND COMPUTING PROGRAM

Bulk samples was prepared by calcinating a mixture of oxides or carbonates of the cations in an appropriate ratio. Since the crystal is inclined to cleave along two basic planes, the sample was cut into thin slices and then thinned by mechanical polishing and ion-milling for electron microscope observation.

A series of [100] electron diffraction patterns (EDPs) were taken from the same region of the same sample with a JEM-1000 electron microscope operated at accelerating voltage 1000 kV with folded exposure time, such as 1, 2, 4, ... 64 seconds. A focal series of corresponding high resolution electron microscope images were obtained by using a JEM-2010 electron microscope operated at 200 kV. The focus step is about 120 Å.

Negatives of the EDPs were scanned by the Perkin-Elmer PDS microdensitometer with a step of $30 \times 30\mu^2$. The density-intensity characteristic curve (D-log I) was constructed from the measurement results of the EDPs. Only the data within the linear range of the curve were used for intensity calculation. Two images with defocus difference about 240 Å in the focal series were chosen and scanned by a Genius ColorPage Sp scanner at a resolution of 300 dpi.

All the works of two-stage image processing were done by using the ECC program [14] except the electron diffraction intensity integration and correction.

3.2. IMAGE DECONVOLUTION

Fig. 3a and b are experimental underfocus images projected along the a-axis. The inset shows the corresponding 1000 kV electron diffraction pattern. The symmetry of the projected structure is 2mm. Hence, the projected unit cell of size 23.5 穋 34.0 Å can be divided into four identical subcells as shown in the bottom left.

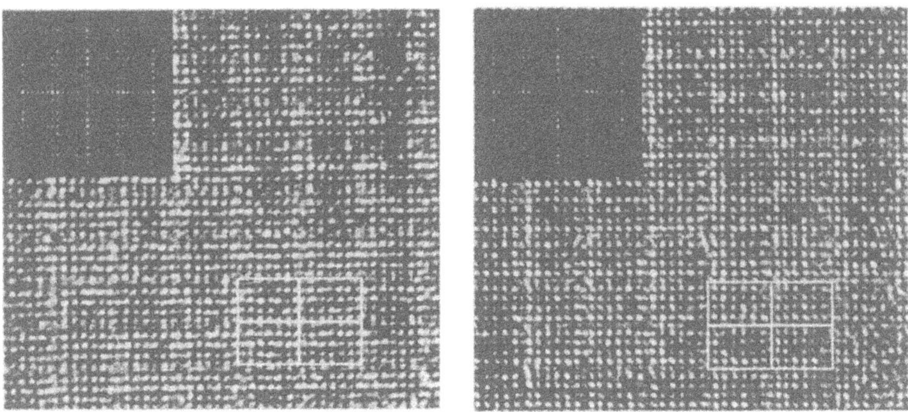

Fig. 3 Two experimental [100] high resolution electron microscope images with focus difference about 240 Å. The insets correspond to [100] electron diffraction patterns taken under accelerating voltage 1000 kV.

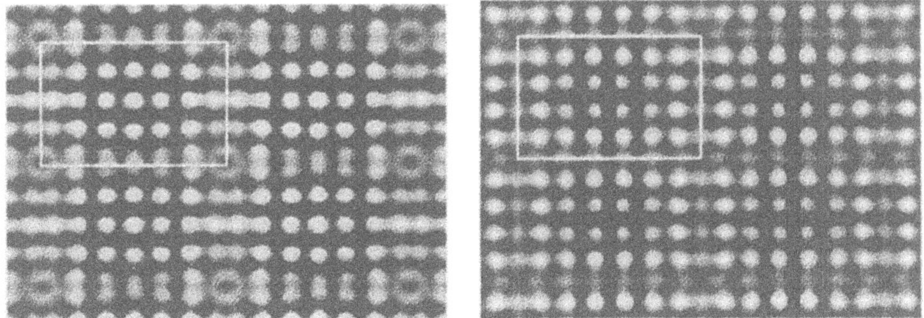

Fig. 4 Images after noise filtering and symmetry averaging. (a) and (b) corresponds to Fig. 3(a) and (b), respectively. The rectangle indicates one fourth of the unit cell.

Fig. 4a and b are images enlarged from Fig. 3a and b, respectively, after Fourier filtering and symmetry averaging. The two images are almost reverse in contrast with each other. The experimental defocus value of each image is searched by using DIMS program [15] in ECC package based on the principle of direct method. The correct defocus is selected from among a series of trial defocus values assigned from -1000 to 0 Å. For each trial defocus a set of structure factors is calculated with the amplitudes equal to square-roots of electron diffraction intensities. Then the weighted average of three figures of merit [7, 16] was computed dealing with Sayre equation [5] or tangent formula [6]. The defocus value corresponding to the set of structure factors which possesses the greatest figure of merit was adopted as the right one. The defocus values of Fig. 4a and b were determined as -480 Å and -710 Å and the corresponding deconvoluted images are shown in Fig. 5a and b, respectively. It can be seen that Fig. 5a and b are almost identical. All the metallic atoms Bi, Sr (La) and Cu appear black with the blackness almost proportional to the atomic weight. Fig. 6 is the schematic diagram showing the positions of metallic atoms in accordance with Fig. 5a and b.

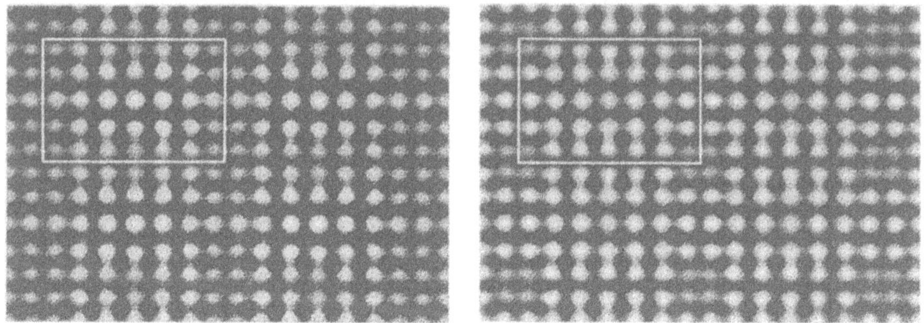

Fig. 5 (a) Image deconvoluted from Fig. 2(a) at defocus -480 Å with resolution 2 Å, (b) Image deconvoluted from Fig. 2(b) at defocus -710 Å with resolution 2 Å.

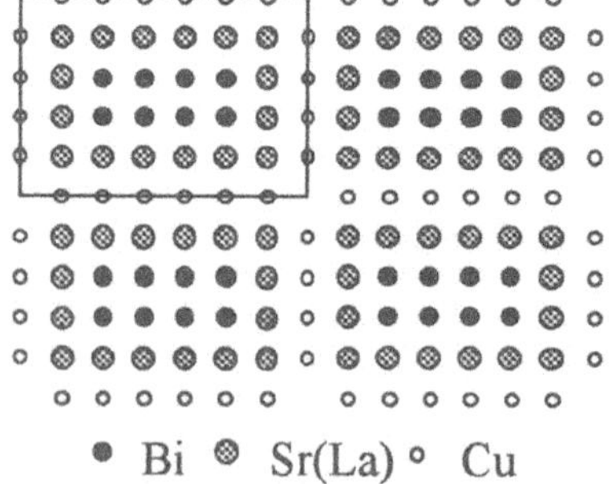

Fig. 6 Projected structure model showing positions of heavy atoms.

3.3. PHASE EXTENSION

Phase extension was carried out also by using the DIMS program in ECC package. The

Fourier transform of the deconvoluted image yields 132 reflections with 2 Å resolution. Their phases are employed as initial phases in the phase extension, while amplitudes of structure factors were obtained from experimental diffraction data. There are altogether 632 observed diffraction intensities up to resolution 1 Å. Among them 172 intensities are symmetrically independent. The 132 starting phases were kept fixed in the process of tangent-formula refinement until last two cycles of iteration. 632 phases within 1 Å resolution were obtained. The potential map (Fig. 7a) calculated with the phases of structure factors obtained from the result of phase extension and the amplitudes from the experimental diffraction data is of poor quality, although the resolution becomes much higher. All atoms are split or diffused. This seems mainly due to the unqualified diffraction data. Therefore, the diffraction intensity was corrected by using the method above-mentioned.

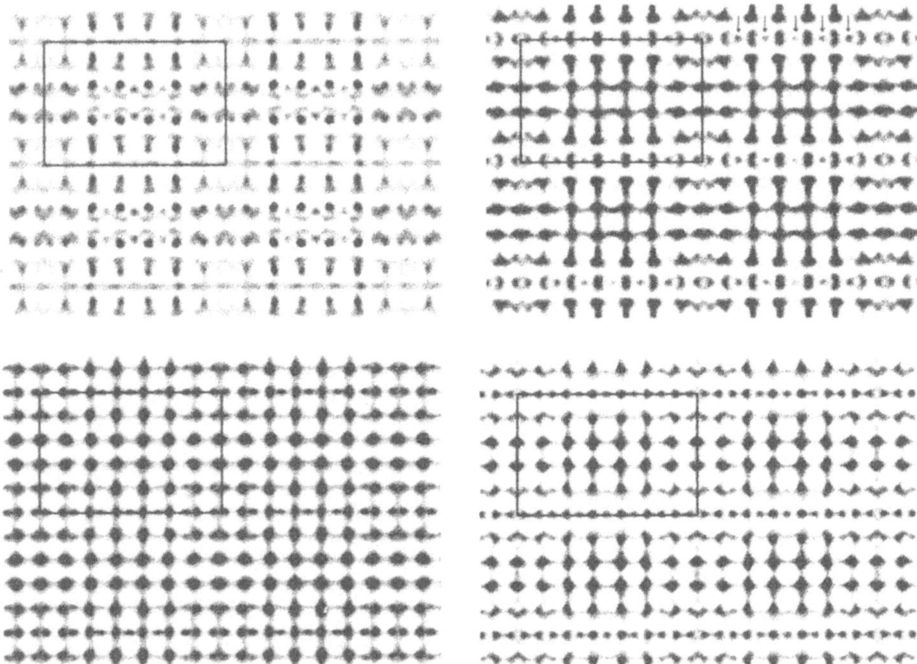

Fig. 7 PPMs of resolution 1 Å obtained after the phase extension. (a) based on original electron diffraction data, (b) diffraction data corrected for one time, (c) for two times and (d) for four times.

3.4. PPMs OBTAINED AFTER DIFFRACTION INTENSITY CORRECTION

Fig. 7a, b, c and d are PPMs obtained after the first, second, third and fourth cycle of diffraction intensity correction. The image quality is improved gradually after each cycle. Comparing with Fig. 7a all black dots become more concentrated in Fig. 7b except those located at the vertices of subcell, which split into two dots. In addition some new black dots appear along the horizontal Cu layers in Fig. 7b as pointed by arrows. These newly appearing black dots should represent oxygen atoms which form the horizontal Cu-O layers together with Cu atoms. Hence, they were included in the calculation of new partial structure factors, while the split black dots at the vertices are ignored. By using the newly corrected diffraction intensities a potential map was calculated with the phases obtained by Fourier synthesis based on Fig. 7b. The resulted image becomes better than before and is shown in Fig. 7c where all atoms appear even more concentrated, but dots located at vertices of subcell remain split. The latter phenomenon indicates that the corresponding oxygen atoms might randomly occupy one of the two split positions. They were included in the third cycle of diffraction intensity correction, and the new potential map was calculated. The phase extension yielded the image containing weak extra dots appeared along vertical Cu columns. These extra dots became slightly stronger after the fourth cycle calculation (see Fig. 7d). Obviously they represent the oxygen atoms forming the vertical Cu-O layers together with Cu atoms. Their contrast is weaker than that of oxygen atoms in the horizontal Cu-O layers. This can be interpreted that the perfection of vertical Cu-O layers is broken down in the structure of present sample. The final PPM map was obtained from Fig. 7d by Fourier synthesis and is shown in Fig. 8.

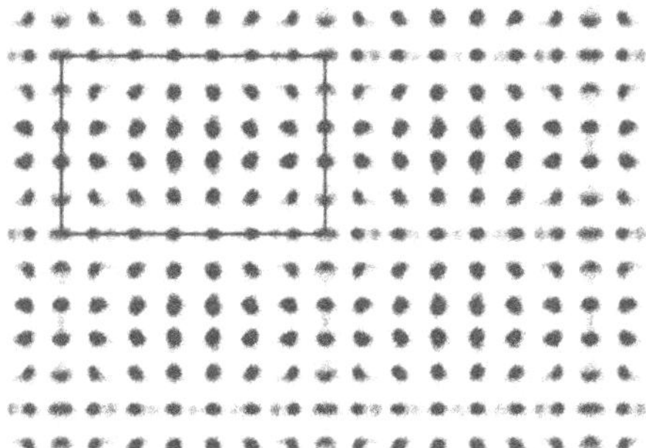

Fig. 8 PPM obtained by Fourier synthesis from Fig. 7d.

4. Conclusion

The [100] projection structure of $Bi_4(Sr_{0.75}La_{0.25})_8Cu_5O_y$ has been determined by using the two-stage image processing technique based on the combination of HREM and electron diffraction. All atoms including oxygen are more or less resolved in the final image. This indicates that this technique is available in *ab initio* crystal structure determination up to the resolution about 1 Å. Although the original observed electron diffraction intensity can be used directly in image deconvolution, it must be corrected for the phase extension. The empirical method of electron diffraction correction introduced in Ref. [11] has been proved to be an efficient way to correct various perturbations, especially that caused by Ewald sphere curvature. The disturbance introduced by dynamic effect can also be attenuated partially with this method.

References

[1] Li, F. H. (1994) Two-stage image processing in HREM, in: *Proc. 13th Int. Congr. on Electron Microscopy*, Paris, Vol. 1, 481-484.

[2] Han, F. S., Fan, H. F. and Li, F. H. (1986) Image processing in high-resolution electron microscopy using the direct method. II. Image deconvolution. *Acta Cryst.* **A42**, 353-356.

[3] Hu, J. J. and Li, F. H. (1991) Maximum entropy image deconvolution in high resolution electron microscopy. *Ultramicroscopy* **35**, 339-350.

[4] Fan, H. F., Zhong, Z. Y., Zheng, C. D. and Li, F. H. (1985) Image processing in high-resolution electron microscopy by using the direct method. I. Phase extension. *Acta Cryst.* **A41**, 163-165.

[5] Sayre, D. (1952) The squaring method: A new method for phase determination. *Acta Cryst.* **5**, 60-65.

[6] Karle, J. and Hauptman, J. (1956) A theory of phase determination for the four types of non-centrosymmetric space groups 1P222, 2P22, $3P_12$, $3P_22$. *Acta Cryst.* **9**, 635-651.

[7] Woolfson, M. M. and Fan, H. F. (1995) *Physical and Non-physical Methods of Solving Crystal Structures*, Cambridge University Press.

[8] Wilson, A. J. C. (1949) The probability distribution of X-ray intensities. *Acta Cryst.* **2**, 318-321.

[9] Uyeda, N., Kobayashi, T., Ishizuka, K. and Fujiyoshi, Y. (1978-1979) High voltage electron microscopy for image discrimination of constituent atoms in crystals and molecules. *Chem. Scripta* **14**, 47-61.

[10] Scherzer, O. (1949) The theoretical resolution limit of electron microscope. *J. Appl. Phys.* **20**, 20-29.

[11] Huang, D. X., Liu, W., Gu, Y. X., Xiong, J. W., Fan, H. F. and Li, F. H. (1996) A method of electron diffraction intensity correction in combination with high resolution electron microscopy. *Acta Cryst.* **A52**, 152-157.

[12] Fuertes, A., Miravitlles, C., Gonzalez-Calbet, J., Vallet-Regi, M., Obradors, X., Rodriguez-carvajal, J. (1989) The tubular crystal structure of new phase $Bi_4Sr_8Cu_5O_{19+x}$ related to the superconducting perovskites. *Physica C* **157**, 525-530.

[13] Lu, B., Li, F. H., Chen, H., Liu, W., Mao, Z. Q. and Zhang, Y. H. (1995) Transmission electron microscopy study of La-Doped Bi-Sr-Cu-O compound. *Jpn. J. Appl. Phys.* **34**, L425-L428.

[14] Wan, Z. H., Li, F. H. and Fan, H. F. Electron Crystallographic Computing (ECC) program, unpublished.

[15] Fu, Z. Q. and Fan, H. F. (1994) DIMS - a direct-method program for incommensurate modulated structures. *J. Appl. Cryst.* **27**, 124-127.

[16] Debaerdemaeker, T., Tate, C. and Woolfson, M. M. (1985) On the application of phase relationships to complex structures. XXIV. The Sayer Tangent Formular. *Acta Cryst.* **A41**, 286-290.

SUCCESS IS NOT GUARANTEED - PRACTICAL MATTERS FOR DIRECT PHASE DETERMINATION IN ELECTRON CRYSTALLOGRAPHY

Douglas L. Dorset,
Hauptman-Woodward Medical Research Institute,
73 High Street, Buffalo, NY 14203-1196 USA

1. Introduction

The success of direct phasing methods for solving crystal structures from electron diffraction intensity data is not only determined by the adherence of the observed intensities to the single scattering approximation. While manipulation of experimental parameters such as crystal thickness (minimum) and electron wavelength (also minimum) are desirable goals, there have been successful determinations based on data collected at just 50 kV. The type of preparation also has a role to play in the collection of intensity data adequate for ab initio structure analyses. For example, is the area irradiated a 1 mm diameter instead of a 1 nm diameter? The former case might minimize some dynamical interactions due to averaging over many crystal orientations while the latter might emphasize the multiple scattering component if the effective sample is a perfect flat crystal.

2. Quasi-kinematical Criterion

In general, it is most important to collect an intensity data set where the experimental Patterson function bears a close resemblance to the autocorrelation function of the actual crystal structure[1]. Dynamical and secondary scattering perturbations are always present. They can be tolerated as long as an extensive overlap between these vector maps can be maintained (Figure 1). This is the so-called 'quasi-kinematical criterion'. A good test to make of the initially collected intensity data is a Wilson plot[2]:

$$\ln\left\langle I_h^{obs}/\sum_i f_i^2\right\rangle = \ln C - 2B_{iso}(\sin^2\theta/\lambda^2)$$

If the recovered value for B_{iso} is typical of those found in typical x-ray crystal structures of similar materials, then the data might be quite favorable for a structure analysis. The presence of systematic absences must also be verified carefully. While forbidden reflections can appear due to secondary scattering contributions[3], collecting a number of diffraction patterns will establish whether or not suspected violations are consistently present. This, of course, is important for determination of space group symmetry.

3. Requirements for Data Collection

In fact, there should be rigorous tests made of the intensity data sets themselves. Diffraction patterns should be recorded at several exposure times so that all of the intensities can be measured on the linear scale of the recording medium. Are the intensities consistent for several

D. L. Dorset et al. (eds.), Electron Crystallography, 213–217.

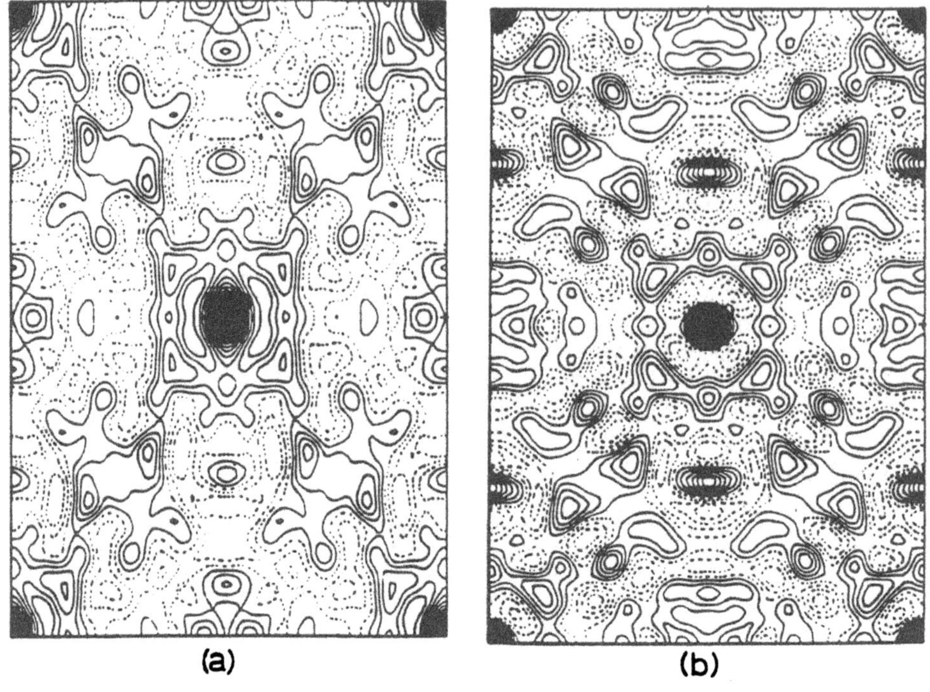

Figure 1. Comparison of Patterson functions for copper perbromophthalocyanine calculated from: (a) 1200 kV electron diffraction intensities and (b) $|F_{calc}|^2$ for the ideal crystal structure.

independently sampled crystals in the same orientation? This can be calculated via $R_{merge} = \sum||F_h^{cryst\ 1}| - k|F_h^{cryst\ 2}||/\sum|F_h^{cryst\ 1}|$. If there is symmetry in the pattern are all equivalent reflections observed at consistent intensities ($R_{sym} = \sum||F_h| - k|F_h^{equiv}||/\sum|F_h|$)? What is the statistical variance of the intensity data? Averaging over numerous equivalent diffraction intensities should give a standard deviation for all reflections that is proportional to Poisson counting statistics ($\sigma(I) = I^{1/2}$) (correlation coefficient greater than 0.95) but with a slope much less than 1.0 (e. g. 0.2) for the whole data set. If there is too much spread in these measured data there may be some source of error, including the preparation of a poor sample (e. g. deformed by irregular elastic bends). When separate zones are combined to assemble a three-dimensional data set, there should be strong common reflections between two patterns so that the scale can be accurately determined.

The sampling of reciprocal space measured by data collection must be adequate to determine the crystal structure. Nature is often perverse. Well-formed crystal faces will most likely correspond to projections down the longest unit cell axes[4], frustrating the identification of individual atomic positions. For organic compounds, optimal sample orientations can be prepared sometimes by the use of epitaxial substrates[5] so that the observed projection will be normal to a molecular plane or long axis. Crystallization from dilute solutions (by evaporation) can be dangerous since shear forces will cause the thin lamellae to be elastically bent. In this case the microcrystals will not be useful for finding symmetrical diffraction patterns useful for intensity measurement. This was demonstrated by control experiments on triphenylene[6], crystallizing in space group $P2_12_12_1$ (**a** = 13.20, **b** = 16.84, **c** = 5.28 Å [lath direction]). The data observed from thin crystals grown on naphthalene layers (cooling a molten solution to a eutectic solid) were superior to those measured from crystals formed by rapid evaporation of dilute benzene solutions. There may be the advantage of epitaxial orientation of a needle projection orthogonal in addition to the one found from solution. Collection of 86 unique data from five projections shows a good agreement to the x-ray crystal structure (R = 0.25). All patterns have good symmetry properties and show internal consistency. On the other hand, since the optimal projection of the structure would be along [001], the existing data are still insufficient for structure determination by direct methods since the most intense reciprocal lattice planes (e. g. hk0) are undersampled. The missing cone of data excluded in most tomographic samplings of single crystal habits can also be a serious limitation, particularly if a very important feature of the molecular packing is expressed by the missing data (e. g. the chain axis repeat in chain folded polymer lamellae)[7]. Reconstructions of the missing data by e. g. the Sayre equation have been only partially successful.

4. Direct Methods Approaches

To date, a number of different types of direct methods have been employed for structure determination from electron diffraction intensities[7]. These include symbolic addition, the tangent formula[9], the Sayre equation[10], random structure generators (e. g. Shake and Bake[11]), as well as maximum entropy and likelihood[12]. Methods making use of Σ_2-phase invariants seem to be very useful. (See Hauptman[13] for a discussion of phase invariants.) This is particularly interesting since experimental measurement of Σ_1- and Σ_2-invariants in convergent beam patterns have produced accurate phase values when dynamical scattering effects are taken into consideration[14]. Part of the reason is that these invariants depend on the accurate estimate of strong normalized structure factor magnitudes. Therefore, convolutional perturbations (multiple scattering) will not affect these values too much, particularly since the absolute ordering of these amplitudes does not seem to be required for the phase determination to succeed. On the other hand, negative quartet invariants are not very useful since predicted values of small $|E_h|$ 'cross-terms' are not always satisfied by the experimental data. Figures of merit based on these invariants will also be similarly affected, of course. Thus, a favorite FOM used in x-ray crystallography, NQEST[15], is not suitable for electron crystallographic determinations. On the other hand, the minimal principle[16] i a good FOM when only the Σ_2-relation is used for its calculation.

Electron diffraction data sets are always smaller than those collected on an x-ray diffractometer. Even when a three-dimensional data set can be measured from the most informative projections of the crystal structures, the vectorial linkage of Miller indices between these various views may be somewhat tenuous. For example, it is always advisable to use a convergence procedure[17] to determine the best path of determining the largest number of phases from a small basis set. Often algebraic unknowns must be incorporated into this basis set to acces a final set of phases large enough to lead to informative maps. (When available, phases from the Fourier transform of high-resolution electron micrographs should not be overlooked. However, care must be taken that the origin definition for image averaging is identical to the one used to evaluate phase invariants[7].) This implies the generation of multiple solutions (most conveniently by the tangent formula or the Sayre equation by traditional means). Here, other techniques for pruning branches such as maximum likelihood can be very effective for limited data sets[12]. It is advisable to build in algebraic phase unknowns so that a number of trial maps can be generated. Often knowledge of conservative structural features in the chemical substance will aid in identification of partial fragments in initial maps so that the complete structure will be found after Fourier refinement. Other figures of merit related to the Cochran condition[18] of 'maximum 'peakiness' for the best phase solution may be useful for narrowing the number of maps to be inspected.

5. References

1. D. L. Dorset (1995) Comments on the validity of the direct phasing and Fourier methods in electron crystallography, Acta Cryst. A51, 869-879.

2. A. J. C. Wilson (1942) Determination of absolute from relative x-ray intensity data, Nature 150, 151-152.

3. J. M. Cowley, A. L. G. Rees, and J. A. Spink (1951) Secondary elastic scattering in electron diffraction, Proc. Phys. Soc. (London) A64, 609-619.

4. L. H. Jensen (1970) Molecular packing in organic crystals, J. Polym. Sci. C29, 47-63.

5. J. C. Wittmann and B. Lotz (1990) Epitaxial crystallization of polymers on organic and polymeric substrates, Progr. Polym. Sci. 15, 909-948.

6. P. R. Pinnock, C. A. Taylor, and H. Lipson (1956) A re-determination of the structure of triphenylene, Acta Cryst. 9, 173-178.

7. D. L. Dorset (1995) Structural Electron Crystallography, Plenum, NY.

8. J. Karle and I. L. Karle (1966) The symbolic addition procedure for phase determination for centrosymmetric an noncentrosymmetric crystals, Acta Cryst. 21, 849-859.

9. J. Karle and H. Hauptman (1956) A theory of phase determination for the four types of noncentrosymmetic space groups: 1P222, 2P22, $3P_12$, $3P_22$, Acta Cryst. 9, 635-651.

10. D. Sayre (1952) The squaring method: a new method for phase determination, Acta Cryst. 5, 60-65.

11. R. Miller, G. T. DeTitta, R. Jones, D. A. Langs, C. M. Weeks, and H. A. Hauptman (1993) On the application of the minimal principle to solve unknown structures, Science 259, 1430-1433.

12. G. Bricogne and C. J. Gilmore (1990) A multisolution method of phase determination by combined maximization of entropy and likelihood. I. Theory, algorithms, and strategy, Acta Cryst. A46, 284-297.

13. H. A. Hauptman (1972) Crystal Structure Determination. The Role of the Cosine Seminvariants, Plenum, NY.

14. J. M. Zuo, R. Høier, and J. C. H. Spence (1989) Three-beam and many-beam theory in electron diffraction and its use for structure-factor phase determination in non-centrosymmetric crystal structures, Acta Cryst. A45, 839-851.

15. G. T. DeTitta, J. W. Edmonds, D. A. Langs, and H. Hauptman (1975) Use of negative quartetet cosine invariants as a phasing figure of merit: NQEST, Acta Cryst. A31, 472-479.

16. H. A. Hauptman (1993) A minimal principle in x-ray crystallography: starting in a small way, Proc. Roy. Soc. (London) A442, 3-12.

17. G. Germain, P. Main, and M. M. Woolfson (1970) On the application of phase relationships to complex structures. II. Getting a good start, Acta Cryst. B26, 274-285.

18. W. Cochran (1952) A relation between the signs of structure factors, Acta Cryst. 5, 65-67.

19. Research was funded by a grant from the National Science Foundation (CHE-9417835). The first part of the title is borrowed from a subsection of a chapter on direct phasing by M. F. C. Ladd and R. A. Palmer in their 1980 book, Theory and Practice of Direct Methods in Crystallography (Plenum, NY).

CRYSTAL STRUCTURE REFINEMENT INCORPORATING CHEMICAL INFORMATION

GEORGE M. SHELDRICK
University of Göttingen
Institut für Anorganische Chemie, Tammannstraße 4,
D-37077 Göttingen, Germany

Abstract

Crystal structure refinement using the new release of the SHELX program is discussed with emphasis on the incorporation of chemical information via constraints and restraints. It would be desirable to use refinement techniques in the structure solution as well, so as to incorporate chemical information into the phasing process. One such approach that looks very promising is ARP

1. Introduction

A common feature of macromolecular refinements and electron crystallography is the comparatively poor typical data to parameter ratio (compared to small-molecule X-ray structure analysis). In both cases it is desirable to supplement the diffraction data with chamical information in the form of constraints or restraints. A new version of the SHELX refinement program (SHELXL-97) is presented here, with the emphasis on the use of chemical information.

Recent advances in cryogenic techniques, area detectors, and the use of synchrotron radiation enable macromolecular data to be collected to higher resolution than was previously possible. In practice this tends to complicate the refinement because it is possible to resolve finer details of the structure; it is often necessary to model alternative conformations, and in a few cases even anisotropic refinement is justified. Although SHELXL [1] provides a number of other features not found in many macromolecular refinement programs (e.g. refinement against data from twinned crystals), it is probably the restrained anisotropic refinement, the flexible treatment of disorder and the estimation of esds that are most likely to be of interest to macromolecular crystallographers.

SHELXL is designed to be easy to use and general for all space groups, but uses a conventional structure-factor calculation rather than a FFT summation; the latter would be faster, but in practice involves some small approximations and is not very suitable for the treatment of dispersion or anisotropic thermal motion. The price to pay

D. L. Dorset et al. (eds.), Electron Crystallography, 219–230.

for the extra generality and precision is that SHELXL is much slower than programs written specifically for macromolecules, but this is to some extent compensated for by the good convergence properties, reducing the amount of manual intervention required (and also the *R*-factor). Two input files are required to run SHELXL: an intensity data file name.hkl and a file name.ins that contains crystal data, atoms, restraints, instructions, etc. All the information in the .ins file is given by four-letter keywords followed by numerical data and references to atoms etc. in free format.

An auxiliary program SHELXPRO is provided as an interface to other protein programs. SHELXPRO is able to generate an .ins file for input to SHELXL from a file in PDB format, including the appropriate restraints etc. SHELXPRO can also generate map files for O and other map interpretation programs, and can display the refinement results in the form of Postscript plots, as well as including the updated coordinates in the .ins file for the next refinement. SHELXL produces PDB and CIF format files that can be read by SHELXPRO and used for archiving.

Macromolecular structures are conventionally divided up into residues, for example individual amino-acids. In SHELXL residues may be referenced either individually, by '_' followed by the appropriate residue number, or as all residues of a particular class, by '_' followed by the class. For example 'DFIX 2.031 SG_9 SG_31' could be used to restrain a disulfide distance between two cystine residues, whereas 'FLAT_PHE CB > CZ' would apply planarity restraints to all atoms between CB and CZ inclusive in all PHE (phenylalanine) residues. Plus and minus signs refer to the next and previous residue numbers respectively, so'DFIX_* 1.329 C_- N'applies a bond length restraint to all peptide bonds ("_*' after the command name applies it to all residues). This way of referring to atoms and residues is not restricted to proteins; it is equally suitable for oligonucleotides, polysaccharides, or to structures containing a mixture of all three. It enables the necessary restraints and other instructions to be input in a concise and relatively self-explanatory manner. These instructions are checked by the program for consistency and appropriate warnings are printed.

Although the unique features of SHELXL are primarily useful for refinement against very high resolution data, tests on SHELXL-93 indicated that only small changes would be required to extend its range of applicability to medium resolution data (say 2.5Å or better). The most important of these changes, implemented in SHELXL-97, were improved diagnostics and more sophisticated anti-bumping restraints, and the addition of non-crystallographic symmetry (NCS) restraints.

2. Constrained and Restrained Refinement

A *constraint* is an exact mathematical condition that leads to the elimination of one or more least-squares variables; a *restraint* is an additional piece of information that is not exact but is associated with an esd. Restraints are normally applied as additional observational equations, i.e. they improve the data to parameter ratio by increasing the number of data but leaving the number of parameters unchanged. The function to be minimized then becomes:

$$M = \Sigma\, w_x(F_o^2 - F_c^2)^2 + \Sigma\, w_r(t - t_c)^2$$

where the first term is summed over all the X-ray data and the second over all the restraints. t is a target value for a restraint and t_c is the current value of the corresponding quantity. In SHELXL the restraint weights w_r are treated as absolute, and are given the values $1/\sigma^2$, where σ is the esd assigned to the restraint, and the X-ray weights w_x are then brought onto an approximately absolute scale by dividing the values set by the X-ray weighting scheme (e.g. $w=[\sigma^2(F_o^2)+(0.1P)^2]^{-1}$ where $P=(F_o^2+2F_c^2)/3$) by the mean value of $w(F_o^2-F_c^2)^2$. This has the desirable side-effect of weighting down the X-ray data releative to the restraints in the early stages of refinement when the agreement between the obseved and calculated intensities is poor, and increasing the contribution of the X-ray data as the agreement improves. In practice the restraint esds are independent of the resolution etc.

2.1 CONSTRAINTS

Typical constraints are the fixing of parameters or the imposition of s strict linear relation between parameters, e.g. coordinates or displacement parameters of atoms on special positions, and the refinement of a single parameter p to describe the occupancies of the two components of a disordered side-chain as p and 1-p. Rigid group constraints enable a structure to be refined with very few parameters, especially when the (thermal) displacement parameters are held fixed (using the BLOC instruction in SHELXL). After a structure has been solved by molecular replacement using a rather approximate model for the whole protein or oligonucleotide, it may well be advisable to divide the structure up into relatively rigid domains (using a few AFIX 6 and AFIX 0 instructions) and to refine these as rigid groups, initially for a limited resolution shell (e.g. SHEL 8 3), then stepwise extending the resolution (using the new STIR instruction). Restraints may still be required to define flexible hinges and prevent the units from flying apart. In view of the small number of parameters and the high correlations introduced by rigid group refinement, L.S. (full-matrix refinement) should be used for this stage (but CGLS will be necessary for the subsequent refinement). After this initial step, which exploits the large convergence radius of rigid-group refinement, in general the more flexible restraints will be used in preference to constraints for the rest of the refinement.

2.2 GEOMETRICAL SIMILARITY AND NCS RESTRAINTS

Very often there are chemically equivalent units in the structure, but it is not easy establish reliable target values for their geometrical parameters. A good approach is to restraint chemically but not crystallographically equivalent 1,2- and 1,3-distances to be equal, but without the use of target values. For example, chemically equivalent bonds in a heme unit can be restrained to be the same; with 4- or 8-fold redundancy this is a powerful and sensible restraint. The SADI instruction in SHELXL sets individual distances to be equal, and the SAME instruction may be used to generate SADI instructions to restrain complete molecules to have the same 1,2- and 1,3-distances etc.

We can take this idea one stage further to generate local NCS (non-crystallographic symmetry) restraints. The use of *NCS restraints* considerably improves the effective data to parameter ratio, and the resulting Fourier maps often look as though they were calculated with higher resolution data than were actually used (because the phases are more accurate). Two types of NCS restraint may be generated automatically with the help of the NCSY instruction. The first type uses the connectivity table to define equivalent 1,4-distances, which are then restrained to be equal. The second restrains the isotropic *U*-values of equivalent atoms to be equal. It is not normally necessary to restrain equivalent 1,2- and 1,3-distances to be equal because the DFIX and DANG restraints will have this effect anyway; but SAME may be used to add such restraints in the absence of DFIX and DANG. The use of *restraints* rather than applying NCS as an exact *constraint* (e.g. in the structure factor calculation) is more flexible (but slower) and does not require the specification of transformation matrices and real-space masks. Experience indicates that NCS restraints should be used wherever possible; it is not difficult to relax them later (e.g. for specific side-chains involved in interactions with other non-NCS related molecules) should this prove to be necessary. Note that restraining equivalent 1,4-distances to be equal is not quite as restrictive as restraining equivalent torsion angles: the 1,4-distances are equal for gauche$^+$ and gauche$^-$ conformations. On the other hand such conformational differences are chemically plausible for exposed side-chains.

2.3 OTHER GEOMETRICAL RESTRAINTS

SHELXL provides distance, planarity and chiral volume restraints, but not torsion angle restraints or specific hydrogen bond restraints. For oligonucleotides, distance restraints [2] may be used, but for reasonably high resolution data it is probably better to assume that for the sugars and phosphates the chemically equivalent 1,2- and 1,3-distances are equal (using the SAME and SADI restraints) without the need to specify target values. In this way the effect of the pH on the protonation state of the phosphates and hence the P-O distances does not need to be predicted, but it is assumed the whole crystal is at the same pH. For proteins, since some amino-acid residues occur only a small number of times in a given protein, it is probably better to use 1,2- and 1,3-target distances based on the study of Engh and Huber [3]; these are employed in the restraints added by SHELXPRO to the .ins file.

The three bonds to a carbonyl carbon atom may be restrained to lie in the same plane by means of a *chiral volume restraint* [4] with a target volume of zero (e.g. 'CHIV_GLU 0 C CD'). Chiral volume restraints are also useful to prevent the inversion of α-carbon atoms and the β-carbons of Ile and Thr, e.g. 'CHIV_ILE 2.5 CA CB' Chiral volume restraints may even be applied to non-chiral atoms such as CB of valine and CG of leucine in order to ensure conformity with conventional atom-labeling schemes (from the point of view of the atom names, these atoms could be considered to be chiral!). The FLAT instruction restrains planar groups by restraining the (chiral) volumes of a sufficient number of atomic tetrahedra to be zero; this works well but is somewhat unconventional, and so the program also prints the r.m.s. deviations from the planes.

2.4 ANTI-BUMPING RESTRAINTS

Anti-bumping restraints are distance restraints that are only applied if the two atoms are closer to each other than the target distance. They can be generated automatically by SHELXL, taking symmetry equivalent atoms and disorder into account. Since this step is relatively time-consuming, in the 1993 release it was performed only before the first refinement cycle, and the anti-bumping restraints were generated automatically only for the solvent (water) atoms (however they could be inserted by hand for any pairs of atoms). In practice this proved to be too limited, so in SHELXL-97 the automatic generation of anti-bumping restraints was extended to all C, N, O and S atoms (with an option to include H...H interactions) and was performed each refinement cycle. Anti-bumping restraints are not generated automatically for (a) atoms connected by a chain of three bonds or less in the connectivity array (unless separated by more than a specified number of residues), (b) atoms with different non-zero PART numbers, and (c) pairs of atoms for which the sum of occupancies is less than 1.1. The target distances for the O...O and N...O distances are less than for the other atom pairs to allow for possible hydrogen bonds. The H...H anti-bumping restraints are applied to all pairs of hydrogen atoms not bonded to the same atom; they help to ensure chemically reasonable conformations for flexible side-chains.

2.5 RESTRAINED ANISOTROPIC REFINEMENT

There is no doubt that macromolecules are better described in terms of anisotropic displacements, but the data to parameter ratio is very rarely adequate for a free anisotropic refinement. Such a refinement often results in 'non-positive definite' (NPD) displacement tensors, and at the best will give probability ellipsoids that do not conform to the expected dynamical behavior of the molecule. Clearly constraints or restraints must be applied to obtain a chemically sensible model. It is possible to divide a macromolecule up into relatively rigid domains, and to refine the 20 TLS parameters of rigid body motion for each domain [5]. This may be a good model for the bases in oligonucleotides and for the four aromatic side-chains in proteins, but otherwise macromolecules are probably not sufficiently rigid for the application of TLS constraints, or they would have to be divided up into such small units that too many parameters would be required. As with the refinement of atomic positions, restraints offer a more flexible approach.

The *rigid-bond restraint* (DELU) [6] assumes that the components of the anisotropic displacement parameters (ADPs) along bonded (1,2-) or 1,3-directions are zero within a given esd. This restraint should be applied with a low esd, i.e. as a 'hard' restraint. For many non-planar groups of atoms, rigid-bond restraints effectively impose TLS conditions of rigid body motion [7]. Although rigid-bond restraints involving 1,2- and 1,3-distances reduce the effective number of free ADPs per atom from 6 to less than 4 for typical organic structures, further restraints are often required for the successful anisotropic refinement of macromolecules.

The *similar ADP restraint* (SIMU) restrains the corresponding U_{ij}-components to be approximately equal for atoms which are spatially close (but not necessarily bonded because they may be in different components of a disordered group). The isotropic version of this restraint has been employed frequently in protein refinements. This restraint is consistent with the characteristic patterns of thermal ellipsoids in many organic molecules; on moving out along side-chains, the ellipsoids become more extended and also change direction gradually.

Neither of these restraints are suitable for isolated solvent (water) molecules. A linear restraint (ISOR) restrains the ADP's to be *approximately isotropic*, but without specifying the magnitude of the corresponding equivalent isotropic displacement parameter. Both SIMU and ISOR restraints are clearly only approximations to the truth, and so should be applied as 'soft' restraints with high esds. When all three restraints are applied, structures may be refined anisotropically with a much smaller data to parameter ratio, and still produce chemically sensible ADP's. Even when more data are available, these restraints are invaluable for handling disordered regions of the structure.

An ensemble distribution created by molecular dynamics is an alternative to the harmonic description of anisotropic motion [8,9], and may be more appropriate for structures with severe conformational disorder that do not diffract to high resolution.

2.6 ADVANTAGES OF RESTRAINED REFINEMENT

Constraints and restraints greatly increase the radius and rate of convergence of crystallographic refinements, so they should be employed in the early stages of refinement wherever feasible. The difference electron density syntheses calculated after such restrained refinements are often more revealing than those from free refinements. In large small-molecule structures with poor data to parameter ratios, the last few atoms can often not be located in a difference map until an anisotropic refinement has been performed with geometrical and ADP restraints. Atoms with low displacement parameters that are well determined by the X-ray data will be relatively little affected by the restraints, but the latter may well be essential for the successful refinement of poorly defined regions of the structure. Premature removal or softening the restraints (to improve the R-value !) often impedes further progress.

3. The Free *R*-factor

The question of whether the restraints can be removed in the final refinement, or what the best values are for the corresponding esds, can be resolved elegantly by the use of R_{free} [10]. To apply this test, the data are flagged (using SHELXPRO) as a working set (90 to 95% of the reflections) and a reference set (the remaining 5 to 10%). The reference set is only used for the purpose of calculating a conventional R-factor that is called R_{free}. It is very important that the structural model is not in any way based on the reference set of reflections, so these are left out of *all* refinement and Fourier map calculations. If the original model was in any way derived from the same data, then many refinement cycles are required to eliminate memory effects. This ensures that the R-factor for the reference set provides an objective guide as to whether the introduction of additional parameters or the weakening of restraints has actually improved the model, and not just reduced the R-factor for the data employed in the refinement ('R-factor cosmetics').

R_{free} is invaluable in deciding whether a restrained anisotropic refinement is significantly better than an isotropic refinement. Experience indicates that both the resolution and the quality of the data are important factors, but that restrained anisotropic refinement is unlikely to be justified for crystals that do not diffract to better than 1.5 Å

Despite the overwhelming arguments for using R_{free} to monitor macromolecular refinements, it is only a single number, and is itself subject to statistical uncertainty because it is based on a limited number of reflections. Thus R_{free} may be insensitive to small structural changes, and small differences in R_{free} should not be taken as the last word; one should always consider whether the resulting geometrical and displacement parameters are *chemically reasonable*. The final refinement and maps should always be calculated with the *full* data, but without introducing additional parameters or changing the weights of the restraints. R_{free} is most useful for establishing refinement protocols; for a series of closely similar refinements (e.g. for mutants to similar resolution) the R_{free} tests only need to be applied to the first.

4. Disorder Made Simple

To obtain a chemically sensible refinement of a disordered group, we will probably need to constrain or restrain a sum of occupation factors to be unity, to restrain equivalent interatomic distances to be equal to each other or to standard values (or alternatively apply rigid group constraints), and to restrain the displacement parameters of overlapping atoms. In the case of a tight unimodal distribution of conformations, restrained anisotropic refinement may provide as good a description as a detailed manual interpretation of the disorder in terms of two or more components, and is much simpler to perform. With high-resolution data it is advisable to make the atoms anisotropic *before* attempting to interpret borderline cases of side-chain disorder; it may well be found that no further interpretation is needed, and in any case the improved phases from the anisotropic refinement will enable higher quality difference maps to be examined.

Typical warning signs for disorder are large (and pronounced anisotropic) apparent thermal motion (in such cases the program may suggest that an atom should be split and estimate the coordinates for the two new atoms), residual features in the difference electron density and violations of the restraints. This information in summarized by the program on a residue by residue basis, separately for main-chain, side-chain and solvent atoms. In the case of two or more discrete conformations, it is usually necessary to model the disorder at least one atom further back than the maps indicate, in order that the restraints on the interatomic distances are fulfilled. The different conformations should be assigned different PART numbers so that the connectivity array is set up correctly by the program; this enables the correct rigid bond restraints on the anisotropic displacement parameters and idealized hydrogen atoms to be generated automatically even for disordered regions (it is advisable to model the disorder before adding the hydrogens).

Several strategies are possible for modeling disorder with SHELXL, but for macromolecules the simplest is to include all components of the disorder in the same residues and use the same atom names, the atoms belonging to different components being distinguished only by their different PART numbers. This procedure enables the standard restraints etc. to be used unchanged, because the same atom and residue names are used. No special action is needed to add the disordered hydrogen atoms, provided that the disorder is traced back one atom further than it is visible (so that the hydrogen atoms on the PART 0 atoms bonded to the disordered components are also correct). The following extract from an .ins file illustrates the action necessary:

```
RESI 38 SER
N  3 0.7714 0.9267 0.0062   11.0 0.1093
CA 1 0.7887 0.9740 0.0744   11.0 0.1370
PART  1
CB 1 0.8386 1.0427 0.0551   41.0 0.1188
OG 4 0.8994 1.0027 0.0230   41.0 0.1820
PART  2
CB 1 0.8414 1.0366 0.0653  -41.0 0.1493
OG 4 0.8368 1.1036 0.0102  -41.0 0.1732
PART  0
C  1 0.7414 1.0167 0.1038   11.0 0.0840
O  4 0.7072 1.0231 0.0690   11.0 0.1018
```

Atoms in PART 1 may bond to other atoms in PART 1 and also to those in PART 0, but not to those in PART 2 etc. Component (PART) 1 has been assigned an occupancy equal to free variable number 4, and component 2 has been assigned an occupancy equal to one minus free variable 4 (this is specified by the codes 41 and –41), so that a single occupancy parameter is refined, and the occupancies of the disordered atoms sum to unity. All other instructions are the same as for non-disordered residues. The last column contains the (isotropic) U-values ($B = 8\pi^2 U$). The program works out itself how to apply the restraints, add H-atoms etc. Note that this very simple and effective treatment of disorder was not available in SHELXL-93.

5. The Solvent Model

It is relatively common practice in the refinement of macromolecular structures to insert water molecules with partial occupancies at the positions of difference electron density map peaks in order to reduce the R-factor (another example of 'R-factor cosmetics'). Usually when two different determinations of the same protein structure are compared, only the most tightly bound waters, which usually have full occupancies and smaller displacement parameters, are the same in each structure. The refinement of partial occupancy factors for the solvent atoms (in addition to their displacement parameters) is rarely justified by R_{free}, but sometimes the best R_{free} value is obtained for a model involving some water occupancies fixed at 1.0 and some at 0.5.

Regions of diffuse solvent may be modeled using *Babinet's principle [11]*; the same formula is employed in the program TNT [12], but the implementation is somewhat different. In SHELXL it is implemented as the SWAT instruction and usually produces a significant but not dramatic improvement in the agreement of the very low angle data. Anti-bumping restraints may be input by hand or generated automatically by the program, taking symmetry equivalents into account. After each refinement job, the displacement parameters of the water molecules should be examined, and waters with very high values (say U greater than 0.8 Å^2, corresponding to a B of 63) eliminated. The F_o-F_c map is then analyzed automatically to find the highest peaks that involve no bad contacts and make at least one geometrically plausible hydrogen bond to an electronegative atom. These peaks are then included with full occupancies and oxygen scattering factors in the next refinement job. This procedure is repeated several times; in general R_{free} rapidly reaches its minimum value, although the conventional R-index continues to fall as further waters are added. It should be noted that the automatic generation of anti-bumping restraints is less effective when the water occupancies are allowed to have values other than 1.0 or 0.5. This approach provides an efficient way of building up a chemically reasonable (but not necessarily unique) network of waters that are prevented from diffusing into the protein, thus facilitating remodeling of disordered side-chains etc. The occupancies of specific waters may also be tied (using free variables) to the occupancies of particular components of disordered side-chains where this makes chemical sense. This procedure may be facilitated by using SHELXPRO to convert the .res output file from one refinement job to the .ins file for the next, or fully automated using the program SHELXWAT that calls SHELXL repeatedly. A related but much more sophisticated approach (ARP) described by Lamzin and Wilson [13] may also be used for automatic irrigation (ARP will also be discussed later in a more general context).

6. The Radius of Convergence

A crucial aspect of any macromolecular refinement program is the radius of convergence. A larger radius of convergence reduces the amount of time-consuming manual intervention using interactive graphics. There are probably a number of contributing factors to the good convergence typically observed for SHELXL, e.g. the refinement against properly weighted F^2 values for *all* data, the inclusion of important off-diagonal terms in the least-squares algebra [4], the ability to refine all parameters at once (i.e. coordinates and displacement parameters in the same cycle), and the restriction to unimodal restraint functions; multimodal restraint functions such as torsion angles or hydrogen bonds tend to increase the number of spurious local minima. It is much better to reserve the multimodal chemical information such as torsion angles for verifying the structure with an independent program such as PROCHECK [14], and to use the unimodal information as restraints. The errors in the FFT calculation of derivatives are larger that those in the structure factors (for the same grid intervals); this would also impede convergence.

Many claims that SHELXL gives R-factors one or two percent lower than other programs have been tracked down either to subtle differences in the model or to not getting trapped in local minima. The differences in the model include the treatment of diffuse solvent and hydrogen atoms, and the ability to refine common occupancies for disordered groups. The inclusion of dispersion terms and the use of a conventional rather than a FFT structure factor summation are also more precise; the approximations in the FFT summation may become significant for high resolution data and atoms with small displacement parameters.

7. Estimated Standard Deviations

No small molecule crystallographer would contemplate publishing a structure without estimated standard deviations, but they are rarely quoted for macromolecules, and then usually only in the form of a Luzzati plot (which is rather inappropriate and was never intended for the purpose [15]!). Provided that there are appreciably more data than parameters, it is in fact possible to invert the full least-squares normal matrix (or at least large blocks of it) from the refinement of a macromolecule, and so derive the esds in all parameters by small-molecule methods. SHELXL uses the full covariance matrix for the estimation of the esds in all dependent parameters such as bond lengths, torsion angles etc.

The structure should be refined to convergence by conjugate gradient least-squares (CGLS) so that the matrix needs to be inverted only once, at the end of the refinement. It turns out that the inversion produces sensible esds even when the calculated shifts would lead to instability. The esds take the restraints into account (in a Bayesian sense) so all restraints should be switched off for this final full-matrix cycle, which is performed with L.S. 1 and DAMP 0 0. This DAMP instruction specifies zero damping (which would otherwise artificially reduce the esds) and zero shift multipliers. All the reflection data should of course be used.

If the full-matrix cycle would take longer than a week or require the purchase of extra memory, an adequate compromise is to use BLOC 1 N_1 > LAST (or something similar) to set up a full-matrix block consisting of all positional but no thermal displacement parameters.

SHELXPRO can be used to plot the atomic positional and bond length esds (a BOND instruction is needed for SHELXL to generate the latter) against the B or B_{eq} values. Preliminary tests suggest that a formula recently proposed by Durward Cruickshank [15] models the dependence of the esds on B_{eq}, effective atomic number, the $R1$-value and the completeness of the data rather well (much better that the Luzzati method).

8. Structure Refinement Techniques in Structure Solution

Since current techniques for solving small-molecule and macromolecular structures by *ab initio* methods appear to require data to almost atomic resolution, yet - with the help of restraints incorporating chemical information - there is no problem in refining structures at much more modest resolution, it is worth asking if refinement algorithms provide a way of incorporating the necessary chemical information into the structure solution process itself. One promising approach is the ARP method [13,16]. In this method, alternate refinement cycles and Fourier maps are calculated; at each stage, a few atoms with high B-values and negative difference electron density are eliminated, and replaced by the most promising candidate atoms found by a sophisticated analysis of the resulting electron density maps. Recent improvements in this method [16] are the use of maximum-entropy as well as least-squares refinement, and sigma-A weighting for the maps. Although geometrical information is taken into account in selecting the new atoms from the maps, the refinement is performed without restraints. At the end of the procedure the final electron density maps have to be refitted; it appears to be best to run several ARP jobs in parallel with slightly different criteria for placing the new atoms, and to calculate a weighted average density from all of them.
The method works well for phase extension (sometimes much better than density modification methods) provided that the resolution is fairly high; for example it is possible to expand the structure of rubredoxin from the positions of the iron and sulfur atoms with data truncated to 1.5Å, and larger fragments (e.g. from molecular replacement with poor sequence identity) can be improved at resolutions as low as 2.0Å.

A possible area for future development of ARP would be to include some form of general restraints, e.g. based on a predicted radial probability distribution function, into the refinement part of the procedure.

Acknowledgement

I am very grateful to a large number of SHELX users, too many to name here, who made many helpful comments and suggestiongs that have found there way into the current SHELXL refinement program. We also thank the the Eropean Commission for support: contract number ERBCHBGCT940731.

9. References

1. Sheldrick, G.M. and Schneider, T.R. (in press) SHELXL: High-resolution refinement, in C.W. Carter and R.M. Sweet (eds.), *Methods in Enzymology*, Academic Press, New York.

2. Parkinson, G., Voitechovsky, J., Clowney, L., Brünger, A.T., and Berman, H.M. (1996) New parameters for the refinement of nucleic acid-containing structures, *Acta Cryst.* **D52**, 57–64.

3. Engh, R.A. and Huber, R. (1991) Accurate bond and angle parameters for X-ray protein structure refinement, *Acta Cryst.* **A47**, 392–400.

4. Hendrickson, W.A. (1985) Stereochemically restrained refinement of macromolecular structures, in H.W. Wyckoff, C.H.W. Hirs and S.G. Timasheff (eds.), *Methods in Enzymology*, Academic Press, New York, Vol. **115B**, pp. 252–270.

5. Driessen, H., Haneef, M.I.J., Harris, G.W., Howlin, B., Khan, G., and Moss, D.S. (1989) RESTRAIN: Restrained structure-factor least-squares refinement program for macromolecular structures, *J. Appl. Cryst.* **22**, 510–516.

6. Rollett, J.S. (1970) Least-squares procedures in crystal structure analysis, in F.R. Ahmed, S.R. Hall and C.P. Huber (eds.), *Computing in Crystallography*, Munksgaard, Copenhagen, pp. 167–181.

7. Didisheim, J.J. and Schwarzenbach D. (1987) Rigid-link constraints and rigid-body molecules, *Acta Cryst.* **A43**, 226–232.

8. Gros, P., van Gunsteren, W.F., and Hol, W.G. (1990) Inclusion of thermal motion in crystallographic Structures by restrained molecular dynamics, *Science*, **249**, 1149–1152.

9. Clarage, J.B. and Phillips, G.N. (1994) Cross-validation tests of time-averaged molecular dynamics refinements for determination of protein structures by X-ray crystallography, *Acta Cryst.* **D50**, 24–36.

10. Brünger, A.T. (1992) Free R value: a novel statistical quantity for assessing the accuracy of crystal structures, *Nature*, **355**, 472–475.

11. Moews, P.C. and Kretsinger, R.H. (1975) Refinement of the structure of carp muscle calcium-binding parvalbumin by model building and difference Fourier analysis, *J. Mol. Biol.* **91**, 201–228.

12. Tronrud, D.E., Ten Eyck, L.F., and Matthews, B.W. (1987) An efficient general-purpose least-squares refinement program for macromolecules, *Acta Cryst.* **A43**, 489–501.

13. Lamzin, V.S. and Wilson, K.S. (1993) Automated refinement of protein models, *Acta Cryst.* **D49**, 129–147.

14. Laskowski, R.A., MacArthur, M.W., Moss, D.S. and Thornton, J.M. (1993) PROCHECK: a program to check the stereochemical quality of protein structures, *J. Appl. Cryst.* **26**, 283–291.

15. Cruickshank, D.W.J. (1994) Protein precision re-examined: Luzzati plots do not estimate final errors, Poster Abstract, I.U.Cr. Meeting, Seattle.

16, Perrakis, A.,Sixma, T.A., Wilson, K.S., and Lamzin, V.S. (1997) The world according to wARP: improvement and extension of crystallographic phases, CCP4 Meeting, York, January 1997.

LEAST SQUARES REFINEMENT OF STRUCTURES FROM DYNAMIC ELECTRON DIFFRACTION DATA

H.W. ZANDBERGEN[1] and J. JANSEN[1,2]
1. National Centre for HREM, Laboratory of Materials Science, Delft University of Technology, Rotterdamseweg 137, 2628 AL Delft, The Netherlands
2. Laboratory for Crystallography, University of Amsterdam, Nieuwe Achtergracht 166, 1018 WV Amsterdam, The Netherlands.

1. Introduction

Structure determination from X-ray and neutron diffraction data is a standard procedure. Starting with a rough model, the accurate structure is determined using a least-squares structure refinement, which is based on kinematic diffraction and in which the differences between calculated and experimental intensities are minimized. With kinematic diffraction the intensities of the reflections increase linearly with thickness. Dynamic scattering, which occurs in electron diffraction (ED), will change the intensities of all reflections with respect to each other as a function of the specimen thickness. An example of this change for $Ce_5Cu_{19}P_{12}$, the example structure of this contribution, is shown in Figure 1. Therefore the kinematic refinement software can only be used for electrons in the regime where the dynamic scattering is small, which is in the case of $Ce_5Cu_{19}P_{12}$ for specimen thicknesses smaller than about 7 nm. For example, we have used kinematic refinement [1] on data sets of very thin $La_3Ni_2B_2N_3$ (about 4 nm thick and in [100] and [110] orientations).

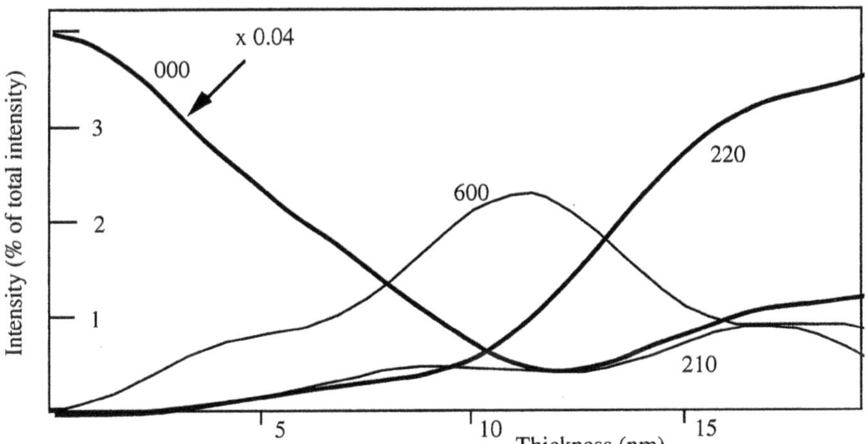

Figure 1. Intensities of several reflections including the central beam of $Ce_5Cu_{19}P_{12}$ as a function of thickness. Note that the intensities do not show a linear increase and that the central beam drops from 100% a zero thickness to about 10% at a thickness of 12 nm.

D. L. Dorset et al. (eds.), Electron Crystallography, 231–241.

Compared to the structure obtained by refinement of neutron powder diffraction data, the positions of La and Ni were quite accurately determined but the positions obtained for N and B were not correct (0.02 and 0.05 nm off). Since it is almost impossible to prepare specimens thinner than 7 nm over a relatively large area, it is essential in the structure refinement from ED that the dynamic diffraction is taken into account.

Although the dynamic diffraction has the disadvantage that the intensities of the reflections depend on the thickness, which make the conventional kinematic diffraction software only valid for very thin crystals, is has on the other hand a major advantage. The advantage is that the dynamic scattering results in a thickness dependence of the scattering of atom columns. The effect of the thickness on the scattering of a single columns of different composition is depicted in Figure 2, where calculated exit waves are shown for a range of thicknesses from 2 to 22 nm. As can be seen from this figure, the scattering amplitude of the atom columns for the hypothetical structure LiNaKRbCs varies strongly over the thickness. Thus if one is able to perform a refinement taking into account the dynamic scattering one can highlight certain atom columns by using a certain thickness. Moreover, if one does simultanuous refining of ED data sets of several thicknesses, the positions of all atoms - including the weakly scattering ones - can be determined quite accurately.

Recently we have developed a software package MSLS [2], in which multislice calculation software is combined with least squares refinement software used in X-ray crystallography. With multislice calculations which are standardly used for image calcultations of HREM images, dynamic diffraction is taken into account explicitely.

2. Experimental

We performed the electron microscopy with a Philips CM30ST electron microscope with a field emission gun operated at 300 kV. The field emission gun has a major advantage that with very small spot sizes the convergence angle is still small, such that the illumination is similar to a plane wave, resulting in sharp diffraction spots. If due to the convergence angle the spots are disc-like, one can use the diffraction lens of the electron microscope for focussing the discs to sufficiently small dots, but this leads to distortions as will be discussed below. High resolution images and electron diffraction patterns were recorded with a 1024x1024 pixel Photometrix CCD camera having a dynamic range of 12 bits. Exposure times were 0.5 to 1.0 second per HREM image. Electron diffraction performed with spot sizes between 6 and 15 nm. Exposure times for the electron diffraction ranged from 0.4 to 2 seconds. To reduce the electron microscope induced contamination and amorphisation, the specimens were cooled to about 100 K.

A small spot size for electron diffraction is used for three reasons: i) to have a relatively small variation of thickness since most crystals are wedge shaped, ii) to reduce the amount of unwanted information like that of the matrix around a small precipitate and iii) to have a little variation in the crystal orientation. The latter reason is quite important which one can appreciate by moving the electron beam in nanodiffraction mode over the specimen: although the crystal is well aligned according to the selected area diffraction, fluctuations in orientation over 1 to 2° in all directions occur, even for areas which are very close to each other (10-50 nm). Such orientation variations should be considered as normal rather than an exception. Calculations have shown [2] that a variation in thickness of the illuminated area leads to an increase in the R-value and a less reliable structure determination.

A typical diffraction pattern is shown in Figure 3. The misorientation of the crystal (0.45°) is evident from the asymmetry of the diffraction spots around the central spot. A misorientation has a strong advantage that it results an increase in the high order reflections on the other side, such that even reflections with d-values well below 0.05 nm can be significant, as can be seen in Figure 3.

The refinements were done with the recently developed software package MSLS of which more details are given below. To allow a comparison between dynamic and kinematic diffraction the MSLS program was also used to simulate a kinematic refinement. In order to approximate a kinematic refinement with the MSLS program, a very small thickness (e.g. 1 nm) or a very low occupancy should be taken. The latter approach was used because in this way the thickness and orientation dependence of the shape of the diffraction spots is properly taken into account. In the calculation of the kinematic R values a 0.1% occupancy of all atom sites and the thicknesses obtained for the dynamic refinement were taken.

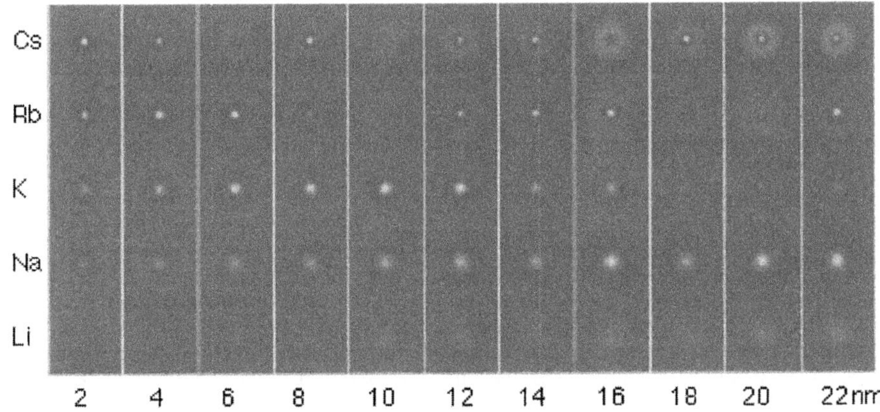

Figure 2. Calculated exit waves (amplitude) for a hypothetical structure with Li, Na, K, Rb and Cs for thicknesses from 2 to 22 nm as indicated. The Cs atoms are most dominant for a thickness of 2 nm, but it is almost invisible for a thickness of 6 nm. This cycle of increase and decrease in visibility has a beat of about 5 nm for Cs. For Rb this beat is about 9 nm and for K 19 nm. Thus by combining electron diffraction data from several thicknesses light elements can be easily determined.

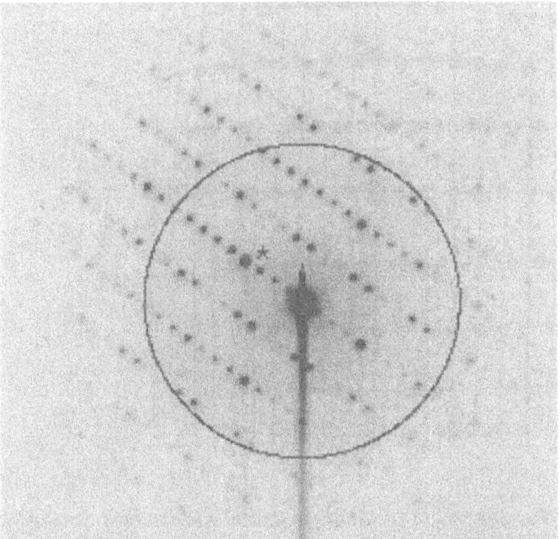

*Figure 3. Electron diffraction pattern taken with a CCD camera. Due to overexposure and the way the CCD is read out, the central beam has a strong tail. The circle indicated 1 Å^{-1}. The * indicated the refined position of the centre of the Laue circle.*

A typical diffraction pattern is shown in Figure 3. The misorientation of the crystal (0.45°) is evident from the asymmetry of the diffraction spots around the central spot. A misorintation has a strong advantage that it results an increase in the high order reflections on the other side, such that even reflections with d-values well below 0.08 nm can be significant, as can be seen in Figure 3.

The indexing procedure

An essential part of the data reduction is the indexing. In this procedure all positions on the recorded image where intensity from reflected beams can be expected have to be is determined irrespectively whether the reflection is strong or weak. In the two-demensional reciprocal space every reflection can be indicated by a vector $\boldsymbol{H}$ which has two integer elements, h and k, the indices. All position in reciprocal space can be described as:

$$\boldsymbol{P} = \boldsymbol{O} + \boldsymbol{A}\ \boldsymbol{H} \qquad (1)$$

where $\boldsymbol{O}$ is the origin and the matrix $\boldsymbol{A}$ consists of the two basis vectors of the reciprocal space (see figure 4). The matrix $\boldsymbol{A}$ describes the orientation of the crystal in respect of the recorded image. Since one uses lenses to image the diffraction pattern to the recording plane, immage distortions can occur. Luckily these errors do not influence the relative intensity of the diffracted beams, but only their positions. Figure (2) shows an example where the effect is a little overdone. However in most diffraction patterns, even taken with the greatest care small deviations from straight lines for the rows of diffraction spots can be observed. Due to the non-linearities expression (1) is not valid.

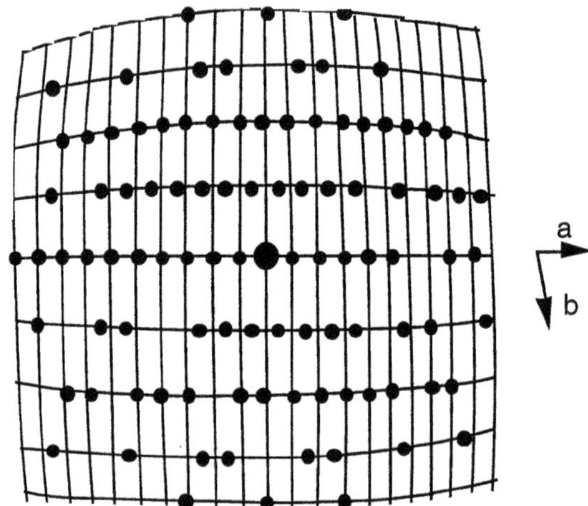

Figure 4. Example of the distortion of the diffraction pattern due to a focusing of the diffraction disks into relatively sharp spots.

To overcome this problem one can introduce correction terms to the righthand side of formula (1).

$$P = \sum_{n=0}^{N} \sum_{m=0}^{N-m} a_{nm} h_1^n \; h_2^m \qquad (2)$$

Where a_{nm} contains the origin and and the reciprocal vectors:

$$a_{00} = 0 \qquad (3)$$

$\mathbf{a}_{01}$ and $\mathbf{a}_{10}$ contain the elements of the matrix **A**. The N in formula (2) indicates the degree of the indexing function. if N equals 1 the linear indexing function of formula (1) is obtained. Since formula (1) is normally a close approximation, one expects that the higher order terms in (2) will be small. To have some measure for the goodness of the resulting indexing a Figure of Merit (FOM) is defined

$$\sum_{i=1}^{2} \sum_{H} (H_i - \mathbf{int}(H_i))^2 \; / \; \#\text{refl.} \qquad (4)$$

This FOM is based on the fact that all reflections can be described by integer indices. If the found positions give rise to rational indices then the FOM starts to rise. A perfect indexing gives a FOM of zero. Table 1 gives typical FOM's for different degree's N.

Table 1. Figure of merit for the fit between the calculated and experimental positions of the reflections using a polynome of the N-th power

N	FOM
1	.084
2	.072
3	.035
4	.033
5	.032

Determining integrated intensities Once the indexing is known all positions in the diffraction pattern where reflections are to be expected can be scanned to obtain indexed intensities. The integrated intensity can be determined by including this position in a box or circle, the size of which is to be determined by the background level. If by enlarging the box or circle no extra intensity above the background is found the optimal size is reached. A larger box would contain the same net intensity with larger calculation errors. Since inelastic scattered electrons should not be included in the resulting intensity the best way is to measure the actual background of a peak close to its position. In practice we take the boundary of the box or circle.

The Choice of Intensity calculation algorithm

If one has an atomic model one has to prove it is correct according to the observed data. In the field of crystallography normally a Least-squares program is used to perform this task and to improve the starting model. The choice of the algorithm which calculates the observed data from the atomic model is of utmost importance for the correctness of the method. The easietst way to calculate intensities is by the kinematic diffraction theory. For X-rays this works quite well. However since electrons interact strongly with the crystal, the electrons tend to be diffracted more than once. Therfore the kinematic theory is valid for very thin crystals only. In the example we will show that the kinematic refinements will lead to very poor results. A better way is to incorporate the dynamical scattering theory into the calculation of the reflection intensities. In literarure several algorithms to perform this task can be found, which lead to the same reflection intensities. Amongst them are 1) Bloch-waves 2) Multi-slice calculations 3) Column approximation. The algorithm for our program was selected by two criteria : It should be not too much computer time consuming and it should be able to handle all types of crystals and orientations. The latter criterion rules out the collumn approach, which allows very fast computing times but fails if the atom columns are too close together. The choice between Bloch waves and Mult-slice was decided on computing time in favour of the latter. Therefore our Least-squares program is called MSLS (Multi-Slice Least-Squares).

The Least-squares algorithm

The least-squares algorithm used is basically the linearised non-linear algorithm which is of common use in crystallographic structure refinement. The main part of the algorithm is a set of linear equation for parameter shifts **s**:

$$\mathbf{M}\ \mathbf{s} = \mathbf{v} \tag{5}$$

where the refinement matrix is given

$$M_{ij} = \sum_m w_m \frac{\partial I_m}{\partial p_j} \frac{\partial I_m}{\partial p_i} \tag{6}$$

and the vector by

$$v_i = \sum_m w_m \quad (I_m - I_{obs_m}) \frac{\partial I_m}{\partial p_i} \tag{7}$$

I_{obs_m} are the observed reflection intensities and I_m are the calculated reflection intensities; p_i is the ith parameter to be refined and w_m are the weights of the reflections. In theory these weights should be $1/\sigma(I)$, in which $\sigma(I)$ is the standard deviation of the intensity. The derivative of the intensity I_m to the parameters p_i raises some problems. The multi-slice algorithm is an iterative process for which no analytical function is available. So, the derivatives cannot be calculated analytically. In MSLS these derivatives are numerically calculated using the definition of the derivative

$$I'(p) = \frac{I(p+\delta) - I(p)}{\delta} \tag{8}$$

with δ tending to zero. Numerically one can only take a finite value for δ. However, δ cannot be too small, due to the accuracy of a computer. A too small δ will cause a random difference between $I(p+\delta)$ and $I(p)$ in the order of the numerical accuracy of the computer. In practice this difference will tend to zero and leads to a singular least-squares matrix M. The optimal values of δ were found by trial and error with simulated data. Just as in a traditional structure refinement program, the parameters in MSLS include atomic positions, Debye-Waller factors, scaling factors etc. As a result of using the multi-slice method, some more parameters may be involved: the crystal thickness, the crystal misalignment and absorption parameters. The misalignment of the crystal is expressed in terms of the centre of the Laue circle in the pattern. A crystal tilt results in a corresponding shift of this centre. Note that crystal tilt and beam tilt are equivalent for diffraction patterns. Therefore, we do not consider beam tilt in this paper.

The refinement of the thickness requires a special trick. The derivatives, I' (formula (8)), imply that the dependency of the intensity to the thickness is continuous. However, by dividing the crystal in as many slices as in the Multi-slice calculation, the thickness becomes a discrete parameter. If the δ in formula (8) is smaller than the slice size the resulting derivative will be zero in many cases since $I(p+\delta)$ and $I(p)$ are equal due to the calculation method. To overcome this problem, in MSLS a

crystal is divided in a certain number of equal slices and a last slice having a thickness of a fraction f of the other slices. It is assumed that this slice contains the same scattering potential as that of the other slices but multiplied by f.
Diffraction patterns may always be multiplied by a constant factor without changing the physics of the system. In MSLS this is a scaling
factor which is needed to scale the observed and calculated intensities to the same order of magnitude. This scaling factor is one of the refinable parameters. However, one has to take care with the definition of this factor. It should be defined in such a way that it is not dependent on the change of any other parameter in the refinement procedure, in particularly the crystal thickness and the absorption parameter. Therefore the actual scaling factor, C, was expressed as function of c, the parameter which was used in the refinement process:

$$C = c \sum_{H=0} I_H{}^{obs} \;/\; \sum_{H=0} I_H{}^{calc} \tag{9}$$

This results in the ideal case in a scaling factor C to be 1.0. In order to get a 3-dimensional crystal structure data from an electron diffraction pattern one zone is not enough in general, though for high symmetric spacegroups one special zone may be sufficient. MSLS allows for simultaneous refinement of several diffraction patterns, each with its own parameters for scaling, thickness and crystal alignment.

Correlation between crystal thickness and Debye-Waller factors
A small problem arises when the crystal thickness and temperature factors are refined simultaneously, because these parameters are highly correlated. Raising both the thickness and the temperature factors results in almost the same least-squares sum. This is not an artefact of the calculation method but lies in the behaviour of nature. Increasing the Debye-Waller factor of an atom means a less peaked scattering potential, which in turn results in a less sharply peaked interaction with the incident electron wave. It can be shown that a thickness of 5 nm and B=2 $Å^2$ will give about the same results as a thickness of 10 nm and B=6 $Å^2$.

3. $Ce_5Cu_{19}P_{12}$, an example

$Ce_5Cu_{19}P_{12}$ is an intermetallic compounds of which we wanted to know the atomic structure. This compound was available as polycrystalline material, which allows the use of X-ray or neutron powder diffraction methods to determine the structure, provided one has a good starting model. We explored another route of structure determination: by electron diffraction. This has as major advantages that secondary phases are not at all a problem (one can simply select small crystals of the right phase), the crystals can be very small, it does not have the problem of overlapping reflections and the composition can be verified by EDX element analysis. In order to obtain a good starting model for the least squares refinement of the electron diffraction data a special form of high resolution electron microscopy was used: through focus holography.

The through focus holography uses the focus dependence of the image distortion by the electron microscope to reconstruct the exit wave. Using algoritms developed by Van Dyck and Coene [3], all useful information is extracted from a series of high resolution electron microscope images taken at different defocus values with known defocus increments. The result is the exit wave function which contains amplitude as well as phase information up to the information limit of the microscope. The exit wave function - or in short exit wave - is the electron wave at the exit plane of the specimen. Since the electron wave passing through the crystal is continuously changed, the exit wave function is still thickness dependent, but not dependent on the distortion introduced by the electron

microscope. As such it provides a major advantage over conventional high resolution electron microscopy (HREM), yielding a more reliable starting model for the electron diffraction refinement.
Electron transparent areas were obtained by crushing. The unit cell of $Ce_5Cu_{19}P_{12}$ was determined by electron diffraction. Theis was done by scanning reciprocal space by tilting a crystal over about 90° and taking a series electron diffraction patterns for different tilts. The unit cell of $Ce_5Cu_{19}P_{12}$ was determined to be hexagonal with a = 1.275 nm and c= 0.396 nm. Because an analogous compound $La_5Cu_{19}P_{12}$ (with a = 1.2773(1) nm and c= 0.39876(3) nm) [4] adopts the space group P-62m we choose this spacegroup for the refinement. However, the spacegroup could have been determine easily by means of convergent beam electron diffraction [see lecture of Tanaka].

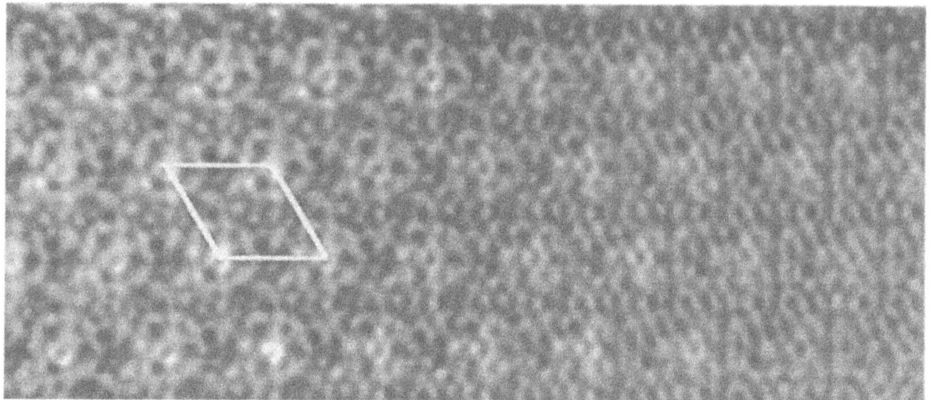

Figure 5 Reconstructed exit wave of $Ce_5Cu_{19}P_{12}$. Note the change in the image from left to right (thinner).

Through focus series of HREM images were taken with crystals in the [001] orientation. A typical example of the resulting exit waves is shown in Figure 5. One can observe in this exit wave that the thickness of the specimen increases rather strongly over the field of view (about 11 nm). This was also observed in other exit waves and conventional HREM images. This shape of the particles is inherent to the method of sample preparation. Transparent areas with a much smaller thickness variation over for instance 20 nm could have been obtained by ion milling. This method was not chosen in our investigation because $Ce_5Cu_{19}P_{12}$ is rather sensitive to moisture, but with proper precautions ion milling is certainly feasible.

Figure 6 shows a part of the exit wave shown in Figure 5 but after applying three fold symmetry. The white dots in this image represent atom columns or groups of atom columns. One can observe several (almost) round white dots and several elongated ones. Since the c axis is short (0.396 nm) and given the resolution of the exit wave, the elongated ones are assumed to be due to two atoms columns. This gives a number of unique atom positions as indicated in the figure. These estimated atom positions are given in Table 1.

Given the thickness dependence of the contrast of the atom columns in the exit waves as shown in Figure 5, the contrast of the dots in the exit waves cannot be used for a asignment of Ce, Cu or P to the various dots in the exit wave. Therefore the MSLS refinement was started by putting Cu atoms at all sites. First the scale factor, crystal (mis)orientation and thickness for each data set were refined. Nine data sets were taken to have a range of thicknesses. All subsequent refinements were done with these nine data sets simultanously. Next a refinement was done of the atom positions but keeping the Debeij Waller factors (B) of all atoms at zero and the ocupancy at 1. This resulted in an overall R value of 10.5 %.

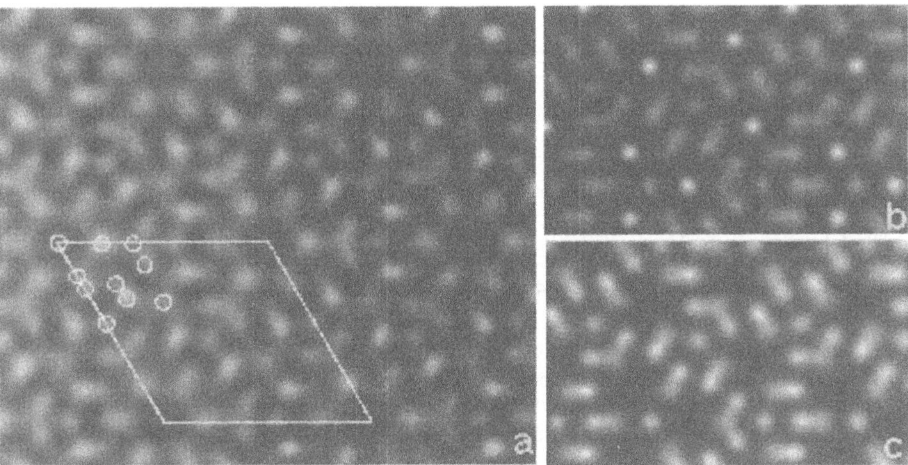

Figure 6. a) experimental exit wave (phase) after averaging (three fold symmetry and mirror plane). The unit cell is indicated. b) and c) are calculated exit waves for thicknesses of 3 nm and 6 nm.In b) and c) the Cu at 0,0,0 is in the left upper corner.

The refined atom positions were used as model for a refinement in which the atom positions were fixed and the occupancies of all atom positions were refined as well as the orientation and thickness of each data set. The resulting overall R value of this refinement was 4.8 %. The resulting occupancies are given in Table 2. One can distinghuish three distinct groups of occupancies: about 1.23 (1.20 and 1.26), about 1 (0.94, 0.95, 0.97 and 1.02) and about 0.68 (0.67, 0.67 and 0.69). These positions can be asigned to Ce, Cu and P respectively. One position (Cu4) has a rather deviating value of 0.74; from the composition it can be concluded that this should be Cu. In the following we will see that this position is indeed extraordinary.

The final refinement was done with Ce, Cu and P on the positions as discussed above. All atom positions and B's, including the crystal orientation and thickness of each data set were refined. The final atomic parameters are given in Table 3. The refined parameters of each diffraction set and the individual R values are given in Table 4.

All atoms have normal B's except Cu(5), which is denoted as Cu4 in Table 2. The B of Cu(5) is about 7 times higher than that of the other Cu positions. This, however, is in very good agreement with the structure data reported for $La_5Cu_{19}P_{12}$, which structure was determined with X-ray single crystal diffraction. Also for $La_5Cu_{19}P_{12}$ the Cu(5) position has a much higher B (6.5) as can be seen from Table 3, in which both the atomic positions and the B's of $La_5Cu_{19}P_{12}$ are given. The larger B for Cu(5) obtained for $La_5Cu_{19}P_{12}$ can be attributed to the anisotropy of B, since the X-ray refinement shows the strongest movement along the c axis. Since the MSLS refinement of $Ce_5Cu_{19}P_{12}$ is only done on [001] diffraction patterns the z component of B is not influencing the diffraction data and thus not measurable. The atomic positions obtained for $Ce_5Cu_{19}P_{12}$ are very close to those of $La_5Cu_{19}P_{12}$. The differences in the atomic positions can be mainly attibuted to the different size of Ce and La.

Table 2
Atomic positional parameters estimated from the reconstructed exit wave, those obtained after the refinement with all Cu atoms and the subsequently refined occupancies.

	Estimated from exit wave		Only positions refined		Only occupancies refined		
atom	x	y	x	y	x	y	n
Cu1	1/3	2/3	1/3	2/3	1/3	2/3	1.20(2)
Cu2	0.80	0	0.817(3)	0	0.8167	0	1.26(1)
Cu3	0.23	0	0.289(4)	0	0.2889	0	0.94(1)
Cu4	0.45	0	0.447 2)	0	0.4465	0	0.74(1)
Cu5	0.18	0	0.173(3)	0	0.1734	0	0.67(1)
Cu6	0.67	0	0.646(4)	0	0.6457	0	0.67(1)
Cu7	0.37	0.16	0.377(2)	0.177(3)	0.3768	0.1769	0.95(1)
Cu8	0.32	0.84	0.317(3)	0.833(3)	0.3174	0.8329	0.69(1)
Cu9	0.51	0.86	0.517(4)	0.875(3)	0.5172	0.8748	0.97(1)
Cu10	0	0	0	0	0.0000	0	1.02(2)

Table 3
Atomic positional parameters and isotropic temperature factors for $Ce_5Cu_{19}P_{12}$ as determined by MSLS and the atomic positional parameters and isotropic temperature factors for $La_5Cu_{19}P_{12}$ [4] as determined by X-ray single crystal diffraction. Except for the notation of the Ce, Cu and P atoms, the corresponding notation for the refinement with only Cu atoms as given in Table 5 is given for easy comparison.

Atoms		MSLS refinement for $Ce_5Cu_{19}P_{12}$				Results X-ray single crystal refinement for $La_5Cu_{19}P_{12}$		
	all Cu	x	y	z	B	x	y	B
Ce(1)	Cu1	1/3	2/3	0	0.06(5)	1/3	2/3	0.65(3)
Ce(2)	Cu2	0.8141(9)	0	1/2	0.06(4)	0.8067(1)	0	0.62(4)
Cu(1)	Cu10	0	0	0	0.44(8)	0	0	0.8(1)
Cu(2)	Cu3	0.2905(9)	0	1/2	0.63(5)	0.2878(2)	0	1.0(1)
Cu(3)	Cu7	0.3812(6)	0.1784(7)	0	0.80(5)	0.3783(2)	0.1725(2)	1.2(1)
Cu(4)	Cu9	0.5203(7)	0.8812(6)	1/2	0.68(4)	0.5174(2)	0.8815(2)	1.2(1)
Cu(5)	Cu4	0.4536(6)	0	0	4.10(11)	0.4508(2)	0	6.5(3)
P(1)	Cu5	0.1782(8)	0	0	0.63(6)	0.1768(2)	0	0.6(1)
P(2)	Cu6	0.6415(5)	0	0	0.99(8)	0.6300(5)	0	0.9(2)
P(3)	Cu8	0.3168(4)	0.8302(4)	1/2	0.95(7)	0.3074(4)	0.8276(4)	0.9(2)

Table 4
Data on the [001] electron diffraction sets used for the structure reported in Table 2. The overall R value in the final MSLS refinement is 2.7 % for the significant reflections and 3.2 % for all reflections. Three R values are given: R(I) being $\Sigma(I(obs)-I(calc))^2/\Sigma I(obs)^2$ and R(√I) being $\Sigma(\sqrt{I(obs)}-\sqrt{I(calc)})^2/\Sigma I(obs)$, to show that the R values do not change drastically when one uses R(√I) instead of R(I). $R(I)_{kin}$ gives the R value with a quasi-kinematic refinement.

thickness	number obs. refl.	crystal misorientation h	k	R(I)	R(√I)	$R(I)_{kin}$
96(2)	305	2.90(13)	-1.43(9)	1.69	2.08	26.9
118(2)	348	1.56(5)	-0.76(6)	2.09	2.66	63.3
118(2)	264	-0.23(5)	-1.71(7)	4.18	5.07	23.0
128(2)	238	-0.18(4)	-0.94(4)	1.09	1.78	33.1
130(2)	306	3.20(8)	0.61(6)	6.74	8.34	25.5
131(2)	154	1.64(6)	-1.01(5)	0.96	0.99	32.1
157(2)	156	0.07(3)	-0.44(3)	1.59	1.81	49.6
163(2)	137	-0.20(3)	-0.62(3)	3.31	3.34	40.6
174(2)	330	1.84(4)	-0.92(3)	4.04	3.58	63.5

The z parameters are not determined in the MSLS refinement because only [001] diffraction patterns were used. With the knowledge that the atoms are located at z=0 or z=1/2 and taking the Cu(1) atom as origin, the z coordinates of all other atoms can be simply determined when one considers the plausible interatomic distances. Although the structure of $Ce_5Cu_{19}P_{12}$ is rather complicated, the interatomic distances are still within a narrow range, e.g. Ce-Cu 0.308-0.326 nm, Ce-P 0.295-0.305 nm, Cu-Cu 0.279-0.286 nm and Cu-P 0.227-0.249 nm. No short Ce-Ce or P-P distances occur.

The R values assuming kinematic diffraction are also given in Table 4. For the calculation of these R values the MSLS program was used with occupancies of 0.1%; only the scalew factors were refined. Evidently these R values are much higher then the ones taking into account the dynamic scattering. A kinematic refinement starting with the atoms at the positions estimated from the exit wave or at the positions given in Table 2 does not lead to R values below 40% nor reliable atom positions.

4. Conclusion

It is possible to determine structures from dynamic electron diffraction data. In most cases low R-values are obtained, which are comparable to those of X-ray single crystal diffraction data. The presence of strong dynamic scattering makes the structure refinement more complex and certainly more time consuming. As such it is a disadvantage. However, it also has a large advantage, because the scattering of a column of atoms depends not only on the scattering potential of the atoms but also on the thickness, resulting in an oscillation in the scattering potential of the column with thickness. Thus for certain thicknesses a column containing only weakly scattering atoms show a higher scattering than a column containing strongly scattering atoms. Consequently by using various thicknesses the weakly scattering atoms can be determined with a higher accuracy. Furthermore it can be used to discriminate between atom types as has been done in the refinement of $Ce_5Cu_{19}P_{12}$, which was started with a model with only Cu atoms on all sites, where by refining the occupancies the locations of the Ce and P atoms were determined.

References

1. Structure of 13K superconductor $La_3Ni_2B_2N_3$ and the related phase LaNiBN, H.W.Zandbergen, J.Jansen, R.J.Cava, J.J.Krajewski, W.F.Peck Jr, E.M. Gyorgy, *Nature* **372**, 759 (1994)
2. MSLS, a least-squares for accurate crystal structure refinement from dynamical electron diffraction patterns, J. Jansen, D. Tang, H.W. Zandbergen and H. Schenk, submitted to Acta Cryst A.
3. Maximum likelihood method for focus-variation image reconstruction in high resolution transmission electron microscopy. W.M.J. Coene, A. Thust, M.Op De Beeck, D.Van Dyck, *Ultramicroscopy* 64, 109 (1996)
4. Crystal srtructure and physical properties of $La_5Cu_{19}P_{12}$ and $Ce_5Cu_{19}P_{12}$, R.J. Cava, T. Siegrist, S.A. Carter, J.J. Krajewski, W.F. Peck and H.W. Zandbergen, *J. Solid State Chem*, in press.

FOURIER REFINEMENT IN ELECTRON CRYSTALLOGRAPHY

Douglas L. Dorset,
Hauptman-Woodward Medical Research Institute,
73 High Street, Buffalo, NY 14203-1196 USA

1. Introduction

As also experienced in x-ray crystallography, a successful direct phase determination with electron diffraction data might only reveal a structural fragment in the initial potential map. This map $\rho(\mathbf{r})$ is calculated from the partial $|F_h|, \alpha_h$ of the accessed reflection set, via:

$$\rho(\mathbf{r}) = V^{-1} \sum_{hkl} |F_o| \exp(i\alpha_{hkl}) \exp(-2\pi i \mathbf{h} \cdot \mathbf{r}) \qquad (1)$$

Recognition of a partial structure solution often requires some knowledge by the experimentalist of molecular (or group) geometry, in addition to the stoichiometry of atomic constituents.

2. A Typical, Nearly-Kinematical, Example

A typical example is the structure analysis of poly (*p*-oxybenzoate)[1] in the [100] projection of the phase I structure, grown as whiskers during polymerization[2]. (The unit cell constants are $\mathbf{a} = 7.45$, $\mathbf{b} = 5.64$, $\mathbf{c} = 12.47$ Å in an orthorhombic cell; the projected symmetry was assumed to be pg due to systematic absences for $\ell = 2n+1$.) The initial map in Figure 1a was generated with phases derived from the Sayre expansion of a basis set. (Since algebraic unknowns were used to generate multiple solutions, anticipated chain geometry, again, was very important for identification of the most promising solution.) Note that only a few atomic positions were observed. Completion of the structure was achieved by Fourier refinement, a procedure that is found to be useful when only two-dimensional, or incompletely-sampled three-dimensional electron diffraction intensities are available.

Trial atomic positions were first ascertained not to violate allowed bond distances and angles (beyond some initial relaxation parameter). These were then used to calculate the first complete set of trial crystallographic phases α_h via the kinematical structure factor expression (where $F_h = |F_h|\exp(i\alpha_h)$):

$$F_h = \sum_j f_j' \exp(2\pi i \mathbf{h} \cdot \mathbf{r}) \qquad (2)$$

where f_j' are electron scattering factors corrected for thermal motion. (An overall B_{iso} Debye-Waller factor is obtained from a Wilson plot[3] before calculating normalized $|E_h(\text{obs})|$. These phases are combined with the $|F_h(\text{obs})|$ values in expression (1) in order to caclulate a new potential map. For the polymer structure (Fig. 1b), correctly identified atomic positions were found to be reinforced. (It is not a good sign if they are not!) Suggested new atomic positions were also

D. L. Dorset et al. (eds.), Electron Crystallography, 243–246.

identified (i. e., other density sites) and these could be used to generate a new phase set via (2). This cyclic procedure was continued until all atoms in the structure could be located at reasonable positions. During this refinement, the crystallographic residual was also minimized.

3. Examples Including Multiple Scattering Perturbations

There can be difficulties with these refinements if the intensity data are significantly perturbed by multiple scattering. An example of an organic structure with heavy atom substituents copper perbromophthalocyanine, can be cited, based on electron diffraction intensities collected at 1200 kV. While the experimental Patterson function was sufficiently similar to the autocorrelation function of the actual crystal structure to permit an ab initio structure analysis, an unconstrained Fourier refinement from initial atomic positions was found to lead to a final potential map that was chemically incorrect (distorted bond distances and angles). In particular, the carbon-halogen bond lengths were far too short and could not be improved in successive cycles of Fourier refinement[4]. The distorted final map coordinates gave an R-value lower than that calculated from more reasonable positions - i. e. the correct structure was not found at a global minimum of the crystallographic residual. If, however, the final map was used to guide the placement of an idealized molecular model with chemically meaningful bonding parameters, a correct solution could be found in a rotational search, even when the kinematical residual was the figure of merit. A 26° tilt value of the molecular plane around **b** (preserving the mm symmetry of the projection) corresponded to the correct crystal structure and the solution could be verified by convergence of a multislice dynamical calulation[5].

Multiple scattering perturbations have also restricted three-dimensional structure refinements[6]. For example, direct phasing of 50 kV texture electron diffraction data from two thiourea polymorphs was successful. However, only the sulfur positions could be located accurately. Attempts to move carbon and nitrogen postions to sites required by known bond distances and angles in successive cycles of Fourier refinement were not very successful, even though the overall phase error was not very large (e. g. 16° for the noncentrosymmetric ferrolectric form[7]).

When it is possible to visualize heavy atom position in high-resolution electron micrographs - e. g. copper perchlorphthalocyanine[8] - a re-scaling[9] of the $|F_h(\text{obs})|$ could be made based on the heavy atom sites via partial kinemetical structure factor calculations (equation (2)). The re-scaling was carried out in zones of $\sin\theta/\lambda$ via $k = |F'_{calc}|^2/|F_{obs}|^2$ to give $|F'_{obs}| = (kI_{obs})^{1/2}$. This procedure greatly improved the refinement of the phthalocyanine derivative[5].

4. The Detection of Light Atom Positions

Finally, it was once imagined that Fourier refinement in electron crystallography would lead to more accurate hydrogen atom positions than were possible in x-ray determinations[10], because of the more favorable ratio of f_H/f_X, where X = C,N,O. Difference maps could be calculated after all heavy atom positions were identified by:

$$\Delta\rho(\mathbf{r}) = V^{-1}\sum_{hkl} |F_o|-|F_c| \exp(i\alpha_h) \exp(-2\pi i\mathbf{h}\cdot\mathbf{r}) \qquad (3)$$

While reasonable hydrogen atom positions have been located for n-paraffin layers or boric acid from electron diffraction data[11,12], some other other reports of similar success for other small organic molecules (diketopiperazine, urea, two forms of thiourea) could not be confirmed, even though direct phase determination verified that the heavier atomic components could be found in ab initio analyses[6]. The difficulty may have been partly due to multiple scattering perturbations since the Wilson plots for the 50 kV texture data sets indicated that $B_{iso} \approx 0.0$ Å^2, a sign that such a data perturbation had occurred. Another contributing factor may have been radiation damage since hydrogen is preferentially abstracted from aliphatic molecules. Higher voltage might lead to a successful determination of hydrogen positions. For example, texture diffraction data from the mineral brucite, collected at 250 kV, permitted the location of hydrogen sites very near those found earlier from neutron diffraction data[13].

5. References

1. J. Liu, B. L. Yuan, P. H. Geil, and D. L. Dorset (1997) Chain conformation and molecular packing in poly (p-oxybenzoate) single crystals at ambient temperature, Polymer, in press.

2. F. Rybnikar, J. Liu, and P. H. Geil (1994) Thin film melt polymerized single crystals of poly (p-oxybenzoate), Macromol. Chem. Phys. 195, 81-104.

3. A. J. C. Wilson (1942) Determination of absolute from relative x-ray intensity data, Nature 150, 151-152.

4. D. L. Dorset, W. F. Tivol, and J. N. Turner (1992) Dynamical scattering and electron crystallography - ab initio structure analysis of copper perbromophthalocyanine. Acta Cryst. A48, 562-568.

5. D. L. Dorset (1997) The accurate electron crystallographic refinement of organic structures containing heavy atoms, Acta Cryst. A53, 356-365.

6. D. L. Dorset (1995) Structural Electron Crystallography, Plenum, NY.

7. D. L. Dorset (1992) Automated phase determination in electron crystallography: thermotropic phases of thiourea, Ultramicroscopy 45, 357-364.

8. N. Uyeda, T. Kobayashi, K. Ishizuka, and Y. Fujiyoshi (1978-1979) High voltage electron microscopy for image discrimination of constituent atoms in crystals and molecules, Chem. Scripta 14, 47-61.

9. D. X. Huang, W. Liu, Y. X. Gu, J. W. Xiong, H. F. Fan, and F. H. Li (1996) A method of electron diffraction intensity correction in combination with high-resolution electron microscopy, Acta Cryst. A52, 152-157.

10. B. K. Vainshtein (1964) Fourier synthesis of potential in electron diffraction structure analysis and its applications to the study of hydrogen atoms, Advances in Structure Research by Diffraction Methods (Vol. 1), R. Brill, ed., Wiley Interscience, NY, pp. 24-54.

11. B. K. Vainshtein, A. N. Lobachev, and M. M. Stasova (1958) Electron diffraction determination of the C-H distances in some paraffins, Sov. Phys. Crystallogr. 3, 452-459.

12. D. L. Dorset (1992) Direct methods in electron crystallography - structure analysis of

boric acid, Acta Cryst. A48, 568-574.

13. B. B. Zvyagin, personal communication.

14. Supported by a grant from the National Science Foundation (CHE9417835)

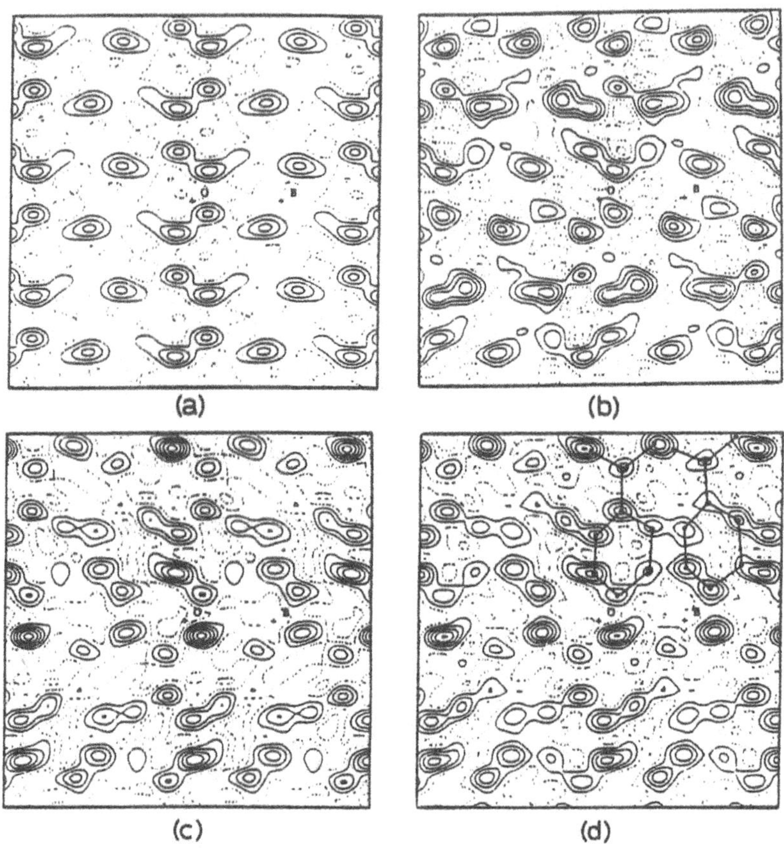

Figure 1. Fourier refinement of poly (*p*-oxybenzoate) from intial map (a) to final map (d), where trial structural model is indicated for the asymmetric unit.

STRUCTURE REFINEMENT THROUGH MATCHING OF EXPERIMENTAL AND SIMULATED HRTEM IMAGES

ROAR KILAAS
National Center for Electron Microscopy
Lawrence Berkeley National Laboratory,
Berkeley, CA 94720, USA

1. Introduction

The goal of performing simulation of HRTEM images is to compare these with the experimental data in order to determine the structure. In practice this means that various models are proposed and that images are calculated until a match is found. At that point, the structure is presumed to be known (atomic positions and atomic numbers) with some given uncertainty. Alternatively, one starts with a given model and varies the model in a systematic fashion searching for a global maximum in the fit between experiment and simulation. This entails that one needs an efficient method to compare the experimental and the calculated image. It also requires knowledge of the uncertainty in the measurement (image intensities in the experimental image) and a way to relate this uncertainty to the uncertainty in chemical composition and atomic positions. This area of quantitative electron microscopy is fairly new and most images are still compared visually. However, it is an active area of research and many techniques from statistics are just now beginning to be used in HRTEM.

2. Acquiring quantitative data

In order to extract quantitative information from electron micrographs, the data must be represented by a set of numbers. Usually, images from TEMs are brought into a digital representation by one of the following methods.

i) Recording the image on a photographic plate and using a scanner to convert the film density into numbers which are stored in a computer.

ii) Recording the image on an image plate.

D. L. Dorset et al. (eds.), Electron Crystallography, 247–260.

iii) Recording the image on a slow scan CCD camera with readout of deposited charge into a computer.

The first approach yields data that is not directly comparable to computer calculations because of the non-linear response of the film. It is however possible to calibrate the response of the film based upon a sequence of controlled exposures using varying exposure times and mapping the resulting scan values versus electron dosage[1]. The image plate and the CCD camera both yield numbers that are linear with respect to the electron dosage and only require a scaling of the data in order to compare to computed values[2]. There has been much discussion about the relative merits of the various recording media above and each has its own advantage. The CCD camera is currently limited to 2K by 2K pixels, although it may be possible to go to 6K by 6K by using multiple chips in the near future. Since its Modulation Transfer Function (MTF) can be characterized, it is straight forward to use deconvolution to compensate for the drop in high frequency response due to spread of electrons and due to spill over of charge to neighboring cells[3]. The image plate has many of the advantages of the CCD camera and covers a larger image area. However, the imaging plate is not gaining as much popularity as the CCD camera. Many laboratories are now starting to do much of the recording on CCD cameras, while still retaining the use of film.

3. Pre-processing of data

Once the image data has been converted to numbers, any necessary processing or transformation of the data can take place. The required pre-processing of the data depends on the nature of the information that is sought and thus there is no one optimal method, but rather a number of possible options.

3.1 SAMPLING AND RESAMPLING OF DATA.

If the image is distorted over the image field of view, either by the action of the imaging system or the recording system, the data can be re-transformed by a warping transformation of the image. This can be done if the transformation can be determined by imaging a perfectly crystalline material and noting deviations from where the atoms are known to be and where they are imaged[4]. On some systems, i.e. the Gatan Imaging Filter, the distortions are measured by recording the image of a square grid of circular holes.

An image of a crystalline material can be resampled onto lattice relative coordinates, such that the unit cell dimensions are represented by an integer number of pixels and commensurate with the dimensions of the final image. This will eliminate streaking in the Fourier transform of the image which is due to the truncation of the image by the edges on boundaries that do not represent a periodic continuation of the image. Streaking can be reduced by multiplying the image with a circular mask. The mask is represented by a circle of pixels with value 1 up to a specified radius and then falling off gradually to 0 within 5 to 10 pixels close to the borders of the image. A side-effect of masking is the increase in noise in the Fourier transform which is discussed below.

If only the image of a single unit cell of the crystalline material need to be determined and compared with the image obtained through an image simulation calculation, the image of the unit cell can be resampled onto the coordinate system and sampling interval used in the computation. This is equivalent to determining the matrix M defined through the equations

$$\mathbf{a}_S = \mathrm{M}\mathbf{a}_e \qquad (1)$$

$$\mathbf{b}_S = \mathrm{M}\mathbf{b}_e$$

Two steps are necessary; the rotation /scaling required to make the lattice base vectors identical ; and secondly the determination of a common origin.

Finding the common origin between experimental and calculated image is determined by cross-correlation between the simulated and the experimental image[5].

3.2 FOURIER TRANSFORMS AND MASKING

The Fourier transform of HRTEM images of crystalline materials provides useful information about lattice spacings and can also be used to compare experimental Fourier amplitudes with theoretical calculations. Because the image being transformed is rarely a periodic function in W (width) and H (height), the Fourier transform of the image of a pure crystalline material is the convolution of the Fourier transform of a perfectly periodic signal (the crystal) with the transform of a window the size of the image dimensions, making a Bragg peak take on the shape of the transform of the window.

name	**window**	**function**	**transform**	**peak profile**	**falloff**	**rel. noise level**
none	1	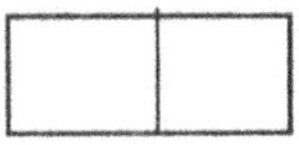	$\frac{\sin(\pi k)}{\pi k}$	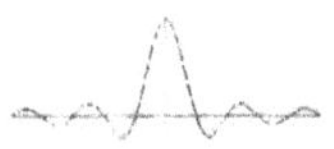	$\frac{1}{k}$	1
cosine	$\frac{\pi}{2}\cos\left(\frac{\pi x}{2}\right)$		$\frac{\cos(\pi k)}{1-4k^2}$		$\frac{1}{k^2}$	1.23
von Hann	$1+\cos(\pi x)$		$\frac{\sin(\pi k)}{\pi k(1-k^2)}$		$\frac{1}{k^3}$	1.5

The use of a mask changes the transform of the window and can be used to make the peak profiles decay faster, but at the expense of increasing the noise level. This is illustrated above, showing the effects of employing masks on a 1-dimensional signal[6]. This also has an effect on locating peak positions in order to determine lattice spacings and on the estimate of the amplitude of the Fourier component. The standard error in both estimates increases as a function of applying a mask, with the cosine window being a good compromise.

3.3 NOISE REDUCTION

In addition to reducing noise, it is also important to have an estimate for amount of noise present and to quote a signal to noise ratio. From two equivalent regions, the noise can be estimated from obtaining the cross-correlation coefficient for two regions. Given a cross-correlation coefficient ccf, the signal to noise ratio can be estimated as

$$S/N \approx \frac{CCF}{1 - CCF} \tag{2}$$

In order to reduce noise and to obtain a statistical average of the image of a single unit cell (motif), the positions of individual motifs can be determined by cross-correlation. Once these are found, statistically equivalent regions can be averaged to find the average motif and to determine the signal to noise ratio associated with individual pixels as a function of position within the unit cell. This determines a standard deviation for each pixel i and can be used to set confidence levels associated with matching of the experimentally averaged image with a calculated image[7].

$$\sigma^2(i) = \frac{1}{M-1} \sum_{j=1}^{M} \left(I_j^2(i) - \langle I(i) \rangle^2 \right) \tag{3}$$

where M is the number of equivalent regions being averaged.

Using a low pass filter to perform a smoothing of the image may be effective depending on the noise level present, particularly when averaging over statistically equivalent regions can not be performed. Smoothing helps the eye see features more clearly; but has the disadvantage that it causes correlation between image pixels, which may distort the significance threshold of simulation mismatch criteria.

Averaging can also be performed through symmetrization which is to average the motif with copies of itself to which symmetry operations known to be present are performed. This will reduce noise levels by a further factor of $1/\sqrt{M}$ when M symmetry related copies are averaged, but may also just disguise defects in imaging conditions.

4. Matching experimental and simulated images

There are a number of various ways to measure similarity or mismatch between two images. Below are a few of these[8].

The mean square difference :

$$D^2 = \left\langle (I_1 - I_2)^2 \right\rangle = \frac{1}{N}\sum (I_1 - I_2)^2 \tag{4}$$

The Root Mean Square Difference:

$$D_{rms} = \sqrt{D^2} \tag{5}$$

The mean modulus difference:

$$D_{mmd} = \left\langle |I_1 - I_2| \right\rangle = \frac{1}{N}\sum_i |I_1 - I_2| \tag{6}$$

The Cross-correlation Coefficient:

$$CCF = \frac{\sum_i (I_1(i) - \bar{I}_1)\cdot(I_2(i) - \bar{I}_2)}{\sqrt{\sum (I_1(i) - \bar{I}_1)^2 \sum (I_2(i) - \bar{I}_2)^2}} \tag{7}$$

The brackets $\langle\ \rangle$ all indicate the mean of the enclosed quantity.

In each of these equations, the sum is over all the pixels i in the image and N is the total number of pixels. The cross-correlation coefficient above is a normalized coefficient where the images are normalized to zero mean.

The CCF (which measures similarity rather than difference) can also be interpreted as the cross-product between two n-dimensional vectors (n being the number of pixels in the image). In that case, one can associate an angle with the CCF, $CCF = \cos\vartheta$, in the general interpretation of an inner product between two vectors as $I_1 \bullet I_2 = |I_1| \cdot |I_2| \cos\vartheta$ with the angle being $\vartheta = \cos^{-1}(CCF)$ This angle is zero for identical images. If the images are normalized to zero mean and unit length as in the definition of the normalized cross-correlation coefficient above, the angle is 180 deg. for a reversal in contrast between the two images I_1 and I_2.

4.1 SIGNIFICANCE AND NOISE.

Each of the above criteria must be tested for the significance of the measured value.

D^2 can be compared to the mean square intensity (or intensity deviation due to noise) in either image.

D_{rms} can be compared with standard deviation of the intensity in either image

A good way to test for the mismatch between two images, is to use a statistical measure for the probability of two images being equal given knowledge of the noise in the images. If one assumes Gaussian uncorrelated noise for each pixel in the experimental image, the optimum statistical measure is given by

$$\chi^2 = \frac{1}{N} \sum \frac{(I_1(i) - I_2(i))^2}{\sigma^2(i)} \tag{8}$$

where N is the number of pixels in the image[9]. The value $\sigma^2(i)$ is the standard deviation associated with the pixel i and can be found as described above from a number of equivalent regions. If an experimental image I_e is compared to a calculated image I_c and there are M adjustable parameters in the calculation, the equivalent expression becomes[7]

$$\chi^2 = \frac{1}{(N-M)} \sum \frac{(I_e(i) - I_c(i))^2}{\sigma_e^2(i)} \tag{9}$$

A mismatch by one standard deviation adds one to the sum in the expressions above and a value of χ^2 of 1 implies that the two images are identical within the uncertainty given by the noise. The expected value for statistically equivalent images consisting of N points is 1 and random deviations from this value by more than $2/\sqrt{N}$ are considered unlikely.

Writing

$$\chi^2 = \frac{1}{N} \sum \frac{(I_1(i) - I_2(i))^2}{\sigma^2(i)} = \frac{1}{N} \sum f^2(i) \tag{10}$$

leads to the definition of a Residual Image $f(i)$[10]

which is used to visualize and to quantify the (mis)match between two images. It has the advantage that instead of presenting a single number for how well two images match, it is a two-dimensional mapping of the local fit. Thus a difference image will more clearly reveal areas of greater mismatch. The optimum match is still defined by minimizing χ^2.

It is important to note that the fitting parameters can also be applied to the Fourier transforms of the images which sometimes will lead to a reduction in the number of the data-points to be compared[11]. In the case of images of crystalline material containing no defects, the Fourier components will be non-zero only for frequencies corresponding to Bragg-reflections of the lattice, although this is strictly only true if the motif has been averaged over many repeating regions and resampled onto lattice coordinates such that streaking due to discontinuities at the boundaries is eliminated. The complex values for the Fourier coefficients take the place of the image intensities.

It is interesting to note that the use of different matching criteria can lead to slightly different values for optimized parameters[12]. An example of this is shown in the table below where the parameters for a relaxation vector of a terminating layer of Nb atoms in a niobium / sapphire interface has been optimized using a range of different image matching parameters.

Image Matching Parameter	**R_x [Å]**	**R_y [Å]**
nip	0.1	-0.1
xcf	0.2	0.1
fmad	-0.1	-0.3
nmad	-0.2	-0.4
rfac	0.1	0.0
nqd	0.0	-0.3
mrd	-0.2	-0.4
disf	0.1	0.1

4.2 ADJUSTING FOR DIFFERENT MEANS AND CONTRAST LEVELS

Since absolute values for image intensities are not known and an experimental image may be linearly related to a calculated image, a useful way of normalizing the image intensities is to subtract the mean and divide by the standard deviation. This ensures that $D^2 = 0$ for linearly related images and a value of around 2 for unrelated data.

Similarly, the Cross-correlation coefficient will lie in the range from -1 to 1, taking the extreme values when the two images are linearly related and being near 0 for unrelated data.

Another approach is to scale the images to the same mean. This is done as follows:

$$I_{calc} = \frac{I_{calc}}{\langle I_{calc} \rangle} \langle I_{\exp} \rangle \tag{11}$$

where the calculated image is scaled to the mean of the experimental image.

In order to understand how the mean value, contrast and image pattern affect the image matching criteria, it is useful to consider how the Root Mean Squared Difference D_{rms} can be separated into three terms[13].

$$D_{rm} = \sqrt{\left\langle (I_1 - I_2)^2 \right\rangle} = \left(\langle I_1 \rangle - \langle I_2 \rangle \right)^2 + \left[\sigma_1 - \sigma_2 \right]^2 + 2(1-\rho)\sigma_1\sigma_2 \qquad (12)$$

where

$$\sigma_{1,2} = \sqrt{\left\langle I_{1,2}{}^2 \right\rangle - \left\langle I_{1,2} \right\rangle^2} \qquad (13)$$

and

$$\rho = \frac{\langle I_1 I_2 \rangle - \langle I_1 \rangle \langle I_2 \rangle}{\sigma_1 \sigma_2} \qquad (14)$$

The first term measures the difference in the mean of the two images and vanishes if both images are normalized to the same mean value. The second term measures the difference in contrast between the two images, while the third term (where ρ is the same as the normalized cross correlation coefficient) measures the difference (similarity) in the pattern of the two images. Thus it is important to note that the normalized cross correlation coefficient CCF only measures similarity in patterns and ignores variation in contrast and differences in mean levels. It is generally found that most of the mismatch between experimental and computer simulated images is due to the difference in contrast[14]. The difference in contrast can be an order of magnitude and the cause is generally attributed to the following factors:

- misalignments
- specimen vibration
- inelastic scattering
- specimen damage

There is however an ongoing debate as to the nature of the discrepancy in contrast as calculations indicate that the factors above are not sufficient to resolve the disparity. A possible explanation is that there is a general background in experimental images that is not accounted for.

4.3 EFFECT OF NOISE ON MATCHING CRITERIA

In order for two images to be considered equal, we need to consider the effect of the uncertainty or error in the matching criteria due to noise and the parameters determining the image.

A study of the effect of noise on the cross-correlation factor reveals that in the presence of noise, the cross-correlation coefficient CCF for the two images I_1 and $I_2+\eta$, where η represent random noise superimposed on image I_2, can be written as[13]

$$CCF(I_1,I_2,\eta)=CCF(I_1,I_2)\Big/\sqrt{1+\frac{\sigma^2(\eta)}{\sigma^2(I_2)}}=CCF(I_1,I_2)\Big/\sqrt{1+\vartheta_n^2} \tag{15}$$

with

$$\vartheta_\eta^2 \equiv \frac{\sigma^2(\eta)}{\sigma^2(I\ )} \tag{16}$$

The effect on the hyper-angle $\vartheta=\cos^{-1}(CCF)$ is in the small angle approximation

$$\vartheta(I_1.I_2,\eta)\approx\sqrt{\vartheta^2(I_1.I_2)+\vartheta_\eta^2} \tag{17}$$

If two images are identical except for a small error in one of the image-formation parameters (defocus, thickness, etc.) the error in the angle ϑ is proportional to the parameter error. The error in the angle due to independent parameter errors is

$$\theta=\sqrt{\sum\vartheta_i^2} \tag{18}$$

Typical mismatches in CCF (pattern matching) due to parameter errors are

Parameter	**Error**	**theta(mrad)**
Noise		0.06
Composition	±0.03	0.02
Thickness	±2nm	0.2
Defocus	±15nm	0.4
Beam Tilt	<1.5mrad	0.8
Astigmatism	<15nm	0.2
Crystal Tilt	<2mrad	0.6
Beam Diverg.	<0.3mrad	0.1
Focal Spread	<5nm	0.15
Vibration	<0.04nm	0.2

4.4 CHI-SQUARE OR CHI-BASED CRITERIA

Although all the methods above measure either the match or mismatch between two images, the important questions is not to what degree do they match, but how well do they match given systematic and non-systematic errors. Thus the fitting parameter must take into account the statistical nature of the data and the accuracy to which we know the data-points. Thus the fitting parameter should depend on a maximum-likelihood (probability) model and be a measure of the probability that A is equal to B, given knowledge of the probability distribution of the data-points. In the presence of Gaussian distribution of uncorrelated noise, each data point has a Gaussian probability distribution with the noise in one pixel uncorrelated to the noise in adjacent pixels, which leads to a χ^2 criteria. The criteria takes into account the number of adjustable parameters and the error in each data point.

As mentioned above, any data point lying one sigma away from the expected value will add 1 to the sum in χ^2.

Similarly, any data point which has only 1% probability of being measured given A = B, adds a value of 6.63 to the sum in χ^2. Thus values of χ^2 greater than about 6 states that there is less than 1% probability that A is equal to B.

The fitting parameter depends on the model of the distribution of data-points due to statistical noise with a Gaussian distribution of uncorrelated noise leading to the χ^2 criteria. However, it is important to determine the statistical nature of the noise in the image. This can be done by examining the noise distribution determined from a large number of image regions considered to be equivalent except for noise. A non-Gaussian distribution will lead to a modified criteria, but still based upon χ [7].

5. Structure determination

In order to determine the "unknown" structure, it is necessary to perform a comparison between calculated images, exit wavefunctions or diffraction patterns and experimentally obtained data. As described above the comparison can be done using different matching/mismatching criteria. Ideally, the determination of the structure is done by modifying the structure until the mismatch between the experimental and calculated data is within the error in the experimental data. In principle the imaging parameters themselves can be allowed to vary together with the atomic coordinates. However, in practice the imaging parameters are optimized separately if possible. This reduces the complexity of the problem and reduces the number of steps involved in the search for a solution which optimizes the matching criteria. In cases involving unknown defects in the presence of a "known" structure, the imaging parameters and specimen thickness are first determined from the known structure.

Determination of an unknown set of input parameters requires the following:

1) An image (in real or reciprocal space) obtained from the experimental data.

2) A computational method yielding an image to be compared to the 1).

3) A method for comparing 1) and 2)

4) A criteria based upon 3) for when 1) and 2) are statistically equivalent.

5) An initial set of adjustable input parameters which are to be optimized so that the final configuration results in satisfying 4)

6) A method for varying the adjustable parameters so that the final configuration is found within finite time.

There is an essential assumption being made above, which states that the computational method used in 2) will produce the image in 1) given the correct choice of input parameters. This is a separate issue which will not be addressed here. The validity of this assumption can be debated and it is acknowledged that computational methods are in need of further refinement. However, in what follows, the assumption is presumed to be valid.

5.1 MATCHING IMAGES OR EXIT WAVEFUNCTIONS

In order to compare calculation with experiment, one can compare either images or diffraction patterns. For perfect structures it may be beneficial to compare diffraction patterns since the number of data points to compare are given by the possible Bragg reflections of the structure[11]. However, for defect structures, the information that describes the defect is located in the diffuse scattering between Bragg spots, and it is more efficient to compare images. The entire discussion relates to both real space and reciprocal space, although only real space images will be referred to.

5.2 SIMULATED THERMAL ANNEALING

Simulated thermal annealing is a relatively new technique for finding the global minimum of a multivariate function[15]. The algorithm is based upon assigning an energy to the system which is a function of the parameters being varied, with the optimum configuration of the system being the minimum energy state, the ground state. A temperature is also assigned to the system and the temperature is slowly being reduced as the configuration is changed. From the initial configuration $E_0(x_1,x_2,....x_n)$, the parameters are varied in a random fashion and for each variation the new energy $E_j(x_1,x_2,....x_n)$ is calculated. The new configuration is always accepted if $\Delta E = E_j(x_1,x_2,....x_n) - E_{j-1}(x_1,x_2,....x_n) < 0$. Otherwise the new configuration has a probability P of being accepted, where

$$P = e^{-\Delta E / T} \qquad (19)$$

E and T being dimensionless quantities.

For each temperature, the system undergoes a given number of variations, accepting or rejecting the new configuration based upon the above criteria. When a specified number of successful transitions have taken place, the temperature is lowered by a certain amount, and the parameters are changed again. As a function of iterations, the energy of the system decreases towards what is hoped to be the minimum energy configuration and the process is terminated when either no more successful variations are made for a given number of attempts or the temperature reaches a lower limit.

When comparing calculated and experimental images, the energy of the system can be chosen to be χ^2 or any of the other quantities that measures image-mismatch. When basing the comparison on the cross-correlation coefficient, the energy can be taken as 1-CCF.

Simulated thermal annealing is a straight forward technique that has proven to be very powerful for finding global minimum without getting trapped in local minima. It is sensitive to the starting conditions and the choice of starting temperature and some experimentation may be required. Near the minimum, it tends to be less optimal than search techniques based upon gradient methods and switching to a different search algorithm may be an alternative once the simulated annealing algorithm has terminated.

5.3 SIMULATED EVOLUTION

Simulated evolution is another technique for obtaining the global minima which is modeled after Darwin's principle of "survival of the fittest"[16] It starts with an initial configuration of all the variables to be fitted and produces a number of sets λ from the initial set using a random generator (mutation generator). This set λ represents the first generation of children. The algorithm proceeds in the following way:

i) Evaluate a quality function Q (goodness of fit) for all λ children.

ii) Select a subset ($\mu < \lambda$) of survivors which will be he parents of the next generation.

iii) Create a new generation by applying the random generator, after selecting and mixing a part of the parent's parameter vectors.

iv) Loop back to i) until one of the following criteria are met: a) a maximum number of generations have been reached or b) a critical goodness-of-fit has been reached.

5.4 OTHER TECHNIQUES

There are other ways to do the refinement which is based upon changing the input parameters so that the system moves in a path where the gradient with respect to the fit is the largest[17]. Each method has its advantages. Simulated thermal annealing and simulated evolution are good techniques for getting close to the optimum fit. Once close to the minimum, gradient methods may be used for further refinement until the match is within the uncertainty of the measurement.

6. References

1. Völkl E. et al (1994) Density correction of photographic material for further image processing in electron microscopy, *Ultramicroscopy* **55**, 75-89
2. Ruijter W.J. de and Weiss J.K. (1992) Methods to measure properties of slow-scan CCD cameras for electron detection, *Rev. Sci. Insts.* **63**, 4314
3. Mooney P.E. et al. (1993) MTF restoration with slow-scan CCD cameras, *Proc. Annual Meeting of the Microsc. Soc. of America* **51**, 262-263
4. Ruijter W.J. de and Weiss J.K. (1993) Detection limits in quantitative off-axis electron holography, *Ultramicroscopy* **50**, 269-283
5. Frank J. (1972) Two-dimensional correlation functions in electron microscope image analysis, *Electron Microscopy*, The Institute of Physics, 622
6. Saxton W.O. (1996) Pre-processing of data, registration, distortions,resampling,noise removal, and noise estimation, NCEM workshop on quantitative HRTEM, April 18-20, NCEM/LBNL Berkeley CA, USA
7. Zhang H. et al (1995) , Structure of planar defects in $(Sr_{0.9}Ca_{0.3})_{1.1}CuO_2$ infinite-layer superconductors by quantitative high-resolution electron microscopy, *Ultramicroscopy* **57**, 103-111
8. Smith A.R. and Eyring L. (1982) Calculation, display and comparison of electron microscope images modeled and observed, *Ultramicroscopy* **8**, 65-78
9. Press, W.H. et al (1986) Numerical Recipes, The art of scientific computing. Cambridge University Press 1986 ISBN 0 521 30811 9, p. 502
10. King W.E. and Campbell G.H. (1994) Quantitative HREM using non-linear least-squares methods, *Ultramicroscopy* **56**, 46-53
11. Thust A and Urban K. (1992) Quantitative high-speed matching of high-resolution electron microscopy images, *Ultramicroscopy* **45**, 23
12. Möbus G. and Rühle M. (1994) Structure determination of metal-ceramic interfaces by numerical contrast evaluation of HRTEM micrographs, *Ultramicroscopy* **56**, 54-70
13. Hÿtch M.J. and Stobbs W.M. (1994) Quantitative criteria for the matching of simulations with experimental images, *Microsc. Microanal. Microstruct.* **5**, 133-151
14. Hÿtch M.J. and Stobbs W.M. (1994) Quantitative comparison of high resolution TEM images with image simulations, *Ultramicroscopy* **53**, 191-203
15. Thust A., Lentzen M. and Urban K. (1994) Non-linear reconstruction of the exit plane wave function from periodic high-resolution electron microscopy images, *Ultramicroscopy* **53**, 101-120
16. Möbus G. (1996) Retrieval of crystal defect structures from HREM images by simulated evolution I. Basic technique, *Ultramicroscopy* **65**, 205-216

17. Press, W.H. et al (1986) Numerical Recipes, The art of scientific computing. Cambridge University Press 1986 ISBN 0 521 30811 9, p. 521

DIRECT METHODS VERSUS ELECTRON DIFFRACTION : THE FIRST EXPERIENCES BY SIR97

R. CALIANDRO, G. CASCARANO, C. GIACOVAZZO, & A. MELIDORO
Istituto di Ricerca per lo Sviluppo di Metodologie Cristallografiche - CNR
c/o Dipartimento Geomineralogico - Campus Universitario
Via E. Orabona, 4 - 70125 Bari - Italy

1. Introduction

Crystal structure analysis *via* electron diffraction data is relatively young : the first Fourier map, of Ba $Cl_2 \cdot H_2O$, was published by Vainshtein & Pinsker [1]. The same kinematical approach was used later on by Cowley [2],[3],[4],[5] for solving a few structures. A more close description of the physical phenomena involved in the electron diffraction was available some years later [6]. This contribution generated two effects : on one side it showed the importance of the dynamical effects, often predominant with respect to kinematical scattering, on the other side reduced the interest for the structure analysis *via* electron data (because of the inadequacy of the kinematic theory). Further contributions to the formulation of the dynamical theory followed ; the reader is referred to the other papers in this book for the basic references and for the description of the main results.

We are interested here only to the applications of direct methods to electron diffraction data. This topic has recently received larger attention, probably due to the pioneering results by Dorset (see [7]) who, using direct methods, solved various organic and inorganic crystal structures. Some years ago direct methods were successfully combined with image processing methods and maximum entropy methods (see other contributions in this book).

Direct methods rely on diffraction magnitudes : any perturbation causing deviation of the observed magnitudes from the moduli of the Fourier transform of the potential field weakens the efficiency of the methods. This is the reason why crystal structure solution *via* electron diffraction data require some special care. As stated elsewhere in this book there are several phenomena which perturb the magnitudes : we quote, among others, dynamical scattering, secondary scattering, incoherent scattering and radiation damage. To reduce their effects several experimental cautiousnesses (e.g., thin crystal samples, high voltage, etc.) are used : however the quality of the data (evaluated on the basis of

D. L. Dorset et al. (eds.), Electron Crystallography, 261–272.

the kinematical theory) is often so poor that the crystal structure solution and analysis are difficult. Additional difficulties arise because :

a) electron diffraction patterns usually provide a subset of the reflections within the reciprocal space. Two-dimensional data are frequent : three-dimensional data are obtained by tilting about appropriate axes, but are often relegated to a few zones.

b) The resolution of the diffraction data is often not very high, mostly for organic samples.

Because of a) and b) a large percentage of reflections is out of the data ; the number of strong phase relationships is then small and the solution is difficult.

The question is now : can direct methods programs, suitably optimized for electron diffraction data, routinary solve most of the small structures ? We will show that SIR97[8], the heir of SIR92[9], is able to face several of these problems. We will focuse our attention on *ab initio* approaches ; i.e., no use will be made of the phase information provided by electron diffraction images.

2. Electron diffraction calculations : from SIR92 to SIR97

The SIR92 package aims at :

1) solving crystal structures *via* direct methods (X-ray data only)

2) refining the molecular model obtained in 1).

The steps 1) and 2) are sequentially and automatically executed : often a partial and quite imperfect model is transformed into the complete crystal structure *via* repeated cycles of structure factor, least squares and electron density calculations, without any user intervention. Isotropic vibrational motion is assumed, and diagonal least squares techniques are employed. SIR92 is however unsuitable for dealing with electron diffraction data. Several modifications have been introduced into SIR97 to allow the use of electron diffraction data. We notice :

a) the kinematic electron scattering factors for neutral atoms, calculated by Doyle & Turner [10] with the use of relativistic Hartree-Fock atomic fields, were introduced in the form

$$f(sin\vartheta / \lambda) = \sum_{i=1}^{4} a_i \exp\left(-b_i sin^2\vartheta / \lambda^2\right) + c.$$

The scattering factor of H has been added according to Cooley[11].

Parameters for ionized atoms can also be used if provided by the user.

b) when a potential map has been calculated, SIR97 selects the most suitable peaks to construct a reasonable molecular fragment. For X-ray data peaks are labelled (in the chemical sense) mostly on the basis of the peak height (large peaks are assumed to correspond to heavy atoms). For electron diffraction data this is no more valid : scattering factors do not assume the value of Z (number of electrons per atom) at $\sin\theta/\lambda = 0$, and may present a variety of forms. (see Fig.1 for an example). The result is that peaks in the potential map are weakly correlated with Z, so that their automatic labeling requires some additional considerations.

The measured intensities are divided in shells (*nshell* be their number), each shell characterized by a representative sinθ/λ value. For each atomic species SIR97 calculates the value

$$Z_{eff} = \sum_{n=1}^{nshell} |f(n)| q(n)$$

where f(n) is the value of f at the sinθ/λ corresponding to the n-th shell, and q(n) is the ratio "number of reflections in the n-th shell" / "total number of measured reflections". For the example described in Fig.1, we have :

$Z_{eff} = 0.679$ for B

$Z_{eff} = 0.860$ for O

$Z_{eff} = 0.165$ for H

Rather than to the value of the scattering factor at sinθ/λ = 0, the peak height is expected to be proportional to Z_{eff} : thus, for the case depicted in Fig.1, the largest peak is expected to correspond to an oxigen rather than to a boron atom. It may be worthwhile noticing that the Z_{eff}'s have to be calculated per each experimental set of data : different data of the same structure would give different Z_{eff} values.

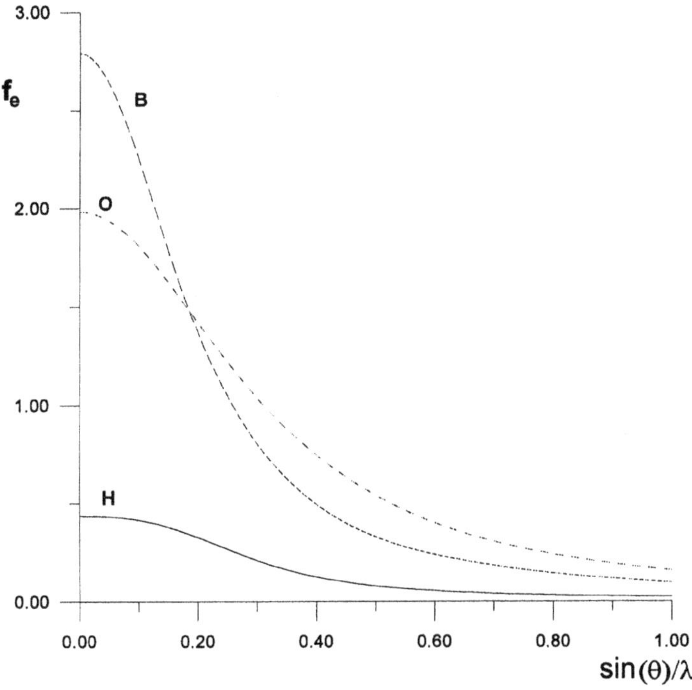

Figure.1 - Electron scattering curves for B, O and H.

c) Because of experimental limitations, electron diffraction data are often bidimensional. SIR97 has been designed to manage both bidimensional and threedimensional data ;
d) the location of hydrogen atoms was disregarded in SIR92, which was designed for X-ray data only. H-atoms are considered by SIR97, which is able to refine their coordinates.

3.The test structures

We used for our applications the ten test structures quoted in Table 1. Three of them belong a non-centrosymmetric space group symmetry.

TABLE 1 - List of test structures and main crystallochemical data

Code	Space group	Unit cell content	a	b	c	α	β	γ
COPPER[12]	P 2_1/m	$Cu_4\ Cl_2\ O_6\ H_6$	5.73	6.12	5.63	90	94	90
DIKE[13]	P 2_1/a	$C_8\ N_4\ O_4$	5.20	11.45	3.97	90	82	90
PARAFFIN[14]	P b a 2	$C_4\ H_8$	7.42	4.96	95.14	90	90	90
PERBRO[15]	C 2/m	$Cu_2\ Br_{32}\ N_{16}\ C_{64}$	19.89	26.46	3.76	90	116	90
THIOUREAF[16]	P m c 2_1	$S_4\ C_4\ N_8$	8.52	5.49	7.54	90	90	90
TI11SE4[17]	C 2/m	$Ti_{66}\ Se_{24}$	25.52	3.45	19.20	90	118	90
TI2SE[18]	P n n m	$Ti_{24}\ Se_{12}$	11.74	14.55	3.45	90	90	90
TI8SE3[19]	C 2/m	$Ti_{64}\ Se_{24}$	25.56	3.44	19.70	90	122	90
UREA[20]	P $\bar{4}$ 2_1 m	$C_2\ N_4\ O_2\ H_8$	5.66	5.66	4.71	90	90	90
CYANO[21]	P 2_1/c	$C_{116}\ N_4$	14.70	9.47	15.42	90	112	90

[12] Voronova,A.A. *et al.* (1958)
[13] Vainshtein,B.K. (1955)
[14] Dorset,D.L. (1976)
[15] Dorset,D.L. *et al.* (1992)
[16] Dvoryankin,V.F. *et al.* (1962)
[17] Weirich, T.E. *et al.*(1996)
[18] Weirich, T.E. *et al.*(in the press)
[19] Weirich, T.E. *et al.*(in the press)
[20] Lobachev,A.N. *et al.*(1961)
[21] Voigt-Martin,I.G. *et al.* (1995)

For each test structure we give in Table 2 :

a) the data resolution in Ångstroms (RES)

b) the number of symmetry independent reflections with measured diffraction intensity (NREF), and their type (i.e. (hkl) means "threedimensional data"). NOMIT is the number of symmetry independent reflections (up to RES resolution) not in the data set : i.e. those for which no measured intensity is available.

TABLE 2 - Data resolution (RES), number of symmetry independent reflections with measured diffraction intensity (NREF), and number of symmetry independent reflections (up to RES) which are not in the data set (NOMIT)

Code	RES (Å)	NREF	Type	NOMIT
COPPER	0.7552	120	hkl	286
DIKE	0.7861	312	hkl	138
PARAFFIN	0.8057	42	hk0	1
PERBRO	0.9618	168	hk0	27
THIOUREAF	0.7630	240	hkl	168
TI11SE4	0.7492	411	h0l	185
TI2SE	0.3627	753	hkl	167
TI8SE3	0.7500	2115	hkl	7
UREA	0.5620	60	h0l	7
CYANO	1.3993	143	hkl	532

It may be observed that in several cases NOMIT is not negligible with respect to NREF. This is usually expected for threedimensional data because of the limited angular range of the tilting technique, but it also occurs in several bidimensional data. For COPPER and CYANO it is NOMIT>NREF.

4. The SIR97 procedure

SIR97 may need a weak interaction with the user : all the phasing process may be automatically performed provided some basic information is supplied. In Table 3 the typical "default" set of commands and directives is shown for processing electron diffraction data.

The commands *%structure* and *%job* are used to define the name of the structure and to introduce a caption on the output file. The *%data* command is used to supply cell parameters, space group symbol, unit cell content and the name of the file containing h,k,l, F_{obs} and $\sigma(F_{obs})$. The directive *electrons* is necessary to inform the program that electron diffraction data are used. The *%continue* command generates all necessary commands to run the program in default way.

TABLE 3 - Example of SIR97 input file

```
%structure copper
%job copperchloride  Sov. Phys. Crystallogr. 3,445-451 (1958)
%data
  electrons
  cell  5.73  6.12  5.63  90.00  93.75  90.00
  spacegroup P 21/m
  content  Cu  4  Cl  2   O  6  H 6
  reflections  copper.hkl
%continue
```

The phasing process is a multisolution one : the default procedure uses the magic integers approach [22] [23], otherwise the supplementary directive "random" in the PHASE modulus activates the random starting set [24]. The correct solution is picked up *via* suitable figures of merit (FOM's) which are combined in the conclusive criterion CFOM. The program automatically analyses the solution with the highest value of CFOM and refines the structural parameters *via* cyclic "structure factor calculation - least squares - electron density calculation" process, labels (in the chemical sense) the electron density peaks and calculates the crystallographic residual index R.

The user can also choose a personal pathway for the crystal structure solution : numerous directives can be fixed to allow a variety of possible pathways.

5. The statistical analysis of the structure factor moduli

The experimental diffraction intensities are put on the absolute scale by the classical Wilson plot technique [25] : as soon as the overall isotropic thermal factor (BISO) becomes available, the normalization of the structure factor moduli is performed. SIR97 routinely analyses the distribution of the normalized moduli to discover the possible presence of pseudotranslational symmetry [26][27]. The results for each test structure are shown in Table 4. We observe :

a) BISO is often negative. The result has no physical meaning and is probably due to the dynamical nature of the diffraction intensities, and/or to the measurement errors, and/or to the high percentage of unobserved reflections. A more careful analysis is however necessary to perfectly identify the source of this effect. SIR97 automatically assumes BISO=0.001 when BISO<0.

b) the pseudotranslational symmetry is often present. In Table 4 we show the most important pseudotranslational vector and the estimated percentage of scattering power of the electron density obeying pseudotranslational symmetry. When this is present one of basic postulates of the direct methods (i.e., the atoms are randomly distributed in the unit cell) is violated, and this adds supplementary difficulties against the success of the phasing process.

TABLE 4 - Overall thermal factor (BISO) and pseudotranslational vectors (u) and percentage of electron density suffering by pseudotranslational symmetry (%)

Code	BISO	u	%
COPPER	-1.32	**b**/4+**c**/2	68
DIKE	-0.36	**a**/4+**b**/4+**c**/2	18
PARAFFIN	0.77	-------	-
PERBRO	0.36	-------	-
THIOUREAF	1.80	**a**/2+**b**/2+**c**/2	60
TI11SE4	6.50	**b**/3+**c**/4	15
TI2SE	-0.99	**a**/2+**b**/2+**c**/3	14
TI8SE3	-1.60	**c**/2	21
UREA	2.72	**a**/2+**c**/2	17
CYANO	-6.95	**a**/2+**b**/4+**c**/4	19

6. About the efficiency of the figures of merit

The combined CFOM ranks the various solutions in order of their expected reliability. Since the set of the structure moduli is affected by several sources of errors and by paucity of data, it is expected that the CFOM criterion will not work as effectively as for the usual X-ray data. Thus CFOM will be frequently far from unity (the expected value for the ideal case) ; furthermore the largest CFOM value might not mark the correct solution. In Table 5 we show, for each test structure, the value of CFOM corresponding to the correct solution and its rank. When rank=i, the correct solution is the i-th in order of CFOM : the most favourable situation occurs when rank=1. A star close to CFOM indicates that the solution has not been found in default, but *via* the random approach.
Table 5 shows that frequently the highest CFOM does not correspond to the correct solution ; however, for the test structures, the solution is always among the most probable trials.

TABLE 5 - Some figures characterizing the phasing process and the subsequent atomic parameters refinement

Code	CFOM	RANK	$<\|\Delta\Phi\|>^{\circ}$	RFINAL	NFOUND /NAT	<DIST(Å)>
COPPER	0.761	4	0.0	0.15	5/5	0.09
DIKE	0.754	2	4.0	0.23	4/4	0.09
PARAFFIN	1.000(*)	1	10.0	0.20	1/3	0.02
PERBRO	0.903	8	58.0	0.25	7/16	0.32
THIOUREAF	0.457	1	14.0	0.19	6/6	0.29
TI11SE4	0.845(*)	1	5.0	0.46	22/23	0.09
TI2SE	0.652(*)	5	0.0	0.22	9/9	0.05
TI8SE3	0.547	3	0.0	0.23	22/22	0.05
UREA	0.39(*)	3	0.0	0.32	3/5	0.16
CYANO	0.0	0	0.0	0.00	0/0	0.00

7. About the quality of the solutions

In the first step of SIR97 a small subset of reflections (those with the largest normalized structure factor moduli) are phased. In the second step a phase expansion process *via* the application of the tangent formula is applied, in the final step a Fourier least squares procedure is applied to refine and complete the structural model. The results are shortly summarized in Table 5 : we denote by $<|\Delta\Phi|>^{\circ}$ the average phase error corresponding to phases determined *via* direct methods (first step). In most of the cases $<|\Delta\Phi|>^{\circ}$ is close to zero : in one case, PERBRO, the crystal structure (or a non-negligible part of it) is eventually found *via* the subsequent refinement procedure. This result suggests that the correct structural model may be identified (among those corresponding to the various trial solutions provided by direct methods), by considering the crystallographic residual

$$R = \frac{\sum_{n} \left| |F_{obs}| - |F_{calc}| \right|}{\sum_{n} F_{obs}}$$

like a figure of merit. In the Table 5 RFINAL is the is the minimum residual value automatically found for the processed trial solutions. Correspondently, the number of atoms found (NFOUND) is given, and is compared with the number of atoms in the asymmetric unit (NAT). The average distance of the SIR97 atomic positions from the published positions is also shown (<DIST>).

The SIR97 performances are quite satisfactory for some structures : e.g., for COPPER, DIKE, TI2SE, TI8SE3, where all the atoms in the asymmetric unit are correctly located, and a small value of <DIST> is determined. As an example, we show in Fig.2 the contoured potential map of TI8SE3, provided by SIR97, to which the projected atomic positions of the published model (indicated by black balls) are superimposed.

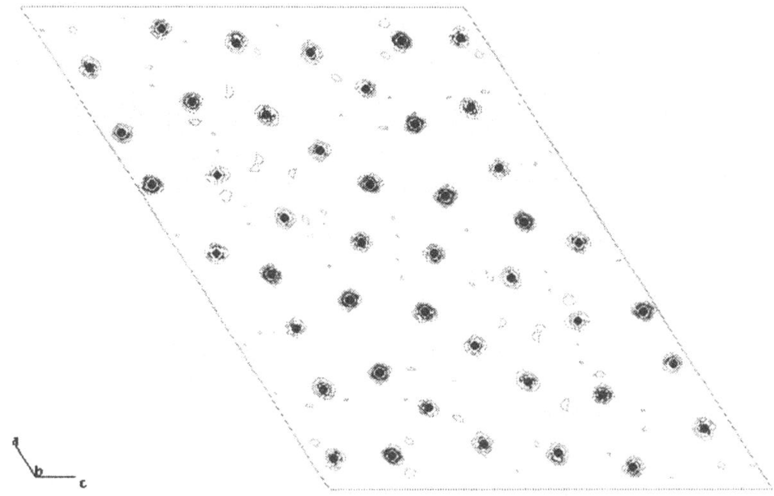

Figure 2 - TI8SE3 - The projections of the published positions, indicated by black balls, are superimposed to the contoured potential map provided by SIR97.

The results are satisfactory also for PARAFFIN, THIOUREAF, TI11SE4, UREA, PERBRO and CYANO. The following remarks might be useful :

a) PARAFFIN - SIR97 finds the carbon atom but misses the two hydrogens. It should be worthwhile noting that in the first E-map all the three atoms are found, but the hydrogen get lost during the automatic least square refinement.

b) THIOUREAF - All the atoms are correctly located by SIR97, but <DIST>=0.29 (distances calculated with respect to the X-ray structure). Dorset [7] observes that for this structure dynamical scattering does not compromise direct phase determination seriously, but leads to some geometrical inaccurancies in the final refined structure.

c) TI11SE4 - 22 over 23 atoms are correctly located, with low geometrical distorsion (<DIST>=0.11). SIR97 ends with RFINAL=0.47 which is particulary high. Subsequent refinement easily identifies the position of the missed atom and lowers the R value.

d) UREA - Three over five atoms are found by SIR97 (the two H atoms are missed). The partial model presents a little geometrical distortion (<DIST>=0.16 Å) that well agrees with Dorset results (see [7]), who obtained a final R value of 0.34 and did not find "any convincing evidence for hydrogen atom positions in the experimental potential map".

e) PERBRO - Seven of the 16 symmetry independent atoms are found by SIR97 (1 Cu, 2 Br, 1N, 3C), with some geometrical distortion in the model. The corresponding potential map is shown in Fig.3 : the complete structure is also sketched for reader usefulness. The qualitative inspection of the Figure shows that the potential map is more informative than the conclusive indications of SIR97. In the original trials for solving the crystal structure (see Dorset et al. [15]), the most reasonable model corresponded to R=0.41. It was found that dynamical effects and other data perturbations, such as the secondary scattering, were present in the data. Dorset [7] concluded that "this structure analysis represent an extreme case for accepting data to be used for *ab initio* determinations".

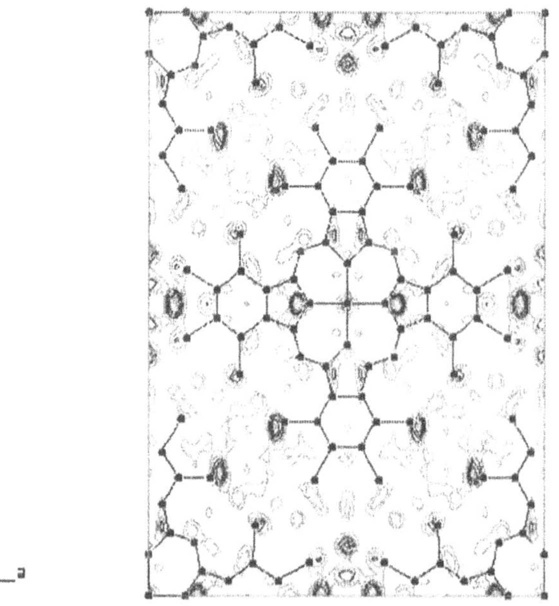

FIGURE 3 - PERBRO - The published structural model (stick and balls) superimposed to the contoured potential map obtained by SIR97.

f) CYANO - The trials with CYANO are just starting. This is a very difficult test both for direct methods and for maximum entropy techniques [21]. 100 trials to phase 51 reflections (those with the largest |E|'s) were generated: only four of them were able to assign phase values to the reflections. The solution ranked 2 by the combined figure of merit provides phases for 46 reflections with $<|\Delta\Phi|> = 51°$ (weighted error = 41°). For reader usefulness, we show in Fig. 4 two potential maps; in a) the map corresponding to the published phases (51 reflections only), in b) that corresponding to our phases (46 reflections only). The correlation factor between the two maps is 0.55, which indicates that informative maps are available *via* direct methods.

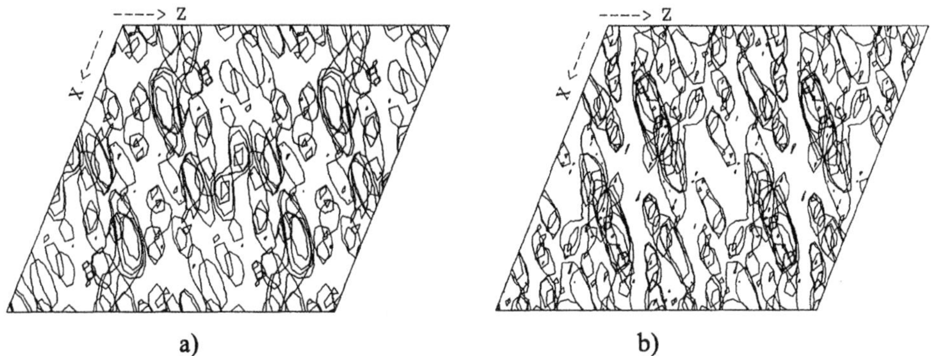

FIGURE 4 - CYANO - a) potential map projected on the (**a**,**c**) plane (51 reflections, published phases) ; b) potential map projected on the (**a**,**c**) plane (46 reflections, direct methods phases)

8. Acknowledgement

The authors would like to thank D. Dorset, S. Hovmöller, I.G. Voigt-Martin, U. Kolbe, T.E. Weirich for making their data available.

9. References

1. Vainshtein, B.K. & Pinsker, Z.G. (1949) Journ. Phy. Chem. SSSR **23**, 1058 (in Russian)
2. Cowley, J.M. (1953a) Acta Cryst. **6**, 516
3. Cowley, J.M. (1953b) Acta Cryst. **6**, 522
4. Cowley, J.M. (1953c) Acta Cryst. **6**, 846
5. Cowley, J.M. (1955) Acta Cryst. **9**, 391
6. Cowley, J.M. & Moodie, A.F. (1957) Acta Cryst. **10**, 609
7. Dorset, D.L. (1995) Structural Electronic Crystallography, Plenum Press, New York
8. Altomare, A.,Cascarano, G., Giacovazzo, C., Guagliardi, A., Moliterni, A.G.G., Siliqi, D., Burla, M.C., Polidori G. & Camalli, M. XVIIth Congress of International Union of Crystallography, Seattle, 8-17 August 1996
9. Altomare, A., Cascarano, G., Giacovazzo, C., Guagliardi, A., Burla, M.C., Polidori, G. & Camalli, M. (1994) Journ. Appl. Cryst. **27**, 435
10. Doyle, P.A. & Turner, P.S. (1968) Acta Cryst. A**24**, 390-399
11. Cowley, J.M. (1992) International Tables for Crystallography, ed. by Wilson, A.J. Kluwer Acad. Publishers, Dordrecht, vol C, 223-373

12. Voronova, A.A. & Vainshtein, B.K. (1958) Sov. Phys. Crystallogr. **3**, 445-451
13. Vainshtein, B.K. (1955) Zh. Fiz. Khim. **29**, 327-344
14. Dorset, D.L. Acta Cryst. (1976). A**32**, 207-215
15. Dorset, D.L. , Tivol, W.F. and Turner, J.N. Acta Cryst. (1992). A**48**, 562-568
16. Dvoryankin, V.F. & Vainshtein, B.K. (1962) Sov. Phys. Crystallogr. **6**, 765-772
17. Weirich, T.E., Ramlau, R., Simon, A., Hovmöller, S. & Zou, X.D. Nature, (1996) **382**, 144-146
18. Weirich, T.E., Poettgen, R., & Simon, A. Z. Kristallogr. (in the press)
19. Weirich, T.E., Poettgen, R., & Simon, A. Z. Kristallogr. (in the press)
20. Lobachev, A.N. & Vainshtein, B.K. (1961) Sov. Phys. Crystallogr. **6**, 313-317
21. Voigt-Martin, I.G., Yan, D.H, Yakimansky, A., Schollmeyer, D., Gilmore, C.J. & Bricogne, G. Acta Cryst. (1995). A**51**, 849-868
22. White, P. & Woolfson, M.M. (1975) Acta Cryst. A**31**, 53-56
23. Declercq, J.P., Germain, G. & Woolfson, M.M. (1975) Acta Cryst. A**31**, 367-372
24. Yao, J.X. (1981) Acta Cryst. A**37**, 642-644
25. Wilson, A.J.C. (1942) Nature (London) **150**, 151-157

STRUCTURE DETERMINATION BY ELECTRON CRYSTALLOGRAPHY USING A SIMULATION APPROACH COMBINED WITH MAXIMUM ENTROPY WITH THE AIM OF IMPROVING MATERIAL PROPERTIES

I.G.VOIGT-MARTIN and U. KOLB
Institut für Physikalische Chemie, Universität Mainz
Jakob Welder Weg 11, D-55099 Mainz

1 Crystal Engineering

Solving a crystal structure is only one of the many problems involved in the process of improving material properties. Because it is difficult to obtain large single crystals from most polymeric and many monomeric organic materials, it is essential to develop electron crystallography to make reliable crystal structure analysis possible.

Crystal engineering is used to produce new organic materials with specific properties such as second harmonic generation, ferroelectricity, triboluminescence, piezoelectricity and pyroelectricity. If these properties are to be induced in the solid state, the crystals must have a resultant polarisability, dipole moment or optical susceptibility. Therefore the constituents of the crystals, the molecules, must be designed to possess a molecular polarisability α, dipole moment μ or appropriate hyperpolarisability tensor components β_{ijk} and the crystals must be non centro-symmetric.

Centro symmetry in organic crystals may be destroyed by using one or several of the following methods:

a) Synthesis of chiral molecules
b) Hydrogen bonding to produce chiral arrays
c) Reduction of ground state dipole to prevent antiparallel stacking
d) Crystal growth in electric fields
e) Production of Langmuir-Blodgett films
f) Use of liquid crystals

D. L. Dorset et al. (eds.), Electron Crystallography, 273–284.

2 Design of the molecule

The properties of a molecule depend not only on the constituent atoms but also on its conformation. Several computer programs are available to calculate this, involving different levels of complexity and accuracy.

2.1 FORCE FIELD CALCULATIONS

Force field programs such as DREIDING [1] use very simple formulae to calculate the lowest energy conformation of a molecule, based on an estimation of the energy of bond-stretching, bond angle deformation, torsional energy as well as the non-bonded Van der Waals-, Coulomb- and hydrogen bond energies. The constants in the equations are included as fixed parameters related to specific atom pairs. These calculations are used to obtain an initial estimate of the molecular geometry, which is subsequently improved by semi-empirical quantum mechanical calculations [2].

$$E = \underbrace{E_s + E_b + E_t}_{\text{bonded}} + \underbrace{E_{VdW} + E_{Coul} + E_{hb}}_{\text{non-bonded atoms}} \qquad (1)$$

E_s = bond-stretching energy
E_b = bond-angle bending energy
E_t = torsional energy
E_{VdW} = non-bonded interaction energy Van-der-Waals
E_{coul} = Coulomb energy
E_{hb} = Hydrogen bond energy

2.2 AB INITIO CALCULATIONS

Ab initio quantum mechanical calculations are the most demanding of these methods and also the most time consuming [3]. These are not based on a large set of parameters obtained from experiment; instead the orbitals of the atoms in the molecule and their overlap integrals are calculated. Programs which can be used are GAUSSIAN [4]or TURBOMOL [5].

In view of the fact that the molecular conformation is adjusted by the crystal field, the semi-empirical method usually suffices for the purpose of determining molecular geometry. However semiempirical programs such as MOPAC6.0 [6] does not calculate the off-diagonal elements of the hyperpolarisability tensor. When these are required it may be necessary to resort to ab initio methods.

2.3 SEMI-EMPIRICAL QUANTUM MECHANICAL CALCULATIONS

Conventional semi-empirical methods (CNDO, INDO, MNDO) to calculate molecular conformations in the gas phase have been used in quantum chemistry for several years. This involves molecular orbital calculations in which a number of simplifying assumptions are made in order to avoid calculating all the integrals involved. The loss of accuracy is compensated by the availability of parameters for atom pairs which are obtained from experiments (18 parameters for each element:U_{ss}, U_{pp}, $ß_s$, $ß_p$, ξ_{ss}, ξ_p, α, G_{ss}, G_{sp}, G_{pp}, G_{p2}, H_{sp}, K_i, L_i, M_i, K_j, L_j, M_j).

The most advanced of them, MNDO, is known to suffer from an overestimation of the repulsion between atoms when they are approximately a Van der Waals distance apart. Good results were obtained with AM1, in which the core repulsion function was modified by additional Gaussian terms. In a further optimisation of parameters (heat of fomation, dipole moments, ionisation potential) the progam MNDO-PM3 was developed [7].

In order to check whether the MOPAC calculations for a specific molecule under investigation are reasonable, spectroscopic information about polarisabilities or dipole moments should be obtained and compared with the calculated values. If these values are sufficiently large to make the molecule a possible candidate for a specific application, single crystals or oriented films can be grown by one of the techniques mentioned in §1.
On the basis of our molecular calculations, the following molecules were chosen for crystal structure determination [8-10]

I) 4-Dimethylamino-3-cyanobiphenyl (DMACB)
II) 9,9'-bianthryl-10-carbonitrile (CNBA)
III) 1-Nitrophenylurethane (NPHU)

As an example Table 1 shows a typical Z-matrix used for MOPAC calculations. It describes the molecular geometry of the CNBA molecule in Figure 1.

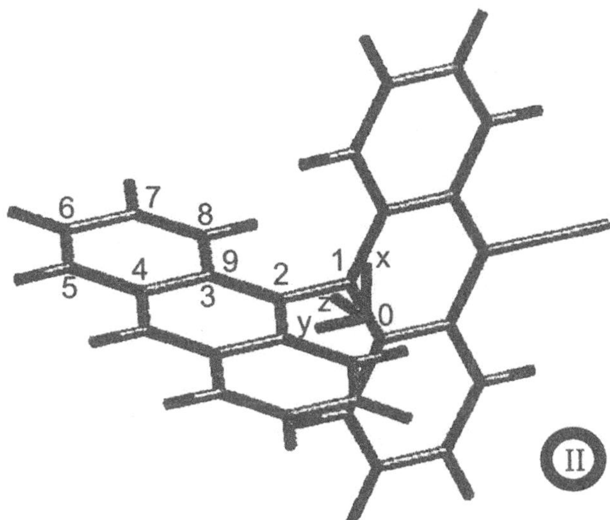

Figure 1: CNBA molecule, labels as used in MOPAC input

Table 1: Head of MOPAC input file for CNBA (labels according to Fig.ure 1)

```
                                                              distance  torsion
                                                                   angle
PM3 1SCF T=24.0H DUMP=60.0M NOXYZ NOINTER POLAR                  ↓     ↓     ↓

CBA gas phase calculation
X  0.0000000  0   0.0000000    0   0.0000000     0    0    0    0     origin
X  1.0000000  0   0.0000000    0   0.0000000     0    1    0    0     x-axis
X  1.0000000  0   90.0000000   0   0.0000000     0    1    2    0     y-axis
X  1.0000000  0   90.0000000   0   -90.0000000   0    1    2    3     z-axis
C  0.6363970  1   44.9032339   1   90.0000000    1    4    1    2     atom 1
C  1.4733800  1   90.0000000   1   -90.0000000   1    5    4    1     atom 2
C  1.4085078  1   119.8060139  1   -51.7480367   1    6    5    4     atom 3
C  1.4285006  1   119.5849459  1   179.9952098   1    7    6    5     atom 4
C  1.4332620  1   119.0851995  1   179.9965863   1    8    7    6       •
C  1.3642967  1   120.8617805  1   -0.0064602    1    9    8    7       •
C  1.4253602  1   120.2047321  1   0.0317701     1    10   9    8       •
C  1.4343728  1   122.1440837  1   179.9942960   1    7    6    8       •
H  1.1014300  1   118.0565306  1   -179.9592963  1    12   7    11      •
H  1.1002566  1   118.6444710  1   -179.9351831  1    11   10   12      •
H  1.1001794  1   121.0719912  1   179.9601577   1    10   9    11      •
..........and so on
```

Subsequently the dipole moments μ and vector components of the hyperpolarisability β can be calculated. The results for the above molecules I-III are shown in Figure 2.

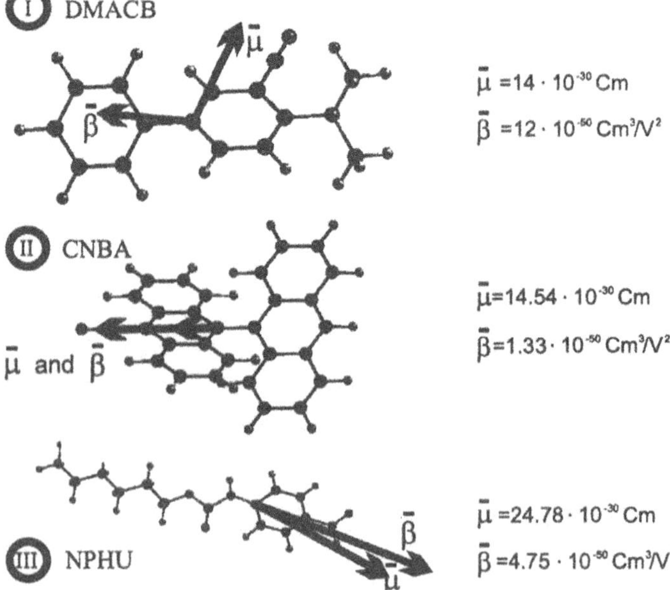

Figure 2: Dipole moment μ and hyperpolarisability β calculated in the gas phase

3 Experimental electron diffraction patterns

Organic crystals are often very thin (100 Å) and are therefore most suitable for analysis by electron diffraction. A eucentric goniometer stage and low dose facilities are absolutely essential to collect diffraction tilt series from beam sensitive compounds. To estimate the space group the following procedure is performed:

a) The basic zone is established by determining the positions corresponding to the largest d-spacing.
b) An axis with strong reflections, preferably the main axis, is chosen as tilt axis.
c) Tilting is performed to the right and to the left and the newly appearing zones are established. An example is shown for NPHU ($P2_1$) exemplarily in Figure 3.
d) A second axis is chosen as tilt axis and the procedure repeated.

From these experiments the unit cell dimensions are determined and from the observed extinctions and density considerations, a number of possible space groups established. Examples of such tilting series for the above examples are shown in references [8-10].

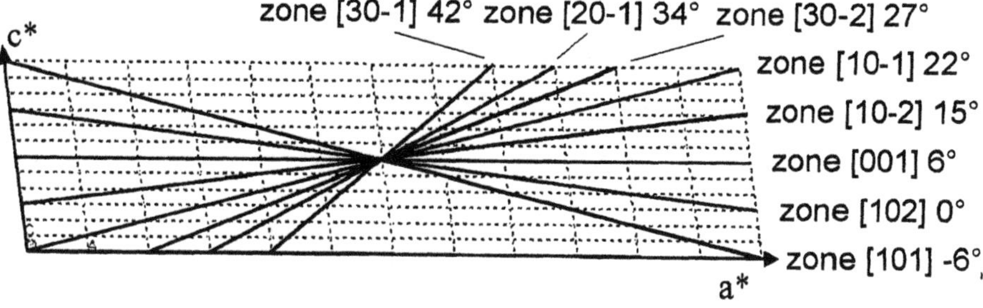

Figure 3: View of the reciprocal space down the tilt axis b*, axis in d* [1/Å], λ=0.0370 Å

4 Powder X-ray diffraction

X-ray powder patterns are easy to measure and should be used to improve the values of d-spacings obtained by electron diffraction and to check whether any of the reflections observed in the electron diffraction pattern are caused by dynamic or secondary scattering. Unfortunately the number of reflections obtained from organic crystals is rarely sufficient for a structure determination by Rietveld methods, so that electron crystallography is the only route to success.

5 Simulation of electron diffraction patterns using packing energy calculation

A relevant question to ask is: Why use simulation methods when direct methods are so successful in solving crystal structures? The reason is that for organic crystals the data set is often extremely sparse and inaccurate, so that direct methods simply do not work. In recent years therefore, considerable effort has been expended in attempts to predict cystal structures from a knowledge of their molecular architecture. Attempts to calculate the potential energy hypersurface have been quite successful using the following methods: Dreiding force field [1], PCK-method [11], Cluster method [12].

In all of these methods several local minima with good negative packing energies are predicted. The intermolecular forces involved are

1) atom-atom potential E_{nb} [13,14] (Van der Waals energy) $E_{nb}(r) = \frac{A}{r^{12}} - B\ \exp(-Cr)$

   A,B,C,D parameters derived from experiment

2) Dipolar forces E_{DD} [15] (Coulomb energy) $E_{DD} = (1/8\ \pi\varepsilon) \sum\sum \frac{q_i q_j}{\varepsilon r_{ij}}$

3) Hydrogen bonding E_{hb} [13,16-18] classified according to their topology as intra-, intermolecular or bifurcated and according to their energy:

   a) Weak H-bonding: donors with small intrinsic electronegativities e.g. C-H---A
   b) Medium H-bonding: typical in water, alcohols, amines, amides and carboxylic acids
   c) Medium-strong H-bonding:
      - charge assisted e.g. $-COO^{-}$---H_3N^{+}
      - resonance assisted e.g. O=C-C=O $\leftrightarrow$ $^{-}$O-C==C-C=OH^{+}
   d) Strong H-bonding: ionic systems e.g. $[F{-}{-}H{-}{-}F]^{-}$

The three energy terms E_{nb}, E_{DD} and E_{hb} add up to the free lattice energy of the molecular crystal

$$U = \tfrac{1}{2} \sum\sum E_{nb}(r_{ij}) + E_{DD} + E_{hb}. \qquad (2)$$

Often the packing energies differ by only a few kcal/mol, so that prediction without any experimental evidence is difficult. Therefore, we use the information obtained from electron diffraction analysis. The MOPAC6.0 calculated molecule is packed into the unit cell of known dimensions using a molecular modelling program such as CERIUS [19] so that the density and symmetry requirements are satisfied and the crystal packing energy is minimised. The required changes in molecular geometry due to crystallisation generally involve only subrotations. The information available at this stage is indicated in Table 2.

Table 2: Available information to build a crystal

MOLECULE:	CRYSTAL:
	Lattice type
Number and type of atom	Size of unit cell
Length and shape of molecule	Number of molecules per unit cell and density
Conformation	Space group
Symmetry	Angle between consecutive zones
Direction of dipole	Diffraction intensities for consecutive zones

Using space filling strategies and symmetry considerations the packing energy U as given in equation (2), is minimised to obtain position and orientation of the molecules with respect to crystal axes.

Finally the values of the polarisation tensor α, the dipole vector μ and the hyperpolarisability vector β are calculated for the modified molecular conformation in molecular coordinates i, j, k and then related to the crystallographic coordinates I, J, K by applying a suitable rotation matrix and using crystal symmetry conditions. The results for DMACB, CNBA and NPHU are shown in Figure 4 [8-10].

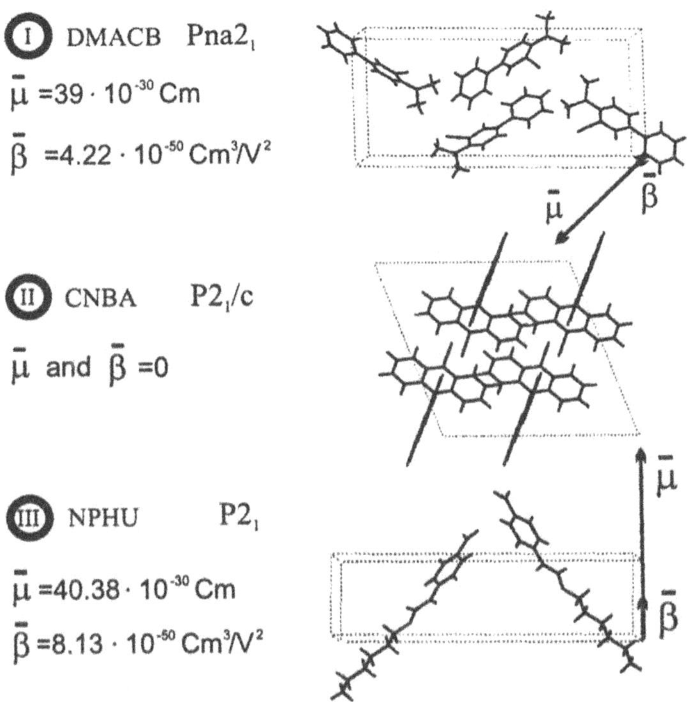

Figure 4: Dipolemoment and hyperpolarisability for molecules I-III in the cell

From these models, electron diffraction patterns are calculated, compared with all experimentally obtained zones and R-factors are calculated. Typical simulated diffraction patterns are shown for NPHU in Figure 5.

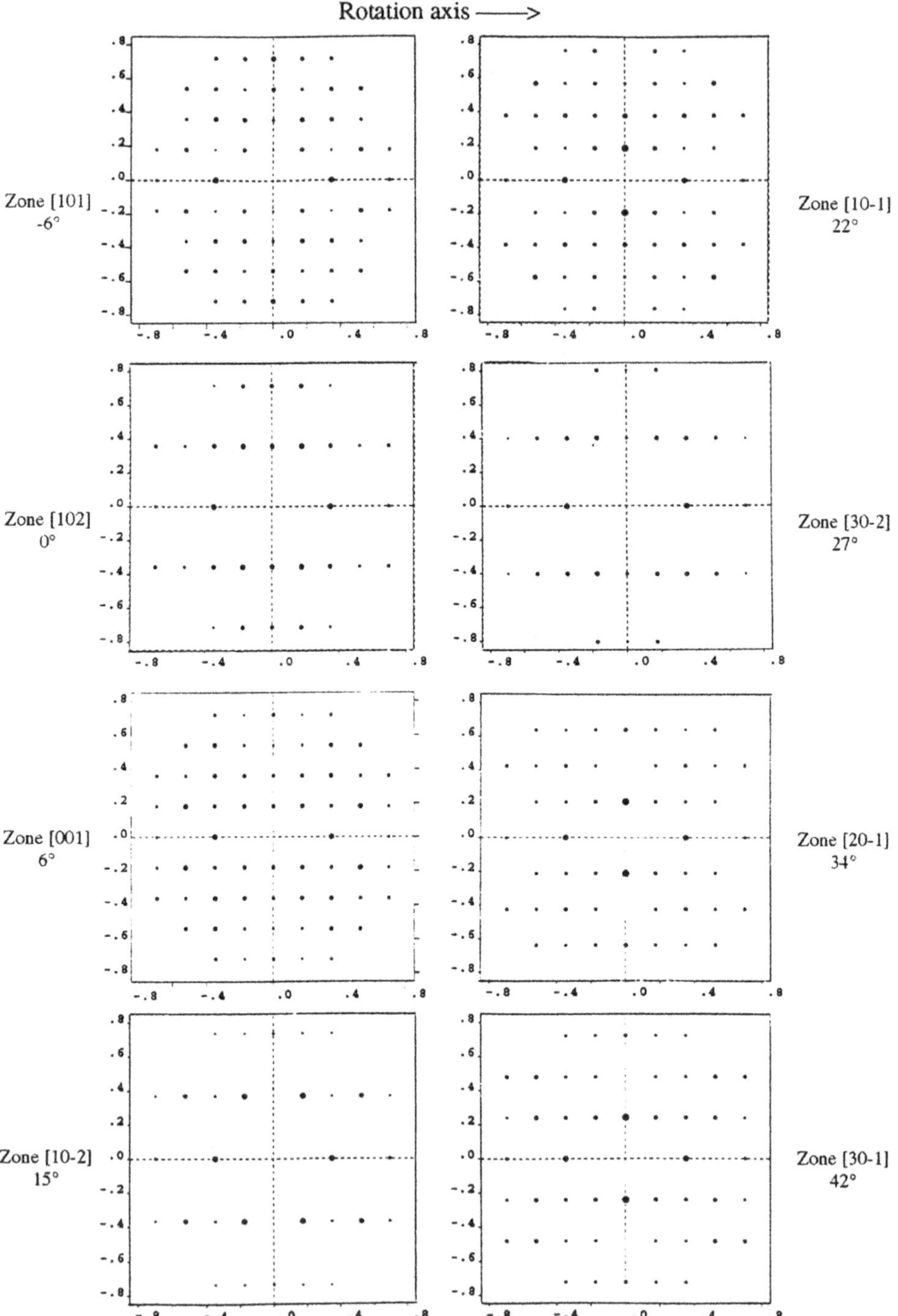

Figure 5: Simulated diffraction pattern for NPHU (III) according to obtained zones; see Figure 3.

6 Quantitative analysis of electron diffraction patterns.

Electron diffraction patterns can be analysed on-line using a suitable slow scan camera. These cameras need to have a large dynamic range, good resolution and a sufficiently large chip so that the diffraction pattern does not have to be demagnified to such an extent that intensities overlap. On-line cameras are very expensive and it may be more convenient to register the diffraction pattern on film emulsion and to analyse the negatives using a CCD camera or a densitometer, although this inevitably leads to a loss of information.

The electron diffraction intensities can then be quantified using ELD [20]. The procedure is as follows:

(a) Correct the calculated diffraction patterns for dynamic and secondary scattering [21]

(b) Calibrate linear range of camera, film emulsion and ELD intensity evaluation.

(c) Ensure that exposure series are obtained to extend linear range and check intensity evaluation

Figure 6 shows a typical set of calibration curves indicating the intensity of individual reflections as a function of exposure time.

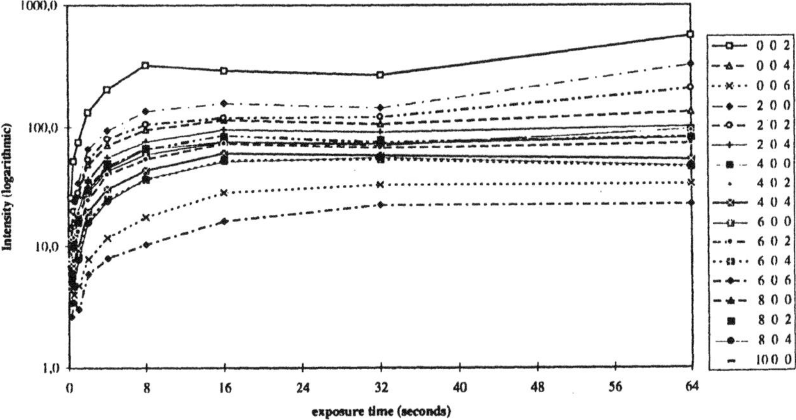

Figure 6: Dependency of intensity on exposure time

When all the zones have been analysed they are merged into a single set by normalising the common axis. This data set is compared with the kinematic values obtained from the model and an R-factor evaluated as:

$$R = \frac{\sum \left| |F_O| - |F_c| \right|}{\sum |F_O|} \qquad (3)$$

where F_0 is the structure factor obtained from the square root of the obseved intensities and F_C is the structure factor calculated from the model.

From electron diffraction data R-values in the range of 0.15-0.35 can be obtained by off-line analysis of the electron diffraction patterns [8-10]. However, it is expected that improved values will be obtained by using on-line slow scan CCD cameras with a large dynamic range.

7 Maximum Entropy and log likelihood statistics

In order to determine the structure independently, the maximum entropy method can be used. Unfortunately the data set from organic molecules is usually very sparse, so that, at best, only low resolution maps can be obtained. In addition, in the early stages of analysis it is extremely difficult to recognise features in the potential maps, especially with an unknown structure. Therefore, it is often very helpful if a model from the simulation procedure delivers some previous indication of projections which might have prominent features.

The formal procedure is as follows [10, 22, 23, 24]:

a) The experimental diffraction intensities are normalised in MITHRIL using electron scattering factors to give unitary structure factors $|U_h|_{obs}$

b) An origin is defined by fixing the phases of suitable reflections using MICE. These phased reflections, which form a basis set H, are used as constraints in entropy maximisation. This generates the root node of a phasing tree. The properties of the maximum entropy map $q^{ME}(x)$ is such that it is capable of generating new phases.

c) New phase information is incorporated into the basis set by adding new strong reflections, permuting their phases and representing each phase choice as a node on the phasing tree.

d) Each node is subjected to constrained entropy maximisation. The log likelihood gain criterion is used to rank the most likely phase choices.

The results of such an analysis for DMACB and CNBA are shown in Figs 7a,b.

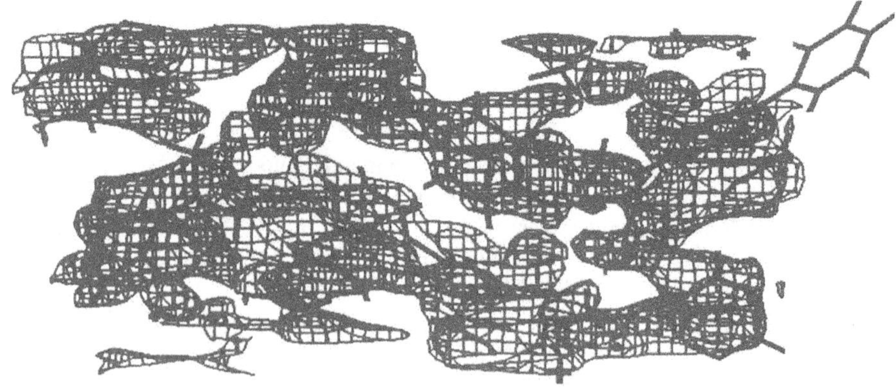

Figure 7a: Centroid map obtained from DMACB (I) with fixed phenyl rings

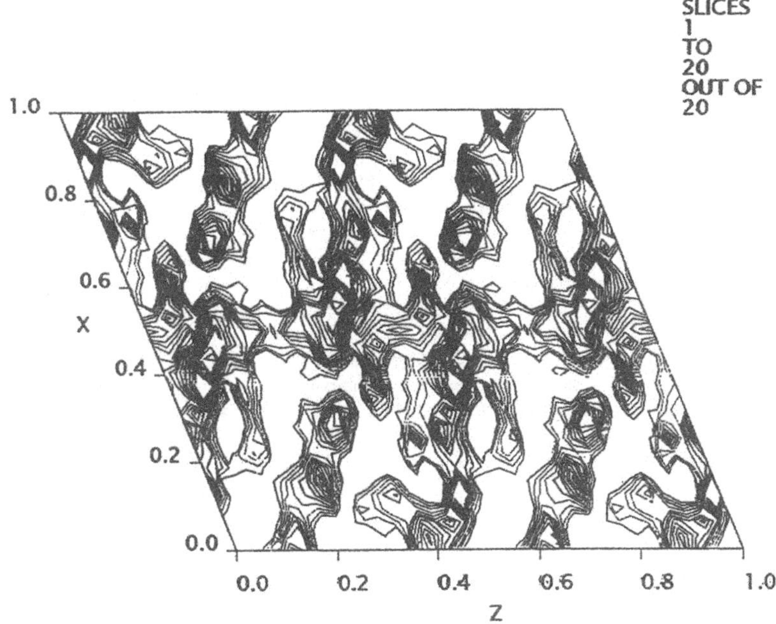

Figure 7b: Potential map obtained from CNBA (II)

Acknowledgement

The authors are indepted to Dipl. Minerologe Hans Kothe for the intensity data of Figure 6 and financial support by the Deutsche Forschungsgemeinschaft is gratefully acknowledged.

References

1. Mayo, S.L., Olafson, B. and Goddard, W.A. (1990) *J.Chem.Phys.* **94** 8897 - 8913.
2. Stewart, J.J.P. (1989) *J.Comput.Chem.* **10** 209 - 231.
3. Richards, W.G. and Cooper, D.L. (1983) *Ab initio Molecular Orbital Calculations for Chemists*, 2 edition, Oxford University Press, New York.
4. Frisch, M.J., Trucks, G.W., Schlegel, H.B., Gill, P.M.W., Johnson, B.G., Robb, M., Cheeseman, J.R., Keith, T.A., Peterson, G.A., Montgomery, J.A., Raghavachari, K., Al-Laham, M.A., Zakrzewski, W.G., Ortiz, J.W., Foresman, J.B., Ciolowski, J., Stefanov, B.B., Nanayakkara, A., Challacombe, M., Peng, C.Y., Ayala, P.Y., Chen, W., Wong, M.W., Andres, L.J., Replogle, E.S., Gomperts, R., Martin, R.L., Fox, D.J., Binkley, J.S., Defrees, D.J., Baker, J., Stewart, J.P., Head-Gordon, M., Gonzalez, C., People, J.A. (1995) GAUSSIAN 94, (Revision B.2) Gaussian Inc., Pittsburgh, PA.
5. TURBOMOL v235, Biosym Technologies (1993), San Diego.
6. Stewart, J.J.P. (1990) MOPAC6.0, *A General Purpose Molecular Orbital Package*, QCPE.
7. Kurtz, H.A., Stewart, J.J.P. and Dieter, K.M. (1990) *J.Comput.Chem.* **11** 82.
8. Voigt-Martin, I.G., .Zhang, Z. X., Yan, D.H., Yakimanski, A., Matschiner, R., Krämer, P., Glania, C., Schollmeyer, D., Detzer, N. and Wortmann, R. (1997) *Colloid.Polym.Science* **275** 18 - 37.
9. Voigt-Martin, I.G., Yan, D.H., Yakimanski, A., Schollmeyer, D., Gilmore, C.J. and Bricogne, G. (1995) *Acta Cryst.* **A51** 849 - 868.

10. Voigt-Martin, I.G., Yakimanski, A., Kolb, U., Matschiner, R. and Wortmann R. (1997)*Ultramicroscopy.(*accepted).
11. QuantumChemistry Exchange Program,Program 548,Indiana University.
12. Gavezotti, A. (1995) *J.Amer.Chem.Soc.* **117** 12299 - 12305.
13. Gilli, G. (1994) *Fundamentals of Crystallography* ed. C. Giacovazzo, Oxford Science Publications.
14. Gavezotti, A. and Filippini (1994) G., *J.Phys.Chem.* **98,** 4831 - 4837.
15. Gasteiger, I. and Marsili, M. (1980) *Tetrahedron* **36,** 3219 - 3228.
16. Lifson, S., Hagler, A.T. and Dauber, P. (1979) *J.Am.Chem.Soc.* **101**, 5111 - 5121.
17. Spackman, M.A. (1986) *J.Chem.Phys.* **85**,6579 and 6587 - 6601.
18. Etter, M.C. and Reutzel, S. (1991) *J.Am.Chem.Soc.* **113,** 2586 - 2598.
19. CERIUS2.0, Molecular Simulations, Oxford.
20. Hovmöller, S. (1992) *Ultramicroscopy* **41** 121 - 135.
21. Dorset, D. (1995) *Structural Electron Crystallography* Plenum Press.
22. Bricogne, G. (1984) *Acta Cryst.* **A40** 410 - 445.
23. Gilmore, C.J. and Bricogne, G. (1992) *Crystallographic Computing 5:From Chemistry to Biology*, ed. Moras, D., Podjarny, A.D., Thierry, J.J., Oxford University Press.
24. Gilmore, C.J., Bricogne, G. and Bannister, C. (1990) *Acta Cryst.* **A46** 297 - 308.

MULTI-DIMENSIONAL ELECTRON CRYSTALLOGRAPHY OF Bi-BASED SUPERCONDUCTORS

FAN HAI-FU [a], WAN ZHENG-HUA [b], LI JIAN-QI [c], FU ZHENG-QING [a], MO YOU-DE [a], LI YANG [a], SHA BING-DONG [a], CHENG TING-ZHU [d], LI FANG-HUA [a] and ZHAO ZHONG-XIAN [c]
[a] *Institute of Physics, Chinese Academy of Sciences, Beijing, China*
[b] *Department of Physics, Zhongshan University, Guangzhou, China*
[c] *National Laboratory for Superconductivity, Chinese Academy of Sciences, Beijing, China*
[d] *Laboratory of Structure Analysis, University of Science and Technology of China, Hefei, China*

1. Introduction

For the structure analysis of crystalline materials, electron crystallographic methods are in some cases superior to X-ray methods. First, many crystalline materials important in science and technology, such as high *Tc* superconductors, are too small in grain size and too imperfect in periodicity for an X-ray single crystal analysis to be carried out, but they are suitable for electron microscopic observation. Secondly the atomic scattering factors for electrons differ greatly from those for X-rays and it is easier for electron diffraction to observe light atoms in the presence of heavy atoms (see *figure* 1). Finally the electron microscope is the only instrument that can produce simultaneously for a crystalline sample a micrograph and a diffraction pattern corresponding to atomic resolution. In principle either the electron micrograph (EM) or the electron diffraction (ED) pattern could lead to a structure image. However the combination of the two will make the procedure much more efficient and powerful [1].

The widespread occurrence of incommensurate modulations in the high-*Tc* superconducting phases and related compounds requires the multi-dimensional crystallographic methods [2] for their structure analysis. On the other hand, the modulation in the high-*Tc* superconductors involves both the metal and the oxygen atoms. The modulation of the latter in the Bi-O layer is important for understanding the mechanism of superconductivity, since it plays an important role in the incorporation of extra O atoms in the Bi-O layer and hence contributes to the hole concentration in the Cu-O layer. Owing to the dominating effect of heavy atoms in X-ray diffraction, electron diffraction may be a better technique to study the oxygen modulation.

D. L. Dorset et al. (eds.), Electron Crystallography, 285–294.

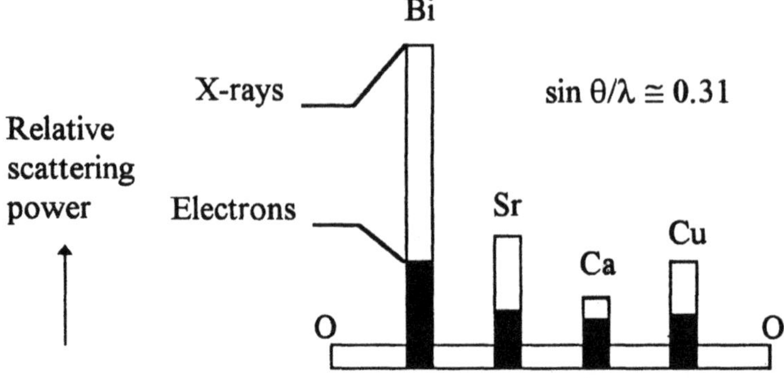

Figure 1. Comparison of the relative scattering power of the elements Bi, Sr, Ca, Cu and O for X-rays and for electrons. Assuming that the relative scattering power of oxygen is the same for X-rays and for electrons, the ratio of relative scattering power between Bi and O is much higher for X-rays (shown as white vertical bar) than for electrons (shown as black vertical bar).

2. The incommensurate modulation in Bi-based superconductors

The chemical formula of the Bi-cuprates can be expressed as $Bi_2Sr_2Ca_{n-1}Cu_nO_{2n+4}$ with n = 1, 2 and 3 corresponding respectively to the Bi-2201, Bi-2212 and Bi-2223 superconducting phases. The basic structure of these compounds can be regarded as consist of alternately a bismuth-oxygen bi-layer and a perovskite layer as shown schematically in *figure* 2. Electron microscopic studies [3, 4] found that there exist various kinds of incommensurate modulations in the structure of Bi-cuprates (shown schematically in *figure* 3). However with the conventional electron microscopic technique it is difficult to observe the modulated structure at atomic resolution. The superconducting phase Bi-2212 has been extensively studied by X-ray and neutron diffraction methods (see [2] and the references there in). The Bi-2201 phase has also been studied by similar methods [5, 6]. However, the published results are not completely consistent with each other, especially on the oxygen atoms of the Bi-O layer. The Bi-2223 superconductor has the highest *Tc* value in the Bi-cuprate family. On the other hand it is most difficult to prepare good-quality single crystals suitable for X-ray analysis for this compound. Hence so far there are no reports on single-crystal structure studies of the Bi-2223 phase by either X-ray or neutron diffraction. Finally an important problem in the study of superconductivity is: whether and how the structure of a superconductor changes when the temperature passes through the *Tc*? The multidimensional electron crystallographic methods have been used for solving the above problems.

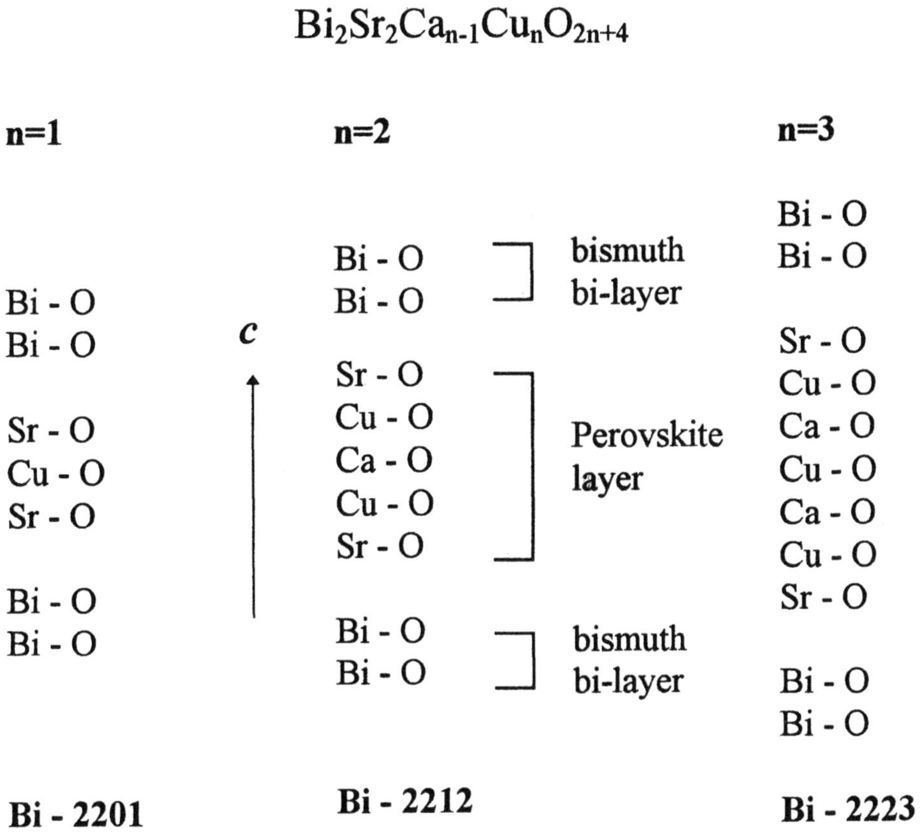

$$Bi_2Sr_2Ca_{n-1}Cu_nO_{2n+4}$$

n=1		**n=2**		**n=3**
				Bi - O
		Bi - O	bismuth	Bi - O
Bi - O		Bi - O	bi-layer	
Bi - O	*c* ↑			Sr - O
		Sr - O		Cu - O
Sr - O		Cu - O	Perovskite	Ca - O
Cu - O		Ca - O	layer	Cu - O
Sr - O		Cu - O		Ca - O
		Sr - O		Cu - O
Bi - O				Sr - O
Bi - O		Bi - O	bismuth	
		Bi - O	bi-layer	Bi - O
				Bi - O
Bi - 2201		**Bi - 2212**		**Bi - 2223**

Figure 2. Structure types in the Bi-cuprate family.

2.1. THE INCOMMENSURATE MODULATION OF THE Pb-DOPED Bi-2223 SUPERCONDUCTING PHASE [7]

This work was based on the preliminary electron diffraction study of Li *et al.* [8] and the average structure obtained by Sequeira *et al.* [9]. The diffraction intensities used in our study were measured from the electron diffraction pattern normal to the ***a*** axis. Since the ***a*** axis is rather short (5.49Å) and is perpendicular to the modulation vector ***q***, no attempt to use 3-dimensional diffraction data was made. The symmetry of the sample belongs to the superspace group P [B bmb] 1-11 with the 3-dimensional unit cell $a = 5.49$, $b = 5.41$, $c = 37.1$Å; $\alpha = \beta = \gamma = 90°$ and the modulation vector $\mathbf{q} = 0.117\mathbf{b}^*$. Five photographs were taken with different exposure times for the same *0klm* electron diffraction pattern. This is an analogue of the multifilm method in X-ray

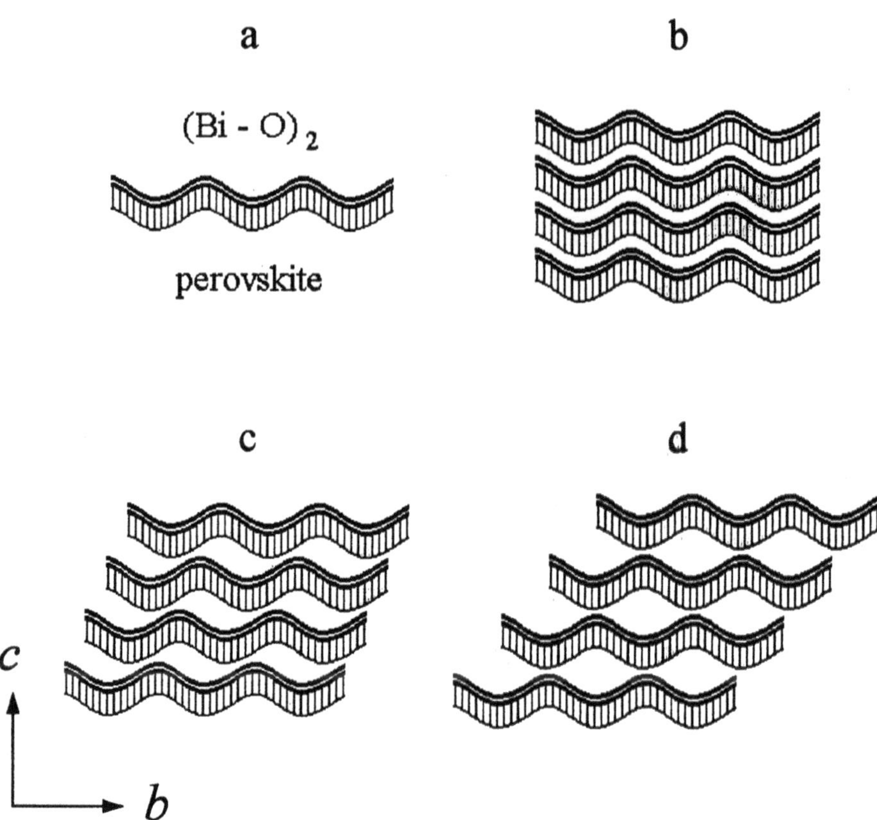

Figure 3. Different modulation modes in the Bi-cuprate family revealed by electron microscopic observation

crystallography for collecting diffraction intensities. A microdensitometer was used to measure the integrated intensities. Structure factor magnitudes were obtained as the square root of diffraction intensities. The R factor for the discrepancy of symmetrically related reflections is 0.12 for the 42 main reflections and 0.13 for the 70 first-order satellite reflections. A few second-order satellites were also observed; however, they are much weaker than the first-order ones and were neglected in the structure analysis.

The structure analysis was carried out in 4-dimensional space, in which the real and the reciprocal unit cells are defined respectively as

$$\boldsymbol{a}_1 = \boldsymbol{a}, \quad \boldsymbol{a}_2 = \boldsymbol{b} - 0.117\boldsymbol{d}, \quad \boldsymbol{a}_3 = \boldsymbol{c}, \quad \boldsymbol{a}_4 = \boldsymbol{d}$$

and

$$\boldsymbol{b}_1 = \boldsymbol{a}^*, \quad \boldsymbol{b}_2 = \boldsymbol{b}^*, \quad \boldsymbol{b}_3 = \boldsymbol{c}^*, \quad \boldsymbol{b}_4 = 0.117\,\boldsymbol{b}^* + \boldsymbol{d}$$

where $\boldsymbol{d}$ is the unit vector normal to the 3-dimensional space, i.e. a unit vector simultaneously perpendicular to the vectors $\boldsymbol{a}$, $\boldsymbol{b}$, $\boldsymbol{c}$, $\boldsymbol{a}^*$, $\boldsymbol{b}^*$ and $\boldsymbol{c}^*$. An atom in the 4-dimensional space without modulation will be something like an infinite straight bar parallel to the fourth dimension $\boldsymbol{a}_4$. Occupational modulation will change periodically the width, while positional modulation will change periodically the direction of the bar. Our task is to find out such a periodic change. This can be accomplished by solving the phase problem and calculating the 4-dimensional potential distribution, the 4-dimensional Fourier map. The incommensurate modulated structure in the 3-dimensional physical space can be obtained by cutting the 4-dimensional Fourier map with a 3-dimensional hyperplane perpendicular to the direction $\boldsymbol{a}_4$. While the modulation wave of all the atoms can be measured directly on the 4-dimensional Fourier map. For details of the multidimensional representation of incommensurate modulated structures the reader is referred to the original papers [10-13].

The phases of the main reflections *0kl0* were calculated from the known average structure. While phases of the satellites *0klm* were derived by the multidimensional direct method [2]. A Fourier map in multidimensional space was then calculated, from which modulation waves of all metal atoms were measured directly. On this basis least-square refinement ended at an *R*-factor of 0.16 for the main and 0.17 for the first-order satellite reflections. *Figure* 4 shows the modulation waves of the metal atoms. *Figure* 5 shows the incommensurate modulation of the Pb-doped Bi-2223 phase in the 3-dimensional physical space projected along the $\boldsymbol{a}$ axis.

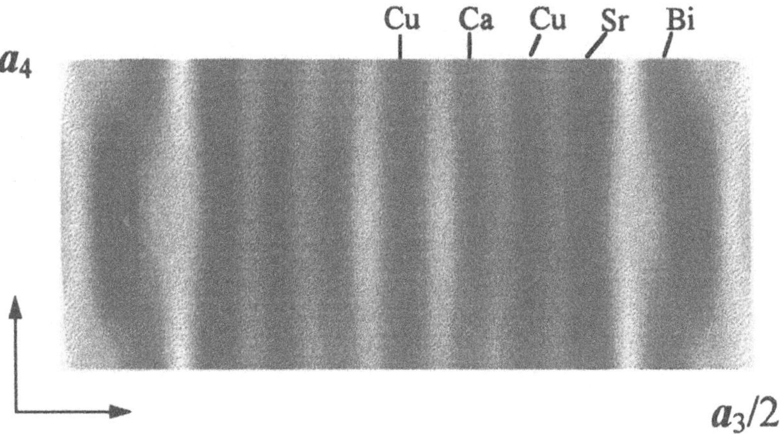

Figure 4. A section, $\int \varphi(x_1,0,x_3,x_4)dx_1$, of the 4-dimensional Fourier map of the Pb-doped Bi-2223 superconductor projected along the $\boldsymbol{a}$ axis

As is seen, both occupational and positional modulations are evident for most metal atoms. An other prominent feature is that the oxygen atoms on the Cu-O layers move towards the Ca layer, forming a disordered oxygen bridge across the layers of Cu(2)-Ca-Cu(1)-Ca-Cu(2) (see *figure* 5).

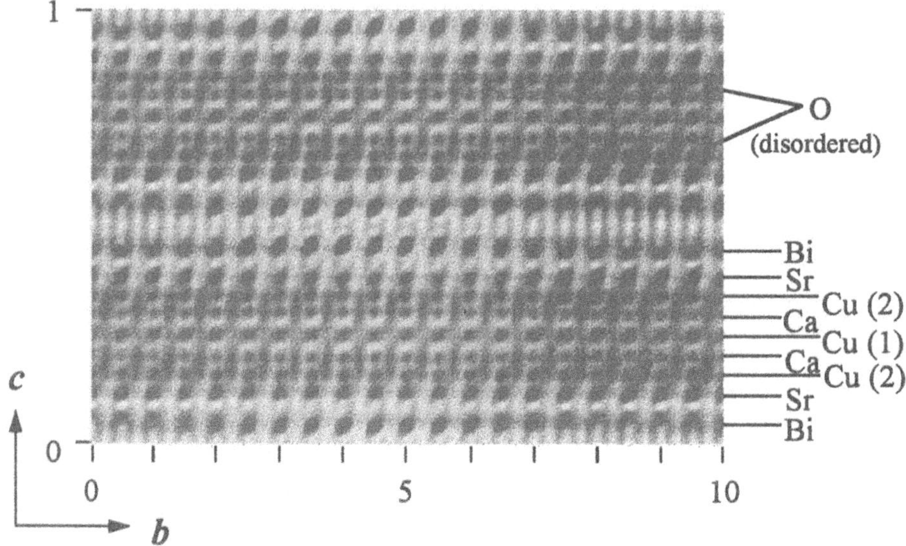

Figure 5. The 3-dimensional potential distribution function of the Pb-doped Bi-2223 superconductor projected along the ***a*** axis. Ten unit cells are plotted along the ***b*** axis, showing the period of modulation to be approximately 8.5 times the length of ***b***.

2.2. THE INCOMMENSURATE MODULATION OF THE Bi-2212 PHASE [14]

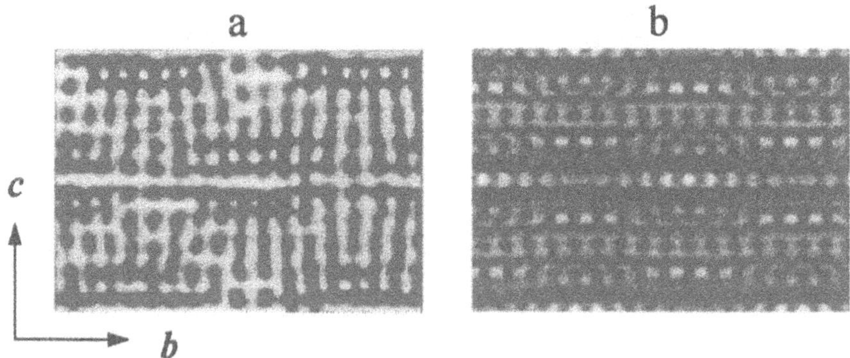

Figure 6. The modulation of the Bi-2212 superconducting phase. (a) experimental EM taken with the incident electron beam parallel with the ***a*** axis [3]; (b) the 3-dimensional potential distribution projected along the ***a*** axis, calculated using structure-factor magnitudes from the corresponding ED and phases from the EM and the direct-method phase extension.

The superconducting phase Bi-2212 has been extensively studied as described in reference [2]. However no electron crystallographic methods at atomic resolution were used in the previous studies. We provide here such an example and show how the combination of EM and ED can be used to determine the incommensurate structure even assuming the average structure is unknown. The sample belongs to the superspace group N [B bmb] 1 -1 1 with $a = 5.42$, $b = 5.44$, $c = 30.5$Å, $\alpha = \beta = \gamma = 90°$ and the modulation wave vector $\boldsymbol{q} = 0.22\,\boldsymbol{b}^* + \boldsymbol{c}^*$. We started with an EM at 2Å resolution (*figure* 6a) and the corresponding ED at 1Å resolution. A set of structure-factor magnitudes with indices *0klm* was measured from the ED. The phases of the main reflections within 2Å resolution were obtained from the Fourier transform of the deconvoluted EM. While phases of satellite reflections and of the main reflection beyond 2Å resolution were derived by the direct-method phase extension. Finally a Fourier map was calculated (*figure* 6b) which reveals the incommensurate modulation of the structure. This procedure can be regarded as an image processing technique in HREM [1] applied to *figure* 6a by making use of the information from the corresponding ED.

2.3. THE INCOMMENSURATE MODULATION OF THE Bi-2201 PHASE

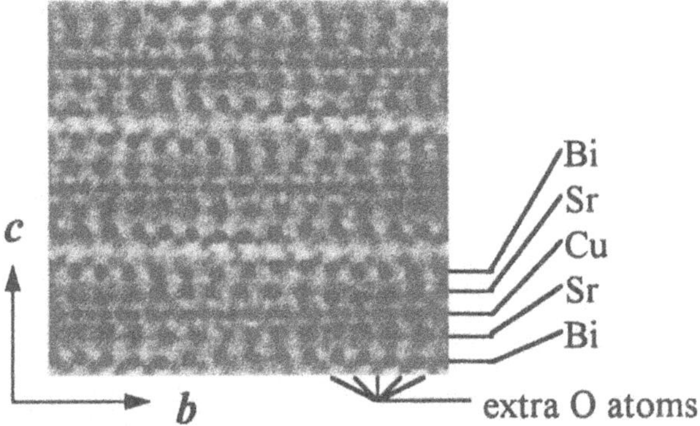

Figure 7. Three-dimensional potential distribution of the Bi-2201 phase projected along the ***a*** axis, showing the incommensurate modulation and extra oxygen atoms.

Crystals of Bi-2201 belong to the superspace group P [B 2/b] -11 with $a = 5.41$, $b = 5.43$, $c = 24.6$Å, $\alpha = \beta = \gamma = 90°$ and the modulation wave vector $\boldsymbol{q} = 0.217\boldsymbol{b}^* + 0.62\boldsymbol{c}^*$. The structure analysis was based on the known average structure [5]. Only electron diffraction intensities of the *0klm* reflections were used. Since inconsistent results have been reported on the oxygen atoms [5, 6], special care was taken for their determination. The 4-dimensional Fourier map projected along the ***a*** axis was calculated with phases from the average structure and the direct-method phase extension. Apart from two oxygen atoms, which are overlapped with metal atoms respectively on the Bi-O and Sr-O layers, the modulation waves of all symmetrically independent atoms were measured directly from this Fourier map. Up to the fourth-order harmonics were included in the expression of the modulation function. Fourier recycling and least-squares refinement led to the R factors: $R_T = 0.32$, $R_M = 0.29$, $R_{S1} = 0.29$, $R_{S2} = 0.36$ and $R_{S3} = 0.52$. Here the R factor is defined as $R = \Sigma| |Fo|-|Fc| |/\Sigma|Fo|$, R_T denotes the R factor for the total reflections, R_M for the main reflections, R_{S1}, R_{S2} and R_{S3} respectively for the first-order, second-order and third-order satellite reflections. The resulting Fourier map clearly shows the main feature of the modulation. However, it contains a number of small additional peaks near some of the Bi and Sr sites. After we failed to eliminate them, they were treated as extra oxygen atoms. By including extra oxygen atoms in the Bi-O layer, after a few cycles of refinement, the R factors dropped to 0.23, 0.20, 0.24, 0.27 and 0.30 for R_T, R_M, R_{S1}, R_{S2} and R_{S3} respectively. Further inclusion of extra oxygen atoms in the Sr-O layer, the least-square refinement ended at the R factors of 0.18, 0.13, 0.19, 0.25 and 0.26 for R_T, R_M, R_{S1}, R_{S2} and R_{S3} respectively. *figure* 7 is the final Fourier map.

2.4 THE INCOMMENSURATE MODULATION OF THE Bi-2212 PHASE AT THE TEMPERATURES ABOVE AND BELOW *Tc*

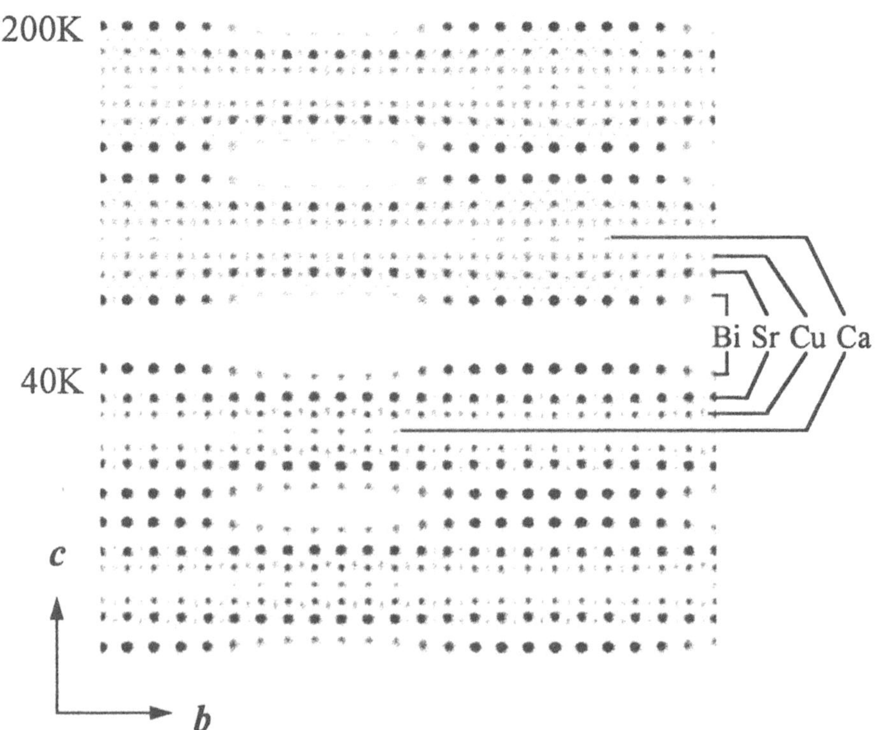

Figure 8. Three-dimensional potential distribution function of the Bi-2212 phase projected along the ***a*** axis. Upper: at the temperature above *Tc*; Lower: at the temperature below *Tc*.

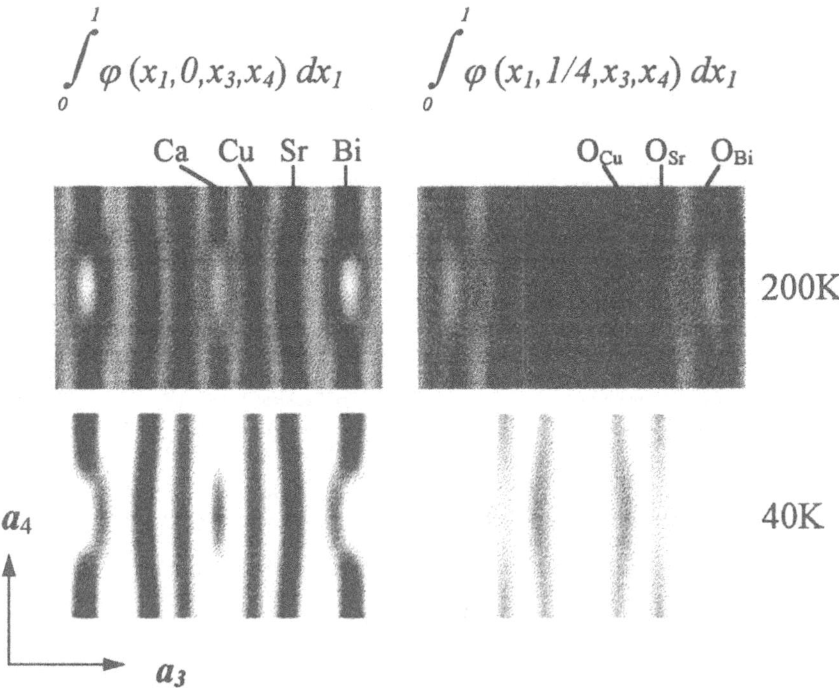

Figure 9. Sections of the 4-dimensional Fourier map of the Bi-2212 phase projected along the ***a*** axis. Upper: at the temperature above *Tc*; Lower: at the temperature below *Tc*. Left: sections showing the metal atoms; Right: sections showing oxygen atoms near $x_2 = 1/4$.

The sample belongs to the superspace group P [B bmb] 1 -1 1 with $a = 5.40$, $b = 5.38$, $c = 30.80$Å; $\alpha = \beta = \gamma = 90°$ and the modulation vector $\boldsymbol{q} = 0.117\boldsymbol{b}^*$. Note that the modulation mode of this sample is different with that of the Bi-2212 sample described in Section 2.2. Electron diffraction patterns were taken respectively at the temperature of 200K (above *Tc*) and 40K (below *Tc*) and with the incident electron beam parallel to the ***a*** axis. The multidimensional direct method was used to determine the incommensurate modulation.

Figure 8 shows the resultant 3-dimensional potential distribution projected along the ***a*** axis. *figure* 9 shows Fourier hyper-sections through the average position of the metal atoms and the unoverlapped oxygen atoms. As is seen, the variations of atoms in both occupancy and position are on average weaker at 40K (below *Tc*) than that at 200K (above *Tc*). A prominent feature can be seen in *figure* 8 is that the modulation wave of the Ca atoms shifts one half of its period along the ***b*** direction. Another prominent, and perhaps more important, feature in *figure* 9 is that the 'extra' oxygen atoms, which distribute on the Bi-O, Sr-O and Cu-O layers at 200K, move towards the Cu-O layer at the temperature below *Tc* (40K).

References

1. Li, F. H. (1997) Image restoration - crystal structure determination by image deconvolution and resolution enhancement, (in this *Proceedings*).
2. Fan, H. F. (1997) Multi-dimensional direct methods, (in this *Proceedings*).
3. Matsui, Y. and Horiuchi, S. (1988) Geometrical relations of various modulated structures in Bi-Sr-Ca-Cu-O Superconductors and related compounds, *Jpn. J. Appl. Phys.* **27**, L2306-L2309.
4. Ikeda, S., Aota, K., Hatano, T. and Ogawa, K. (1988) A new mode of modulation observed in the Bi-Sr-Ca-Cu-O system, *Jpn. J. Appl. Phys.* **27**, L2040-L2043.
5. Gao, Y., Lee, P., Ye, J., Bush, P., Petricek, V. and Coppens, P. (1989) The incommensurate modulation in the $Bi_2Sr_{2-x}Ca_xCu_2O_6$ superconductor, and its relation to the modulation in $Bi_2Sr_{2-x}Ca_xCu_2O_8$, *Physica C* **160**, 431-438.
6. Yamamoto, A., Takayama-Muromachi, E., Izumi, F., Ishigaki, T. and Asano, H. (1992) Rietveld analysis of the composite crystal in superconducting $Bi_{2+x}Sr_{2-x}CuO_{6+y}$, *Physica C* **201**, 137-144.
7. Mo, Y. D., Cheng, T.Z., Fan, H. F., Li, J. Q., Sha, B. D., Zheng, C. D., Li, F. H. and Zhao, Z. X. (1992) Structural features of the incommensurate modulation in the Pb-doped Bi-2223 high *Tc* phase revealed by direct-method electron diffraction analysis, *Supercond. Sci. Technol.* **5**, 69-72.
8. Li, J. Q., Yang, D. Y., Li, F. H., Zhou, P., Zheng, D. N., Ni, Y. M., Jia, S. L. and Zhao, Z. X. (1989) Structure Symmetry and microstructures of Bi(Pb)-Sr-Ca-Cu-O system, *Progress in High Tc Superconductivity* **22**, 441-443.
9. Sequeira, A., Yakhmi, J. V., Iyer, R. M., Rajagopal, H. and Sastry, P. V. P. S. S. (1990) Novel structural features of Pb-stabilised Bi-2223 high-*Tc* phase from neutron-diffraction study, *Physica Scripta* **C167**, 291-296.
10. Janner, A. and Janssen, T. (1977) Symmetry of periodically distorted crystals, *Phys. Rev. B* **15**, 643-658
11. De Wolff, P. M. Janssen, T. and Janner, A. (1981) The superspace groups for incommensurate crystal structures with a one-dimensional modulation, *Acta Cryst.* **A37**, 625-636.
12. Janner, A. Janssen, T. and De Wolff, P. M. (1983) Bravais classes for incommensurate crystal phases, *Acta Cryst.* **A39**, 658-666.
13. Jansen, T., Janner, A., Looijenga-Vos, A. and De Wolff, P. M. (1992) Incommensurate and commensurate modulated structures, in Wilson, A. J. C. (ed.), *International Tables for Crystallography*, Kluwer Academic Publishers, Dordrecht, pp. 797-835; 843-844.
14. Fu, Z. Q., Huang, D. X., Li, F. H., Li, J. Q., Zhao, Z. X., Cheng, T. Z. and Fan, F. F. (1994) Incommensurate modulation in minute crystals revealed by combining high-resolution electron microscopy and electron diffraction, *Ultramicroscopy* **54**, 229-236.

STRUCTURE DETERMINATION BY MAXIMUM ENTROPY AND LIKELIHOOD

C.J.Gilmore
Department of Chemistry
University of Glasgow
Glasgow G12 8QQ, Scotland, UK.

1. Introduction

This chapter will illustrate the process of solving a crystal structure using the maximum entropy (ME) method via the Bricogne formalism [1,2,3]. The structure in question is the membrane protein Omp F porin [4] from a set of two-dimensional electron diffraction data at *ca.* 6Å resolution. It follows the method outlined by Gilmore, Nicholson & Dorset [5]

2. Data Normalisation

We start with the process of normalisation in which the raw intensities are converted to unitary structure factors via the equation:

$$\left(\left|U_{\mathbf{h}}\right|^{obs}\right)^2 = k\sigma_2\left(\left|F_{\mathbf{h}}\right|^{obs}\right)^2 \Big/ \sigma_1 \exp\left(-2B\sin^2\theta/\lambda^2\right)\varepsilon_{\mathbf{h}}\sum_{j=1}^{N} f_j^2 \tag{1}$$

where B is an overall, isotropic temperature factor, k a scale factor (both obtained by a Wilson plot), f_j is the electron scattering factor for atom j, the summation spans the N atoms in the unit cell, θ is the Bragg angle for radiation of wavelength, ε is the staistical weight λ and:

$$\sigma_n = \sum_{j=1}^{N} z(eff)_j^n \tag{2}$$

where $z(eff)_j$ is the effective atomic number of atom j. Each U magnitude has an associated phase angle $\varphi_{\underline{h}}$ which is to be determined. Table 1 shows the unit cell and symmetry information needed for normalisation using MITHRIL [6,7], and Table 2 lists the raw structure factors before normalisation along with the experimentally observed phase angles. We are not going to use the latter directly, but they will act as a useful

D. L. Dorset et al. (eds.), Electron Crystallography, 295–304.

check on how well the *ab initio* phasing is progressing. Notice also that we employ the standard deviations of the structure factors in the ME calculations.

```
CELL  72  72  5  90 90 120  !Cell parameters
LATT   A  P           !Acentric, primitive  cell
ELECTRON                         ! Electron data
SYMM -Y,X-Y,Z       !Symmetry operations
SYMM -X+Y,-X,Z
SYMM Y,X,Z
SYMM X-Y,-Y,Z
SYMM -X,-X+Y,Z
CONT  1500 C        ! 1500 C atoms in cell
DATA 1                           ! Data format
      (3I3,F7.2,F5.2,I5)
```

Table 1. The Normalisation Data

The Wilson plot is shown in Figure 1.

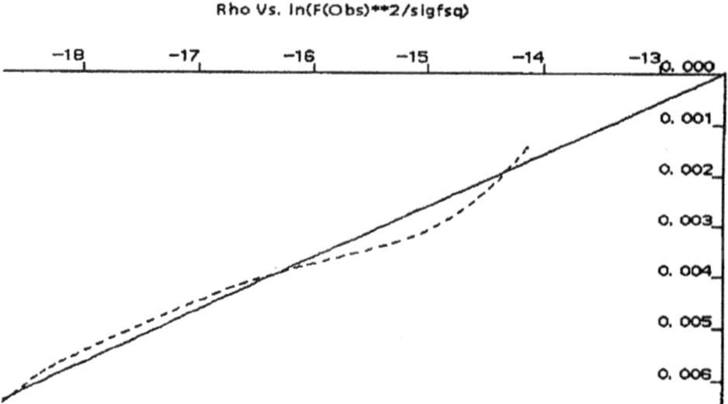

Figure 1. The Wilson plot for Omp F Porin

At first sight all seems well except that the overall temperature factor is B=327±57Å^2. This is clearly unlikely, but is a common feature when carrying out normalisation with such a sparse data set. Negative B's are also very common. To overcome the problem, a temperature factor of B=0.0Å^2 is imposed on the data to give a set of U-magnitudes listed in Table 3.

3. Getting started with the phasing

Normally one would begin by defining the origin, but the plane group is p31m and all the reflections are structure seminvariants so no such reflections are available. To overcome this the strongest 5 centric reflections: (1) 5 0 0 (4) 4 0 0 (8) 3 0 0 (10) 2 0 0 (14) 7 0 0 are given permuted phases. This gives 2^5 nodes on a rootless phasing tree. Each of these phase combinations is subjected to constrained entropy maximisation and a log-likelihood gain (LLG) computed for each calculated for each one as described in a previous chapter by both Bricogne and Gilmore.

h	k	l	|F(obs)|	σ(|F(obs)|)	Phase
1	0	0	0.78	0.08	180
2	0	0	7.44	0.70	180
3	0	0	10.78	1.10	180
4	0	0	10.76	1.10	0
5	0	0	22.95	2.30	180
6	0	0	1.19	0.11	0
7	0	0	5.32	0.50	0
8	0	0	1.54	0.15	0
9	0	0	2.55	0.25	180
10	0	0	1.77	0.18	0
1	1	0	3.30	0.33	324
2	1	0	5.78	0.58	43
3	1	0	2.84	0.28	214
4	1	0	5.69	0.60	237
5	1	0	4.58	0.45	58
6	1	0	3.95	0.39	104
7	1	0	2.45	0.24	346
8	1	0	1.52	0.15	184
9	1	0	1.27	0.13	34
10	1	0	0.77	0.08	217
2	2	0	9.37	0.94	230
3	2	0	9.89	0.99	118
4	2	0	7.42	0.74	305
5	2	0	3.52	0.35	36
6	2	0	4.61	0.46	259
7	2	0	.00	0.30	98
8	2	0	1.88	0.20	348
9	2	0	0.82	0.08	173
3	3	0	14.93	1.50	261
4	3	0	6.81	0.68	159
5	3	0	5.01	0.50	33
6	3	0	5.90	0.59	70
7	3	0	3.02	0.30	297
8	3	0	0.72	0.07	245
4	4	0	5.78	0.58	42
5	4	0	3.53	0.35	290
6	4	0	1.48	0.15	195
7	4	0	2.82	0.28	129
8	4	0	0.88	0.09	272
5	5	0	2.24	0.22	16
6	5	0	2.46	0.25	30
6	6	0	1.46	0.14	264

Table 2. The raw intensity data for Omp F porin

No.	h	k	l	|U(obs)|	σ(|U(obs)|)	d	Constraint
1	5	0	0	0.10330	0.01035	12.47	180/360
2	4	-2	0	0.06775	0.00679	18.00	None
3	6	-3	0	0.05809	0.00583	12.00	None
4	4	0	0	0.04797	0.00490	15.59	180/360
5	5	-2	0	0.04424	0.00442	14.30	None
6	8	-4	0	0.04341	0.00435	9.00	None
7	5	-1	0	0.04154	0.00438	13.61	None
8	3	0	0	0.04115	0.00419	20.78	180/360
9	6	-2	0	0.03350	0.00333	11.78	None
10	2	0	0	0.03275	0.00307	31.18	180/360
11	7	-3	0	0.03104	0.00310	10.25	None
12	7	-2	0	0.02619	0.00260	9.93	None
13	3	-1	0	0.02552	0.00256	23.57	None
14	7	0	0	0.02456	0.00230	8.91	180/360
15	9	-3	0	0.02383	0.00238	7.86	None
16	2	-1	0	0.02363	0.00236	36.00	None
17	8	-3	0	0.02313	0.00230	8.91	None
18	11	-4	0	0.02217	0.00219	6.47	None
19	8	-2	0	0.02135	0.00212	8.65	None
20	6	-1	0	0.02075	0.00203	11.20	None
21	8	-1	0	0.01857	0.00181	8.26	None
22	7	-1	0	0.01812	0.00178	9.51	None
23	10	-5	0	0.01730	0.00170	7.20	None
24	9	-4	0	0.01650	0.00163	7.98	None
25	10	-2	0	0.01465	0.00155	6.81	None
26	10	-3	0	0.01438	0.00142	7.02	None
27	9	-2	0	0.01411	0.00141	7.62	None
28	4	-1	0	0.01262	0.00124	17.29	None
29	11	-5	0	0.01185	0.00120	6.54	None
30	9	0	0	0.01049	0.00102	6.93	180/360
31	10	0	0	0.00861	0.00087	6.24	180/360
32	8	0	0	0.00722	0.00069	7.79	180/360
33	9	-1	0	0.00719	0.00070	7.30	None
34	10	-4	0	0.00702	0.00070	7.15	None
35	12	-6	0	0.00617	0.00059	6.00	None
36	11	-1	0	0.00616	0.00063	5.92	None
37	10	-1	0	0.00612	0.00062	6.54	None
38	6	0	0	0.00467	0.00042	10.39	180/360
39	12	-4	0	0.00433	0.00043	5.89	None
40	11	-2	0	0.00400	0.00038	6.14	None
41	11	-3	0	0.00349	0.00033	6.33	None
42	1	0	0	0.00342	0.00034	62.35	180/360

Table 3. The U-magnitudes, their standard deviations, phase restrictions and resolution (Å) for Omp F porin.

The phasing tree is listed in Table 4. It can be seen that there is initially little discernible variation in LLG, but the phase analysis using the t-test correctly obtains the correct pattern of phases as in shown in tables 5 and 6. Analysis gives no significant indications for main effects, but, for the double and triple sign analysis of likelihood, there are significant and correct indications which can be used to derive single phase angles correctly.

Node	Entropy(x10^{-2})	LLG	NS + L	Error
1	-0.031	0.01	-0.4604	52
2	-0.023	0.01	-0.3433	126
3	-0.029	0.02	-0.4303	17
4	-0.023	0.00	-0.3468	92
5	-0.025	0.00	-0.3703	81
6	-0.023	-0.01	-0.3466	156
7	-0.028	0.00	-0.4212	47
8	-0.023	0.00	-0.3459	121
9	-0.028	0.00	-0.4231	75
10	-0.024	0.00	-0.3527	150
11	-0.024	0.02	-0.3768	41
12	-0.024	0.00	-0.3573	115
13	-0.030	0.01	-0.4437	105
14	-0.023	0.01	-0.3433	180
15	-0.029	0.01	-0.4410	70
16	-0.023	0.01	-0.3504	145
17	-0.030	0.02	-0.4481	34
18	-0.024	0.01	-0.3530	109
19	-0.029	0.02	-0.4311	0
20	-0.023	0.01	-0.3446	74
21	-0.025	0.00	-0.3698	64
22	-0.023	0.00	-0.3516	138
23	-0.025	0.00	-0.3678	29
24	-0.023	0.00	-0.3514	104
25	-0.023	0.00	-0.4344	58
26	-0.024	0.00	-0.3563	132
27	-0.028	-0.01	-0.4236	23
28	-0.028	0.00	-0.3546	98
29	-0.032	0.01	-0.4821	87
30	-0.024	0.01	-0.3493	162
31	-0.029	0.01	-0.4363	53
32	-0.023	0.00	-0.3437	127

Table 4. The phasing tree for permuting 4 centric reflections without an origin.

No	**No**	**No**	**PosAv**	**NegAv**	**Signif**	**Sign**
1	10		0.006	0.002	0.153	+
8	10		0.010	-0.002	0.000	+
10	14		0.002	0.006	0.173	-
1	8	10	0.002	0.006	0.126	-

Table 5. The phase relationships from the t-test.

No	h	k	l	Deduced phase	Correct?	Σ_1	Prob
1	5	0	0	180	Yes	0	.776
4	4	0	0	0/180			
8	3	0	0	180	Yes	0	.527
10	2	0	0	180	Yes	180	.505
14	7	0	0	0/180			

Table 6. The derived phases and a comparison with the Σ_1 formula.

No	h	k	l	|U(obs)|
2	4	-2	0	0.068
3	6	-3	0	0.058
4	4	0	0	0.048

Table 7. The permuted reflections for level 2 of the phasing tree.

Node	Entropy	LLG	NS+L	Error
1	-0.025	0.00	-0.3690	0
2	-0.038	0.01	-0.5676	85
3	-0.038	-0.01	-0.5715	99
4	-0.038	0.01	-0.5654	77
5	-0.037	0.02	-0.5529	63
6	-0.038	-0.01	-0.5695	85
7	-0.038	0.01	-0.5721	99
8	-0.038	0.02	-0.5753	77
9	-0.038	0.01	-0.5635	63
10	-0.038	0.01	-0.5637	66
11	-0.038	0.02	-0.5753	80
12	-0.038	0.01	-0.5721	58
13	-0.038	-0.01	-0.5695	44
14	-0.037	0.02	-0.5527	66
15	-0.038	0.01	-0.5658	80
16	-0.038	-0.01	-0.5715	58
17	-0.038	0.01	-0.5676	44
18	-0.038	0.01	-0.5714	53
19	-0.038	0.03	-0.5707	68
20	-0.038	0.01	-0.5770	46
21	-0.038	-0.02	-0.5755	31
22	-0.038	0.03	-0.5666	53
23	-0.038	0.01	-0.5790	68
24	-0.039	-0.02	-0.5817	46
25	-0.038	0.01	-0.5716	31
26	-0.038	0.01	-0.5716	35
27	-0.039	-0.02	-0.5818	49
28	-0.038	0.01	-0.5791	27
29	-0.038	0.03	-0.5666	13
30	-0.038	-0.02	-0.5755	35
31	-0.038	0.01	-0.5770	49
32	-0.038	0.03	-0.5707	27
33	-0.038	0.01	-0.5714	13

Table 8. The second level phasing tree.

Note that the Σ_1 indications are wrong 2 out of 3 times whereas likelihood is always correct. The indications for reflection (1) are very strong so we define the phase of (1) as 180^0. This becomes node 1 of the phasing tree and we now permute the reflections listed in Table 7. This gives 4^2x2=32 nodes i.e. nodes 2-33. The resulting LLGs after entropy maximisation are given in Table 8.

Analysis of the LLGs gives 4 nodes which pass the t-test. They are listed in table 9. One has a negative LLG and so is discarded.

Node	LLG	Entropy	Score	No. of violations
29	0.032	-0.038	1.000	0
30	-0.017	-0.038	1.000	0
31	0.007	-0.038	1.000	0
32	0.033	-0.038	1.000	0

Table 9. Analysis of the second level nodes.

Nodes 29 (13), 31(49), and 32(27) are kept (the mean absolute phase errors are in parentheses). Now let us examine the centroid map for node 29. For **k** acentric the Fourier coefficients are:

$$|U_{\mathbf{k}}|^{obs}\left[I_1(X_{\mathbf{k}})/I_0(X_{\mathbf{k}})\right]\exp\left(i\varphi_{\mathbf{k}}^{ME}\right) \tag{3}$$

where:

$$X_{\mathbf{k}} = (2N/\varepsilon_{\mathbf{k}})|U_{\mathbf{k}}|^{obs}\left|U_{\mathbf{k}}^{ME}\right| \tag{4}$$

For **k** centric, these coefficients become:

$$|U_{\mathbf{k}}|^{obs}\tanh(X_{\mathbf{k}})\exp\left(i\varphi_{\mathbf{k}}^{ME}\right) \tag{5}$$

with:

$$X_{\mathbf{k}} = (N/\varepsilon_{\mathbf{k}})|U_{\mathbf{k}}|^{obs}\left|U_{\mathbf{k}}^{ME}\right| \tag{6}$$

where $|U_{\mathbf{k}}|^{obs}$ is the observed U-magnitude, $\left|U_{\mathbf{k}}^{ME}\right|$ is that extrapolated from the ME map, $\varphi_{\mathbf{k}}^{ME}$ is the phase angle predicted by the ME process and $\varepsilon_{\mathbf{k}}$ is the statistical weight of reflection **k.** The centroid map is shown in figure 2 for 4 unit cells. For reference the true map using image derived phases is shown in Figure 2. The agreement is excellent with a map correlation coefficient of 0.95.

Now we permute the reflections listed in Table 10. This gives 4^3x2x3=184 nodes. (Remember we kept three nodes from level 2).

No	h	k	l	|U(obs)|
7	5	-1	0	0.042
8	3	0	0	0.041
9	6	-2	0	0.034
10	2	0	0	0.033

Table 10. The reflections permuted in level 3 of the phasing tree.

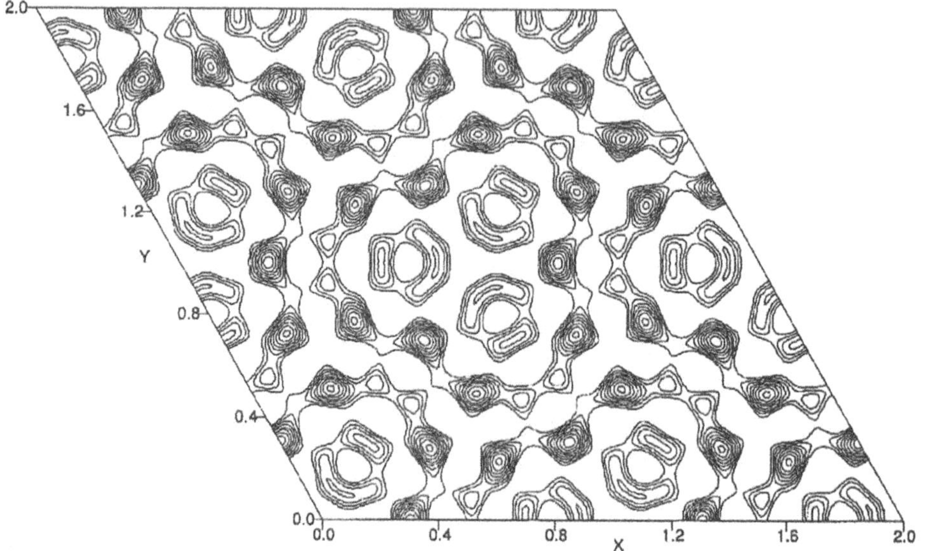

Figure 2. The centroid map for node 29.

After analysis nodes 42(25), 50(18), 53(9), 98(56), 125(40), 189(26) are kept (the mean absolute phase errors are in parentheses). Figure 4 shows the best centroid map (for node 53). For reference the correct map using experimental phases is shown in Figure 4. The map correlation coefficient is 0.94.

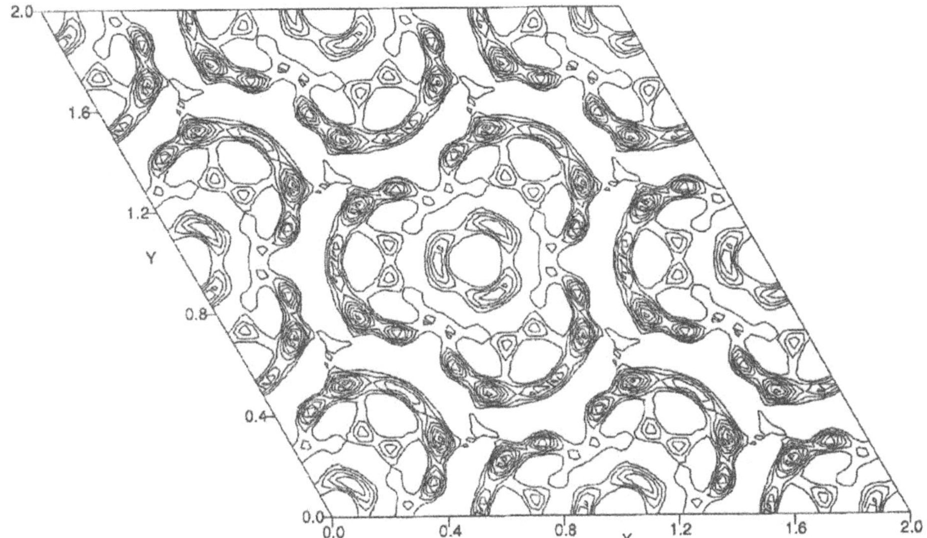

Figure 3. The true map for Omp F porin using image derived phases.

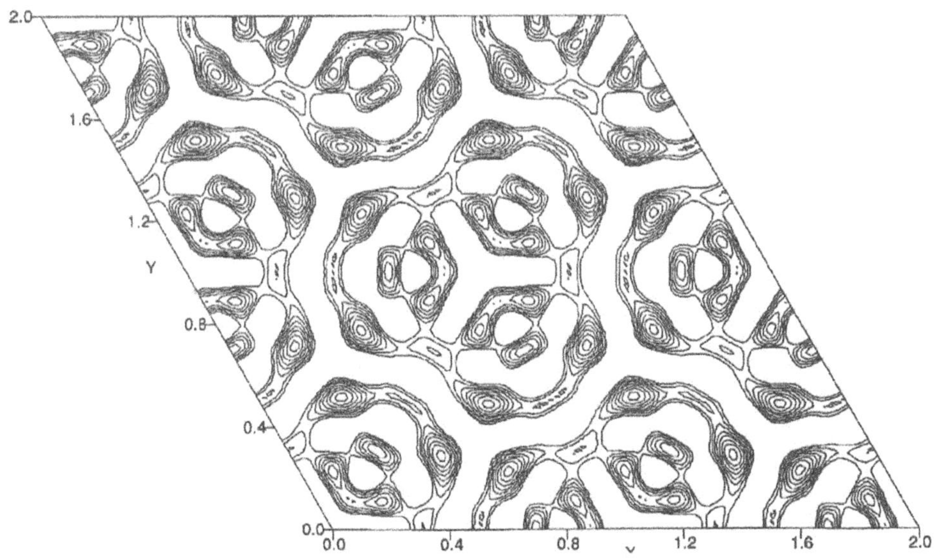

Figure 4. The centroid map for node 53 with a mean phase error of 9^0 and a map correlation coefficient of 0.94.

References

1. Bricogne, G. (1984) Maximum entropy and the foundations of direct methods, *Acta Cryst.* **A40**, 410-445.
2. Bricogne, G. and Gilmore, C.J. (1990) A multisolution method of phase determination by combined maximisation of entropy and likelihood. I. Theory, algorithms and strategy, *Acta Cryst.* **A46,** 284-297.
3. Gilmore, C.J., Bricogne, G. and Bannister, C. (1990) A multisolution method of phase determination by combined maximisation of entropy and likelihood. II Application to small molecules, *Acta Cryst.* **A46,** 297-308.
4. Sass, H. J., Büldt, G., Beckmann, E., Zemlin, F., Van Heel, M., Zeitler, E., Rosenbusch, J. P., Dorset, D. L., and Massalski, A. (1989). *J. Mol. Biol.* **209**, 171-1
5. Gilmore, C.J., Nicholson, W.V., and Dorset, D.L., (1996) Direct methods in protein electron crystallography: the ab initio structure determination of two membrane protein structures in projection using maximum entropy and likelihood, *Acta Cryst.* **A52,** 937-946.
6. Gilmore, C.J. (1984) MITHRIL - an integrated direct-methods computer program *J.Appl.Cryst.* **17**, 42-46
7. Gilmore, C.J. and Brown. S.R. (1988) New developments in the MITHRIL computer program . **22**, 571-572.

CRYSTALLOGRAPHIC IMAGE PROCESSING ON MINERALS:
3D structures, defects and interfaces

X.D. ZOU[1], E.A. FERROW AND D.R. VEBLEN[2]

Mineralogy and Petrology, Institute of Geology, Sölvegatan 13, Lund University, S-223 62 Lund, Sweden

Present addresses:
[1] Structural Chemistry, Stockholm University, S-106 91 Stockholm, Sweden and [2] Earth & Planetary Sciences, Johns Hopkins University, Baltimore, MD 21218, USA

1. Introduction

High resolution electron microscopy (HREM) combined with crystallographic image processing (CIP) is a powerful technique for structure determination of crystals too small to be studied by common techniques and for studying defects and interfaces. This technique is especially useful in mineralogy since low grade metamorphic rocks and clay-rich sediments are composed of fine grain crystals; and since the crystal structures of minerals are very complicated and not well ordered. Here we will present some examples on the application of HREM and CIP to silicate minerals.

2. 3D crystal structure determination of silicate minerals

It is possible to obtain crystallographic structure factor amplitudes and phases from HREM images. Distortions in HREM images, due to for example crystal tilt, wrong focus and astigmatism can be compensated for by crystallographic image processing [1]. The projected crystal potential can be retrieved. If one of the crystal axes is short and atoms do not overlap along this short axis, it may even be possible to resolve individual atoms in the projection. However, the unit cells of silicate minerals are usually so large that atoms overlap in every projection. In these cases atoms cannot be resolved from any single projection. HREM images from several projections are needed for a complete 3D structure determination. The power of 3D reconstruction on minerals was demonstrated on staurolite [2,3]. HREM images from five different projections taken on a JEOL ARM-1000 operated at 800kV were combined to generate a 3D potential map. At the resolution of 1.38Å, all atoms including oxygen could be resolved from the 3D map.

We are applying the 3D reconstruction technique to a modulated layer silicate, bannisterite, in order to see if we can solve this complex structure *ab initio* by electron crystallography. Bannisterite has a monoclinic space group *A2/a* with a=22.9, b=16.4, c=24.6 Å and β=94.36° [4]. The structure of bannisterite is more complicated than

D. L. Dorset et al. (eds.), Electron Crystallography, 305–312.

staurolite and there are 83 unique atoms in the unit cell. The ideal composition is $CaKFe_{20}Si_{29}Al_3O_{76}(OH)_{16}\cdot nH_2O$. HREM images of bannisterite from 20 different zone axes have been collected on a JEOL 4000EX operated at 400kV. Images from four of the zone axes are shown on the left column in Figure 1.

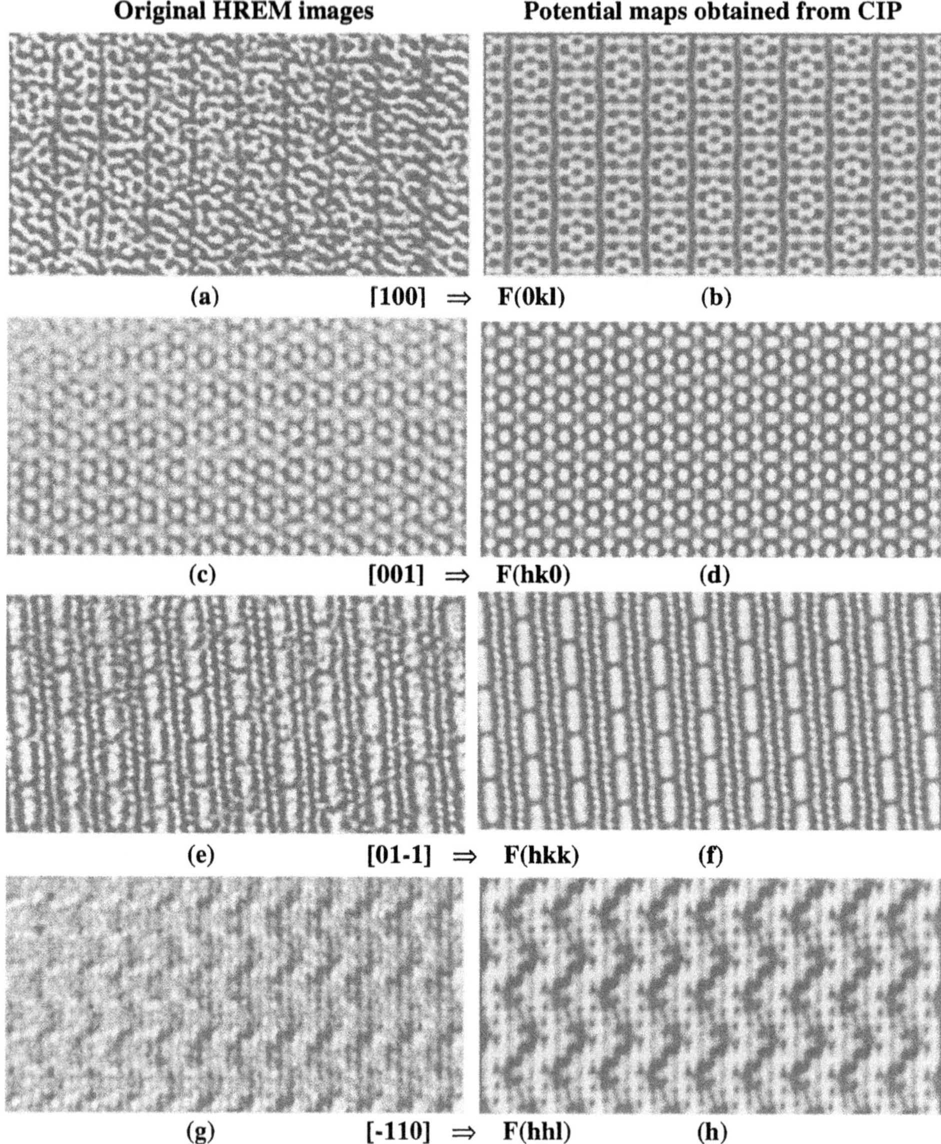

Figure 1 HREM images of bannisterite taken along different zone axes. From the [100] projection (a) the structure factors F(0kl) are obtained. The projected potential map (b) is calculated from the F(0kl). (c) and (d) The [001] projection from which the F(hk0) are obtained. (e) and (f) The [001] projection from which the F(hkk) are obtained. (g) and (h) The [-110] projection from which the F(hhl) are obtained. (g) was taken at a wrong focus which has been corrected for in (h).

Each of the HREM images was studied by crystallographic image processing. Optical distortions were corrected and the symmetry was imposed onto the amplitudes and phases of the Fourier transform (Figure 1). The symmetrized amplitudes and phases should correspond to the structure factors of the crystal. The inverse Fourier transform of the symmetrized amplitudes and phases was then proportional to the projected potential of the crystal.

The structure factors, initially indexed in 2D were reindexed to their proper 3D indices according to the crystallographic zone axis of the projection (Fig. 1) and the phases shifted according to the shift of the origin to be consistent with the 3D space group *A2/a* in the International Table of Crystallography. All structure factors F(hkl) obtained from different zone axes were then combined and an inverse 3D Fourier transform calculated according to:

$$\varphi(x\,y\,z) = \frac{\lambda}{\sigma\,\Omega} \sum_{hkl} F(h\,k\,l) \exp[-2\pi i(hx + ky + lz)]$$

At the resolution of 1.6Å of the HREM images, it should be possible to resolve all individual atoms in the reconstructed 3D potential map if enough projections are included. The 3D reconstruction of bannisterite is still in progress. This procedure was first used for 2D protein crystals [5] and later for 3D inorganic crystals [2,3].

3. Studies of defects and interfaces by crystallographic image processing

All real crystals contain defects and local disorder, both chemical and structural. HREM is one of the most powerful tools in studying structures of defects and interfaces, since we can directly **see** the defects. However, we may ask: do the HREM images really represent the **true** structure of the defects? We know that only those images which are taken under optimum conditions (optimum focus and crystal well aligned) are directly interpretable. These optimum conditions are usually difficult to achieve in practice. Image simulation is then used to interpret the images. However, a model has to be proposed before the simulation, and this may not be possible if the structure of the defect is unknown.

3.1. STUDY OF DEFECTS AND INTERFACES BY QUASI-OPTICAL FILTERING

Using image processing, it is possible to filter away the noise and compensate for the optical distortions and crystal tilt, so that the defects and interfaces in the image after processing can be directly interpreted. It is also possible to select certain parts of information from an image by quasi-optical filtering of its Fourier transform, to enhance their contribution in processed HREM images.

We have found two new types of interfaces in bannisterite using HREM combined with image processing. A model for each of the interfaces was proposed based on images after compensating for the contrast transfer function and noise filtering. Atomic coordinates of the model was obtained using the program system CrystalKit. A HREM image was simulated from the proposed model by MacTempas [6] (Fig. 2c) and agrees very well with the experimental image. One of the interfaces in bannisterite is shown in Figure 2.

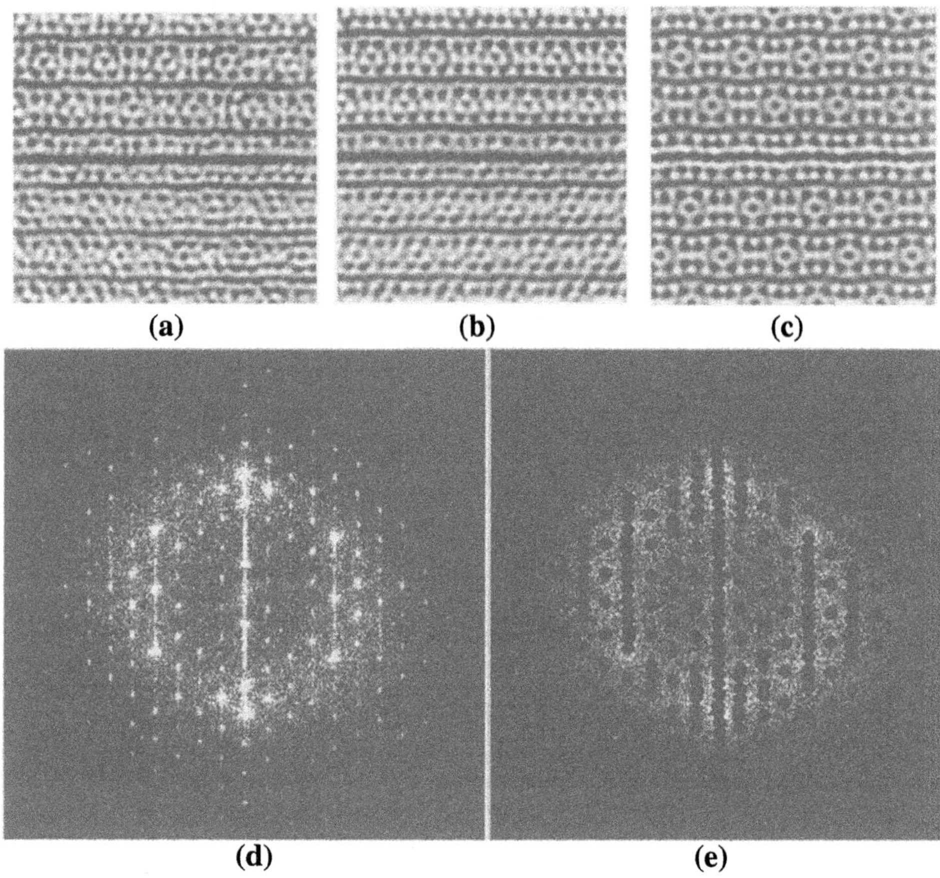

Figure 2. Interfaces in bannisterite studied by quasi-optical filtering. (a) HREM image of an interface. (b) reconstructed map after compensating for the contrast transfer function and quasi-optical filtering. (c) simulated HREM image using an interface model proposed from (b). (d) Fourier transform of (a). Note that the horizontal defect gives rise to vertical streaking in the FT. (e) Fourier transform of (a) showing the quasi-optical filter used for reconstructing (b).

3.2. STUDY OF DEFECTS BY CRYSTALLOGRAPHIC IMAGE PROCESSING

Crystals contain certain symmetry and defects destroy the symmetry. However, the symmetry of the crystal is usually not completely destroyed by the presence of the defect, i.e. crystals with defects and interfaces still contain certain crystallographic symmetry. One type of defects is planar defects with translations parallel to the defect plane. There are only 7 possible symmetries for a HREM image taken parallel to the defect plane. These are the 7 border groups [7], shown in Figure 3.

Symbol	Symmetry description	Corresponding plane groups
t		*p1*
t : 2		*p2*
t · *m*		*pm*
t : *m*	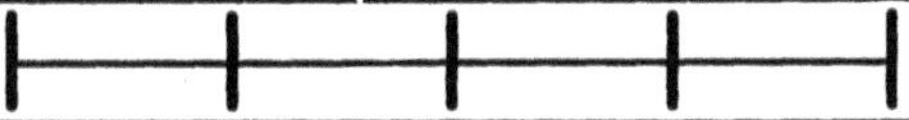	*pm*
t · *a*		*pg*
t : 2 · *a*	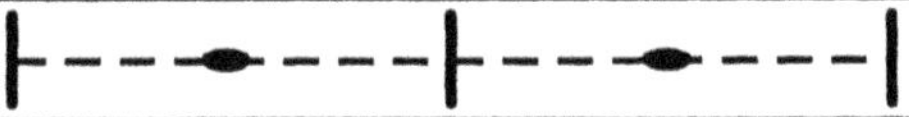	*pmg*
t : 2 · *m*	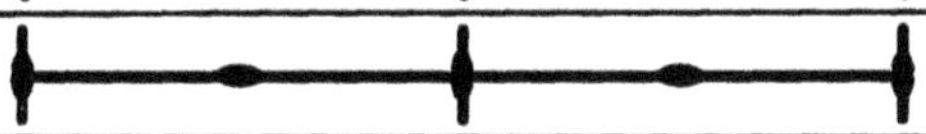	*pmm*

Figure 3. The seven border symmetry groups and their relation with the plane groups of the crystal. (after Vainshtein [7]).

Figure 4 shows HREM images and the corresponding Fourier transforms of two types of planar defects from anthophyllite and bannisterite, respectively. Both crystals have *mm* symmetry. However, the symmetries of the two defects on the border are different. The defect in anthophyllite (Fig. 4a) does not destroy the two mirror planes of the anthophyllite and the border symmetry is *t:2·m*, while the two mirror planes in bannisterite are lost (Fig. 4c) and the border symmetry is only *t:2*. Planar defects give rise to streaking in the Fourier transform, as shown in Fig. 4b and 4d. The symmetry of the defects can be also determined from the Fourier transform.

Recently, we have developed a method for applying crystallographic image processing to defects and interfaces. The symmetry of the defects is used for compensating for the distortions in HREM images, such as those caused by crystal tilt. Here we demonstrate the method, using anthophyllite (Fig. 4a) as an example. More detailed description will be published in a separated paper. Anthophyllite has the space group *Pnma*, with a=18.5863, b=18.0649 and c=5.2895 Å. An HREM image taken along the *c* axis is shown in Figure 5a. Each black dot in the HREM image represents a tetrahedral chain along the *c* axis and each dark vertical line represents an octahedral row (Fig. 4a and Fig. 5a). Due to slight crystal tilt, the image of the tetrahedral chains and octahedral rows are blurred.

The interface has the same periodicity as the anthophyllite in the directions parallel to the interface. In the direction perpendicular to the interface, the translation symmetry is lost. Since dealing with crystals is much simpler than dealing with defects/interfaces, we have generated a crystal from the defect/interface. The first step is to select a super unit cell in which the defect/interface is included (Fig. 5a). An artificial crystal image is then generated from the super cell (Fig. 5b) and its Fourier transform is calculated (Fig. 5c). If the cell width along the ***b*** axis is large enough, the streaking from the interface (Fig. 4b) will be finely sampled (Fig. 5c). The Fourier transform is indexed according to the sampling (Fig. 5c) and amplitudes and phases are extracted from the lattice position of the Fourier transform of the artificial crystal. The symmetry is determined from the phases to be *pmm*, which shows that the symmetry of the interface has the border group $t : 2 \cdot m$ (Fig. 3).

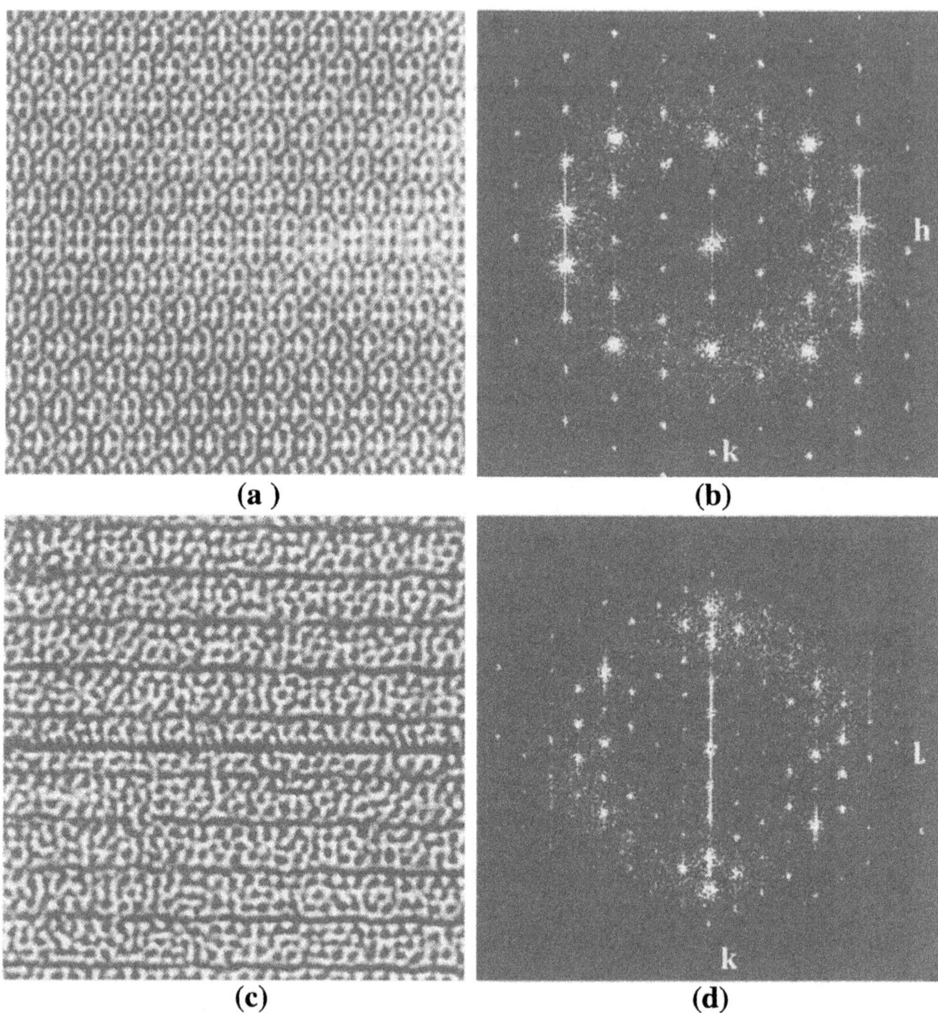

Figure 4. Defects in real and reciprocal space. (a) HREM image of the double chain anthophyllite with intergrowth of a triple chain jimthompsonite projected along the c axis. (b) The Fourier transform of (a). The streaking in the Fourier transform appears vertically, passing through the diffraction spots. Note that the diffraction spots have *cmm* symmetry and the streaks show *mm* symmetry. (c) HREM image of a planar defect in bannisterite projected along the a axis. (d) Fourier transform of (c). Here the diffraction spots have *cmm* symmetry but the streaking only *p2* symmetry.

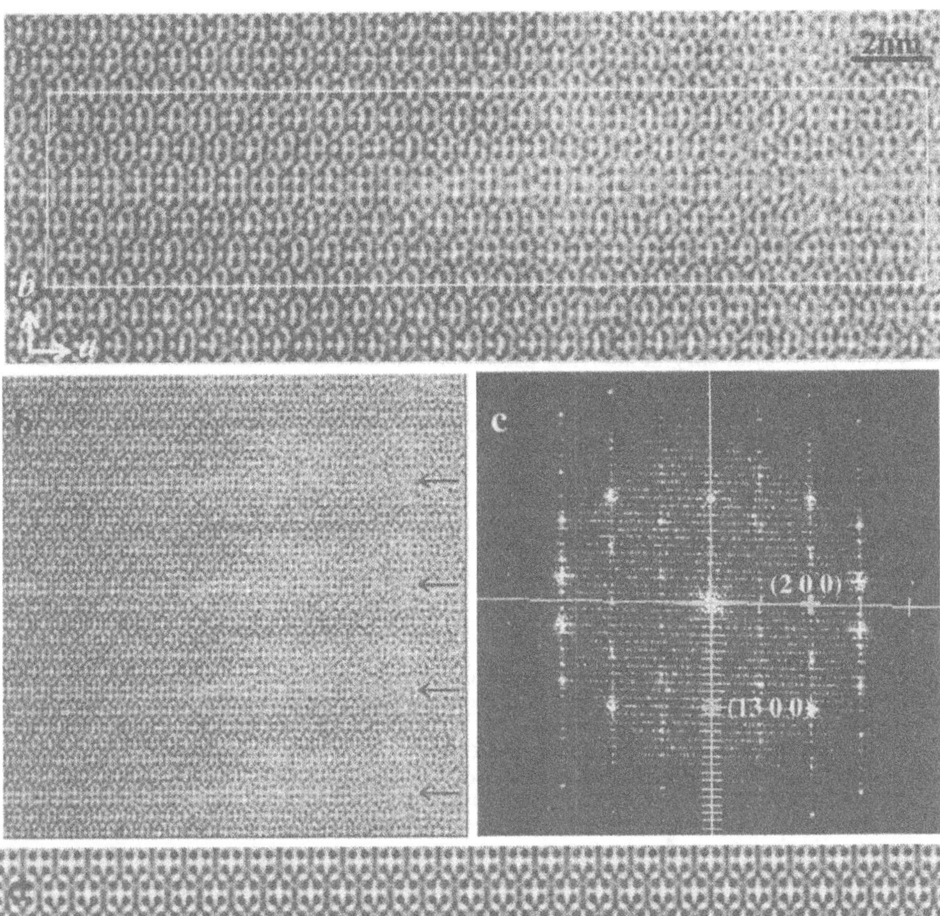

Figure 5. Crystallographic image processing of the planar defect in Fig. 3a. (a) A large unit cell including the defect is selected from the image. (b) A 2D artificial crystal image is generated by repeating this "unit cell" several times vertically. The interfaces are marked. (c) The Fourier transform of the artificial crystal image. The amplitudes and phases are extracted from the lattice points in the FT. The border symmetry of the defect is *t:2·m* while the symmetry of the artificial crystal is *pmm*. The *pmm* symmetry is imposed on the amplitudes and phases and an improved map (d) is obtained by inverse Fourier transformation of the symmetrized amplitudes and phases. In this improved map it is easy to see the atomic arrangement of the triple chain within the double chain crystal.

The *pmm* symmetry is imposed on the amplitudes and phases and thereafter an inverse Fourier transform is calculated. The tetrahedral chains and octahedral rows are clearly resolved in the reconstructed map (Fig. 5d). Groups of 2x2 black dots in the map are the double tetrahedral chains, which are separated by vertical rows of octahedra in anthophyllite crystals. At the interface, groups of 2x3 black dots indicate intergrowth of a triple tetrahedral chain lamellar in the anthophyllite crystal.

The example described here shows that crystallographic image processing can be used to compensate for distortions in HREM images of defects/interfaces. The HREM image is greatly improved. The structural features at the defect/interface are easier to recognise from the improved image. It is possible to build up an atomic model of a defect/interface based on this reconstructed map. This is only the first attempt of utilising the symmetry in studying defects. More developments of the crystallographic image processing method in studying defects and interfaces are in progress.

Acknowledgements

The Swedish Natural Science Research Council are greatly acknowledged for providing Veblen a Tage Erlander professorship and Zou a post doc scholarship.

References

1. Zou, X.D. (1998) Crystal structure determination by crystallographic image processing: II. Compensate for defocus, astigmatism and crystal tilt, in *Electron crystallography,* eds Dorset, Hovmöller and Zou, Kluwer Academic Publishers, Dordrecht.
2. Downing, K.H., Hu, M., Wenk, H.-R. and O'Keefe, M.A. (1990) Resolution of oxygen atoms in staurolite by three-dimensional transmission electron microscopy, *Nature (London)* **348**, 525-528.
3. Wenk, H.-R., Downing, K.H., Hu, M. and O'Keeefe, M.A. (1992) 3D structure determination from electron-microscope images: electron crystallography of staurolite, *Acta Cryst.* **A48**, 700-716.
4. Heaney, P.J., J.E. Post and H.T. Evans (1992) The crystal structure of bannisterite, *Clays and Clay Minerals,* **40**, 129-144.
5. Unwin, P.N.T. and Henderson, R. (1975) Molecular structure determination by electron microscopy of unstained crystalline specimen, *J. Mol. Biol.* **94**, 425-440.
6. Kilaas R. (1990) MacTempas: a program for simulating high resolution TEM images and diffraction patterns. Berkeley: Total resolution.
7. Vainshtein, B.K. (1994) *Modern Crystallography I: Fundamentals of crystals Symmetry, and Methods of Structural Crystallography,* Springer-Verlag, Berlin, Heidelberg, New York, 107.

ELECTRON DIFFRACTION IN POLYMER CRYSTAL STRUCTURE ANALYSIS: SOME EXAMPLES.

Stefano V. Meille,
Dipartimento di Chimica, Politecnico di Milano,
via Mancinelli 7, 20131 Milano, Italy

1. Introduction

The relative merits of a technique depend often on the specificity of the field where it is applied and on the available alternatives. In this perspective it is not surprising that in polymer crystallography electron diffraction has been comparatively more fruitful than for low-molecular weight compounds [1]. This stems from two sets of reasons:

1) X-ray diffraction data for polymers are much poorer than those available for most other systems because:
   a) polymers are only moderately diffracting, generally multiphasic materials with a substantial amorphous fraction coexisting with crystalline phases.
   b) polymer crystals are very small and anisotropic (typically 10 nm thick lamellae with lateral dimensions of the order of 1μm but often much less) and in practice not adequate for single crystal X-ray work. Structural investigations have largely been based on fibers patterns which yield immediately the fiber repeat value and the probable helical symmetry. While this accounts for the popularity of the technique, substantial problems remain. Only more recently some polymer structures have been solved and refined from X-ray "powder data".
   c) the lattice dimensions (except for the axial repeat, if fibers are available) and the space group symmetry, because of the broadness and the overlapping of reflections, are often difficult to identify.
   d) for the same reasons, and due to the irregular shapes often displayed by diffraction maxima in fiber patterns, integrated X-ray diffraction intensities are hard to measure and semi-quantitative estimates are still often used in structural refinements of polymers.

D. L. Dorset et al. (eds.), Electron Crystallography, 313–322.

2) Electron diffraction data from synthetic polymers has obvious advantages as it allows to study individual polymer microcrystals separately, individual crystalline phases in samples where more than one crystalline polymorph coexist, structure morphology relationships and continuos diffraction related to disorder which often characterizes polymer crystals. Furthermore data can be of rather good quality as was understood quite early [2] and subsequently demonstrated in a number of papers identifying the different factors giving rise to the specific situation, namely:
   a) the strong interaction between electrons and matter, i.e. the features of the atomic scattering cross sections for electrons that make this type of radiation particularly suitable for poorly diffracting systems
   b) dynamical scattering is kept at a relative minimum because i) polymer crystals, are very thin ii) generally only low atomic number elements are present in polymers,

The disadvantages of electron diffraction from polymers are in essence the same as for thin organic crystals [1]. Incoherent multiple scattering from stacks of lamellar crystals with melt-crystallized polymer thin films may be somewhat more specific to polymers and bending of lamellar crystals has been demonstrated both for polymers and for low molecular weight systems. Radiation damage is clearly also a major issue but it can be kept at a minimum applying low dose techniques and working at sub-ambient temperature.

It appears important to stress that even electron diffraction data of relatively poor quality, as long the limitations are clear, can be extremely useful in the case of polymers. We have already mentioned that electron diffraction represents the most efficient and reliable technique to determine the crystalline unit cell and its symmetry, a task which can hardly be achieved even with advanced modeling techniques. However, used in conjunction with molecular modeling and packing analysis, which with second row elements are adequate to determine both the conformation for the isolated and the crystalline molecule (see e.g. Ferro et al., ref. [3]), even semi-quantitative electron diffraction intensities can serve as a validation of proposed structures, or as a figure of merit to test also marginally different models.

With good data full refinements can be carried out, taking care to properly constrain the molecular model. In this respect we note that this is the standard procedure in refinements of polymer crystal structures also using X-ray data. In fact the rule is to take standard bond lengths, allow only marginal variations for bond angles and refine torsion angles [4]. In the latter case the deviations from rotational isomeric states seldom exceed 25°. If this approach is accepted, the only truly free variables remain the three molecular "rigid body" translations and the three rotations which often may be further reduced because of lattice symmetry, and by the requirement that the intramolecular translation vector i.e. the fiber repeat, must coincide with a given lattice vector (which in polymer crystallography is usually c).

The problem of "over-refinement" remains a very basic one and traditionally in the case of polymers it has been tackled intuitively but adequately by crystallographers using chemical "sensitivity". The question arises also for single crystal X-ray data and is discussed in the contribution by Sheldrick to the present volume. It is obviously pertinent also to fiber or powder X-ray data or whenever there is a poor data to refined parameters ratio and constraints need to be applied. The issue coincides in practice with the determination of the relative weights of the diffraction data and of the applied constraints and is related to the problems arising when diffraction data sets of different origin are used. Different practical solutions have been adopted but the more convincing is to determine the relative weight of the constraints as well as of the data as a function of their standard deviations which, however, may be difficult to estimate. In this respect, especially with systems for which force-fields are not well established, the usefulness of model compounds is apparent. As we shall see, the solution and refinement of polymer crystal structures affected by extensive disorder may in any case be very difficult to handle.

Since refinements based on SAED data are carried out assuming kinematic scattering while dynamic effects are known to be important in electron diffraction, the validity of this approximation needs to be verified and in various cases [1, 5] this was done calculating n-beam dynamic amplitudes from the refined structure. With a relatively low number of repeat units in the beam direction (i.e. with thin crystals) the improvement was found to be in general very modest. Alternative validation procedures have relied on X-ray data from unoriented specimen again finding remarkable levels of agreement which confirm that properly considering stereochemical constraints polymer electron diffraction data are perfectly adequate also for structural refinements. Still another test of the structure reliability can be comparison of the structural model from the diffraction data refinement with packing energy minimization results. In this context it appears to be helpful to adopt the experimentally determined lattice constants and symmetry while allowing for simultaneous minimization of the inter- and intramolecular energy contributions.

An additional comment is in order with respect to the practice of using, together with electron diffraction data, also X-ray data in separate or combined sets. As noted in the case of fibers and single crystals, the degrees of disorder and the crystal structure itself may in principle [6] be influenced by different sample morphologies. It would appear thus safer when using different diffraction data sets to choose morphologically closely similar samples. This may be difficult to achieve but, in our experience the data from lamellar crystals used for electron diffraction are closely comparable to X-ray data from unoriented polycrystalline samples which can be used for Rietveld type refinements. The use of this procedure, which allows for a simpler treatment of experimental X-ray intensity data than for fibers, is advisable in conjunction with SAED single crystal data while SAED fiber data should be used with X-ray fiber data.

In the subsequent sections some examples of structure determinations using SAED data will be given highlighting some of the issues anticipated so far. For the chosen investigations large amounts of additional structural information is available from different sources (X-ray data, packing energy calculations etc.) allowing to estimate the reliability of the models obtained by refinement from SAED data.

2. The crystal structure of γ-Polypivalolactone.

Polypivalolactone (PVL) is a highly crystalline polyester (Figure 1) that can be useful as a model in polymer crystallization studies. It exhibits polymorphic behavior and in the more common α-modification it was found by a fiber X-ray diffraction study [7] to adopt a chain conformation deviating substantially from the minimum energy proposed

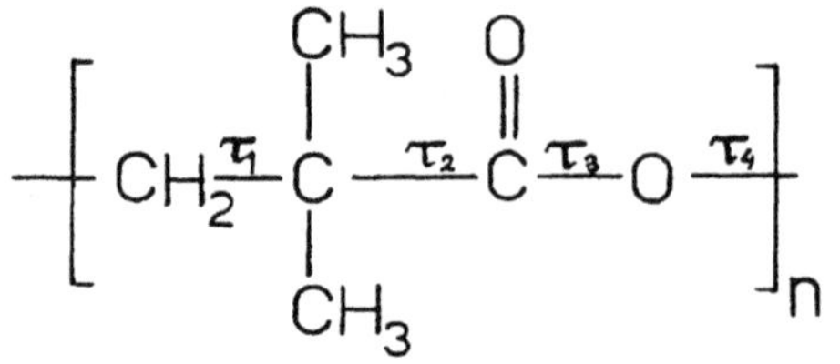

$$\left[CH_2 \overset{\tau_1}{-} C(CH_3)_2 \overset{\tau_2}{-} C(=O) \overset{\tau_3}{-} O \overset{\tau_4}{-} \right]_n$$

Figure 1. The chemical repeat unit in polypivalolactone

on the basis of rough conformational calculations [8]. The fiber diffraction result was in essence confirmed by a powder study [9] and in Table 1 the values of the more relevant torsion angles are given for the three different models. The structure of the γ-modification, which cannot be obtained in pure macroscopic samples, but always co-crystallizes in different amounts with the α-phase, was solved using tridimensional electron diffraction data obtained from melt crystallized thin films [10]. The unit cell and the space group could be established even if space group forbidden reflections (i.e. 010, 030 and 100) were apparent in the hk0 zone but absent in others and in X-ray patterns from individual γ-spherulites. Considering the identity of the axial repeat with the α modification, simple rigid body trial and error procedures allowed to solve the structure achieving encouraging disagreement factors both with the model proposed for the α structure and with the conformational model. Subsequent refinements using the linked atom procedure [4] allowed for relatively small but significant adjustments of bond and torsion angles, which were constrained in an effort to reproduce the bias of the ester group for a planar conformation. A substantial improvement of the disagreement factor was achieved which appeared to be reasonably balanced for data of different zones, even if disagreement appeared to concentrate more on certain hk0 reflection of intermediate intensity. This fact and the space group forbidden reflections suggest that, in the specific case non-kinematic scattering is more important in the hk0-zone, possibly due to incoherent multiple scattering from stacked lamellae.

Stereochemically acceptable rigid body models and the results of the constrained refinement were tested with a Rietveld procedure against X-ray diffraction data of unoriented PVL samples in which both the α and γ-modification coexisted. As shown in Table 1, it was found that the relative fit of the different models with electron diffraction data is well reproduced with the powder data set, once the substantial contribution of the coexisting α phase to powder data is adequately considered.

Model	τ_1	τ_2	τ_3	τ_4	$R_{(e.d)}$	$R_{(Rietveld)}$
I (isolat. chain *)[(b)]	60	34	180	208	0.22	0.112
II (α - phase chain*)[(a)]	46	54	191	182	0.17	0.105
III (e.d. refined)	52	48	178	195	0.14	0.094

Table 1: Main chain torsion angles and disagreement factors of different models with electron diffraction and powder data for γ-PVL (* rigid body).

Detailed conformational an packing energy calculations were also performed on both the α and the γ-crystalline modifications of PVL [3], adopting the lattice dimensions and symmetry determined from diffraction work. The results support the idea that the chain conformations in these two highly crystalline phases differ significantly from each other in a way which is consistently determined for the distinct diffraction data sets, and for the energy calculations. The structural refinement based on diffraction data were however constrained in a way that presently still largely depends on personal chemical sensibility.

Model	τ_1	τ_2	τ_3	τ_4	E(isolated)*	E(crystal)*	r.m.s.d. (Å)
Isolated chain[(d)]	51	50	178	193	-	-	
α - phase chain	51	50	191	182	0.34	0.00	0.10
γ - phase chain	53	46	180	196	0.13	0.07	0.05

Table 2: Main chain torsion angles and relative energies for minimum energy models of the PVL chain respectively isolated and in α- and γ -phase crystals. (Energy in kcal/mol, r.m.s. deviations from models refined with diffraction data)

3. The crystal structure of γ-isotactic polypropylene (γ-iPP)

The crystal structure of γ-iPP is unique as it is the only so far known ordered polymer structure in which non-parallel polymer chains are found within the same crystal [11,12]. While in the original solution and refinement of the structure electron diffraction intensity data were not used, SAED patterns in the literature [13,14] were instrumental in determining the correct lattice. In fact without these data it is very doubtful that the structure would have been solved at all. A projection view of the structure along one of the molecular axis directions is presented in Figure 2.

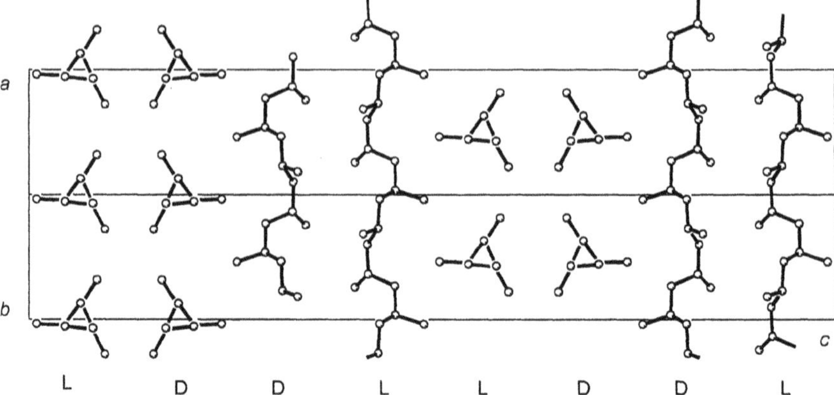

Figure 2. A view of the γ-iPP crystal structure along one of the two molecular orientations.

Also because of its uniqueness the structure was examined in detail with respect to packing energy and compared with the more common α-modification again with simultaneous minimization of intra- and intermolecular energy [15]. In this case it was found that for both structures the minimum energy *crystalline* conformation does not deviate in a perceptible way from the minimum energy 3_1 *isolated helix* conformation. In fact the same conformation is adopted in all three crystalline modification of iPP. These packing energy calculations showed the γ-crystalline arrangement determined from X-ray powder diffraction intensities to correspond closely to a potential energy minimum which was equal or lower than the α-structure with comparable disorder.

An independent electron diffraction investigation [16] confirmed the symmetry and qualitatively the correspondence of electron diffraction patterns to those expected for the orthorhombic γ-iPP structure. A quantitative study [17] presents additional interest as electron diffraction data are the only independent experimental intensity data which can be used for a detailed check of this crystal structure which, for obvious reasons is unstable in fiber morphology. Two zones of the diffraction patterns (taken by Bernard Lotz) were analyzed, namely the hhl corresponding to the chain axis direction and the h0l which is obtained from lamellae lying flat. A total o 22 observed and 10 unobserved reflections were used which, considering the high symmetry of the Fddd space group, should not be considered few. The correctness of the structure results immediately as rigid body treatments yield disagreement factors around 0.25. Further constrained refinement can easily lower this value to 0.15. However in this case the level of agreement with the results of the other studies was poor: the r.m.s. deviations of the carbon atom coordinates were 0.3Å from those of the packing energy study and 0.2Å from those of the powder data refinement [12]. The structure refined from electron diffraction, even after relaxation showed at this point substantially higher energy although the potential energy minimum remains in essence the same one in all three studies. The model from electron diffraction improved significantly if instead of scaling the data of the two zones using the common reflections (i.e. the 00l row) a different scale factor is refined for the two data sets, starting from the approximate value determined from the common 00l reflections. In addition the more reliable common reflection intensities were used in the refinement, i.e. those with a larger spread in intensity values. With this procedure, using essentially a rigid body approach, values of 0.142 of the R factor are obtained with r.m.s deviations of 0.04 Å from the packing energy minimization model. With additional constrained refinement R values as low as 0.128 with r.m.s.d. from the energy minimum of 0.10 Å can result confirming that also in this case, using proper constraints and an adequate treatment of the observed reflections, a structure refinement is indeed feasible and that it leads to one and the same result as refinement from powder data or by packing energy minimization.

4. The structure of form III polybutene

This structure was recently solved by direct methods and refined using 3D electron diffraction data [18]. The work is part of an important attempt due essentially to Dorset [1], to show that a structure solution using direct methods is possible in favorable cases also with a relatively limited set of electron diffraction data. It is clear that the future application of such techniques to structures where molecular modeling is problematic will develop only tackling problems that initially have to be simple and for which results can be double-checked. On the other hand the authors of the investigation noticed that unconstrained least-squares refinement, did not produce a stereochemically acceptable model consistent with our previous discussion. Sensible results could however be achieved by Fourier refinement techniques, yielding a structural model which differs only marginally from that obtained by a constrained refinement in a pioneering powder diffraction study by Cojazzi et al. (ref. [19]) of the same crystalline modification.

The case of the structure solution and refinement of polybutene form III from electron diffraction data, while opening to new possibilities, also implies that constrained powder refinements (coupled if possible with lattice determination from electron diffraction) is probably at present the most expedient way to refine crystal structures of polymers with relatively simple repeat units at the present time. Such a suggestion results considering how much easier it is experimentally to obtain reliable intensity profiles by X-ray powder diffraction rather than by either fiber X-ray or single crystal electron diffraction work. This obviously does not question the importance of developing new structure solution techniques which in the future may lead to solve problems that are presently untreatable.

5. The β-isotactic polypropylene (β-iPP) structure

The interest of this crystal structure lies specifically in the conclusion that suggestions put forward in the previous discussion of other polymer structures may not apply fully to this case.

The β-polymorph of iPP was identified more than thirty years ago: its lattice was found [20] to be hexagonal (or trigonal), with very large 9 or even 12 chain unit cells being invoked to account for the observed reflections.

A recent electron diffraction study [21] concluded that, apart for diffraction effects due to α-phase impurities, all the reflections of β-iPP can be accounted for by a three-chain unit cell with $a = b = 11.03$ and $c = 6.49$Å. Space group $P3_121$ was found to be the more probable option on account of the fit to electron diffraction data, powder data and preliminary packing energy calculations. These suggestions were in essence confirmed by an independent electron diffraction study [22] published approximately at the same time, but differing in the detailed model proposed.

The determination of the lattice of this structure relied upon electron diffraction data. It is notable that in analogy to what found in other cases there is close correspondence in the fit of the different models to X-ray powder diffraction data and to SAED data, indicating that dynamical effects, are rather unimportant also in this case. As already mentioned there is agreement between diffraction data and the packing energy analysis on space group $P3_121$ as a viable representation of the structure. However, if the

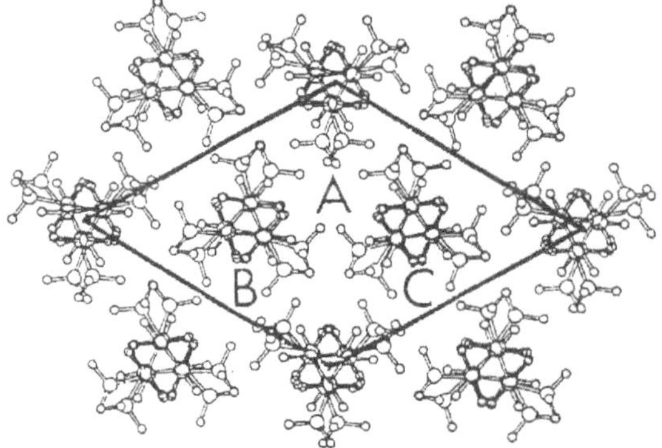

Figure 3. The β-iPP structure obtained from packing energy minimization

"best $P3_121$ models" obtained respectively from diffraction data (see ref. [21], Figure 6) and from the energetic studies (Figure 3) are compared, it is plain to see that they do not correspond. We can add that while it is fairly easy with rigid body chains to achieve good disagreement factors with the hk0 projection data, and apparently stereochemically acceptable models, these need to be modified substantially to fit tilted crystal data.

In short, while the gross features of the models show some correspondence, the details of the structure are far less defined than for the other instances we discussed, suggesting extensive disorder. This is evidenced by 1) the presence of continuos intensity streaks running perpendicular to the a^* (or b^*) direction and by 2) the fact that the reciprocal lattice symmetry is higher (the Laue group is probably P6mm) than any space group that can be reasonably thought to apply locally. It is notable that even a $P6_122$ model fully consistent with the Laue symmetry, implying extensive disorder but still with a comparatively low energy value, yields poor R values with the both the powder and the SAED diffraction data sets.

On the other hand the packing energy study identifies a considerable number of structurally distinct minima whose energy lies within less than one kJ/(mole of trimer) from the absolute minmum. It is possible that a proper combination and weighting of the different relative minima will afford a model matching the diffraction data: however such a model has not been identified yet. We are left at present with very reasonable local models, none of which, taken alone is however able to account for the diffraction data.

6. Concluding remarks

The very high potential of electron crystallography in polymer crystal structure analysis is apparent. SAED data even of modest quality very often allow adequate solution and constrained refinement of polymer crystal structures. It is suggested that molecular modeling and/or supportive X-ray powder profiles in combined approaches with electron diffraction may be more expedient. In this context there is still the need for computer programs in which even energy minimization procedures and fitting of the model to structural data can proceed simultaneously. While in most cases the combined approach is just another option it may be the only practically viable one in order to tackle complex problems involving disorder.

References

1. Dorset, D.L. (1995) "Structural electron crystallography", Plenum Press, New York.
2. Claffey,W., Gardner, K., Blackwell, J., Lando, J. , Geil, P.H. (1974) Structure analysis of polymer single crystals by electron diffraction, *Phil. Mag.* **30**, 1123-1232.
3. Ferro, D. R., Brückner, S., Meille, S.V., Ragazzi, M. (1990) Conformational Study of α and γ poly-pivalolactone based on intra- and intermolecular interactions, *Macromolecules* **23**, 1676-1680.
4. Arnott, S., Campbell Smith, P.J. (1978) LALS: a linked atom least-squares reciprocal-space refinement system incorporating stereochemical restraints to supplement sparse diffraction data, *Acta Cryst.* A**34**, 3-11.
5. Moss B., Dorset, D.L. (1982) Refinement of linear polymer structures determined from electron diffraction data, *J.Polym.Sc.Polym.Phys.Ed.* **20**, 1789-1904.
6. Hasegawa, H., Claffey, W., Geil, P.H. (1977) Analysis of the crystal structure of poly(ethylene sulfide), *J. Macromol. Sc.* **B13**, 89-100.
7. Perego, G., Melis, A., Cesari, M. (1972) The crystal structure of polypivalolactone, *Makromol. Chem.* **157**, 269-278.
8. Cornibert, J., Hien, N.V., Brisse, F., Marchessault, R.H. (1974) Comments on the crystal structure of polypivalolactone and deviation of ester group from *trans* planarity, *Can. J. Chem.* **52**, 3742-3747.
9. Brückner, S., Meille, S.V., Porzio, W. (1988) The structure of α-polypivalolactone. A refinement based on the Rietveld method, *Polymer* **29**, 1586-1589.
10. Meille, S.V., Brückner, S., Lando, J.B. (1989) The structure of γ-polypivalolactone: a combined analysis of single crystal electron diffraction data and powder X-ray diffraction profiles with the Rietveld method, *Polymer* **30**, 786-792.
11. Brückner, S., Meille, S.V. (1989) Non-parallel chains in crystalline γ-isotactic polypropylene, *Nature* **340**, 455-457.
12. Meille, S.V., Brückner, S., Porzio, W. (1990) γ-Isotactic polypropylene. A structure with nonparallel chain axes, *Macromolecules* **23**, 4114-4121.

13. Ferro, D.R., Brückner, S., Meille, S.V. Ragazzi, M. (1992) Energy calculations for isotactic polypropylene: a comparison between models of the α and γ crystalline structures, *Macromolecules* **25**, 5231-5235.
14. Morrow, D.R., Newman B.A. (1968) Crystallization of low-molecular weight polypropylene fractions, *J.appl.Phys.* **39**, 4944-4950.
15. Lotz, B., Graff, S., Wittmann, J.C. (1986) Crystal Morphology of the γ (triclinic) phase of isotactic polypropylene and its relationship to the α-phase, *J.Polym.Sc.Polym.Phys.Ed.* **24**, 2017-2032.
16. Lotz, B., Graff, S., Straupé, C., Wittmann, J.C., (1991) Single crystal γ phase isotactic polypropylene: combined diffraction and morphological support for a structure with non-parallel chains, *Polymer* **32**, 2902-2910.
17. Meille, S.V., Brückner, S., Lotz, B., (manuscript in preparation) Refinement of the γ phase isotactic polypropylene crystal structure from electron diffraction data.
18. Dorset, D.L.; McCourt, M.; Kopp, S.; Wittmann, J.C.; Lotz, B. (1994) Direct determination of polymer crystal structures by electron crystallography - isotactic poly(1-butene) form III, *Acta Cryst.* B**50**, 201-208.
19. Cojazzi, G.; Malta, V.; Celotti, G.; Zannetti, R. (1976) Crystal structure of form III of isotactic poly(1-butene), *Makromol.Chem.* **177**, 915-926.
20. Turner-Jones, A., Aizlewood, J.M., Becket, D.R. (1964) Crystalline forms of isotactic polyoropylene, *Makromol. Chem.* **75**, 134-158.
21. Meille, S.V., Ferro, D.R., Brückner, S., Lovinger, A., Padden, F.J.; (1994) The structure of β-isotactic polypropylene: a long standing structural puzzle, *Macromolecules* **27**, 2615-2622.
22. Lotz, B.; Kopp, S.; Dorset, D.L.,(1994) Sur une structure cristalline originale de polymères en conformation hélicoïdale 3_1 ou 3_2 *C.R..Acad. Sci. Paris* **319**, 187-196.

MEMBRANE PROTEINS SOLVED BY ELECTRON MICROSCOPY AND ELECTRON DIFFRACTION

A. HOLZENBURG
School of Biochemistry and Molecular Biology
School of Biology
University of Leeds, Leeds LS2 9JT, UK

1. Prologue

I have been appointed rather late to give a lecture at this school and for this reason, as all members in the audience will have been aware, there were no lecture notes for my contribution. I tried to compensate for this shortcoming by offering a special tutorial, and a small crowd has actually taken me up on this offer. From these and any subsequent discussions I have learnt a lot for myself and the open, friendly and enthusiastic attitude of most of the students made this school a unique experience.

The original title of this lecture was scheduled as "Membrane proteins solved by electron microscopy and electron diffraction". However, since this is a school aimed at disseminating knowledge of practical approaches as well as the theory behind it, it seems more appropriate to view this contribution under the heading "How to solve membrane proteins ...".

D. L. Dorset et al. (eds.), Electron Crystallography, 323–342.

2. Introduction, overview and some clarifying remarks

While soluble proteins, domains thereof or peptides are ideally suited for 3-D crystallisation and X-ray analysis to a resolution at which individual atoms can be visualised, the situation changes abruptly when one deals with membrane proteins which, by their very nature, prefer a 'two-dimensional environment' and are therefore very difficult to crystallise in three dimensions. For this reason viable alternatives are needed and the most obvious alternative is to grow two-dimensional (2-D) crystals (crystals that are only one unit cell thick). These 2-D crystals can be readily analysed in the electron microscope and are the prerequisite for electron crystallographic studies. At this point we should differentiate between the two main classes of membrane proteins: integral membrane proteins and membrane-associated proteins. Both require a '2-D matrix' in order to fulfill their functional role in a physiological setting, and this matrix is the lipid bilayer. Integral membrane proteins will want to span the entire membrane in order to be active while for membrane-associated proteins the contact with the lipid surface is crucial. As with 3-D crystals, it is the quality of the 2-D crystals that ultimately determines the resolution and several 2-D crystallisation strategies have been developed. For integral membrane proteins, a possible way forward is via reconstitution into 2-D protein (and lipid) crystals: The idea is to mix detergent-protein and detergent-lipid mixed micelles at a suitable lipid-protein and lipid-detergent ratio, pH, ionic milieu and temperature and then deplete the resulting ternary micelles (detergent-lipid-protein) of the detergent component (e.g. by dialysis) in order to induce the formation of 2-D crystals of the protein, ideally within lipid sheets rather than vesicles [1]. Other approaches to 2-D crystals of integral membrane proteins include in situ methods, batch methods etc. and they have been thoroughly reviewed in the literature [2]. The 2-D crystallisation of membrane-associated proteins can be very effectively carried out on planar lipid films [3]: Take a small trough filled with buffer containing the protein to be crystallised and add a dilution lipid and a ligand lipid (a lipid to which the protein to be crystallised has a very high affinity) dissolved in e.g. chloroform/hexane. The lipids will then orientate themselves at the air-water interface, the protein will recognise the ligand lipid, form complexes with it and crystals will eventually form via lateral diffusion and self-organisation of the lipid-protein complexes.

For electron microscopical observation, 2-D crystalline samples can either be embedded in negative stain (e.g. uranyl acetate) for low to intermediate resolution studies or in vitreous water (no ice crystals are present that could damage the protein), tannin or glucose. The latter three embedding media preserve the structure of proteins to high resolution [4]. In keeping with this it is crucial to observe the specimens and record electron micrographs at low doses (< 1000 e^-/nm^2) and low temperature (< -150 °C; "cryo" studies). The electron micrographs are then digitised (unless one has access to on-line image acquisition facilities via e.g. a slow-scan CCD camera), Fourier transforms are calculated, phase and amplitude information is extracted, data sets from different crystals are merged, and finally, via Fourier synthesis, a projection map is calculated. Tilting the specimen relative to the incident beam will give access to 3-D structural information.

At this point, we have not yet talked about electron diffraction and have only dealt with images. With images it is important to remember that the Fourier transform of an image provides both, amplitude and phase information. The latter is related to the phase structure of the object and should not be confused with the propagating wave front phases which are lost in the image. When dealing with images, there is no phase problem as far as the crystallographic structure factor phases are concerned. Just to keep things clear in our minds: when speaking in the following about amplitude and phase information, we can say that the phases are concerned with the positions of centres of mass and the amplitudes describe the volumes of these centres of mass.

In order to be able to record electron diffraction patterns with sufficient signal-to-noise ratios, quite large (in the region of a few 1000 unit cells) and well-ordered 2-D crystals are required. In other words, the biggest challenge (= substitute for the old fashioned term 'problem') is to grow 2-D crystals of sufficient quality, that is size and order. Once electron diffraction patterns have been recorded, one would use the amplitude information from these patterns and combine it with the phase information as obtained from the Fourier transforms of the images in order to calculate the structure. We know that images provide us with both, amplitude and phase information, so why would one want to do that ? This is because the amplitude information from the images is modulated by an 'electronoptical shortcoming' called the contrast transfer function and suffers from an exponential fall-off with increasing resolution so that the amplitude information cannot be taken as read but requires tricky correction and scaling procedures. Furthermore, the amplitude information from images tends to be quite scattered. For these reasons it is very advantageous to record electron diffraction patterns. More specifically, the usage of electron diffraction amplitudes bestows several benefits: there are no disturbing effects due to specimen drift (which can be severely limiting) since the diffraction pattern is translationally invariant, and electron diffraction amplitudes are not affected by the contrast transfer function. These positive effects together with the fact that the accumulated electron dose required for recording electron diffraction patterns is below the dose required for images, it is not astonishing that electron diffraction patterns contain information to higher resolution than the Fourier transform of a corresponding image. However, in contrast to the 'poorer' image amplitudes, image phases can be easily corrected for the contrast transfer function, are very reliable (less scatter when compared to the image amplitudes) and readily extendable using the now well established maximum entropy methodology [5]. In other words, image phases and electron diffraction amplitudes constitute an ideal combination. In biological sciences, it was this combination of information that originally bore the name "electron crystallography". Robert M Glaeser has, in 1985, defined electron crystallography as ..." any use of electron diffraction, electron scattering or direct imaging in which the analysis of data is essentially crystallographic in its execution or in its objectives", and I think this is a very useful definition.

Following on from what has been said above, it should be evident that the situation in electron crystallography is quite different from X-ray crystallography where structures are either solved to high resolution or not solved at all. With electrons, one can solve a structure over a wide resolution range: With regards to membrane proteins, high resolution information (< 0.8 nm) has been obtained for e.g. the well known bacteriorhodopsin [6], LHC-II [7] and porin [8] to name but a few. At intermediate (1.5

to 0.8 nm) or low resolution (above 1.5 nm, typically 2-3 nm) structures have been obtained for a great number of membrane proteins and it should be stressed that establishing structure-function relationships does not always depend on structural details at atomic resolution. This is especially true with big proteins and protein complexes. In the following, I am going to use three examples to illustrate these points.

3. The Examples

3.1. PHOTOSYSTEM II

The first example is concerned with photosystem II (PSII) from higher plants. PSII is a protein-pigment complex located in the thylakoid membrane of plants and cyanobacteria (for an overview see e.g. Nicholson *et al.* [9]). Within chloroplasts, PSII occupies primarily those parts of the thylakoid membrane system known as grana. Upon illumination, it provides the plant with chemical potential and the environment with oxygen allowing us to adopt an aerobic lifestyle. PSII is a transmembrane complex, it is big (approaching 1 MDa), and it is extremely labile, i.e. it truly is an X-ray crystallographer's nightmare. Out of the more than 20 polypeptides, the major ones are

- the reaction centre heterodimer (D1/D2) + Cyt b-559 (= reaction centre),
- the extrinsic oxygen evolution-enhancing subunits of 33, 23 and 16 kDa (OEE-33, OEE-23, OEE-16) constituting the oxygen evolution complex (OEC), and
- the core chlorophyll-binding proteins of 47 and 43 kDa (CP47, CP43).

All of the above polyeptides are part of the so-called PSII core. The peripheral light-harvesting antennae comprise a number of polypeptides of 25 to 30 kDa and complement the core light-harvesting proteins CP47 and CP43 in their function to trap light energy as efficiently as possible and to transfer it to the reaction centres. Between 8 and 20 copies of these peripheral light-harvesting polypeptides are associated with

PSII and the major light-harvesting antenna polypeptide is called light-harvesting complex II (LHC-II, see 3.2.).

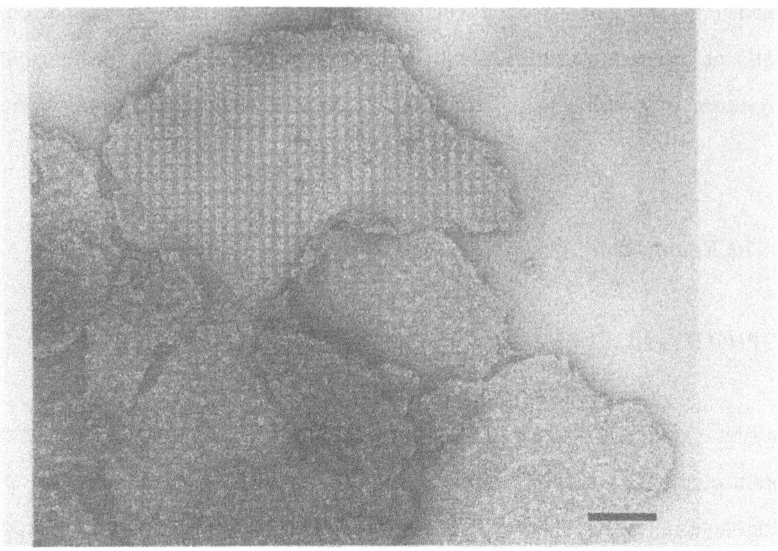

Figure 1. Electron micrograph of a negatively stained (protein appears white) *in situ* 2-D crystal of PSII (a x b = 17.7 x 20.1 nm, $\gamma = 91°$, *p*1) surrounded by membranes containing non-crystalline PSII. The scale bar corresponds to 100 nm.

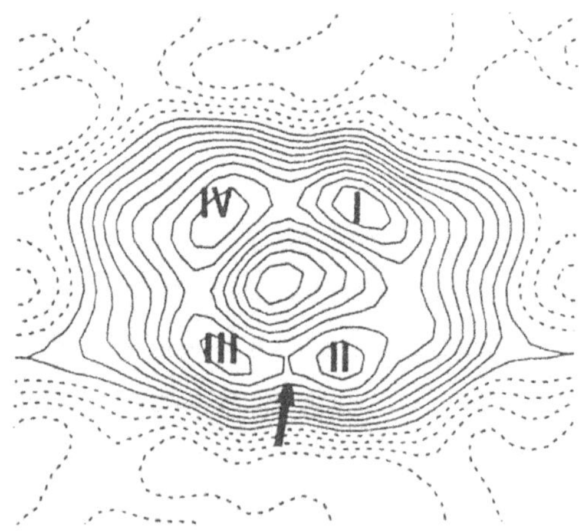

Figure 2. Fourier projection map of native PSII revealing four domains (the location of domain V is indicated by an arrow) and a central protein deficit. One unit cell is shown.

In order to investigate the structure of native PSII, one has to analyse the complex whilst in its native membrane environment. Electron microscopical analysis of negatively stained *in situ* 2-D crystals (Figure 1) resulted in a projection map at low resolution (3 nm) showing the complex to consist of four major domains surrounding a central stain dip (Figure 2).

By tilting the specimen relative to the incident electron beam, as mentioned earlier, one can gain access to 3-D structural information. However, the impracticality to tilt a specimen by as much as 90° means that not all of the 3-D information is retrieved (missing cone problem) and as a result of this the resolution along z is worse compared to the resolution in the a,b plane. By calculating a 'wedge diagram' (extent of data plotted as radius in the x^*y^* plane *versus* z^*; cf. Henderson et al. [6]) one can determine as to whether the information is complete within a given 'wedge'. With this being the case and employing a maximum tilt angle of 60° (87% of the Fourier space has been sampled with % completeness = sin 60 x 100), the resolution along z is worse by a factor of about 1.3.

Looking at a plot (e.g. like the one in Holzenburg *et al.*[10]) showing the modulation of the amplitudes and phases along z^* for each reflection (h,k), one can easily see that the phase information is more reliable than the amplitude information. Slices through the 3-D structure reveal 4 major domains and a central deficit on one side (known as the lumenal side) of the complex while the other side (referred to as stromal side) is mainly characterised by two domains forming a central ridge to the structure (Figure 3e and f). The overall 3-D structure of PSII is highly anisotropic going from top to bottom, left to right or back to front. It also protrudes much more from the lumenal side of the membrane than from the stromal side with which it is flush aiding its organisation within the chloroplast membrane system (Figure 4). Estimating the mass of the complex using the enclosed volume gives a number that agrees well with the unit cell containing one PSII complex, i.e. PSII operates as a monomer *in vivo* and not as an oligomer.

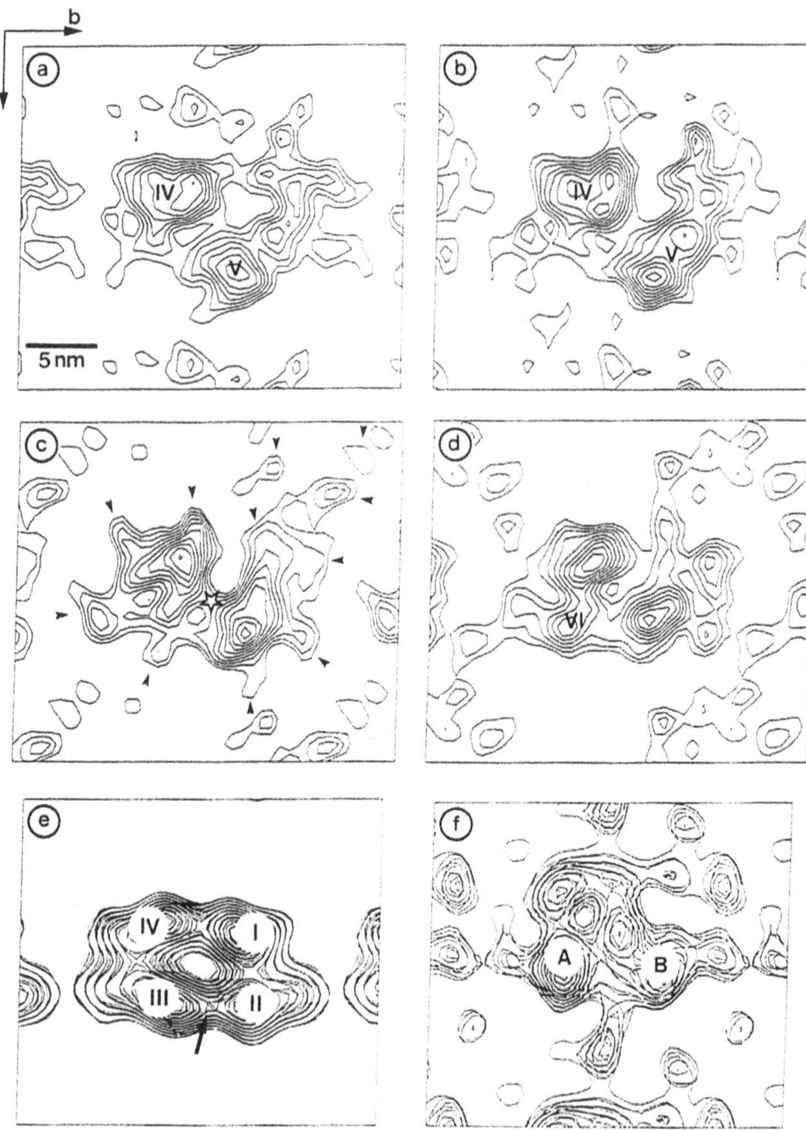

Figure 3. Slices through the 3-D structure of PSII before (e,f; native) and after (a-d) Tris-treatment. The lumenal side of the complex (a,b) is dominated by two large domains IV and V. At the centre of the membrane (c), the star marks a point of pseudo-twofold symmetry. Towards the stromal side (d), another domain (VI) is apparent between domains IV and V. As yet unidentified peripheral features are marked by arrowheads (see [11]). In (e) and (f), three consecutive lumenal and stromal slices, respectively, have been superimposed. The labels in (e) correspond to those in Figure 2, and in (f), the two domains forming a central ridge are labelled (A and B). Reproduced from Ford *et al.* [11].

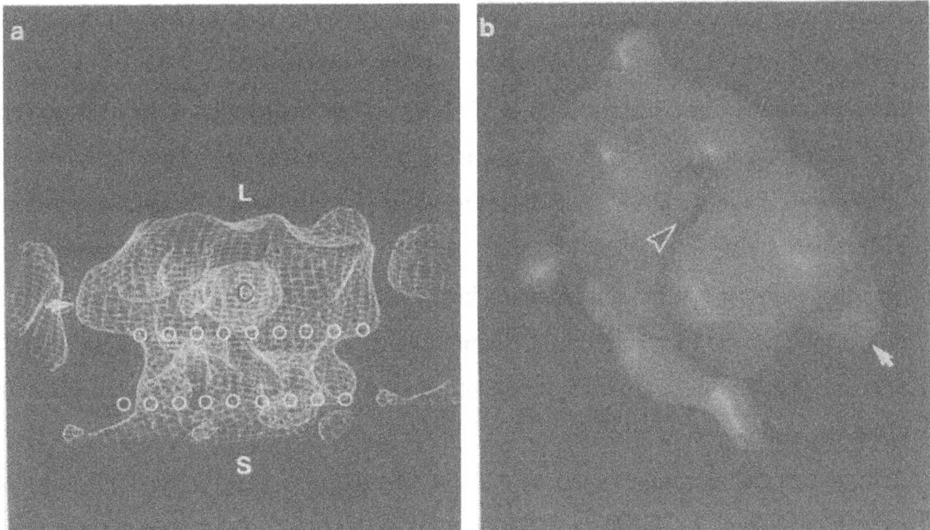

Figure 4. 3-D chicken wire (a) and surface relief (b) presentations of native PSII. The arrows point at identical features in (a) and (b) to facilitate comparison. The intramolecular cavity is highlighted by an arrowhead in (b) and labelled "C" in (a). In (a), lumenal (L) and stromal (S) faces as well as the estimated position of the thylakoid membrane (open circles) are indicated.

Furthermore, the central protein deficit is now discernable as a distinct intramolecular cavity (Figure 4), a biologically extremely important structural feature as we shall see. From the electron microscopical point of view, another observation deserves attention. It is interesting to note that the features in the projection map are identical to those on the lumenal side of the 3-D map. This is due to the fact that these 2-D crystal containing membranes have a tendency to adsorb lumenal face down onto the support film. Together with preferential stain trapping between the lumenal face and the support film this causes a non-uniform stain distribution along *z* with a very strong bias towards lumenal features and very little information on the stromal side when analysed in projection only.

In order to be able to relate some of the observed domains to specific polypeptides of PSII, 2-D crystals were subjected to simple Tris-washing, which removes the three extrinsic OEE-polypeptides, and, after biochemical and functional characterisation, were again analysed in three dimensions [11]. As a result of the OEC removal, the stain could reach the interior parts of the complex much better, the crystals obtained were better ordered and adsorbed both ways (lumenal side up and down) to the support film. These factors allowed the retrieval of a combined data set showing an unbiased and more detailed 3-D structure with a maximum resolution of 1.8 nm. As a result, the 3-D map after Tris-washing revealed a further domain (V) between domains II and III and a domain VI located between IV and V within the boundaries of the membrane (Figure 3). By calculating a 3-D Fourier vector difference map (native minus Tris-washed) and displaying this difference map within the native 3-D map, the location of the OEC could be determined (Figure 5). The data revealed that (i) there is only one copy of each OEE-polypeptide per PSII and (ii) all of them are located on the lumenal side of the complex where they coincide with domains I, II and III and help forming the intramolecular cavity. Washing the OEC off resulted not only in an opening of this cavity but also in a drastic decrease in the oxygen evolution activity, so that a true structure-function relationship could be established. For many years, scientists have proposed a special environment required for oxygen evolution (after all, this process takes place at a redox potential of around +1V): an intramolecular cavity is a perfect solution [11,12].

Following on from this, the question arises which OEE-polypeptide corresponds to which domain. This was solved by treating the 2-D crystals with different concentrations of NaCl inducing a differential removal of the OEE-polypeptides [12]. Since they are all located on the lumenal side and since preferential staining can reveal information specific to the lumenal side (see above), an assignment

Figure 5. 3-D Fourier vector difference map (labelled 1-3) displayed inside the native PSII 3-D map (C = cavity). After differential removal of the OEE-polypeptides [12], 1 could be assigned to OEE-33, 2 to OEE-23 and possibly another 10-kDa extrinsic polypeptide, and 3 to OEE-16. Modified from Ford *et al.* [11].

can be introduced without the need for further 3-D analysis [12]. By comparing the different data sets it became clear that domain III houses OEE-16, II houses OEE-23, and I OEE-33. Due to the relatively large volume in the 3-D difference map asssigned to OEE-23 (Figure 5) and the finding that a further 10-kDa polypeptide is required for the binding of OEE-23 to the lumenal face of the complex, it seems very probable that

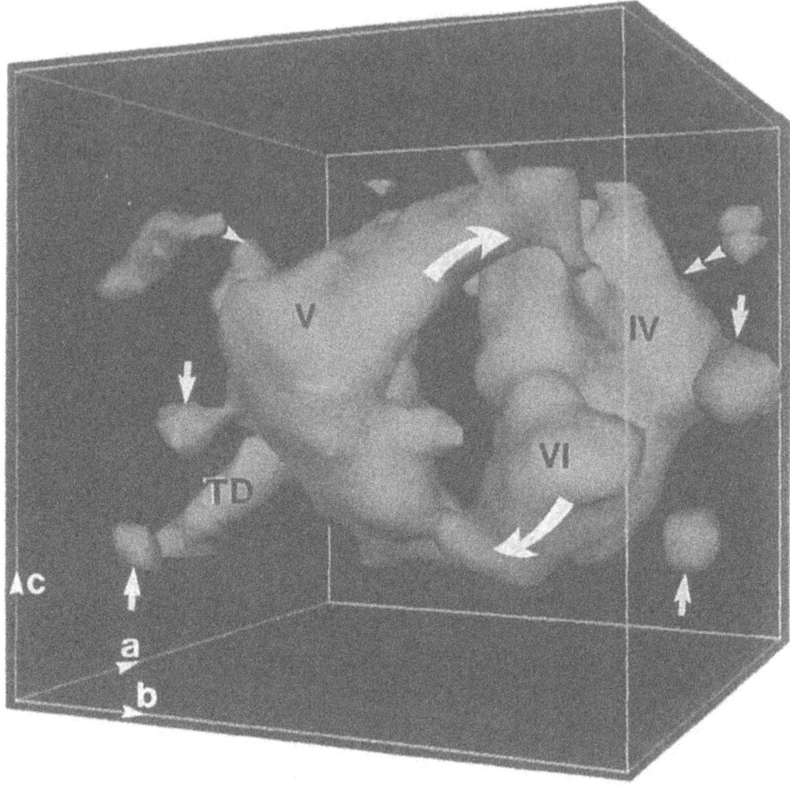

Figure 6. Surface rendered 3-D map of Tris-treated PSII (viewed side-on) revealing domains IV, V, and VI, spherical densities delineating the putative lipid bilayer (arrows), and the embracing architecture of the major domains (curved arrows). The arrowheads point at small unidentified lumenal domains. Reproduced from Ford *et al.* [11].

both the10-kDa polypeptide as well as OEE-23 are part of domain II. Additional information on the identity of domains IV and V was gained from cross-linking and other biochemical and biophysical studies suggesting, in conjunction with the 3-D data after Tris-washing [11], that domain IV corresponds to CP47. This would leave only domain V as possible candidate for CP43 and domain VI, which is sandwiched between domains IV and V, is therefore very likely to house the reaction centre (Figure 6).

The described experiments show how one can tackle the structure of a large and fragile membrane protein complex to a resolution that is not high but nevertheless biologically meaningful. Striving for higher resolution data for the native complex, 2-D crystals of PSII have been embedded in vitreous water and data were recorded under cryo-conditions (see 2.). The crystals, at that point in time, were not large enough to record electron diffraction patterns, so all the information had to come from images necessitating the recording and evaluation of thousands of electron micrographs prior to the averaging over 53 small but well-ordered crystalline areas [13]. Averaging was carried out in reciprocal space using the lattice vectors to facilitate the alignment. The feasibility of this approach has so far only been demonstrated by Perkins *et al.* [14] for purple membrane preparations where the results could be validated by comparing them to a known data set. With PSII, the approach was re-assessed by checking the self-consistency of the final structure factors that allowed the calculation of a projection map to 1.3 nm resolution from data sets that, when judged individually, showed strong reflections only up to about 5 nm resolution [13]. It is therefore important to realise that an effective merging can be carried out as long as there are enough reflections to unambiguously determine the lattice parameters making sure that the crystalline areas to be merged are isomorphous. The general strategy is summarised in Figure 7. The cryo-electron crystallographic data to 1.3 nm resolution represent the highest resolution structure so far for the native complex and confirm earlier estimates of the molecular mass as well as the assignments of three central domains to CP47, CP43 and the reaction centre heterodimer D1/D2 plus cytochrome b-559 (Figure 8). The data also proved valuable for discussing the evolution of reaction centres from anoxygenic to oxygenic photosynthesis [13]. Even though the map shows details to 1.3 nm resolution, the light-harvesting antenna polypeptides (LHCs), which are expected to be located around the core of PSII, are not well defined and their likely positions in the map are characterised by low density features only. This could be due to the LHCs being disordered in the native membrane. However, the major light-harvesting antenna polypeptide LHC-II can be isolated and undergoes oligomerisation into trimeric complexes which lend themselves for the outstanding electron crystallographic studies carried out by da Da Neng Wang, Werner Kühlbrandt and Yoshinori Fujiyoshi [7].

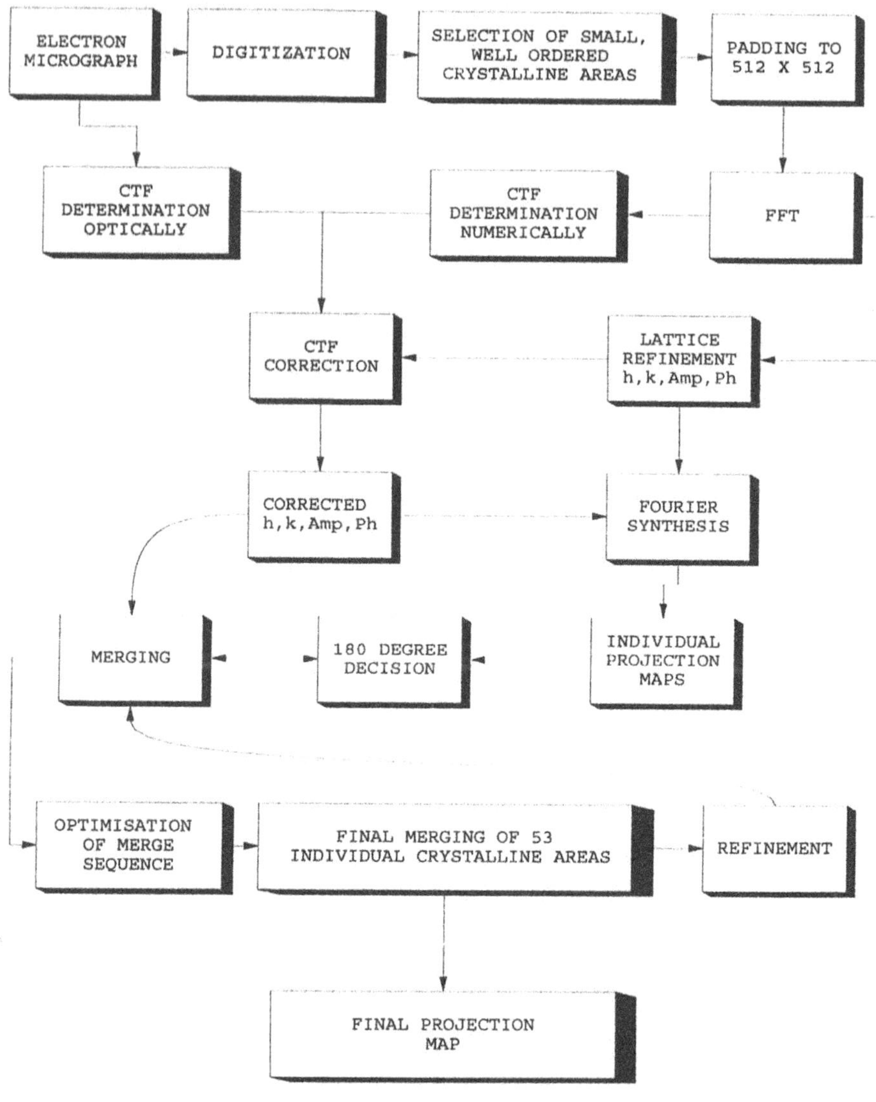

Figure 7. Flow chart describing the stages included in the averaging of the PSII image cryo-data in reciprocal space. The 180° decision relates to the application of appropriate rotation functions for crystals that are rotated with respect to each other (for more details, see Stoylova *et al.* [13]).

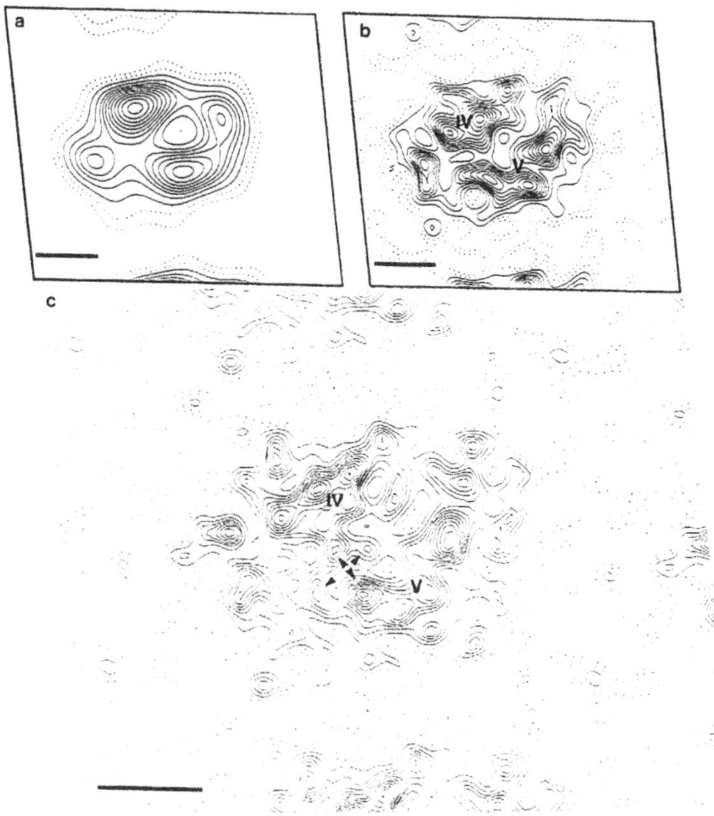

Figure 8. Fourier projection maps of native PSII in the frozen-hydrated state (unstained) after amplitude and phase correction calculated to (a) 5 nm, (b) 2.5 nm, and (c) 1.3 nm resolution. In (c), the central regions within domain VI are highlighted by arrowheads. Other domains are labelled as above. Dashed contour lines are drawn at 1σ and 2σ, and the first solid contour line at 3σ. The scale bar corresponds to 5 nm.

3.2. LHC-II

The trimeric complexes of LHC-II form 2-D protein-lipid crystals of high quality. Due to the plane group being *p*321 ($a = b = 12.95$ nm), fusion of smaller crystals into larger ones did not constitute a stumbling block since there is no need to consider any of the problems arising from crystals which are flipped (rotated through 180° along an axis parallel to the *a,b* plane) and/or rotated with respect to each other. The crystals were initially analysed in different embedding media at liquid nitrogen temperature [4] and electron diffraction patterns showing reflections to 0.346 nm resolution could be recorded. The first high-resolution 3-D data sets were calculated to 0.6 nm resolution in the *a,b* plane and about 0.8 nm resolution in the *z* direction [15] by combining amplitude information from electron diffraction patterns with the phase information from images, i.e. the resolution in the image was inferior to the resolution in the electron diffraction pattern. Secondary structure features were clearly resolved showing three transmembrane helices (A, B, C), two of which (A and B) cross over each other [15]. The positions of 15 chlorophyll molecules were also assigned. When looking at the modulation of amplitudes and phases along z^* [15], it becomes clear how much more reliable the electron diffraction amplitude information is versus the amplitude information from images [e.g. 10]. However,

- at high tilt, electron diffraction maxima can be considerably displaced from their expected positions due to the curvature of the Ewald sphere [16]. This effect is reduced at higher accelerating voltages (i.e. shorter wavelength).
- It is also crucial that the specimens are absolutely flat. Is this not the case, then one can observe a blurring of the reflections at high angles [16].

Finally, analysing the crystals at liquid helium temperature using Yoshinori Fujiyoshi's 400-kV electron cryomicroscope equipped with an extremely stable top entry stage, resulted in Fourier transforms of images which extended to a resolution comparable to that of electron diffraction data (!) [7]. The lower temperature in conjunction with the special stage results in less specimen drift and allows to record images at higher electron doses. Furthermore, the higher accelerating voltage improves coherence, decreases specimen damage (also contrast but that is compensated for by the higher affordable doses), Ewald sphere curvature and the attenuation of the envelope of the contrast transfer function at higher resolution. The resolution of the final 3-D map was 0.34 nm in the *a,b* plane and better than 0.49 nm orthogonal to it, which allowed, the polypeptide chain to be traced, individual amino acids as well as pigment molecules to be identified, and a detailed energy transfer mechanism to be proposed [7].

3.3. HUMAN BLOOD COAGULATION FACTOR IX

The third and last example is concerned with a low resolution study of the blood coagulation factor IX (FIX). Using the lipid layer technique mentioned earlier, FIX was 2-D crystallised under quasi-physiological conditions, bound to the surface of a lipid layer. The 2-D crystals were negatively stained, images were recorded in the electron microscope and a projection map was calculated to 3 nm resolution [17]. The unit cell ($a = b = 14.7$ nm, $\gamma = 97.1°$, *p*2) contained two molecules related by a point of twofold rotational symmetry and each molecule consists of a large flat domain and a small stalk-like domain attached to the large domain off-centre (Figure 9). Because one can deduce from the stain distribution that the small domain is pointing towards the lipid layer, the orientation of the factor relative to the membrane plane can be determined. A piece of information that cannot be obtained by any other method.

By combining this low resolution information with the known X-ray data [18] one can assign the large domain to the catalytic domain and the second epidermal growth factor-like (EGF-2) peptide and the stalk-like domain to the N-terminal γ-carboxyglutamic acid-rich domain (Gla domain; responsible for membrane binding) and EGF-1 (Figure 9). This way a complete picture of the *in vivo* situation is obtained. A very good example of how synergistic a combined X-ray/electron crystallographic approach can be. But this is not to say that the lipid layer technique should only be considered for low resolution studies. Quite the opposite is true, since this approach has in a number of cases led to very high quality 2-D crystals that diffract electrons to a resolution of better than 0.4 nm [e.g. 3].

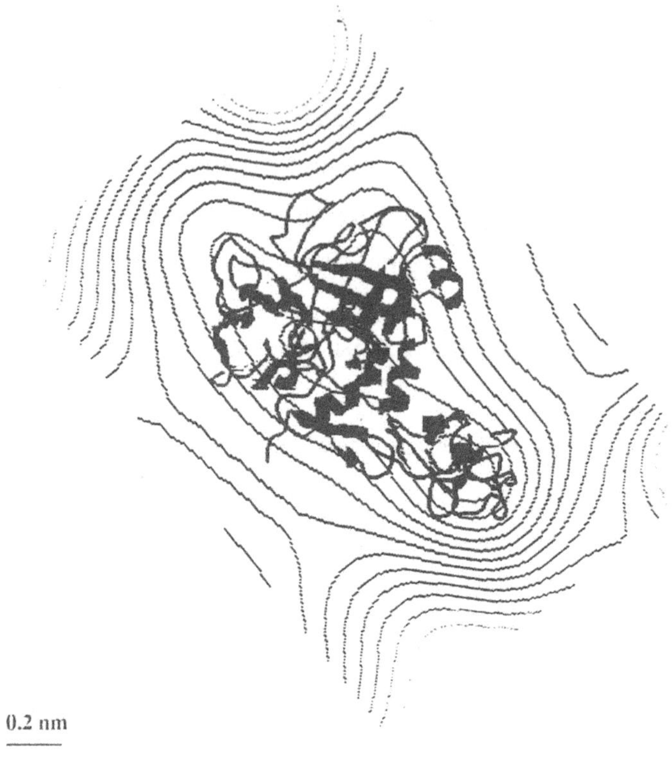

Figure 9. Projection map of lipid-bound FIX with the X-ray structure [18] superimposed (light grey = catalytic domain and EGF-2, dark grey = EGF-1 and Gla domain).

4. Epilogue

Since its begginings in the 1930s, electron micrsopy has come a long way. Over the last years, advances in biological electron crystallography (specifically, the engineering of better microscopes, improvements of specimen preparation and image processing techniques) have made huge leaps forward. While it is to be hoped that the field will continue to flourish, the biggest gains could perhaps be encountered, if the X-ray and electron communities worked together more closely, underpinned by a sound basis of mutual understanding and appreciation.

Acknowledgements

I like to thank Dr. Robert C. Ford and his coworkers at UMIST as well as my past and current laboratory members, particularly Dr. Svetla Stoylova, for sustaining the *tour de force* on PSII. I am also grateful to Dr. Stoylova for providing Figures 7 and 9. The financial support by the UK Biotechnology and Biological Sciences Research Council, the Academic Development Fund of the University of Leeds, the Leeds Centre for Molecular Recognition in Biologial Sciences, NATO, the Nuffield Foundation, and the Royal Society is greatly acknowledged.

References

1. Holzenburg,A. (1995) Electron microscopical analysis of ion channels. In: Ion channels - a practical approach (ed. R.H. Ashley) 269-290, IRL Press, Oxford.
2. Jap,B.K., Zulauf,M., Scheybani,T., Hefti,A., Baumeister,W., Aebi,U. and Engel,A. (1992) 2-D crystallisation: from art to science. *Ultramicroscopy* **46**, 45-84.
3. Brisson,A., Olofsson,A., Ringler,P., Schmutz,M. and Stoylova,S. (1994) Two-dimensional crystallisation of proteins on planar lipid films and structure determination by electron crystallography. *Biol. Cell* **80**, 221-229.
4. Wang,D.N. and Kühlbrandt,W. (1991) High-resolution electron crystallography of light-harvesting chlorophyll *a/b*-protein complex in three different media. *J. Mol. Biol.* **217**, 691-699.
5. Gilmore,C.J., Nicholson,W.V. and Dorset,DL. (1996) Direct methods in protein electron crystallography: the *ab initio* structure determination of two membrane protein structures in projection using maximum entropy and likelihood. *Acta Cryst. A* **52**, 937-946.
6. Henderson,R., Baldwin,J.M., Ceska,T.A., Zemlin,F., Beckmann,E. and Downing, K.H. (1990) Model for the structure of bacteriorhodopsin based on high-resolution electron cryo-microscopy. *J. Mol. Biol.* **213**, 899-929.
7. Kühlbrandt,W., Wang,D.N. and Fujiyoshi,Y. (1994) Atomic model of plant light-harvesting complex by electron crystallography. *Nature* **367**, 614-621.
8. Walian,P.J. and Jap,B.K. (1990) 3-Dimensional electron diffraction of Phoe porin to 2.8 Å resolution. *J. Mol. Biol.* **215**, 429-438.

9. Nicholson,W.V., Ford,R.C. and Holzenburg,A. (1996b) A current assessment of photosystem II structure. *Bioscience Reports* **16**, 159-187.

10. Holzenburg,A., Bewley,M.C., Wilson,F.H., Nicholson,W.V. and Ford,R.C. (1993) Three dimensional structure of photosystem II *Nature* **363**, 470-472.

11. Ford,R.C., Rosenberg,M.F., Shepherd,F.H., McPhie,P. and Holzenburg,A. (1995) Photosystem II 3-D structure and the role of the extrinsic subunits in photosynthetic oxygen evolution. *Micron* **26**, 133-140.

12. Holzenburg,A., Flint,T.D., Shepherd,F.H. and Ford, R.C. (1996) Photosystem II: Mapping the locations of the oxygen Evolution-enhancing subunits by electron microscopy. *Micron* **26**, 121-127.

13. Stoylova,S., Flint,T.D., Ford,R.C. and Holzenburg,A. (1997) Projection structure of photosystem II *in vivo* determined by cryo-electron crystallography. *Micron*, in press.

14. Perkins,G.A., Downing,K.H. and Glaeser,R.M. (1995) Crystallographic extraction and averaging of data from small image areas. *Ultramicroscopy* **60**, 283-294.

15. Kühlbrandt,W.and Wang,D.N. (1991) Three-dimensional structure of plant light-harvesting complex determined by electron crystallography. *Nature* **350**, 130-134.

16. Baldwin,J.and Henderson,R. (1984) Measurement and evaluation of electron diffraction patterns from two-dimensional crystals. *Ultramicroscopy* **14**, 319-336.

17. Stoylova,S., Gray,E., Barrowcliffe,T.W., Kemball-Cook,G. and Holzenburg,A. (1997) Electron crystallographical study of lipid-bound human factor IX. *Thromb. Haemost.* **Suppl.** 590

18. Brandstetter,H., Bauer,M., Huber,R., Lollar,P. and Bode, W. (1995) X-ray structure of clotting factor IXa - active site and module structure related to Xase activity and hemophilia B. *Proc. Natl. Acad. Sci. U.S.A.* **92**, 9796-9800.

THE NEED FOR ELECTRON CRYSTALLOGRAPHY IN MINERAL SCIENCES

M. MELLINI
Dip. Scienze della Terra - Università di Siena
Via delle Cerchia 3, 53100 Siena - Italy

1. Introduction

Electron crystallography may largely contribute to mineral sciences. The present aim is to understand why and how, starting from the already existing examples.

The success of TEM in mineral sciences dates back to the early '70s. Delay with respect to other fields was due to technical and scientific aspects. A technical problem was the specimen preparation, finally solved by ion milling. A scientific constraint was a scale bias, due to the occurrence of X-ray crystallographers (they had no interest, as their objects were smaller than the poor TEM resolution) and of mineralogists interested to microstructures that could be studied using optical microscope and electron microprobe. The current state has changed: crystallographers found that electron diffraction and imaging is useful even as regards structural analysis, mineralogists understand the deal of information contained in a TEM picture.

The growth of electron microscopy in mineral sciences may be schematized with reference to a few main studies. Among them, the pioneristic studies done on chrysotile asbestos: by doing top quality microscopy upon difficult material (fibrous habit, beam damage), Yada [1, 2, 3] revealed the complex curled layers of chrysotile and the wave-like arrangement of antigorite. The first summary about HRTEM was the seminal paper by Buseck and Iijima [4]. A comprehensive book was edited by Wenk [5]; by reporting the main TEM results in different fields of mineralogy; that book was for a long time a database for retrieving information on the TEM arsenal. Within a few years, the number of TEM laboratories steadily arose; a new approach to mineral study emerged, at the same time as the common appraisal of the mineral nature was changing from the static, idealized three-dimensional order to a more interesting dynamic pattern.

The very reason for the quick success of TEM was the impressive aspect of high resolution images: any observer, even if unaware about contrast transfer function or dynamical scattering, accepted the self-explaining evidence of the image. Images were more convincing than electron diffraction patterns, as "seeing is believing". Luckily, microscopists observed that "believing is seing", and work was done to avoid oversimplified interpretations; SHRLI-80 [6] was the first public domain program used to check interpretations, based upon expensive multislice calculations.

Unfortunately, developments in microscope hardware, coupled with new computing hardware, led scientists to overlook alternative uses of electron diffraction.

D. L. Dorset et al. (eds.), Electron Crystallography, 343–352.

Examples are the still limited popularity of CBED (that, in spite of the many possibilities, does not yet find extensive use in mineral sciences), the limited attention paid to the pioneristic work of the russian scientists, who demonstrated how mineral structures might be successfully determined using electron diffraction data only [8, 9, 10].

Early years of extensive HRTEM study of minerals were devoted to the defective microstructures, as opposite to the average structures revealed by X-rays [11]. The number of mineral structures determined by HRTEM and/or electron diffraction is still limited [12], with results often qualitative, sometimes possibly wrong and, in most of the cases, totally unaccurate. Time is now ripe to exploit electrons as tools able to reveal also the local average structure, namely able to determine and refine crystal structures that occur as small volumes, with an accuracy on bond distances better than one tenth Å.

2. Mineral structures from electron diffraction

The following three sections justify the need for electron crystallography. The matter has been artificially divided into: a) structure analysis based mostly upon electron diffraction; b) structure analysis based mostly upon high resolution imaging. The examples given will obviously testify the common occurrence of intermediate cases.

First attempts of structure analysis using electrons were done many years before the advent of HRTEM [8, 13]. Quantitative analyses of layer silicates (kaolinite, biotite, celadonite, muscovite) were performed measuring diffracted intensities by oblique texture electron diffraction and solving the structures starting from models or Patterson analysis. That approach will not be detailed, as elsewhere present in this course.

The case of the manganese and magnesium silicates gageite and balangeroite shows TEM used as an electron diffraction camera [14]. Gageite was a pipe-like structure, where manganese octahedra defined a tunnel occupied by disordered silicate tetrahedra; X-ray data had been refined to R = 0.17, a figure indicating major problems. Electron diffraction showed spots requiring a trebled *c* value, and led to remove disorder in tetrahedral distribution, by postulating a four-repeat chain in the octahedral pipe. The R factor went to 0.155. After a new X-ray data collection, using the trebled parameter on a better crystal, R was 0.049. Gageite examplifies the cooperation between electrons and X-rays: electrons (revealing weak diffractions) usefully integrate X-rays, and overcome pseudosymmetry and modulation problems. As final bonus, lattice imaging and electron diffraction showed different polytypes, clarified the complex fibrous intergrowth between balangeroite and chrysotile, showed Wadsley defects within balangeroite.

Bementite and parsettensite are modulated layer silicates, where a continuous octahedral sheet is linked with tetrahedral sheets that have periodic inversions in the orientation of tetrahedral apices. These minerals have constituted a problem, recently solved by electron diffraction, powder X-ray diffraction, Rietveld refinement and DLS. DLS (= Distance Least Squares) is a refinement procedure that optimizes bond geometry, assuming the general polyhedral connectivity, lattice parameters and crystal symmetry, and imposing that calculated bond distances and angles fit the expected ones. The structure of bementite [15] consists of two sheets of manganese octahedra, rotated by 22° each to the other, and linked through an inverted tetrahedral sheet, containing pairs of six-membered rings interconnected with five- and seven-membered rings. That model was derived starting from electron diffraction and high resolution imaging, building a crystalchemically sound model refined by DLS procedure, comparing the calculated images with the observed ones. Parsettensite shows islands of silicate tetrahedral rings [16]. Tilting experiments showed that structure factors could be derived using a kinematical approximation, owing to use of very thin grains. Therefore, different models for the single layer were compared with the hk0 Fourier transform built using diffracted intensities, and the best one (R = 0.2) was refined. The final three-dimensional model was obtained by expanding the two-dimensional structure using DLS procedure. The final check was finally given by the comparison between observed ad calculated lattice images.

3. Mineral structures from high resolution imaging

High resolution imaging as a tool for structure determination may be traced back to systematic studies started in the '70s. The reader should consult the original papers, if interested to understand the approach. Often, they contain a not straightforward solution of the problem, and build models based upon comparative analysis of different data. They therefore appear appropriate to stress the need for more powerful tools, as indicate procedures far from a general diffraction approach, prone to quantitative data treatment.

Tunnel-structure oxide minerals are built up by edge-sharing manganese octahedra, and normally occur as fine-grained material [17]. The manganese octahedra build columns, linked together to form ribbons 1-, 2-, 3-, 4-, n-columns wide; four ribbons connect each to the other forming the tunnel. Depending on number of octahedral columns in the ribbons, different [m x n] tunnels are possible: e.g., [1 x 2] tunnels in ramsdellite, [2 x 2] in hollandite, [2 x 3] in romanechite. Tunnel oxide minerals have important transport and storage properties, as host large radius cations and water molecules that easily diffuse along the tunnel. From the point of view of electron microscopy, the tunnel is a low potential zone, producing strong contrast at distances of a few octahedral sizes. Therefore, tunnel structures were a good case where to combine known structural principles and early imaging techniques. These revealed new structural types, easily interpreted as [3 x 3], [3 x 4], [3 x 5], [2 x 4], [2 x 7], etc. tunnels. That study demonstrated how complex minerals, difficult to study using conventional analytical approaches, might be easily understood using electron imaging.

The approach was extended to a group of silicates, now known as biopyriboles (the word combines biotite, pyroxene, amphibole). They result from intercalation of the mica, M, and the pyroxene module, P [18]. For instance, amphibole is MP. These phases (structurally related, and colinearly varying in composition among the two end-members) form a polysomatic series. HRTEM showed that jimthompsonite and chesterite were the M_2P and M_2PMP polysomes. Other local structures were discovered and interpreted using multislice simulations. The case of biopyriboles was important also because explained the solid state reaction pyroxene -->amphibole -->intermediate biopyriboles -->mica. Local faults at uncoherent or semicoherent interfaces behave as "highways", able to promote cation and water diffusion throughout the reacting crystal.

Cebaite, $Ba_3Ce_2(CO_3)_5F_2$, occurs as tiny crystals, a few hundreds Å in size [19]. The structure was solved starting from the chemically related mineral huanghoite,

$BaCe(CO_3)_2F$. Observed high resolution images for huangoite (whose structure was known) were compared with the simulated ones, to determine the thickness and defocus conditions producing images having high contrast in the positions of light atoms. Atomic positions for heavy and light elements explained not only the distances between the periodic bright dots, but also the intensity variation. Cebaite was an example of successful structure determination in conditions other than weak phase object approximation, namely for thickness where dynamical scattering was important.

Carlosturanite is a fibrous hydrated magnesium silicate. Electron diffraction was used to derive unit cell parameters. The structure model was proposed starting from a) chemical composition, b) unit cell, c) coherent interfaces with intergrown chrysotile [20]. The model derives from lizardite, modified introducing ordered tetrahedral vacancies and substituting hydroxyl anions or water molecules for oxygen atoms. Exhaustive one-to-one comparison with images calculated under definitely dynamical scattering conditions was limited by experimental difficulties: misalignement (average orientation of the fiber bundle, with individual fibers misaligned up to a few degrees), beam damage, separation of thinnest fibers during ion milling. Carlosturanite was the S_5X polysomes, and the faulted periodicities matched the values of the S_4X, S_6X, S_7X polysomes. As seen from the present viewpoint, the study is a qualitative structural analysis, performed using a poorly resolving microscope. A well equipped and numerous team is now operating to overcome that primitive work [21]. Together with powerful microscopes, crystallographic image processing may contribute in improving the model.

The case of orientite (a calcium and manganese silicate) deals with the relationships between electron and X-ray diffraction. Two teams had proposed different structures, having the same scaffolding of oxygen atoms and differing in the location of some tetrahedral and octahedral cations [22]. Both the models had been refined to acceptable R factors. However, they were unable to fit the HRTEM images; owing to systematic absences, calculated images had 9 Å periodicities, whereas the images strongly indicated the doubled value of 18 Å. This led to reconsider the structure, produce a third model hybrid between the previous ones and consistent with high resolution images. When introduced into the X-ray refinement, the R factor dropped from 0.085 to 0.069.

TiO_2(B) occurs within the polymorph anatase as semicoherently intergrown thin lamellae [23]. Its structure was solved combining AEM, electron diffraction, high-resolution imaging and simulations, DLS techniques. After interpretation of images, the model was allowed to vary under those constraints. The technique was successfully tested in-situ, by redetermining the structure of anatase: atomic coordinates, displaced from known positions, converged to values within 1% of expected.

4. A caveat

This review on one hand emphasizes the skillfulness used, on the other one the absence of systematic, really quantitative approaches to structure determination.

HRTEM structure analysis started for favourable cases, where contrast was easily intepreted as structural units producing contrasted deflecting potential (e.g., the tunnel-oxide structures), and evolved towards complex situations when a preexisting crystalchemical approach was supplying models to be tested (e.g., the biopyriboles, with polysomatism as the key to explain structures). The situation improved with multislice calculations, that offered a quantitative check (a persisting weak point was however constituted by uncertainity in the values of imaging parameters). Images were further exploited emphasizing electron diffraction (e.g., analysis of polytypism in gageite and balangeroite, use of kinematical intensities in parsettensite), using DLS optimization (e.g., TiO_2(B) and parsettensite), utilizing X-ray data either for performing Rietveld refinement (bementite) or single crystal analysis (e.g., gageite and orientite).

Those approaches are more ingenious than systematic; we do not know how reliable are the models; no accurate bond pattern can be given; comparison between observed and calculated diffraction intensities is qualitative; most often, scientists solved structures where a good general knowledge was preexisting; the work was limited to defined cases, such as polysomes; the whole work had to be validated looking for outer constraints (e.g., X-rays); rarely, the real imaging conditions were known (defocus, thickness, misalignement, occurrence of kinematical or dynamical scattering).

For several reasons (the computational and psychological difficulty to deal with dynamical scattering) the quantitative exploitation of electron diffraction data was overlooked, and people using electrons focused attention on "conventional" HRTEM. The experimental approach was consciously limited to the optical resolution of the microscope; only the small fraction of reflections falling within the optimal objective aperture was used, diffractions falling outside were totally disregarded. Paradoxically, if the same approach had been used also in the case of X-rays (that do not have any optical resolution), the structures known with precision better than 0.01 Å would have never been determined.

Work was based upon two-dimensional images. The scientific advancement was, consciously or unconsciously, limited to subjects that could be studied within that limit. This means that, for instance, nowadays we know a lot about modular structures (e.g., polytypes, polysomes), if the structure of the fundamental modules is known and the same principles still hold for the derivative arrangements. As the detailed analysis of heavily faulted specimens was relatively easy, many reports dealt with extended defects; however, point defects (e.g., vacancies and chemical substitutions) are still poorly known, in spite of their importance [24].

Exploratory activity showed the potential of electron crystallographic approaches, but the approach has now to become more quantitative and more systematic.

5. Electron crystallography applications in mineral sciences

The previous studies were mostly focused on the mineral problem; sometimes they also promoted tools that are now in the electron crystallography domain; for this reason, should appear here again. They were not included, as this section is devoted to studies that specifically attempted to develop the methods, that is to the elaboration of raw electron microscopy and electron diffraction data.

Among them, a first example is the revisitation of oblique texture electron diffraction. Those early determinations have been recently duplicated, by combining old experiments and new crystallographic software [25]. Direct methods of phase determination (that is a probabilistic approach developped within the domain of X-ray crystallography) were applied to the diffracted intensities previously measured. The resulting potential maps were sufficient to identify most of the atomic positions, here included the oxygens. The paucity of electron diffraction data available however prevented the "automatic" structure recognition, and still required the active intervention of the crystallographer.

The method was further extended to boric acid [26]. Using fourty years old room-temperature electron diffraction data, as well as new low-temperature data, the structure was solved ab-initio. Reliable bond geometry was derived for oxygen and boron atoms: the B-O distance was found to be 1.36(1) vs. 1.367 Å (neutrons) the O-H distance 1.11(11) vs. 0.97 Å. Although the cases of layer silicates and boric acid may have been favoured by particular conditions of electron scattering (as discussed by the author), they indicate the reliability of ab-initio structure determination, with no information other than the electron diffraction pattern.

3D structure determination has been obtained for the mineral staurolite, $Fe_2Al_9Si_4O_{24}H$ [27]. Images, taken in different projections in thin regions where dynamical scattering was minimal, were Fourier transformed to derive amplitude and phases. The nonperiodic noise was eliminated by Fourier filtering, obtaining images decidedly improved with respect to the recorded ones. The calculated diffraction pattern was used to determine defocus and astigmatism.The 3D potential map was built using 59 independent reflections, after a procedure based upon determination of structure factors, origin assignement by imposing appropriate phase relationships, correction for contrast transfer function, data merging. The map showed all the atoms, with a resolution of 1.38Å. After discussing the experimental requirements, the authors presented major arguments related with the general feasibility of 3D electron crystallography, the possibility to overcome problems of dynamical scattering, the possibility to determine accurate mineral structures from a few unit cells, the possibility to combine 3D electron crystallography with X-ray diffraction (e.g., supplying the starting set of phases needed for phase extension procedure). That study contains many suggestions useful for the advancement in structure analysis; the whole procedure strongly parallels what described as "crystallographic image processing" [28, 29] .

An example of two-dimensional crystallographic image processing is found in the study of the modulated manganese and iron layer silicate bannisterite [30]. After Fourier transform and image enhancement, the calculated image revealed more detail than the experimental one. Gross features and structural details such as pinching and swelling of the sinusoidal octahedral sheet, overlapping regular and inverted octahedra, columns of large interlayer cations could be easily observed.

Tourmaline is a borosilicate mineral; early studies had indicated different plane group symmetries, interpreted as possibly due to medium range ordering after cation diffusion under the electron beam. That issue has been recently reanalyzed [31]; accurate crystallographic image processing confirmed that early observation, and showed compositional dependence in the deviation from the plane group. In particular, Fe-poor elbaites show *p31m* , Fe-rich elbaites *p3m1*, rubellite *p6*, Fe-Mg-Al bearing specimens pseudo-*p6* plane groups.

One of the problems of HRTEM is the knowledge of actual imaging conditions, coupled with the correction of less-than-optimal imaging conditions. We have previously recalled that defocus and astigmatism may be determined in-situ. A further problem arises with local misalignement; namely, owing to the internal texture of the specimen (e.g., small domains randomly rotated) as well as to experiment artifacts (e.g., bending of the thin foil), the imaging conditions may vary from point to point. This is important also when electron sensitive specimens are analyzed; in such cases, common procedure is the careful alignement in a given position, followed by movement to an adjacent fresh area for image recording. In the case of orthopyroxene, the effects of crystal tilt could be easily removed by image processing, for thickness smaller than 100 Å and tilt angles lower than 2.5°. In that way, interpretable images could be reconstructed [32].

6. Some problems to be solved

The previous chapter reported cases solved using electron crystallography. This section will present examples that require reliable electron crystallography techniques. The cases are ranged in an approximate order, starting from the simplest ones.

- Phase identification: identification of mineral phases is common practice of mineralogy. The procedure is based upon macroscopic observation, physical properties, X-ray diffraction, chemical analysis. Electron diffraction enters the procedure from a long time (e.g., the cases of asbestos and clays). However, phase identification is now extended to microphases (e.g., microinclusions of metallic copper in biotite [33] or of hematite in rutile [34]). The problem is analogous to macrophases, but more difficult as restrained to the use of electron diffraction, electron imaging, chemical analysis in the microscope. Possible unknowns are environmental particles, cosmic dusts, synthetic products, possibly in a polyphasic arrangement. The investigator should identify each of them, using a preparation technique that preserves the original textural relationships. The ideal tool would be an analytical microscope, capable to record the X-ray emission, of tilting the specimen over a large range, and modified to automatically record diffractions originating from areas smaller than 1000 Å; the data should be recombined to produce the three-dimensional reciprocal lattice; the orientation matrix should be used for piloting the crystal tilt; the lattice parameters should be known with accuracy and precision sufficient for phase identification; diffracted intensities should be available. Namely, a first exigence is evolution of the TEM towards the state of automated electron diffractometer.

- Modular mineralogy: many minerals are described as commensurate or uncommensurate modulated structures. Namely, a basic module is chemically or structurally modulated, producing a superperiodicity multiple, or not, of the basic periodicity. The simplest cases are polytypes [35] and polysomes [12, 36]. The first aim is the statistical description of the overall arrangement of intercalated modules (this information is available, in a synthetic way, within the shape of diffraction spots; otherwise, spreaded out in the lattice image). The second goal is chemical recognition (some tool able to reveal whether any chemical heterogeneity occurs). The third goal is the structural identification of each module (based upon details of the image contrast, or the diffraction pattern from selected image regions). The fourth goal the accurate structure determination of small, faulted zones. Those targets may be reached only using a combination of different techniques: conventional electron diffraction and imaging;

image analysis in the direct and reciprocal space; imaging techniques based upon chemical contrast generation (e.g., use of electron energy loss or high-angle annular dark-field [37]); capability of in-situ structure determination.

- Reaction microstructures: minerals often reveal microstructures inherited from the reaction mechanisms. Examples are reaction rims among different phases; exsolution microstructures; continuous and discontinuous processes connected with diagenesis and metamorphism; intermediate (stable and metastable) phases produced within quickly reacting systems (e.g., reactions following the Ostwald ripening rule); early stages of crystal growth, such as morphological unstability of nanocrystal germs [38]. These microstructures are important, because related with the mechanism responsable for their production. In some cases, they also have practical implications; an example is metamictization (e.g., a crystal-to-amorphous phase transition induced by isotopic decay) with respect to nuclear technologies such as materials durability or waste disposal.

- Determination and refinement of the average structure: trivial case is the use of electron crystallography in support to other tools, such as X-ray powder and single crystal X-ray diffraction (e.g., offering a set of starting phases for phase generation through probabilistic methods). A less trivial case is structural analysis for cases where X-rays cannot be applied. This situation may arise from habit or size of the crystal, or because crystals are faulted. The problem consists in the application of techniques, hopefully capable to refine structure models, with an accuracy on bond distance better than 0.1 Å. Different approaches may be considered: exploitation of high resolution imaging and Fourier transform (as done in [27, 29]), or CBED methods.

- Poorly crystalline minerals, such as the ones that occur during the sedimentary or alteration processes, or may be found as complex arrangements within cosmic dusts, or are produced by bacterial activity and, more generally, by a biomineralization process, or represent local evolution of a metastable state (e.g., crystal nucleation in a glass): in all those cases the researcher is confronted with tiny objects, poorly diffracting, structurally irregular, difficult even to describe.

- Site population, that is the distribution of crystalchemically similar atoms among similar sites. That problem is solved for a long range order state by X-rays (if a crystal exists), or for a short range order state by spectroscopies. Further exigence is the contemporaneous analysis of short and long range order states by combining microstructural capabilities (e.g., imaging antiphase boundaries and twin planes) together with probes able to deal with local site partitioning. A still partial solution is ALCHEMI [39]; another solution would be the accurate local structure refinement.

- Point defects: as already stated, the main problem is how to see disordered point defects [18, 24]. An example is the study of mullite, devoted to understand what is the oxygen vacancy distribution [40]. However, for the moment, both diffraction and imaging techniques do not offer satisfactory general solutions.

- Oxidation state of the elements: for the moment, high resolution EELS seems to be the most promising one [38, 41].

- Correlation micro-macro: advanced structural and chemical knowledge of small regions explains macroscopic systems. In fact, the microscopic arrangements modify bulk chemical, physical and transport properties; this modified properties, in turn, determines modified bulk reactivity (both because of the modified thermodynamics properties, as well as of more favourable kinetics). On one hand, mineral sciences need advanced electron crystallography tools to understand microscopic features; on the other one, they need quantitative exploitation of that microscopic knowledge at the macroscopic scale.

7. References

1. Yada, K. (1967) Study of chrysotile asbestos by high resolution electron microscopy, *Acta Cryst.* **23**, 704-707.
2. Yada, K. (1971) Study of the microstructures of chrysotile asbestos by high resolution microscopy, *Acta Cryst.* **5**, 119-124.
3. Yada, K. (1979) Microstructures of chrysotile and antigorite by high resolution electron microscopy, *Can. Miner.* **17**, 679-691.
4. Buseck, P.R. and Iijima, S. (1974) High resolution electron microscopy of silicates, *Amer. Miner.* **59**, 1-21.
5. Wenk, H.R. (1976) ***Electron microscopy in mineralogy***, Springer Verlag. Berlin
6. O'Keefe, M.A., Buseck, P.R., Iijima, S. (1978) Computation of high-resolution TEM images of minerals, *Nature* **274**, 322-324.
7. Pinsker, Z.G. (1953) ***Electron diffraction***, Butterworth, London.
8. Zvyagin, B.B. (1967) ***Electron diffraction analysis of clay mineral structures***, Plenum Press. New York.
9. Drits, V.A. (1987) *Electron diffraction and high-resolution electron microscopy of mineral structures*, Springer Verlag, Berlin.
10. Vainshtein, B.K., Zvyagin, B.B. and Avilov, A.S. (1992) Electron diffraction structure analysis, in JM.Cowley (ed.), ***Electron diffraction techniques***, Oxford University Press, Oxford, pp. 216-312.
11. Mellini, M. (1989) High resolution transmission electron microscopy and geology, Adv. Electrons Electr. Phys. 76, 281-326.
12. Allen, F. (1992) Mineral definition by HRTEM: problems and opportunity, in P.R. Buseck (ed.), *Minerals and reactions at the atomic scale: transmission electron microscopy*, Rev. Miner. **27**, 289-334.
13. Zvyagin, B.B. (1994) Electron diffraction analysis, in A.S. Marfunin (ed.), *Advanced Mineralogy*, vol. 1, Springer Verlag, Berlin, pp.50-63.
14. Ferraris, G., Mellini, M. and Merlino, S. (1987) Electron diffraction and electron microscopy study of balangeroite and gageite: crystal structures, polytypism, and fiber texture, *Amer. Miner.* **72**, 382-391.
15. Heinrich, A.R., Eggleton, R.A. and Guggenheim, S. (1994) Structure and polytypism of bementite, a modulated layer silicate, *Amer. Miner.*, **79**, 91-106.
16. Eggleton, R.A. and Guggenheim, S. (1994) The use of electron optical methods to determine the crystal structure of a modulated phyllosilicate: parsettensite, *Amer. Miner.* **79**, 426-437.
17. Turner, S., Gorshkov,A.I. and Buseck, P.R. (1994) Tunnel-structure oxide minerals, in A.S. Marfunin (ed.), *Advanced Mineralogy*, vol. 1, Springer Verlag, Berlin, pp. 90-95.
18. Veblen, D.R. (1992) Electron microscopy applied to nonstoichiometry, polysomatism, and replacement reactions in minerals, in P.R. Buseck (ed.), *Minerals and reactions at the atomic scale: transmission electron microscopy*, Rev. Miner. **27**, 181-230.
19. Li, F.H. and Hashimoto, H. (1984) Use of dynamical scattering in the structure determination of a minute fluorocarbonate mineral cebaite $Ba_3Ce_2(CO_3)_5F_2$ by high resolution electron microscopy, *Acta Cryst.* **B40**, 454-461.
20. Mellini, M., Ferraris, G. and Compagnoni, R. (1985) Carlosturanite: HRTEM evidence of a polysomatic series including serpentine, *Amer. Miner.* **70**, 773-781.
21. Alberico, A., Baronnet, A., Belluso, E., Ferraris,G., Miehe, G., Prencipe, M., Rodewald, M. and Soboleva, S. (1996) Some evidences for revisiting the structural model of carlosturanite, *Assoc. Ital. Crist., Abstracts Alessandria Meeting*, P-67.
22. Mellini, M., Merlino, S. and Pasero, M. (1986) X-ray and HRTEM structure analysis of orientite, *Amer. Miner.* **71**, 176-187.
23. Banfield, J.F., Veblen, D.R. and Smith, D.J, (1991) The identification of naturaly occurring TiO_2(B) by structure determination using high-resolution electron microscopy, image simulation, and distance least squares refinement, *Amer. Miner.* **76**, 343-353.
24. Veblen, D.R. and Cowley, J.M. (1994) Direct imaging of point defects by HRTEM, in A.S. Marfunin (ed.),

Advanced Mineralogy, vol. 1, Springer Verlag, Berlin, pp.172-173.

25. Dorset, D.L. (1992a) Direct phasing in electron crystallography: determination of layer silicate structures, *Ultramicroscopy* **45**, 5-14.
26. Dorset, D.L. (1992b) Direct methods in electron crystallography. Structure analysis of boric acid, *Acta Cryst.* **A48**, 568-574.
27. Wenk, H.R., Downing, K.H., Meisheng, H. and O'Keefe, M.A. (1992) 3D structure determination from electron microscope-images: electron crystallography of staurolite, *Acta Cryst.* **A48**, 700-716.
28. Wang, D.N., Hovmoller, S., Kihlborg, L. and Sundberg, M. (1988) Structure determination and correction for distortions in HREM by crystallographic image processing., *Ultramicroscopy* **25**, 303-316.
29. Zou, X. (1995) Electron crystallography of inorganic structures - theory and practice. Doctoral Dissertation, Stockolm University
30. Ferrow, E.A. and Hovmoller, S. (1993) Crystallographic image processing (CIP) and high-resolution transmission electron microscopy (HRTEM) studies of bannisterite: a 2:1 type modulated layer silicate, *Eur. J. Mineral.* **5**, 181-188.
31. Ferrow, E.A., Wallenberg, L.R. and Skogby, H. (1993) Compositional control of plane group symmetry in tourmalines: an experimental and computer simulated TEM, crystallographic image processing, and Mossbauer spectroscopy study, *Eur. J. Mineral.* **5**, 479-492.
32. Zou, X., Ferrow, E.F. and Hovmoller, S. (1995) Correcting for crystal tilt in HRTEM images of minerals: the case of orthopyroxene, *Phys. Chem. Minerals* **22**, 517-523.
33. Ilton, E.S. and Veblen, D.R. (1988) Copper inclusions in sheet silicates from porphiry Cu deposits, *Nature* **334**, 516-518.
34. Banfield, J.F. and Veblen, D.R. (1991) The structure and origin of Fe-bearing platelets in metamorphic rutile, *Amer. Miner.* **76**, 113-127.
35. Baronnet, A. (1992) Polytypism and stacking disorder,- in P.R. Buseck (ed.), *Minerals and reactions at the atomic scale: transmission electron microscopy*, Rev. Mineral. **27**, 231-288.
36. Veblen, D.R. (1991) Polysomatism and polysomatic series: a review and applications, *Amer. Miner.* **76**, 801-826.
37. Wang, S., Buseck, P.R. and Liu, J. (1995) High-angle annular dark-field microscopy of franckeite, *Amer. Miner.* **80**, 1174-1178.
38. Ricolleau, C., Audinet, L., Gandais, M., Gacoin, T., Boilot, J.P. and Chamarro, M. (1996) Influence of growth conditions on the structural properties of CdS_xSe_{1-x} (x = 0.4 and x = 1) nanocrystals, *J. Crystal Growth* **159**, 861-866.
39. Buseck, P.R. and Self, P. (1992) Electron energy-loss spectrosocpy (EELS) and electron channelling (ALCHEMI), in P.R. Buseck (ed.), *Minerals and reactions at the atomic scale: transmission electron microscopy*, Rev. Mineral. **27**, 141-180.
40. Epicier, T., O'Keefe, M.A. and Thomas, G (1990) Atomic imaging of 3:2 mullite, *Acta Cryst.* **A46**, 948-962.
41. Garvie, L.A.J., Craven, A.J. and Brydson, R. (1994) Use of electron-energy loss near-edge fine structure in the study of minerals. *Amer. Miner.* **79**, 411-425.

ELECTRON DIFFRACTION OF MINERAL STRUCTURES AND TEXTURES

Boris B. Zvyagin
Institute of the Geology of Ore Deposits, Petrography, Mineralogy, and Geochemistry
Russian Academy of Sciences
Staromonetny pr. 35
Moscow 109017, Russia

1. Introductory Remarks

The essence of the lectures given are presented in a list of publications cited at the end of this article. Mostly, the discussion will concentrate on the material given in Ref. 18. In addition, the material below presents some more recent reflections on the subject.

2. Minerals as a Favorable Applications Area for Electron Diffraction

Almost from the very beginning of the technique, it became clear that the world of natural substances, represented also by minerals, contained many examples that are suitable for investigation by electron diffraction. Indeed a case can be made for an actual need of electron diffraction, even applications where electron diffraction serves as the most efficient technique for providing unique information about the sample that, at the same time, is detailed and reliable. For example, in 1928, Kikuchi used mica flakes as the object for first demonstrating the type of electron diffraction pattern named after him. Muscovite, another mica, served as a sample object for pioneering convergent beam electron diffraction experiments by Kossel and Mollenstedt in the period 1938 to 1942. Patterns had been obtained by Aminoff and Broome in Sweden (1933 to 1938) from the edges of talc and graphite crystals. They also observed sphalerite ZnS-ZnO transformations by electron diffraction. Hendricks (USA, 1938) obtained texture patterns of clay minerals while Honjo and Mihama (Japan), as well as Kulbicki (France) were the first to publish (in 1954) selected area electron diffraction patterns from the mysterious clay mineral, halloysite.

Major developments in the electron diffraction studies of minerals occurred during the second half of the century, when the functions and specific features of the various kinds of electron diffraction experiment were further refined. Within this period, investigations could be classified according to which electron diffraction technique was used, as well as the specific objects that were studied and the kind of problem that was solved. Essential structural results were obtained mainly by high resolution electron diffraction and selected area electron diffraction, in combination with electron microscopy (sometimes accompanied by spectroscopy). It was only later that the potential advantages of convergent beam techniques had been fully established.

Typical problems investigated included structure determination and/or refinement, polytype analysis (to be extended in a general way to a modular analysis), as well as substance identification of individual particles, mixtures or regular intergrowth associations. Of special significance were the possibilities not provided by any other diffraction tool - e. g. detailed analyses of structural and lattice variations, including distortions, modulations, the display of sub- and super-periodicities, both in real and reciprocal space, and identification of peculiar kinds of crystals including cylindrical lattices, etc.

D. L. Dorset et al. (eds.), Electron Crystallography, 353–357.

Typically, the objects investigated, from which useful results were obtained, were highly-dispersed (fine-grained) minerals consisting of platy or fibrous microcrystals. Their structures were found to differ widely in terms of order-disorder and symmetry. Studies of phyllosilicates and clay minerals, oxides and hydro-oxides of Fe and Mn, hybrid and interstratified minerals, chain-ribbon silicates, sulfosalts, sappharines, etc., have enriched the field of structural mineralogy. In particular, generalizations from electron diffraction data, as well as a verification of results from other techniques, have done much to establish relationships uniting systems, sets, or series of minerals. Included in these results has been the actual discovery of new minerals, as well as their groups and families.

3. Oblique Textures - a Special Technique for Electron Diffraction Structure Analysis

It is well-known that, at high accelerating voltages (small wavelength), electron diffraction patterns can be considered as a planar sampling of the reciprocal lattice normal to the incident beam direction and passing through the reciprocal lattice origin. Two features of the diffraction pattern may be considered: its geometry and its intensity distribution. True single crystal diffraction patterns image two-dimensional sets of intensity maxima whereas polycrystal diffraction patterns represent one-dimensional sequences of three-dimensional sets of circular reflections. Oblique texture diffraction patterns contain three-dimensional reflection sets regularly distributed in two dimensions. Therefore, oblique texture electron diffraction patterns are more complete than those from single crystals and more regular than those from polycrystals. Accordingly, they provide more definite information about the lattice geometry (unit cells), obtained directly through analysis of the patterns. Such patterns are also a useful source of intensity data, since there are fewer overlaps of reflections than found in polycrystalline ('powder') diffraction patterns. These intensities are quite reliable for structure determination, moreover.

Oblique texture electron diffraction has features superior to other electron diffraction techniques. Obviously, since the interaction of an electron with matter is very strong, much more than say for x-rays or neutrons, the sampled area for diffraction experiments can therefore be very small. It might be said, in general, that the small dimensions of coherent scattering in nonideal crystals explain why the kinematical approximation for interpreting electron diffraction intensities (hence their utility for electron diffraction structure analysis) is usually more valid than expected from ideal crystalline objects of the same dimension. The probability for coherent single scattering of the primary beam is therefore small, and multiple scattering may be neglected since their probabilities are products of correspondingly small numbers. Thus the influence of multiple scattering may not represent an appreciable fraction of the total intensity to prevent reliable structure results from being obtained.

In selected area diffraction, the diffraction pattern is formed by an object area about 1 micrometer in breadth and e. g. 200 to 500 Å thick. In such selected area experiments, deviations from ideal crystallinity (e. g. bending) in different parts of the sample contribute to the intensity of different reflections. Thus, the diffraction intensity depends not only on the scattering power of the unit cell, through the structure factor amplitude $|F(hkl)|$, but also on the effective scattering volumes contributing to individual reflections. Dark field electron microscopy reveals that these volumes can actually vary for different hkl's.

Macrocrystals studied by x-ray diffraction or neutron diffraction also experience an averaging effect over a great number of effective coherent scattering volumes so that there is generally a reliable dependence of measured intensity on the underlying structure factor magnitudes. Microcrystals studied by selected area electron diffraction, on the other hand, do not provide an good analogy to this former situation because of the different volume elements contributing to different reflections. While selected area methods might provide other advantages (due to the visualization of an undistorted reciprocal net), they can be a rather poor source of intensity data.

Oblique texture diffraction provides reliable intensities free of the restrictions mentioned for selected area techniques, since a great number of individual microcrystals contributes intensity to each reflection, thus providing an averaging of the type mentioned for x-ray or neutron studies of macrosamples. The averaging over orientations (due to a larger sampled area - e. g. 1 mm diameter) provides a favorable experimental relationship between intensity and |F(hkl)|. Measurement of these improved intensities can be improved by modern detection systems. In addition, combination with EELS allows the electron diffraction pattern to be decomposed into contributions due to defined energy losses. Isolation of elastically scattered electrons by energy filtering is especially important for providing nearly kinematical intensities. This improvement is very important for ab initio structure analysis - i .e. the solution of the crystallographic phase problem - and may also affect other factors associated with reliable derived structure parameters. Not only are the final results restricted to the determination of atomic coordinates and interatomic distances, but one can also observe directly the distribution of intracrystalline electrostatic potential and thus discern the crystal properties related to it.

4. Certain Aspects of the Solution of Electron Diffraction Problems

For the application of electron diffraction structure analysis to mineral problems, certain specific approaches to the interpretation and calculation of electron diffraction patterns are called for. Minerals are materials with mainly low symmetry (monoclinic or triclinic) so the interpretation of the diffraction patterns should be adjusted to this fact. Even though modern electron microscopes are equipped with goniometer stages, the interpretation of electron diffraction patterns for arbitrary orientations can remain problematic. Electron diffraction studies of minerals have revealed relatively simple relationships between the direct lattice and the reciprocal lattice that are given in reference 18. The connection between the two lattices is simplified by dividing the approach into two steps. Plane **ab** and repeat **c** are characterized along with the normal projection c_n to plane ab for the direct lattice. Next, the two-dimensional set of hk directions and the arrangement of hkl nodes along them are established for the reciprocal lattice. In order to compare interpretations by several authors, as well as to establish the most rational choice of unit cell, the relationships between different coordinate settings becomes very important. Contrary to the practice of many texts and even the latest edition of the *International Tables for Crystallography*, the emphasis should not be placed on the transformation of atomic coordinates. Even in the most simple Cartesian example, they are not obvious and their derivation is not very easy. Instead the transformation of unit cell vectors should be the basis for all other transformations, e. g. for coordinates in the direct lattice, unit vectors and indices in the reciprocal lattice. When the cell vectors are defined by the lattice nodes, the relationships are established in a very simple way, with coefficients as integers or rational fractions.

For example, if $\mathbf{A}_i = p_{ik}\mathbf{a}_k$ and $\mathbf{a}_i=q_{ik}\mathbf{A}_k$ (for the definition of i,k, see ref. 18) are, respectively, the direct and reverse transformation of the direct lattice unit cell vectors, then Det. p_{ik} = N and Det. q_{ik} = 1/N are volume ratios of corresponding cells. (The numbers of cells of one choice are contained in the cell of another choice.) It is known that the transformations of atomic coordinates or direct lattice vector components, X_i, $U_i = q_{ki}x_k$, u_k; x_i, $u_i = p_{ki}X_k$, U_k, are contravariant to the cell vector transformations. The same holds for the corresponding cell vectors of the reciprocal lattice: $\mathbf{A}^*_i = q_{ki}\mathbf{a}^*_k$, $\mathbf{a}^*_i = p_{ki}\mathbf{A}^*_k$. The h_i, H_i indices (coordinates in the reciprocal lattice), which are contravariant with $\mathbf{a}^*_i$, $\mathbf{A}^*_i$, are covariant with $\mathbf{a}_i$, $\mathbf{A}_i$ as x_i, X_i are covariant with $\mathbf{a}^*_i$, $\mathbf{A}^*_i$.

Such relationships can be demonstrated by a simple but important case. Let a subsidiary orthogonal axis $\mathbf{c}_o$ be normal to the **ab** plane replacing the real axis **c** which is tilted to this plane. Thus: $\mathbf{a} = \mathbf{a}_o$, $\mathbf{b} = \mathbf{b}_o$, $\mathbf{c} = \mathbf{c}_o + \mathbf{c}_n = x_n\mathbf{a}_o + y_n\mathbf{b}_o + \mathbf{c}_o$. This coordinate system is useful for modeling crystal structures, also the consideration of polytypes, also revealing equivalence or independence of structure descriptions in different settings, and, finally, for choosing the most suitable unit cell. For the reverse transformation, the matrices are defined: q_{ik} = (1 0 0 /0 1 0/x_n y_n 1); q_{ki} = (1 0 x_n/0 1 y_n/ 0 0 1). Since $\mathbf{a}_o = \mathbf{a}$, $\mathbf{b}_o = \mathbf{b}$, $\mathbf{c}_o = -x_n\mathbf{a} - y_n\mathbf{b} + \mathbf{c}$, p_{ik} = (1 0 0/ 0 1 0/ -x_n -y_n 1); p_{ki} = (1 0 -x_n/0 1 -y_n/0 0 1).

For ideal phylosilicate polytype models, x_n or $y_n = -1/3$ and an orthogonal axis $\mathbf{c}_o$ = (**a** or **b**) + 3**c** may be chosen. The diffraction patterns may thus be described with $h_o = h$, $k_o = k$, L_o = (h or k) + 3l, whereas $h = h_o$, $k = k_o$, $l = (L_o - h_o,k_o)/3$. These relationships are useful for constructing F-tables for identification of polytypic families.

5. Acknowledgement

The latest developments in the electron crystallography of minerals made in the IGEM, Russian Academy of Sciences, were supported by the Russian Foundation of Fundamental Investigations (RFFI), projects 93-05-14248 and 96-05-64983, for which the author is greatly indebted.

6. Recommended References

1. Chukhrov, F. V. (ed.) (1975) *Hypergene Iron Oxides in Geological Processes.* Nauka Press, Moscow

2. Chukhrov, F. V., Gorshkov, A. I. and Drits, V. A. (1989) *Hypergene Manganese Oxides* Nauka Press, Moscow.

3. Drits, V. A. (1987) *Electron Diffraction and High-Resolution Electron Microscopy of Mineral Structures.* Springer Verlag, Berlin.

4. Goodman, P. (ed.) (1981) *Fifty Years of Electron Diffraction.* D. Reidel Publ. Co., Dordrecht.

5. Pinsker, Z. G. (1953) *Electron Diffraction.* Butterworths, London.

6. Vainshtein, B. K. (1964) *Structure Analysis by Electron Diffraction.* Pergamon Press, Oxford.

7. Vainshtein, B. K. (1986, 1994) *Modern Crystallography*, Vol. 1, Springer Verlag, Berlin

8. Vainshtein, B. K., Zvyagin, B. B. (1993) Electron diffraction structure analysis (EDSA). *International Tables for Crystallography*, Vol. B. (U. Shmueli, ed.) Kluwer, Dordrecht, pp. 310-314.

9. Vainshtein, B. K., Zvyagin, B. B., Avilov, A. S. (1992) Electron diffraction structure analysis, In: *Electron Diffraction Techniques*, Vol. 2 (J. M. Cowley, ed.) Oxford Univ. Press, pp. 216-312.

10. Zhukhlistov, A. P. (1993) Electron diffraction techniques in the solution of mineralogical problems. Izvestia Akad. Nauk. Ser. Phys., 57(2) 16-21.

11. Zhukhlistov, A. P., Avilov, A. S., Ferraris, G., Zvyagin, B. B., Plotnikov, V. P. (1997) Evidences for a three-site random distribution of hydrogen atoms in the structure of brucite $Mg(OH)_2$ provided by electron diffractometry. Kristallografiya (Crystallography Reports) 42(2), in press.

12. Zvyagin, B. B. (1967) *Electron Diffraction Analysis of Clay Mineral Structures*. Plenum Press, NY.

13. Zvyagin, B. B. (1989) Structural characteristics of polytypes delivered by texture diffraction patterns. In: *X-ray and Neutron Structure Analysis in Material Science* (J. Hasek, ed.) Plenum Press, NY, pp. 353-360.

14. Zvyagin, B. B. (1992, 1997) Oriented texture patterns. *International Tables for Crystallography*, Vol. C. (A. J. C. Wilson, ed.) Kluwer, Dordrecht, section 4.3.5.

15. Zvyagin, B. B. (1993) Electron diffraction analysis. In: *Advanced Mineralogy* (A. S. Marfunin, ed.) Springer Verlag, Berlin, pp. 50-63.

16. Zvyagin, B. B. (1993) Oblique-texture electron diffraction in powder crystallography. *Proc of EPDIC-2* (R. Delhez and E. J. Mittemeijer, eds.) Trans. Tech. Publ, Switzerland, pp. 125-138.

17. Zvyagin, B. B. (1993) Electron diffraction analysis of minerals. MSA Bull. 23(1) 66-79.

18. Zvyagin, B. B. (1994) Electron diffraction of textures and minerals. Crystallography Reports 39(2), 239-245.

19. Zvyagin, B. B., Gorshkov, A. I. (1969) Electron microscopy and electron diffraction. In: *Techniques of Electron Microscopy of Minerals*, Nauka Press, Moscow, Chapter 6 (pp. 207-310).

20. Zvyagin, B. B., Vrublevskaya, Z. V., Zhukhlistov, A. P., Sidorenko, O. V., Soboleva, S. V., Fedotov, A. F. (1979) *High-voltage Electron Diffraction in the Study of Layer Minerals*. Nauka Press, Moscow.

21. Zvyagin, B. B., Zhukhlistov, A. P., Plotnikov, V. P. (1996) Development of electron diffractometry of minerals. In: *Structural Studies of Crystals* (Collected Works to the 75th Anniversary of the Akademician B. K. Vainshtein). Nauka Physmatlit, Moscow, pp. 225-234.

STRUCTURAL ELECTRON MICROSCOPY CHARACTERIZATION OF THE TERNARY COMPOUND S_4In_2Zn OBTAINED BY CHEMICAL TRANSPORT

D.R. ACOSTA, J.ABASOLO*, A.LOPEZ-RIVERA, M.BRICEÑO**
*IFUNAM, A.P. 20-364, 01000 México D.F., MÉXICO
**Univ. de Los Andes, Fac. de Ciencias,Mérida 5101,VENEZUELA

1. Introduction

Crystals of the ternary semiconductor S_4In_2Zn grown by chemical transport methods presents hexagonal crystallographic configuration. The structure is based upon a close packing of sulphur atoms with tetrahedral and octahedral indium atoms and tetrahedral zinc atoms located in the interstices. Some physical properties like electrical conductivity when measured along the c-axis and also perpendicular to it, are highly anisotropic: $\sigma_{\parallel c} / \sigma_{\perp c} \cong 10^{-3}$ (1). This behavior might be originated by structural differences along the mentioned directions; the presence of stacking faults, superlattices, polytypes, etc, are most of times responsible for structural variations from one to other crystallographic direction. Also there is a controversy related with the crystallographic space group associated with the hexagonal phase of S_4In_2Zn. In this work we present some results of the structural characterization after transmission electron microscopy studies were carried out; bright and dark field images, together with selected area (SAED) and convergent beam electron diffraction (CBED) patterns were used to characterize our material. Also, high resolution electron microscopy (HREM) images and multislice calculations were used in order to improve structural determinations.

2. Experimental

Samples were obtained by chemical transport method and present mica-like configurations; even when several methods were tried to strip-out very thin laminates before mounting in 400 mesh copper grids, it was not possible to obtain individual thin flakes of the sample. Observations were carried out in a Jeol 2010-FX microscope with a double tilt sample holder for CBED studies, and in a Jeol 4000-EX for high resolution characterization.

3. Results

Bright and dark field images revealed the layered configurations of S_4In_2Zn samples, and even when structural stability was observed under beam irradiation, most of times bend contourns like those observed in figure 1 and dislocations nets among other structural intrinsec defects were commonly detected. From transmission diffraction patterns several configurations were detected: perfect hexagonal lattice, stacking faults, polytypes; for all these cases, parameters corresponding to S_4In_2Zn in very close agreement with those reported were found: a = 0.378 nm and c= 3.701 nm from SAED patterns and a= 0.38 nm and c = 3.706 nm from CBED patterns. Also lattice parameters for other ternary zinc sulfide semiconductors were detected in minor quantities. These results were corroborated from HREM images analized with traditional and computer assisted measurements.

D. L. Dorset et al. (eds.), Electron Crystallography, 359–362.

CBED diffractions patterns were obtained in the [001] exact direction at room and liquid nytrogen temperature and in most of cases patterns coming from laminates conformed by several stacking layers were recorded. FOLZ and HOLZ rings togehter wiht Kikuchi bands are visible in most of images. The diffraction group 6mm was readily derived from symmetry considerations in each figure. Following the standard symmetry analysis method reported by Rackman et al (2) the 6mm point group was determined and also using the symmetry tables (2) ,(3), the space group $R\bar{3}m$ was deduced for the S_4In_2Zn crystals. Symmetry breakings in several whole CBED patterns along the c- axis were observed from many crystallites. This might come from loss in periodicity along the c-axis and in turn this can be associated with the existence of stacking faults and polytypes in this direction.

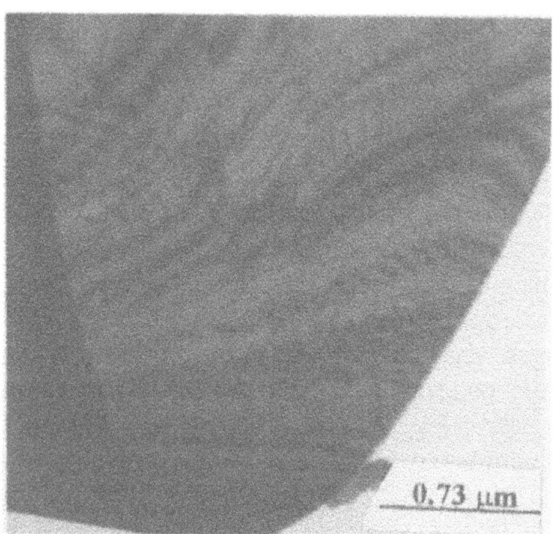

Figure 1. Bright field image of a bent laminate. Bend contourns and net of dislocation can be observed.

Figure 2. TEM difraction pattern from a perfect crystallite very close to the [001] direction.

It was not possible to record HOLZ lines from the direct central disk even when cooling sample holder was used to minimize thermal scattering effecs; this difficulty might be originated in the multiple layered configuration of the S_4In_2Zn sample. From TEM diffraction patterns and from HREM images, the existence of polytypes was confirmed. At least two diifferent polytypes have been observed; crystallographic analysis of polytypes reveals variation in cell parameters along the c-axis direction from c = 3.70 nm to c = 3.81 nm; this kind of disorder might come from the existence of numerous stacking faults and extra planes as could be observed in HREM images. An example of the polytypes observed are presented in figure 6, the sequence of fringes and fringe packets observed in several HREM images is not always completely periodic; this has been attributed to the possible existence of another crystallographic phase (4). The symmetry breaking observed in some CBED patterns might come from irregular lattice periodicities associated with structural and compositional faults.

The origiin of polytypes and modulated structures observed in S_4In_2Zn samples seems to be caused by the differences in the stacking of the layers of sulphur atoms. We have found in most of cases, polytypes with constant number N of lamnate packets, but polytypes with a different number B of laminates also have been observed and this is an evidence that the mixing of polytype forms is a consequence of the crystal growth process. At present time the density of structural faults has not been fully quantified in order to correlate directly, structural properties and electrical conductivity, for instance. In the other hand, also we have found zones of our samples with almost perfect hexagonal characteristics as derived from SAED and CBED patterns and from HREM images.

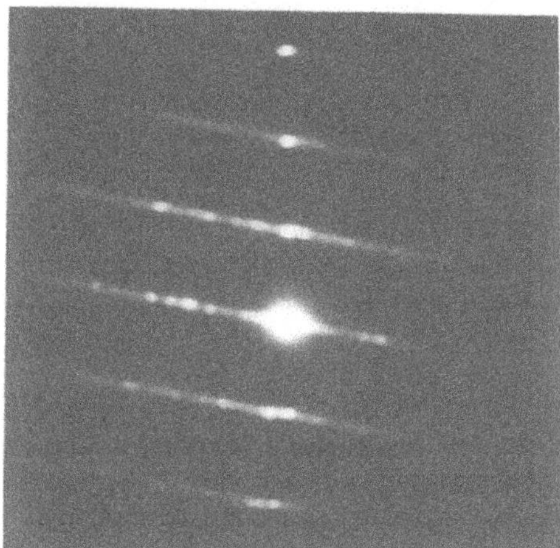

Figure 3. TEM diffraction pattern along b-axis direction. Reflections coming from modulated structures are visible.

Figure 4. TEM diffraction pattern from where the presence of polytypes and other structural defects are deduced.

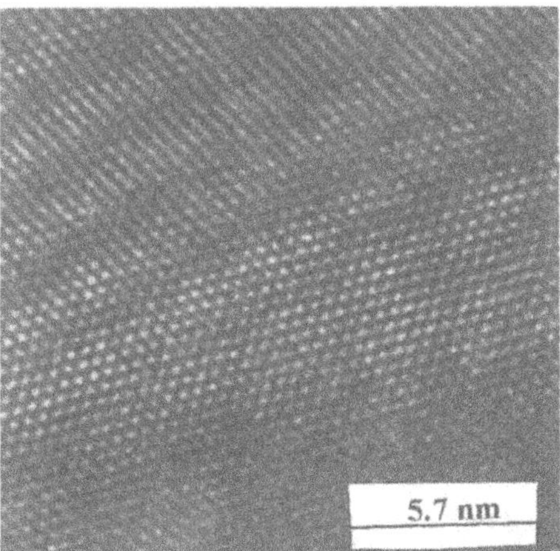

Figure 5. HREM image along the c-axis. Moiré fringes are visible together with structure without defects.

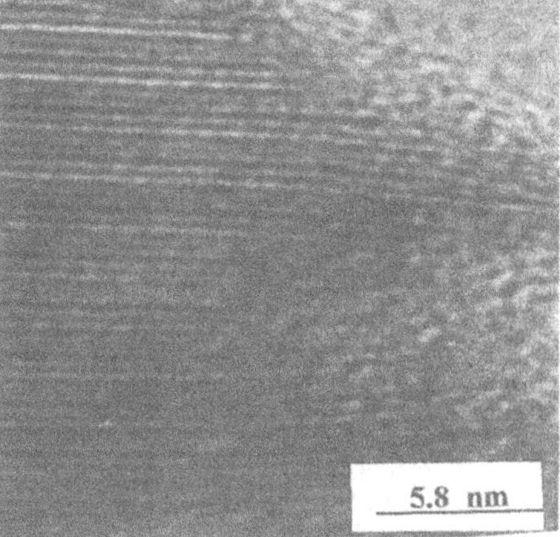

Figure 6. HREM image perpendicular to c-axis. Polytypes, bent planes and vacant lattice planes can be observed.

Figure 7. CBED pattern from a laminate of S_4In_2Zn. The 6mm diffraction group can be readily deduced from the image.

Figure 8. This is a multi slice projected potential images calculated using data reported in the literature.

4. Conclusions

The layered and faulted structure of the ternary S_4In_2Zn semiconductor has been determined from transmission electron microscopy studies. From CBED patterns analysis the space group $R\bar{3}m$ has been assigned to the hexagonal structure studied. The layered structure along the **c**- axis has to do directly with the differences observed by other authors in electrical conductivity., given that conduction electrons have to overpass higher potential barriers than those travelling along **a** and **b** directions

Acknowledgment

One of us (D:R:A) thanks the partial finnancial support from DGAPA UNAM project IN-106295 to participate in the 25th Electron Crystallography School held in Erice, Italy

5. References

1. Anagnostopoulos,A.N.Manolikas, C., Papadoulos, D.and Spyridelis,J: Phys. Status Sol. 72, (1982),731
2. Buxton,B.F., Eades, J.A., Steeds, J.W. and Rackmahn, G.M.; Phil. Trans.R.Soc. London, 281,(1976),171
3. Tanaka, M., Terauchi, M. and Kaneyama, T.; "Convergent Beam Electron Diffraction II"Edit. JEOL Ltd,, 1988
4. Radautsan,S.I., Donika,F.G., Kyosse, G.A. and Mustya, I.G.; Phys. Sta.Sol. 37,K123,(1970),

DIFFRACTION CONTRAST IN TEM IMAGES OF MODULATED SEMICONDUCTOR ALLOYS

S. P. AHRENKIEL
National Renewable Energy Laboratory
1617 Cole Blvd., Golden, Colorado 80401 U.S.A.

1. Introduction

Composition modulation is detected in transmission electron microscope (TEM) images and diffraction patterns of several technologically important semiconductor materials, including $ZnSe_{0.5}Te_{0.5}$ alloys and (AlAs)/(InAs) short-period superlattices (SPSs) [1,2]. Here we examine calculated and experimental diffraction contrast in TEM images.

2. Diffraction and Imaging

We describe the propagation of high-energy electrons in thin crystalline foils using dynamical diffraction theory [3]. We then examine image formation in the TEM.

2.1. DYNAMICAL DIFFRACTION

The wave function $\psi(\mathbf{r})$ of an high-energy electron in the sample is a solution to

$$[\nabla^2+4\pi^2(k^2+U(\mathbf{r}))]\psi(\mathbf{r})=0 \quad (1)$$

where $\mathbf{k}$ is the incident wave vector ($k=|\mathbf{k}|$), and $U(\mathbf{r})$, which is proportional to the crystal potential, is written as a Fourier sum over reciprocal-lattice vectors $\mathbf{g}$ and modulation vectors $\mathbf{g}_m$ (assuming $g_m<g$ for $g\neq 0$, $g_m\neq 0$)

$$U(\mathbf{r})=\sum_{\mathbf{g},\mathbf{g}_m} U_{\mathbf{g}+\mathbf{g}_m} e^{2\pi i(\mathbf{g}+\mathbf{g}_m)\cdot\mathbf{r}} \quad (2)$$

The solutions to (1) are modulated Bloch waves

$$\psi^{(j)}(\mathbf{r})=\sum_{\mathbf{g},\mathbf{g}_m} \psi^{(j)}_{\mathbf{g}+\mathbf{g}_m} e^{2\pi i(\mathbf{k}^{(j)}+\mathbf{g}+\mathbf{g}_m)\cdot\mathbf{r}} \quad (3)$$

Below a sample of thickness T, the beam amplitudes are (with $\mathbf{k}^{(j)}=\mathbf{k}+\gamma^{(j)}$)

$$\Psi_{\mathbf{g}+\mathbf{g}_m}=\sum_j \psi_0^{(j)*}\psi^{(j)}_{\mathbf{g}+\mathbf{g}_m} e^{2\pi i\gamma^{(j)}T} \quad (4)$$

Applying the high-energy approximation, we obtain an eigenvalue problem in terms of the excitation errors $s_{\mathbf{g}+\mathbf{g}_m}$ (with $k=|\mathbf{k}+\mathbf{g}+\mathbf{g}_m+\mathbf{s}_{\mathbf{g}+\mathbf{g}_m}|$) and extinction distances $\xi_{\mathbf{g}+\mathbf{g}_m}=k/U_{\mathbf{g}+\mathbf{g}_m}$ that is solved for the $\gamma^{(j)}$ and $\psi^{(j)}_{\mathbf{g}+\mathbf{g}_m}$ to determine the $\Psi_{\mathbf{g}+\mathbf{g}_m}$.

D. L. Dorset et al. (eds.), Electron Crystallography, 363–366.

2.2. IMAGING

The image intensity is approximately

$$I(\mathbf{r})=\left|\sum_{\mathbf{g},\mathbf{g}_m} O(\mathbf{g}+\mathbf{g}_m)\Psi_{\mathbf{g}+\mathbf{g}_m} e^{2\pi i(\mathbf{g}+\mathbf{g}_m)\cdot\mathbf{r}}\right|^2 \tag{5}$$

where $O(\mathbf{g}+\mathbf{g}_m)$ is the objective aperture masking function. The approximate bright-field (BF) and dark-field (DF) image intensity resulting from long-period modulation ($s_{\mathbf{g}_m}\approx 0$, $s_{\mathbf{g}+\mathbf{g}_m}\approx s_{\mathbf{g}}$) is calculated using six beams ($\mathbf{0}$, $\mathbf{g}$, $\pm\mathbf{g}_m$, and $\mathbf{g}\pm\mathbf{g}_m$)

$$I_{BF}(\mathbf{r})=\left|\Psi_0+\Psi_{\mathbf{g}_m} e^{2\pi i\mathbf{g}_m\cdot\mathbf{r}}+\Psi_{-\mathbf{g}_m} e^{-2\pi i\mathbf{g}_m\cdot\mathbf{r}}\right|^2, \tag{6}$$

$$I_{DF}(\mathbf{r})=\left|\Psi_{\mathbf{g}}+\Psi_{\mathbf{g}+\mathbf{g}_m} e^{2\pi i\mathbf{g}_m\cdot\mathbf{r}}+\Psi_{\mathbf{g}-\mathbf{g}_m} e^{-2\pi i\mathbf{g}_m\cdot\mathbf{r}}\right|^2$$

The diffracted beams form interference patterns that are interpreted as image contrast.

3. Results

Composition modulation in substitutional alloys produces satellites with structure factor proportional to the difference in form factor of the modulated species. In addition, bond-length differences (lattice mismatch) among alloy constituents causes strain modulation. The local Fourier component is written to lowest order as

$$U_{\mathbf{g}}(\mathbf{r})=[1-2\pi i\mathbf{g}\cdot\mathbf{R}(\mathbf{r})]U_{\mathbf{g}}+2[x(\mathbf{r})-x]\Delta U_{\mathbf{g}} \tag{7}$$

where $x(\mathbf{r})$ is the local and x the average composition, respectively, $\mathbf{R}(\mathbf{r})$ is the local displacement, $U_{\mathbf{g}}$ is the average Fourier component, and $\Delta U_{\mathbf{g}}$ (with $\Delta U_0\approx\Delta U_{\mathbf{g}}$) is the maximum potential variation due to composition modulation. The alloy functions are

$$x(\mathbf{r})=x+x_m\cos(2\pi\mathbf{g}_m\cdot\mathbf{r}), \qquad \mathbf{R}(\mathbf{r})=\mathbf{R}_m\sin(2\pi\mathbf{g}_m\cdot\mathbf{r}) \tag{8}$$

We have assumed the lattice parameter varies linearly with composition. Abbreviating

$$a=U_{\mathbf{g}}/2k, \qquad b=x_m\Delta U_{\mathbf{g}}/2k, \qquad c=\pi(\mathbf{g}\cdot\mathbf{R}_m)U_{\mathbf{g}}/2k \tag{9}$$

we write the relevant matrix elements as

$$1/2\xi_{\pm\mathbf{g}}=a, \qquad 1/2\xi_{\pm\mathbf{g}_m}=b, \tag{10}$$

$$1/2\xi_{\mathbf{g}\pm\mathbf{g}_m}=b\mp c, \qquad 1/2\xi_{-\mathbf{g}\pm\mathbf{g}_m}=b\pm c$$

Analytical solutions for the beam amplitudes can be obtained for special cases, and the 6×6 matrix problem is solved numerically in the general case. For simplicity, we assume a two-beam condition ($s_{\mathbf{g}}\approx 0$). Figs. 1 and 2 show BF/DF contrast resulting from composition modulation, while Fig. 3 shows coherent strain contrast only.

Experimental data are shown in Figs. 4 and 5. Vertical modulation is observed in $ZnSe_{0.5}Te_{0.5}$ alloys with a period λ_m of only 30Å, giving strong satellites in diffraction patterns (DPs). Strong lateral modulation with a period of 200Å has been observed in (AlAs)/(InAs) SPSs. Double periodicity is observed in (002) DF, but not in (002) BF, possibly because of the strong (004) contribution to BF.

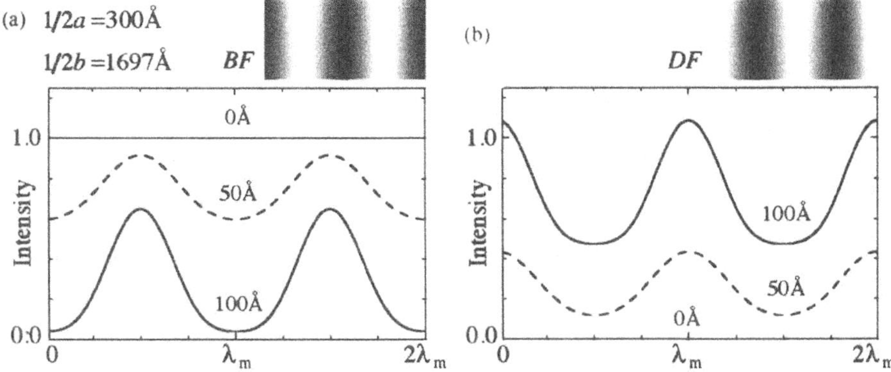

Figure 1. Image contrast for composition modulation using a strong reflection.

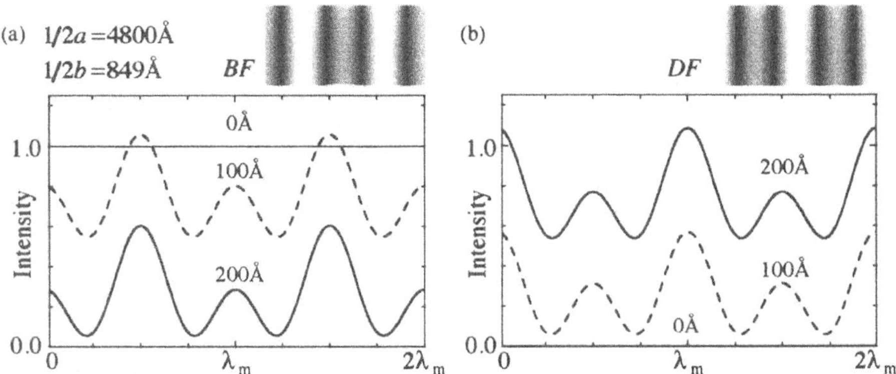

Figure 2. Image contrast for composition modulation using a weak reflection.

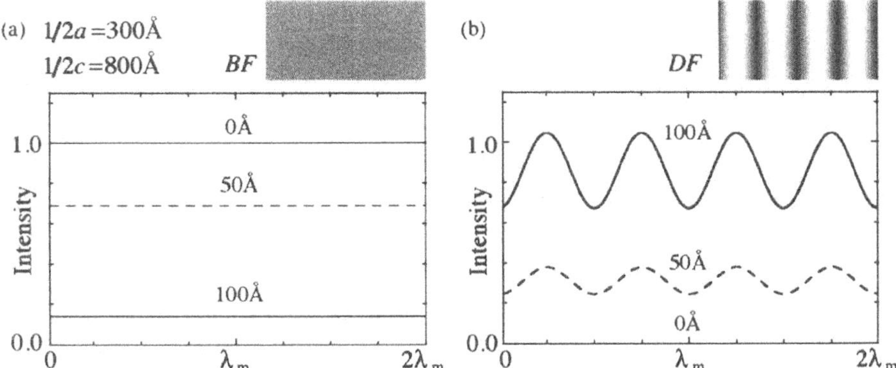

Figure 3. Image contrast for coherent strain modulation only using a strong reflection with $\mathbf{g}\cdot\mathbf{R}_m \neq 0$. *Strained regions appear dark in DF images.*

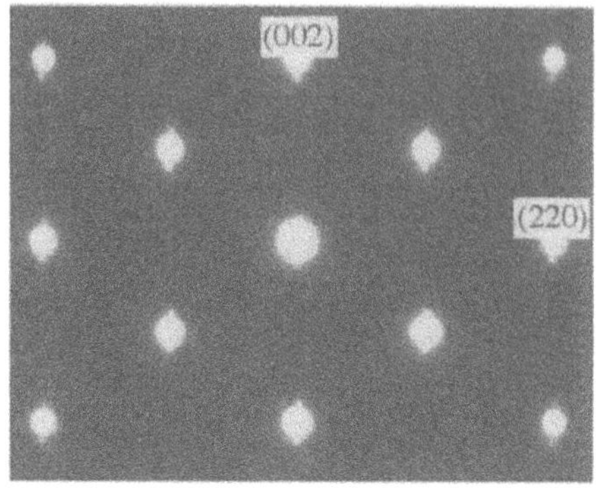

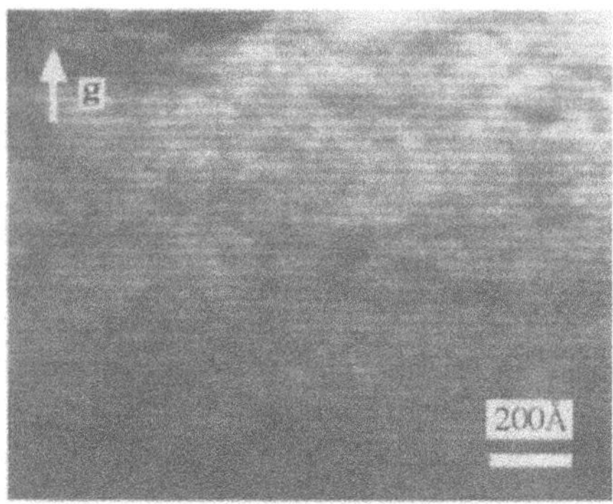

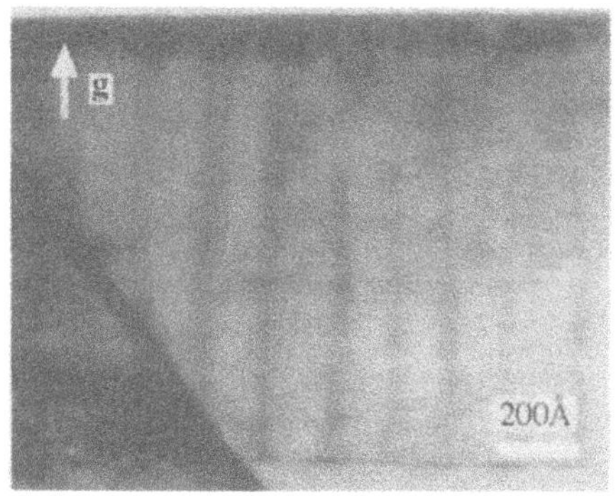

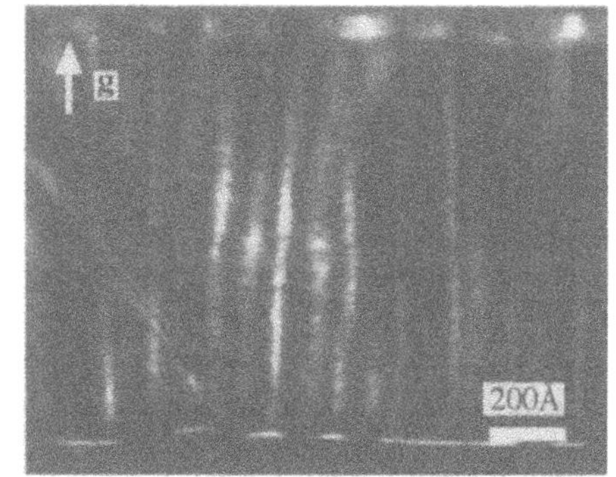

4. References

1. Ahrenkiel, S.P., Xin, S.H., Reimer, P.M., Berry, J.J., Luo, H., Short, S., Bode, M., Al-Jassim, M., Buschert, J.R. and Furdyna, J.K. (1995) Self-organized formation of compositionally modulated $ZnSe_{1-x}Te_x$ superlattices, *Phys. Rev. Lett.* **75**, 1586-1589.
2. Mirecki-Millunchik, J., Twesten, R.D., Follstaedt, D.M., Lee, S.R., Jones, E.D., Zhang, Y., Ahrenkiel, S.P., Mascarenhas, A. (1997) Lateral composition modulation in AlAs/InAs short period superlattices grown on InP(001), *Appl. Phys. Lett.* **70**, 1402-1404.
3. Reimer, L. (1984) *Transmission Electron Microscopy*, Springer-Verlag, New York.

ELECTRON CRYSTALLOGRAPHY OF A METAL AZO SALT PIGMENT

G.BOYCE, J.R.FRYER and C.J. GILMORE.
Department of Chemistry
University of Glasgow,
Glasgow G12 8QQ, Scotland, UK.

1. PIGMENTS

Pigments are coloured or fluorescent particulate solids that are insoluble in the vehicle or medium in which they are incorporated. Among organic pigments, azo compounds constitute the largest group, both with respect to the number of different structures and total production volume. As pigments retain a crystallite structure throughout the colouring process, it is essential to ascertain as much structural information as possible through determination of the crystal structure.

2. STRUCTURE DETERMINATION

In essence, structure determination from electron diffraction data is a two part process. Firstly, the size and shape of the unit cell can be determined from the geometry of the diffraction pattern. Secondly, lattice type and distribution of atoms in the structure can be determined from the relative intensities of diffraction spots.

Crystal structure analyses of these compounds has been hampered by the small crystallite size exhibited by these materials that precludes the use of standard single crystal diffraction techniques. This has led to development of methodology utilising Electron Crystallography.

3. ELECTRON MICROSCOPY

Metal Azo salt pigment under investigation has general molecular structure shown.

D. L. Dorset et al. (eds.), Electron Crystallography, 367–370.

Figure 1.

Crystals of this compound were examined using a JEOL 1200EX TEM. Electron diffraction patterns were obtained for two projections allowing unit cell parameters to be determined. Examples of the patterns are shown in micrographs 1 & 2.

Micrograph 1

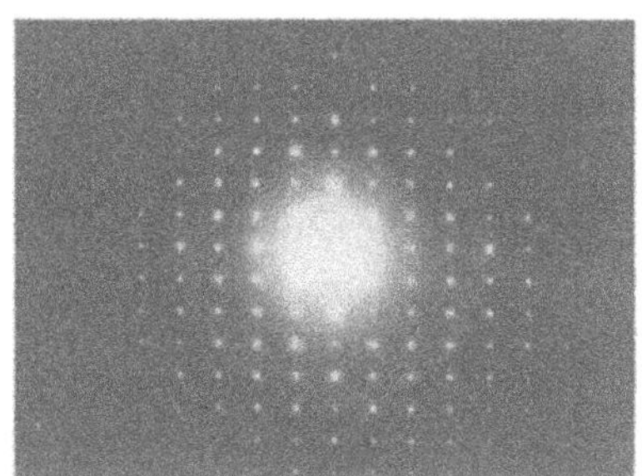

Micrograph 2

4 DATA MANIPULATION

It is possible to retrieve intensity information from a micrograph using digitisation techniques followed by CRISP [1], a crystallographic image processing program. Obtained from this are a list of indexed reflections and their corresponding intensities. As each diffraction pattern could be reproduced many times, it was decided to merge datasets before attempting structure solution.

Using the SG MERGE facility within the **maXus** [2] suite of programs for structure determination, it was possible to correctly scale data and merge equivalent reflections, thus setting data in the correct format for use within MICE [3-5]. Initial R_{init} values on merging for each dataset were,

Projection 1, $R_{init}=0.122$ *Projection 2,* $R_{init}=0.169$

5 MICE

MICE is a multisolution method for direct phase determination. The program uses experimental intensity information and calculates a series of structure factors encompassing all possible phase values [6]. This is carried out by first normalising diffraction intensities to give unitary structure factors $| U_h |^{obs}$. Next, an origin is defined by fixing the phases of suitable reflections. These phased reflections form a basis set, {H}, and are used as constraints in an entropy maximisation procedure [7] which generates the root node of a phasing tree. The properties of the maximum entropy map, $q^{ME}(x)$, produced are such that it is capable of extrapolating new phase and intensity information. At this point, the extrapolation is usually too weak to be of value and new phase information needs to be incorporated into the basis set. This gives rise to a series of phase choices and each choice is represented as a node on the second level of the phasing tree.

Nodes are ranked by a likelihood function [8] which evaluates the agreement between the extrapolated structure magnitudes from the relevant maximum entropy distribution and the experimentally measured values. The centroid maps produced can be interpreted in an analogous manner to electron density maps and from this information on areas of atomic density can be gleaned.

6 RESULTS

Unit cell parameters were determined and results agreed with those from Powder X-ray diffraction, **a=8.74Å** **b=10.92Å** **c=34.98Å** **β=94.2°**
Projection 1 - c2mm **Projection 2 - p2mm**

Unique electron diffraction intensities were normalised to give unitary structure factors, with 38 and 94 reflections for each projection respectively.

MICE calculations were carried out for each independent projection. For both, the root of the phasing tree was generated by carrying out a constrained entropy maximisation. A second level was generated and more phases permuted. This generated 128 nodes for each projection and each individual node was subjected to constrained entropy maximisation, as before. Nodes were ranked according to likelihood values and the maps generated were examined. Examples of maps for each projection are shown in figure 2 and 3. The areas of highest density correspond to the heavy metal atom that is present in these compounds. Also, the coordinates of these atoms can be determined and it has been seen that both projections agree as to the areas of high atomic density.

Figure 2.

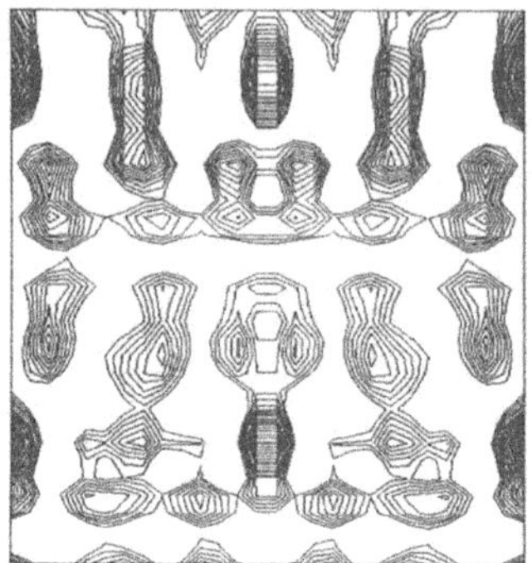

Figure 3

REFERENCES

1. Hövmoller, S. (1992) CRISP- Crystallographic image-processing on a personal computer. *Ultramicroscopy*, **41**, 121-135.
2. maXus - MAC Science Co. LTD, Yokohama, Japan and Nonius B.V., Delft, The Netherlands.
3. Bricogne, G. and Gilmore, C.J. (1990) A multisolution method of phase determination by combined maximisation of entropy and likelihood. I. Theory, algorithms and strategy, *Acta Cryst.*, **A46**, 284-297.
4. Gilmore, C.J., Bricogne, G. and Bannister, C. (1990) A multisolution method of phase determination by combined maximisation of entropy and likelihood. II Application to small molecules. *Acta Cryst.*, **A46**, 297-308.
5. Gilmore, C.J. (1996) Maximum entropy and Bayesian statistics in crystallography: a review of practical applications. *Acta Cryst.* **A52**, 561-589.
6. Voigt-Martin, I.G., Yan, D.H., Yakimansky, A., Schollmeyer, D., Gilmore, C.J. and Bricogne, G. (1995) Structure determination by Electron crystallography using both Maximum entropy and simulation approaches, *Acta Cryst.* **A51**, 849-686.
7. Bricogne, G. (1984) Maximum entropy and the foundations of direct methods, *Acta Cryst.***A40**, 410-445.
8. Bricogne, G. (1988) Maximum entropy methods in the crystallographic phase problem, in N.W. Isaacs and M.R. Taylor (eds.), *Crystallographic Computing 4: Techniques and New Technologies,* Clarendon Press, Oxford, pp. 60-79.

SYSTEMATIC STUDY OF METAL PARTICLES (Pt, Ni) CONTRAST ON AMORPHOUS SUPPORT (SILICA) USING MULTISLICE

A.L.CHUVILIN, T.E.CHESNOKOVA
Institute of Catalysis SB RAS,
av. Lavrentieva 5
Novosibirsk 90, Russia 630090

Metal particles being viewed on the background of amorphous support is the common system when studying catalysts by TEM. Usual problems one deal with studying samples of this kind are:

- particles size is of the same order of magnitude as point spread function of instrument, that causes visible particle size to depend on defocus value;
- amorphous contrast of support suppresses particle images and corrupts them drastically.

In order to study the influence of imaging conditions and support background on particles images we made image calculations of silica supported Pt and Ni particles. Variety of system compositions and imaging conditions were as follows:

- imaging conditions
  - JEM-100CX, JEM-2010 and JEM-4000EX were simulated;

TABLE 1. Parameters of instruments been used

Instrument	JEM-100CX	JEM-2010	JEM-4000EX
HT	100 kV	200 kV	400 kV
Cs	2.9 mm	0.5 mm	1.0 mm
Resolution point-to-point	4 Å	1.97 Å	1.67 Å
Scherzer focus	-1230 Å	-410 Å	-460 Å

  - beam convergence 0 - $3.5*10^{-3}$rad;
- system composition
  - silica thickness: 30, 60, 90, 120 and 150Å
  - particles sizes: 13, 25, 68 135 and 288 atoms
  - particle composition: Pt and Ni;
- particle orientation varies from [111] to [110]
- particle structure: fcc, "ikosahedral", "dekahedral".

Set of model images allow one to investigate contrast behaviour in limits defined. We will highlight here some features that seem to be interesting for practice.

D. L. Dorset et al. (eds.), Electron Crystallography, 371–374.

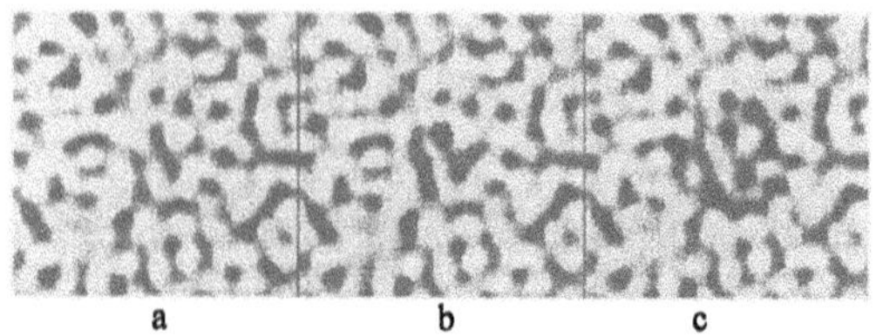

Figure 1. Calculated images of Pt particle on silica support, 100kV, Scherzer defocus. Size of each image is 45x45Å^2. Silica thickness is 150 Å. Particle sizes: a) 7Å, b) 12Å, c) 21Å.

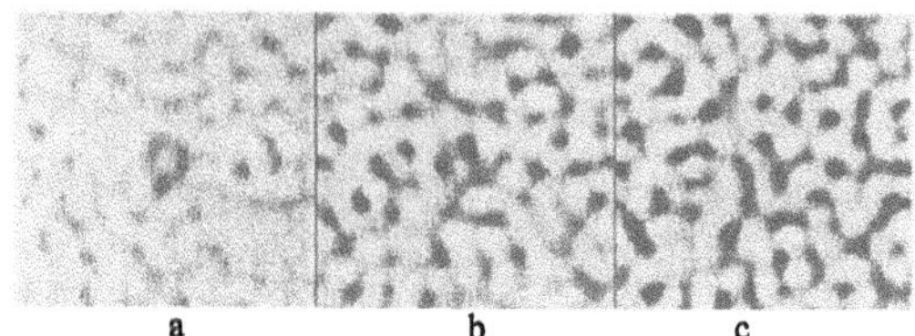

Figure 2. Calculated images of Pt particle on silica support, 100kV, Scherzer defocus. Particle size 10Å. Silica thickness: a) 30Å, b) 90Å, c) 150Å.

Figures 1 and 2 demonstrate that at 100kV (at 4Å resolution) smallest visible particle size is mainly determined by underlying support thickness. In this case visible "internal structure" on the image of a particle does not reflect structure of a particle but this is a noise from support. At these imaging conditions contrast of the particles is mainly "Z contrast" due to high atomic number of Pt. Figure 3(a, b, c) demonstrate that diffraction contrast (due to diffracted beams cut-off by diaphragm) of 12Å particles have

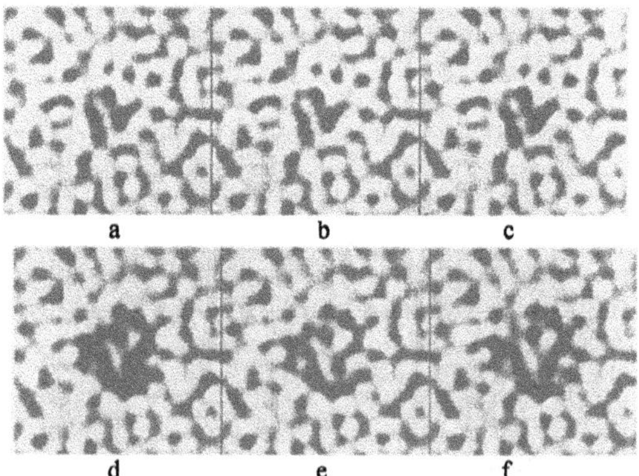

Figure 3. Calculated images of Pt particles on silica support, 100kV, Scherzer defocus. Silica thickness is 150 Å. Particle size: a), b), c) 12Å, d), e), f) 21Å. Particles orientation: a), d) [111] zone, c), f) [110] zone, b), e) intermediate orientation.

undetectable value. At the other hand for larger particle (Figure 3(d, e, f)) contribution of diffraction contrast became significant.

At higher accelerating voltages (and resolution about 2Å) contrast of particles is either the "phase contrast" of lattice planes (Figure 4(c, f)) or diffraction contrast (Figure 4(a, d)). Therefore "visibility" of particles is determined by particle orientation.

Particle as large as 21Å can be just invisible at 400kV when being in "nonreflecting" orientation (Figure 4e), remaining poor contrast been due to cut-off of pair of {220} reflexes. Angle of misorientation at which lattice is still visible depends on particle size and is about 15° for 12Å particle and about 10° for 21Å particle.

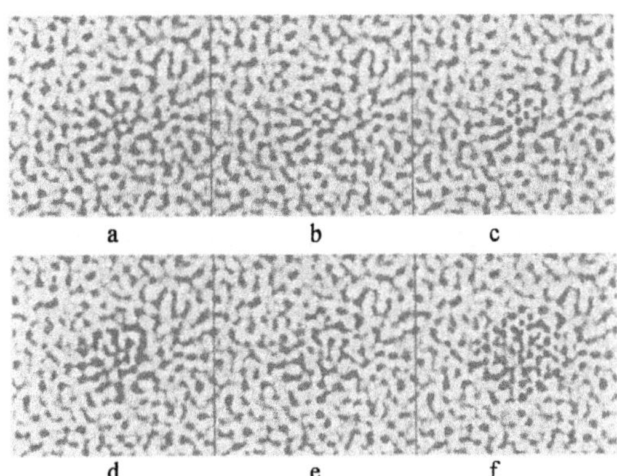

Figure 4. Calculated images of Pt particles on silica support, 400kV, Scherzer defocus. Silica thickness is 150 Å. Particle size: a), b), c) 12Å, d), e), f) 21Å. Particles orientation: a), d) [111] zone, c), f) [110] zone, b), e) intermediate orientation.

As one may expect contrast is lower for lower atomic number. Figure 5 shows comparison of images of Pt and Ni particles for different accelerating voltages. Low kV conditions seems to be preferable for light particles observation.

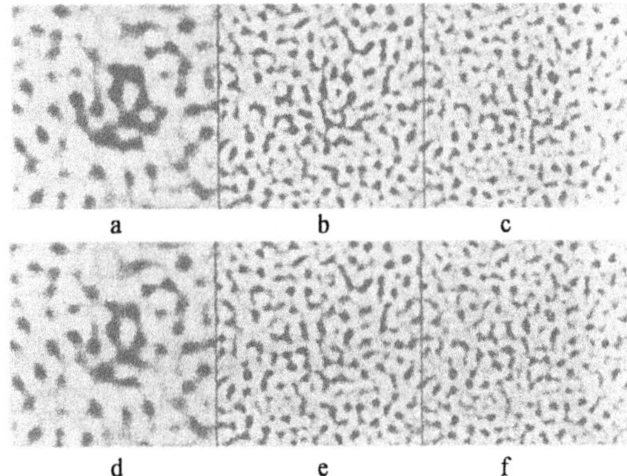

Figure 5. Calculated images of Pt and Ni particles on silica support, Scherzer defocus. Silica thickness is 90Å. Particle size is 21Å. Accelerating voltages: a), d) 100kV, b), e) 200kV, c), f) 400kV.

Influence of beam convergence on the contrast of particles was studied intensively. At Scherzer defocus increase of convergence leads only to little decrease of contrast. At over- and underfocus conditions increase of convergence selectively suppress spatial frequencies and blure the images. Not all spatial frequencies are suppressed equally and conditions can be chosen at which background noise of support goes down and lattice fringes of a particle enhance. Figure 6 illustrate the case.

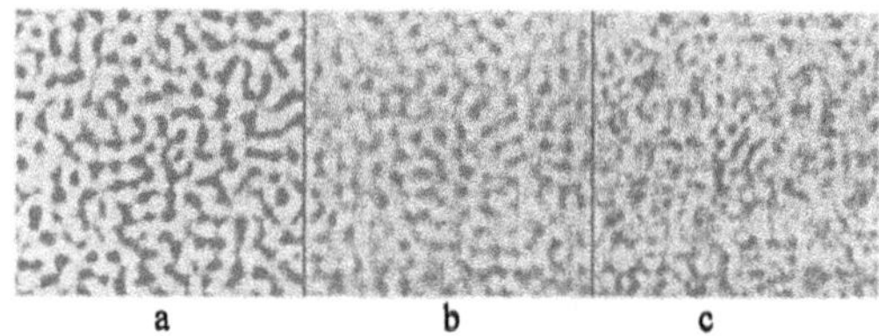

Figure 6. Calculated images of Pt on silica support, 200kV. Silica thickness is 150Å. Particle size is 10Å. Imaging conditions: a) df=-400Å, conv=$0.5*10^{-3}$rad, b) df=-200Å, conv=$3.0*10^{-3}$rad, c) df=-600Å, conv=$3.0*10^{-3}$rad.

We also studied the possibility to detect "ikosahedral" structure of small particles on support. Model was build as an ikosahedr consisting of 6 layers of Pt atoms, with Pt-Pt shortest interatomic distance being the same as in fcc bulk metal. Figure 7(a, b, c) illustrates that all specific features of low resolution images (including "butterfly contrast" at 2-fold orientation) are lost because of support noise. At conditions allowing lattice resolution specific 2-fold, 3-fold and 10-fold contrast is still visible on 150Å support at Scherzer defocus.

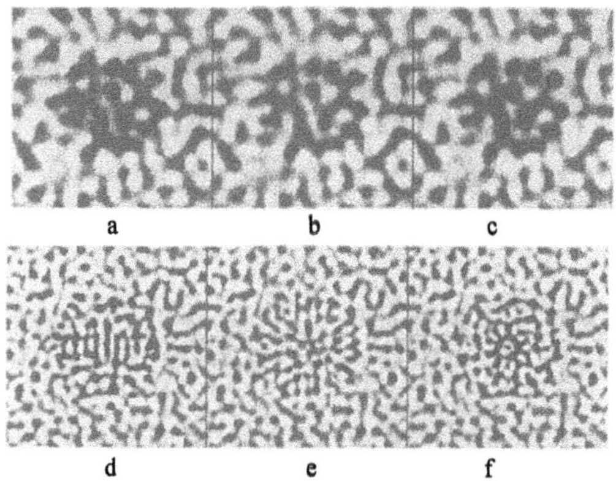

Figure 7. Calculated images of Pt ikosahedral particle on silica support, Scherzer defocus. Silica thickness is 150Å. Accelerating voltages: a), b), c) 100kV, d), e), f) 200kV. Parallel to beam is: a), d) 2-fold axis, b), e) 3-fold axis, c), f) 5-fold axis.

Total set of about one thousand images allows one to study other particular features of contrast and search for optimal experimental conditions to observe small particles on amorphous support.

This work was supported by RFBR grant No.93-03-4836.

FIRST STEPS IN THE STRUCTURE DETERMINATION OF AN OXYCARBONATE SUPERCONDUCTOR FROM ELECTRON DIFFRACTION INTENSITIES

E. GAUTIER, D. TRANQUI, C. CHAILLOUT
Laboratoire de Cristallographie, CNRS, BP166, 38042 Grenoble cedex 9, France

1. Introduction

In 1994, a new high temperature superconducting family was discovered in the Ba-Ca-Cu-O system [1-3]. Subsequently, the presence of carbon forming CO_3 groups was detected by high resolution electron microscopy [4], and confirmed from EELS measurements [5]. The general formula $Cu_{1-y}C_yBa_2Ca_{n-1}Cu_nO_x$ was thus proposed for this family. The samples are prepared at high temperature (≈ 1000°C) and under high pressure (≈ 8GPa) generated in a belt-type apparatus. Members with n=3 to 6 have been observed by electron diffraction, but only the n=3 and n=4 members are obtained as bulk phases.

These compounds present a real interest since their superconducting transition temperatures are among the highest (117K for n=4 and 120K for n=3 at optimal doping) and since they do not contain any poisonous element such as Bi, Tl, or Hg.

In view of understanding the clues for having such high Tc's, it is very important to compare the detailed structures of different superconducting families. Structural determinations either from conventional x-ray [6] or powder neutron [7] diffraction have been reported for the n=4 member. However, only the average structure was obtained. In particular, the superstructure observed on electron diffraction patterns and corresponding to a doubling of both a and c axes was not taken into account. We also collected powder x-ray synchrotron data on a n=3 sample but no significant improvement in the structure determination was achieved. The main difficulties are due to the presence of small amounts of different impurities, some of them being not identified. Furthermore, the amount of sample synthesized in one experiment is very small (≈70mg), which implies to mix several batches not necessarily strictly identical for a neutron experiment. This is why we have used electron microscopy including diffraction data to get informations about the local structure of our samples. Our first results are reported below.

2. Experimental

High resolution electron microscopy images as well as electron diffraction patterns were obtained on a Philips CM300 super-twin microscope operated at 300kV (Cs=1.2mm), and recorded on photographic plates (Kodak SO 163). The diffraction patterns were digitised using a 8-bits CCD camera. Different exposure times were used for the same

D. L. Dorset et al. (eds.), Electron Crystallography, 375–378.

pattern in order to stay within the linear range of the film and of the camera. The diffracted spots were integrated using the IDL software. Image simulations were carried out using the EMS package [8].
Diffraction patterns for a n= 3 sample were collected either in the Selected Area Diffraction mode or in the microdiffraction one which was found to give better results. This is mainly due to the fact that the data come from a zone of the sample which is very small (50-100 Å) and, consequently, of nearly constant thickness. On the other hand, a disavadvantage consists of the possible excitation of high order Laue zones (HOLZ). We selected a thin area of the sample, close to the edge, to reduce the dynamical effects. We collected patterns of different zone axes tilting around [100], [010], and [110] axes of the same crystallite.

3. Results and discussion

3.1. STRUCTURAL MODEL

Electron diffraction patterns taken along the [010] zone axis can be indexed using a cell of lattice parameters a=b≈3.85 Å, and c≈ 8.5 +3.2 (n-1) Å. However, superstructure reflexions indicate that the real cell is given by 2axbx2c (*Figure 1.*). Patterns taken along the [001] zone axis with no reflexion at h/2 k/2 0 confirm that the a and b axes are not doubled simultaneously, but that the crystal present microdomains rotated by 90° with respect to each other (*Figure 2.*). These domains are clearly visible on the HREM image taken along the <100> zone axis (*Figure 3.*). Different cations are imaged as bright dots and most of them can be identified by comparison with simulated images based on the model proposed from the neutron or x-ray results. In one domain, a superstructure (2a x 2c) consisting of very large bright dots is observed. Such a contrast has already been reported and attributed to the presence of carbonate groups [9]. The superstructure can thus be associated with an ordering of Cu and CO_3 groups in the basal plane of the structure. A scheme of the structure is given in *Figure 4.*

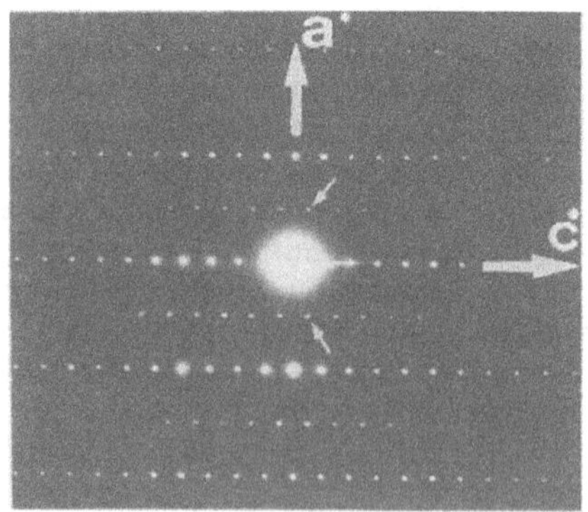

Figure 1. Electron diffraction pattern taken along the [010] zone axis (n=3). Superstructure reflexions are indicated by a small arrow

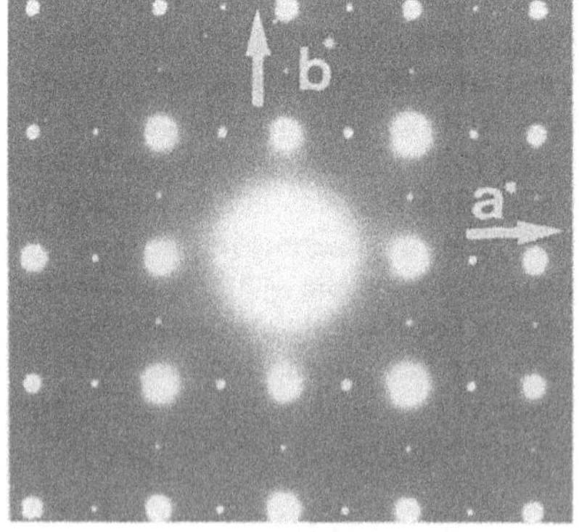

Figure 2. Electron diffraction pattern taken along the [001] zone axis (n=3)

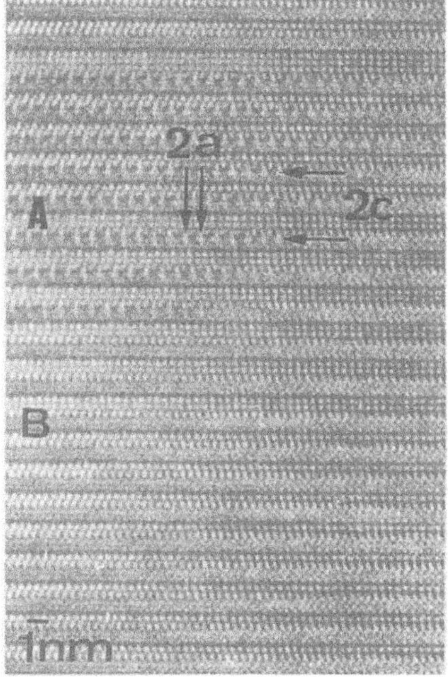

Figure 3. HREM image of an n=3 crystallite taken along the <100> zone axis. The superstructure is only visible in one domain (A) ([010] orientation), the other domain (B) being orientated along [100].

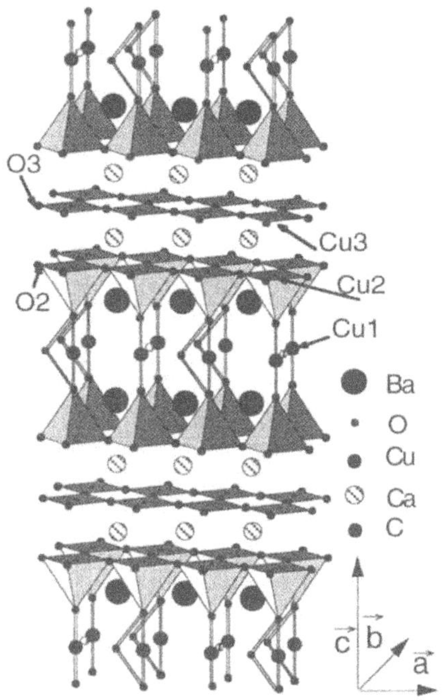

Figure 4. Scheme of the structure of $C_{0.5}Cu_{0.5}Ba_2Ca_2Cu_3O_{8+x}$

3.2. FIRST RESULTS IN STRUCTURE ANALYSIS

From the [001] zone axis, eleven other zone axes were recorded by tilting around [100], [010], and [110]. The superstructure reflexions were not taken into account in this preliminary study. Equivalent reflexions based on the average tetragonal structure of a given zone axis were averaged. Large variations of some reflexions, (2,0,0) for example, measured along different directions were observed. They are due to the strong dynamical effects which have usually the tendency to decrease the diffracted intensities. Accordingly, as a crude approximation, we decided to select the highest value obtained among the different ones. We ended up with 48 independent reflexions. We did not take into consideration the curvature of the Ewald sphere which remained very small as we did not go very far in the reciprocal space, as well as the thickness effect.
A Patterson map was done from the set of the 48 intensities. Positions of five highest peaks were found compatible with the structure obtained from synchrotron x-ray diffraction (Table 1). This enabled us to localize the heaviest atoms, namely the Ba, Ca and the Cu from the CuO_2 planes (*Figure 4.*). We then tried to refine the structure using the SHELXL93 [10] program modified for electron diffraction, but we could not get convergence.

TABLE 1. Patterson search results to compare with the synchrotron powder x-ray diffraction ones

Interatomic vectors from RX-model				obtained from Patterson maps			
x	y	z		x	y	z	peak height
0	0	0.2926	Ba-Ba	0	0	0.2907	379
0	0	0.2809	Cu1-Cu2				
0	0	0.2420	Ba-Ca				
0	0	0.2191	Cu2-Cu3				
0.5	0.5	0	O2-O2	0.5	0.5	0	288
0.5	0.5	0	O3-O3				
0	0	0.4655	Ba-Ca	0	0	0.5	238
0	0	0.5	Cu1-Cu3				
0.5	0.5	0.3309	Ca-Cu2	0.5	0.5	0.34	230
0.5	0.5	0.3537	Ba-Cu3				
0.5	0.5	0.3883	Ca-Cu1				
0.5	0.5	0.5729	Ba-Cu2	0.5	0.5	0.5	176
0.5	0.5	0.6118	Ca-Cu1				

Our first results have shown that, even in the case of a complicated structure with several cations species, it is possible to get reliable informations from a simple Patterson approach. However, to go further in our analysis, more accurate and higher resolution of observed data are needed along with appropriate correction procedures of thickness and dynamical effects.

References

1. Alario-Franco, M.A., Chaillout, C., Capponi, J.J., Tholence, J.L., Soutetie, B. (1994) A new HTSC family: the copper analogs of the single-layer Hg or Tl copper oxide superconductors, *Physica C* **222**, 52- 56
2. Ihara, H., Tokiwa, K., Ozawa, H., Hirabayashi, M., Negishi, A., Natahuta, H., Song, Y.S. (1994) New High-Tc Superconductor Family of Cu-Based $Cu_{1-x}C_xBa_2Ca_{n-1}Cu_nO_{2n+4-\delta}$ with Tc>116K, *Jpn. J. Appl. Phys.* **33**, L503-L506
3. Jin, C.Q., Adachi, S., Wu, X.J.,Yamauchi, H., Tanaka, S. (1994), 117K superconductivity in the Ba-Ca-Cu-O system, *Physica C* **223**, 238-242
4. Kawashima, T., Matsui, Y. , Takayama-Muromachi, E. (1994) New oxycarbonate superconductors $(Cu_{0.5}C_{0.5})Ba_2Ca_{n-1}Cu_nO_{2n+3}$ (n=3,4) prepared at high pressure, *Physica C* **224**, 69-74
5. Alario-Franco, M.A., Bordet, P., Capponi, J.J., Chaillout, C., Chenavas, J., Fournier, T., Marezio, M., Souletie, B., Sulpice, A., Tholence, J.L., Colliex, C., Argoud, R., Baldonedo, J.L., Gorius M.F., Perroux, M. (1994) The superconducting "copper/carbonate cuprates" An electron microscopy study, *Physica C* **231**, 103-108
6. Akimoto, J., Oosawa, Y., Tokiwa, K., Hirabayashi, M., Ihara, H. (1995), Crystal structure analysis of $Cu_{0.6}Ba_2Ca_3Cu_4O_{10.8}$ by single-crystal X-ray diffraction method, *Physica C* **242**, 360-364
7. Shimakawa, Y., Jorgensen, J.D., Hinks, D.G., Shaked, H., Hitterman, R.L., Izumi, F., Kawashima, T. , Takayama-Muromachi , E., Kamiyama, T. (1995) Crystal structure of $(Cu,C)Ba_2Ca_3Cu_4O_{11+\delta}$ (Tc=117K) by neutron powder diffraction analysis, *Phys. Rev. B* **50**, 16008-16014
8. Stadelmann, P. A. (1987), EMS-A Software Package for Electron Diffraction Analysis and HREM Image Simulation in Material Science,*Ultramicroscopy* **21**, 131-146
9. Hervieu, M., Boullay, P., Domengès, B., Maignan, A., Raveau, B. (1993), The Oxycarbonate $Y_{1.6}Ca_{0.4}Ba_4Cu_5CO_3O_{11}$, n=2 Member of the "123"-Type Derivatives $(Y_{1-x}Ca_x)_nBa_{2n}Cu_{3n-1}CO_3O_{7n-3}$, *J. Solid State Chem.* **105**, 300-304
10. Sheldrick, G.M. (1993). SHELXL93, program for the refinement of crystal structures, University of Göttingen, Germany

EFFECTS OF LOCAL CRYSTALLOGRAPHY ON STRESS-INDUCED VOIDING IN PASSIVATED COPPER INTERCONNECTS*

R. R. KELLER
National Institute of Standards and Technology
Materials Reliability Division, 853
325 Broadway
Boulder, CO 80303, U. S. A.

J. A. NUCCI
Cornell University
School of Electrical Engineering, Phillips Hall
Ithaca, NY 14853, U. S. A.

Abstract

We have measured local variations in microtexture and grain boundary misorientation in narrow, passivated copper interconnects, using electron backscatter diffraction. This allowed us to differentiate between the local crystallography associated with voids and that associated with regions remaining intact during thermal treatment, all within the same lines. In general, grain boundaries intersecting voids exhibited structures that provided more favorable kinetic pathways for atom migration than those boundaries present within intact regions of the same line. Specifically, grains near voids showed a locally weaker <111> texture than those in unvoided regions. Boundaries between such grains were more likely to be of twist character and of higher angle character than those in unvoided regions. Such boundaries, when of tilt character, were more likely to have misorientation axes parallel to the film plane. Local variations in crystallography are shown to play an important role in determining interconnect reliability.

1. Introduction

We show in this paper the effects of local variations in thin film crystallography on interconnect reliability. In particular, we consider the crystallographic aspects of microstructural features that are of nearly the same dimensions as interconnect thicknesses and widths, or approximately 1 μm. Such features are believed to play a significant role in determining the susceptibility of metal interconnects to stress-induced voiding and electromigration [1, 2]. Average grain size and texture [3] have been shown to have important effects. Namely, larger grains and stronger fiber texture generally lead to improved reliability. However, localized variations in such microstructural features may be intuitively expected to affect performance more than average structures since interconnect dimensions are now small enough that single grains and grain boundaries often traverse the whole line width or thickness. An example of the importance of local variations was demonstrated by Rodbell et al. [1], where aluminum lines of stronger av-

D. L. Dorset et al. (eds.), Electron Crystallography, 379–382.

erage texture exhibited poorer reliability than those of weaker average texture. Electron backscatter diffraction (EBSD) measurements indicated that voids formed near grains that were oriented significantly away from the <111> texture in the stronger average texture lines. It was suggested that grain-to-grain misorientation differences played a strong role in that behavior. Measurements of local texture variations were also made in copper lines subjected to stress voiding conditions [4, 5]. Results similar to those in reference [1] were obtained, but with greater confidence, since a much larger data set was collected. Namely, grains adjacent to voids showed a wider distribution of orientations about the overall <111> line texture than those in unvoided regions. We discuss in this paper the importance of local crystallography in assessing the reliability of copper lines, relying on the results described in references [5, 6].

2. Experimental

2.1. INTERCONNECT FABRICATION

The specimens consisted of copper lines, 0.75 to 2.0 μm in width, and 0.5 μm in height, deposited by electron-beam evaporation onto thermally-oxidized silicon substrates. A 50 nm layer of tantalum was used as both an adhesion layer and a diffusion barrier, encasing the copper lines. The overlying passivation layer consisted of 1.2 μm of SiO_2. Samples were annealed in vacuum for 1 h at 400°C after passivation, to induce thermal stresses upon cooldown to room temperature. The passivation layer was removed by reactive ion etching in order to expose the bare copper surfaces for subsequent examination.

2.2. ELECTRON BACKSCATTER DIFFRACTION ANALYSIS

We used an EBSD system in a scanning electron microscope (SEM) to collect crystallographic information from the surfaces of the copper lines. EBSD patterns were collected with each specimen at a 70.5° tilt towards a low-light, silicon-intensified target camera. The SEM was operated at 30 kV, with probe currents in the range 0.25 to 0.75 nA. For the capture of each pattern, we averaged 64 frames and performed a flat field correction by normalizing the raw pattern to an image containing no crystallographic information. The sampling area of the copper was of approximate diameter 0.2 μm, which enabled analysis of linewidths as small as 0.75 μm. The data collection scheme consisted of placing the electron beam into arrays of positions on the copper lines, and capturing the diffraction patterns. We distinguished between beam positions immediately adjacent to stress voids and those within regions of the specimens that remained intact after thermal processing, as shown in figure 1. A total of 94 diffraction patterns were collected near voids, and 132 patterns were collected from intact regions. The data were analyzed in terms of (111) pole figures and misorientation angle distributions, also called Mackenzie plots.

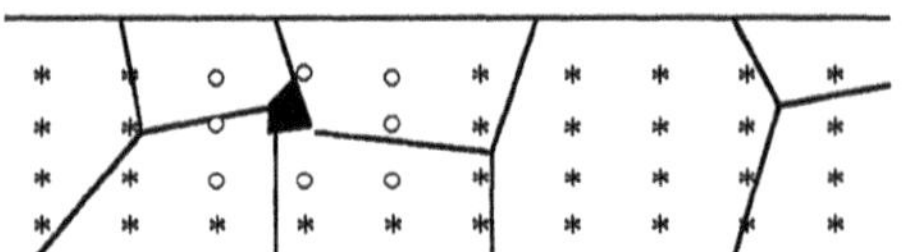

Figure 1. Electron beam positions. Circles represent positions adjacent to voids. Stars are positions in intact regions.

3. Results and Discussion

3.1. LOCAL TEXTURES AND GRAIN BOUNDARY STRUCTURES

Pole plot distributions calculated from orientation distribution functions revealed texture strength differences between data from voided and unvoided regions. Grain orientations near voids showed an approximately 25% weaker texture than those in unvoided regions. Vapor-deposited metal films often develop a columnar grain structure, where grain boundary planes are inclined nearly perpendicular to the film plane. In such cases, a perfect texture necessarily requires that all grain boundaries be tilt boundaries, with misorientation axes lying normal to the film plane. We refer to such boundaries as tilt-A boundaries. A schematic of such a boundary is shown in figure 2(a). Deviation from perfect textures leads to two other types of boundaries, twist and tilt-B. Twist boundaries are those with misorientation axis lying in the film plane and normal to the boundary plane, as shown in figure 2(b). Tilt-B boundaries are those with misorientation axis lying in the film plane, and in the boundary plane, as seen in figure 2(c). These boundary structures determine local diffusivities within the film plane. The tilt-A structure allows atoms to migrate easily via grain boundary diffusion through the boundary, in a direction parallel to the film. The twist and tilt-B structures allow rapid diffusion through grain boundaries, in the film plane, with the tilt-B structure being the most efficient. In terms of void growth, the twist and tilt-B boundaries are expected to lead to more rapid growth than the tilt-A boundary. The net effect of local texture variations is to create a distribution of grain boundary structures, each with a different diffusivity. A further effect is due to misorientation angle. Grain boundaries intersecting voids showed, on the average, a higher misorientation angle than those in unvoided regions. This result can be interpreted in terms of the total number of dislocations within a boundary. Higher angles suggest more dislocations, and therefore greater diffusivities, in general. The characteristic (high diffusivity) boundary associated with voids is then one of twist or tilt-B character, combined with high misorientation angle.

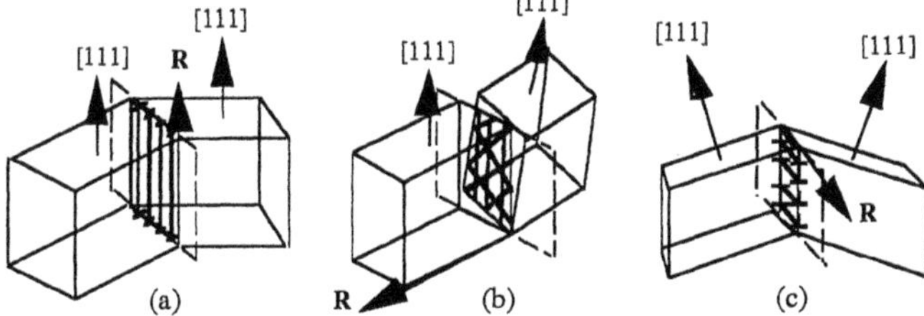

Figure 2. Grain boundaries, with misorientation axes **R** in a thin film. Film plane lies normal to **R** of left figure for each case. Cube edges represent lattice axes. (a) tilt-A boundary, (b) twist boundary, (c) tilt-B boundary.

3.2. STRESS VOIDING CONTRASTED WITH ELECTROMIGRATION

The most favorable configuration for rapid stress void growth at a triple junction is a structure with high diffusivity in many in-plane directions. An example is a triple junction consisting of three twist or tilt-B boundaries. Conversely, the most favorable configuration for rapid electromigration void growth is a structure that creates an atomic flux divergence during electron flow. An example is a triple junction consisting of the intersection of tilt-A and non-tilt-A boundaries, with the tilt-A boundaries positioned "upstream" with respect to electron flow. A line microstructure favorable for stress void growth might not necessarily be favorable for electromigration void growth. Figure 3 illustrates schematically the requirements for growth in these two cases. The local grain boundary crystallography will be the determining factor for rapid void growth. Real microstructures consist of mixtures of tilt and twist boundaries, so distinction between the favorable cases for stress voiding and electromigration will not be quite as clear.

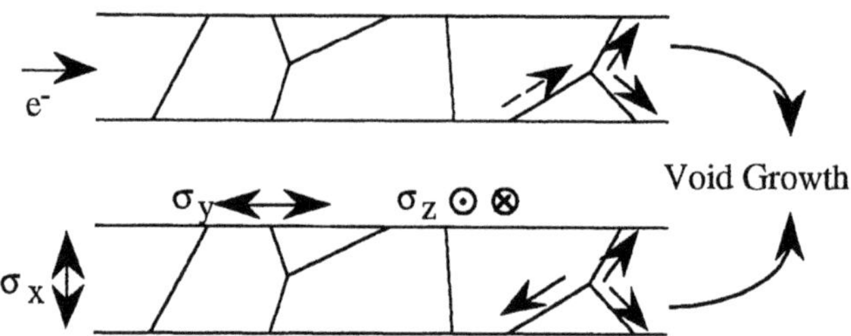

Figure 3. Boundary structures favorable for void growth. Top: electromigration with both low- and high-diffusivity boundaries, amidst high density electron flow. Bottom: stress voiding with three high diffusivity boundaries amidst triaxial tensile stresses.

4. Summary

We have used electron backscatter diffraction to investigate effects of local variations in the crystallography and texture of thin film lines on stress voiding. Voids formed in regions of locally weaker texture. The effects were attributed to variability in grain boundary structures and diffusivities. The variation in boundary structure has different effects for stress voiding as compared to electromigration.

5. References

1. Rodbell, K. P., Hurd, J. L., and DeHaven, P. W. (1996) Blanket and Local Crystallographic Texture Determination in Layered Al Metallization, in H. J. Frost, M. A. Parker, C. A. Ross, and E. A. Holm (eds.), *Polycrystalline Thin Films: Structure, Texture, Properties and Applications II,* Mater. Res. Soc. Symp. Proc. Vol. 403, Mater. Res. Soc., Pittsburgh, pp. 617-626.
2. Sanchez, Jr., J. E., Randle, V., Kraft, O., and Arzt, E. (1992) Morphology and crystallography of electromigration induced transgranular slit failures in aluminum alloy interconnects, *Submicrometer Metallization: The Challenges, Opportunities, and Limitations,* SPIE Vol. 1805, SPIE, Bellingham, WA, pp. 222-231.
3. Knorr, D. B. and Rodbell, K. P. (1993) Effect of Line Width on Electromigration of Textured Pure Al Films, in K. P. Rodbell, W. F. Filter, H. J. Frost, and P. S. Ho (eds.), *Materials Reliability in Microelectronics III,* Mater. Res. Soc. Symp. Proc. Vol. 309, Mater. Res. Soc., Pittsburgh, pp. 345-350.
4. Field, D. P., Nucci, J. A., and Keller, R. R. (1996) Interconnect Failure Dependence on Crystallographic Structure, *Proc. 22nd Int'l Symp. for Testing and Failure Analysis,* ASM International, Metals Park, OH, pp. 351-355.
5. Nucci, J. A., Keller, R. R., Sanchez, Jr., J. E., and Shacham-Diamand, Y. (1996) Local crystallographic texture and voiding in passivated copper interconnects, *Appl. Phys. Lett.* **69,** 4017-4019.
6. Nucci, J. A., Keller, R. R., Field, D. P., and Shacham-Diamand, Y. (1997) Grain boundary misorientation angles and stress-induced voiding in oxide passivated copper lines, *Appl. Phys. Lett.* **70,** 1242-1244.

HOW TO DETERMINE RELIABLE INTENSITIES USING FILM METHODS?

H. KOTHE AND U. KOLB
Institut für Physikalische Chemie, Johannes-Gutenberg Universität Mainz, Jakob Welder Weg 11, 55099 Mainz

Abstract

The basis of a successful structure analysis is to obtain reliable intensities. In the field of electron crystallography intensity data can be collected on-line with a CCD-camera as well as off-line using image plates or film material. Whereas image plates are read out with a laser, film material is analysed via a densitometer or digitised using a CCD-camera or a scanner. Both, CCD-camera and scanner, uses CCD-technology and we show that both systems can be used for intensity evaluation. In order to obtain reliable intensities from film media it is important to define and calibrate the experimental conditions, the digitization process and the evaluation of intensity data exactly. A high optical resolution and a high optical range are necessary for a good evaluation.
For comparable CCD-chips the results obtained are similar. Due to the fast development in computer technology, the systems used here are no longer comparable to the high end products of today with 16 bit optical density and 3000 by 3000 dpi optical resolution but one can expect that both systems lead to similar results. Scanners are more easy to handle and so we prefer a high end transmission-scanner for intensity evaluation in the future.

1. Experimental conditions

In order to avoid the use of saturated peaks and to increase the range of resolution by analysing even very weak reflections at higher angles, exposure series for each zone must be taken. Based on a preliminary exposure series for one zone, the shortest exposure time should deliver the strongest peak not saturated, while for the longest exposure time the weak peaks should not disappear in the background. To control the development conditions (time and quality of developer) it is useful to take a calibration picture containing areas of different exposure times as shown in Fig. 1.

Figure 1 Calibration picture for development conditions

2. Digitisation

As scan tools a CCD-camera and a scanner were compared and optimal scan conditions for both were developed. For the CCD-camera the intensity of the light box, the aperture and the distance should be chosen in such a way that the background is not cut off and maximum magnification is reached.

D. L. Dorset et al. (eds.), Electron Crystallography, 383–387.

2.1. CORRECTION FOR NON-LINEARITIES

It is important to overcome the non-linearity of the film and the CCD-sensors first. This can be performed by using a photographic calibration strip (21 steps). First of all one needs to define a range on the calibration strip, which includes the dynamical range on the negatives. The chosen range should not cut off the background of the diffraction pattern. The strongest peak of the shortest exposed negative must be analyzed properly. Therefore the background of the image and the greylevel of the strongest beam of the diffraction pattern can be used for defining a reasonable range on the calibration strip. The scanning conditions have to be chosen so that one step which is brighter than the background can be recognised on the calibration strip. This correction can be performed in the program package CRISP[1].
The digitisation process is different for CCD-camera and scanner (Fig.2). For both systems the calibration strip and the negatives have to be digitised under the same conditions. For the CCD-camera each negative was digitized first and the CCD-correction was applied in CRISP using the data obtained from the calibration strip in a second step. Using the scanner the information on the calibration strip was analysed via CRISP and used as a tonecurve for the scanner. The tonecurve obtained was applied automatically during the scanning and no further correction was necessary after digitisation.

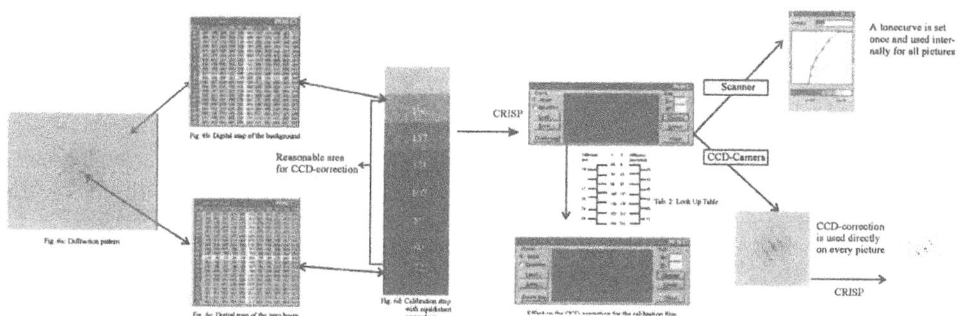

Figure 2 Scheme of CCD-correction and digitisation process

2.2. NOISE OF CCD-CAMERA AND SCANNER

A problem of the CCD-technique is the optical noise of the CCD-chip, which leads to slight differences between two scans. In comparison to the AGFA-scanner, the Sony-CCD-camera exhibited a higher noise level. In the case of electron diffraction patterns, in which all the information is contained in relatively few pixel even a small change of the pixel contents causes a big effect. In order to decrease this error, which is especially a problem of darker regions on the negative, one should digitise the same negative several times to decrease optical noise.
In order to lower the noise we recommend scanning a diffraction pattern five times and adding the data up directly in CRISP. Modern scanners and CCD-cameras provide such a statistical option.

2.3. GREYLEVELS FROM CCD-CAMERA AND SCANNER

To compare the data from CCD-camera and scanner the greylevel of a calibration film (20steps) with exposure times between 0.5 and 640 seconds were measured. For the CCD-camera three different apertures were used whereas for the scanner with 10 bit internal D_{min} and D_{max} were chosen in an optimal range. The resulting curves exhibited different shapes according to the different scanning conditions. For a better comparison all curves were normalised on 256 greylevels using formula 1.

$$\text{new greylevel} = 255 - 255 * \frac{(\text{greylevel of the unexposed film} - \text{greylevel of the exposed film})}{(\text{greylevel of the unexposed film} - \text{greylevel of the longest exposure film})}$$

(1)

After this correction the same results were found for the CCD-camera and the scanner.

2.4. COMPARISON OF DIFFERENT SYSTEMS

Compared to the microdensitometers, the CCD-technique, which includes the CCD-cameras and scanners, is a comparatively cheap technique. The analysis of a diffraction pattern can be performed in one digitisation process. CCD-cameras use a static matrix CCD sensor, whereas scanners contain a lamp, which is moved at a constant speed over the scanning area. A mirror system leads the transmitted light to a linear CCD-sensor. We compared for our work one Sony-CCD-camera and an AGFA-Flatbed scanner with transmission unit (Table 1) with comparable technical specifications. A 12 bit PCO CCD-camera was also tested, but the intensity evaluation is not ready yet because there are still problems with utilising 12 bit files in ELD.

Name	Agfa StudioScanII	Sony	PCO SensiCam
type	Desktop Flachbett scanner	CCD-camera	CCD-camera
optical resolution	400 dpi horizontal* 800 dpi vertical	752*572 Pixel	640 * 480 Pixel
dynamical range internal external	 10 Bit 8 Bit	 8 Bit	 12 Bit
tonecurve	variable (use of the internal 10 Bit)	gamma=1 or gamma=0,45	gamma=1 or gamma=0,45
scan area	Din A4 (30*20 cm)	variable	variable
options	- control via software - batch scan possible		- control via software - statistical scan modus

Table 1Technical data for used systems

3. Evaluation of intensity data

As an example the peak intensities of an exposure series were evaluated using the CRISP module ELD [2,3]. The diffraction patterns were indexed and the axes were refined. The intensities were analysed by shape fitting (integration is also possible), where the parameters of a Gaussian curve are fitted to the peak in 4 cycles.

For our investigations the NLO-active compound, 2,6-bis(4-dimethylamino-benzylidene)cyclohexanone (DMABC) was used. An exposure series was taken for the [010]-zone of DMABC. Each picture was digitised 5 times and analysed. Symmetrically equivalent reflections were merged and a mean value was calculated. Figs. 3a and 3b show the intensities obtained as a function of the exposure time for CCD-camera and scanner.

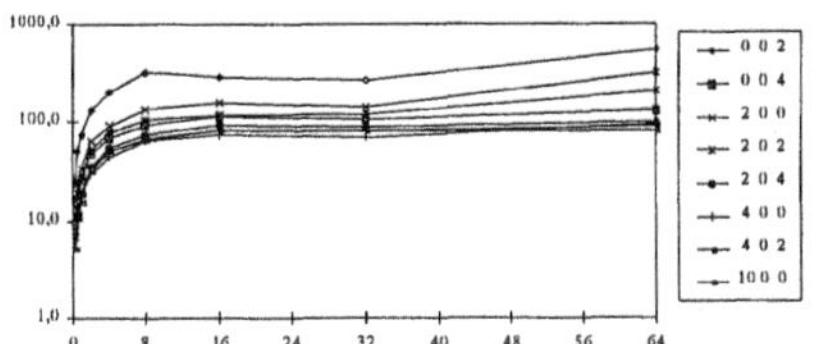

Figure 3a Obtained intensities from CCD-camera

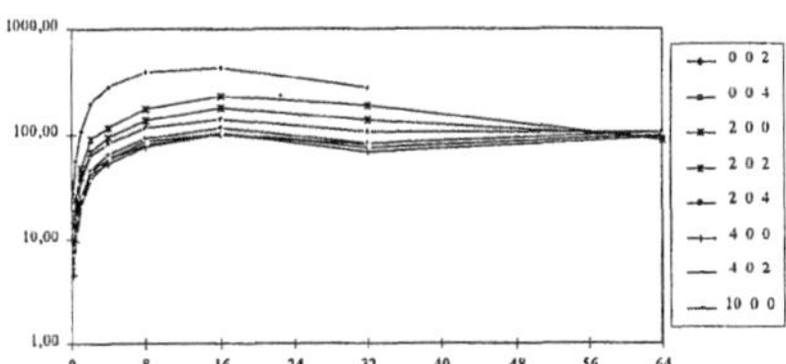

Figure 3b Obtained intensities from scanner

For our recording conditions values up to 16 seconds exposure time turned out to be reasonable.
An analysis of the (10 0 0) reflection as a function of exposure time did not exhibit a big difference in the observed intensity, but the estimated intensities show significantly different values especially at long exposure times.
A cut through the (10 0 0) reflection for different exposure times (Fig. 4a) and a cut through the (10 0 0) reflection for each digitisation process for a long exposure time (Fig. 4b) does not show significant changes in the peak form. If the peak is in the linear range, the estimated and observed intensities are reasonable whereas the analysis of saturated reflections causes considerable problems with the shape fitting in ELD. This clearly shows that the intensity evaluation of nearly saturated peaks in ELD does not work correctly. An advanced peak analysis of ELD is available in the new Version of the CRISP packages and will be tested in the near future.

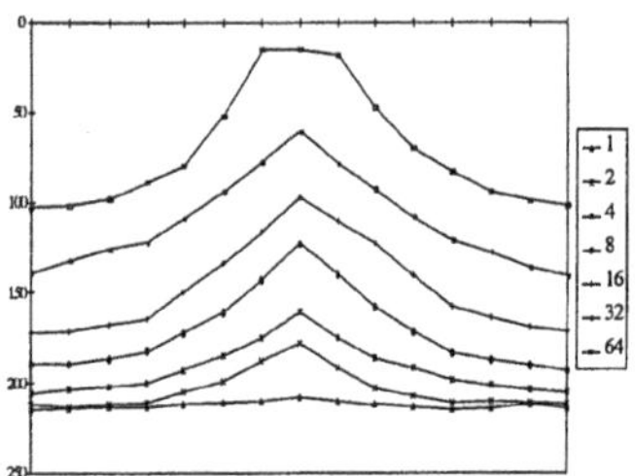

Figure 4a Cut through the (10 0 0) reflection for different exposure times

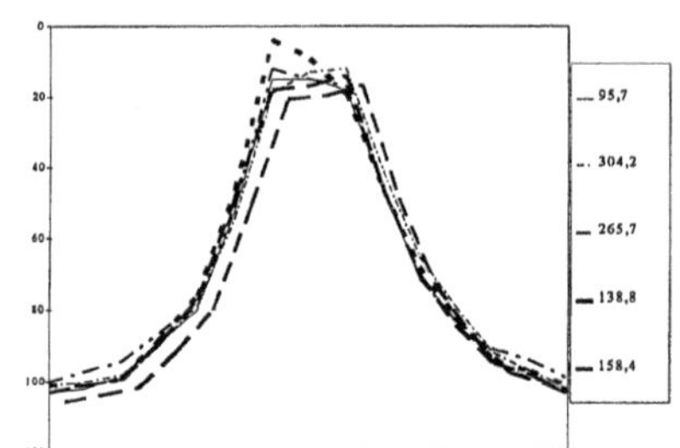

Figure 4b Cut through the (10 0 0) reflection for a long exposure times

To normalise the exposure series of a zone to one data set it is necessary to eliminate non reliable intensities beforehand. The ratio between the intensities of two different exposure times must correspond to the correct mathematical function. To determine the relationship between the intensities of different exposure times for each reflection, the highest intensity in the linear range is used and multiplied by the relation factor I(2t)/I(t) to normalise data for one exposure time. Figs. 5a and Fig. 5b show the result for the CCD-camera and the scanner. Calculation of R-value between scanner and CCD-camera data gives 9,9%.

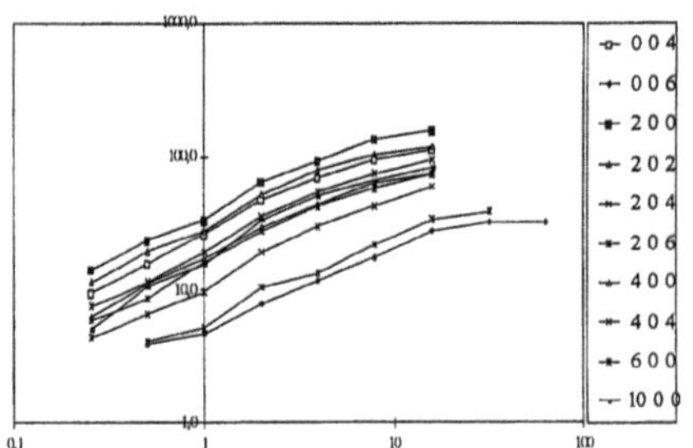

Figure 5a Extrapolation for the CCD-camera

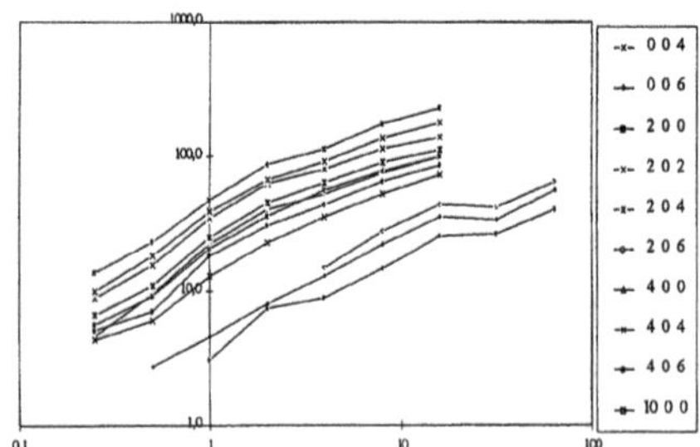

Figure 5b Extrapolation for the scanner

This procedure must be repeated for each zone. The next step in getting a 3-d dataset is to normalize the different zones. For this step you can use some formula [3]

4. Conclusion

Both CCD-camera and scanner can be used to digitise electron diffraction patterns. Both methods give similar results, but scanners are easier to handle. To digitise negatives with the CCD-technique, a high dynamical range and a high optical resolution is necessary. Recently progress in the CCD-technology has made 12 bit or 16 bit CCD-sensors available at reasonable cost. Also the optical resolution of the CCD-sensors is getting better, so that today a point resolution of 7 μ is possible. The combination of a higher optical resolution and a higher optical range should help to get a better intensity starting set.

References

1. Hovmöller, S.(1992) CRISP: crystallographic image processing on a personal computer, Ultramicroscopy **41**, 121-135
2. Zou, Z., Sukharev, Y., Hovmöller, S. (1993) ELD - a computer program system for extracting intensities from electron diffraction patterns, Ultramicroscopy **49**, 147-158
3. Zou, Z., Sukharev, Y., Hovmöller, S. (1993) Quantitative electron diffraction - new features in the program system ELD, Ultramicroscopy **52**, 436-444

SURFACE STRUCTURES SOLVED BY DIRECT METHODS

E. LANDREE*, C. COLLAZO-DAVILA*, D. GROZEA*, E. BENGU*,
L. D. MARKS* AND C. J. GILMORE†
*Department. of Materials Science and Engineering
Northwestern University, Evanston, IL 60208 USA
†Department of Chemistry
University of Glasgow
Glasgow G12 8QQ, Scotland, UK

1. Introduction

For years Direct Methods (DM) have proven to be an invaluable tool for determining bulk crystal structures from diffraction data, and it would now appear that the same can be said for surface structures as well [1]. Surface structures can be determined from surface transmission electron (or X-ray) diffraction data by Maximum Entropy [2] or Minimum Relative Entropy [3] methods. The Minimum Entropy method is combined with a Genetic Algorithm [4,5,6] (GA) global search routine for determining possible solutions. Competition between exploration and optimization is strongly dependent upon how the GA probes the solution space, relying on factors such as mutation rate, population size, number of generated children and the form of the Figure of Merit. Atomic positions are generated using a heavy-atom holography [7] algorithm which includes minor refinement. Solutions are then discriminated based upon physical or chemical considerations, and final atomic positions refined using χ^2 minimization.

This approach has already solved two-dimensional test models, as well as known structures such as the Si(111)-5x2-Au, the Si(111)-√3x√3-Au [1], and the Si(111)-7x7 [8] as a means of confirming its usefulness for solving real surface crystal structures. Several unknown structures have also been solved including the Si(111)-4x1-In [9] Si(111)-6x6-Au [10], and TiO_2(100)-1x3 [11].

1. Diffraction from Surfaces

Two techniques sensitive to surface diffraction are Grazing Incidence X-ray Diffraction (GIXD) and off-zone Transmission Electron Diffraction (TED). There is strong evidence to suggest that TED from surfaces can be treated as a single scattering phenomenon [12] (kinematical) and therefore can be analyzed by DM. From surface diffraction data, one can measure h,k and intensities. The measured reflections are then

D. L. Dorset et al. (eds.), Electron Crystallography, 389–392.

treated identically by DM, with the only difference being whether one uses X-ray or electron scattering factors.

2. Genetic Algorithm

One of the greatest difficulties in solving the "Phasing Problem" is determining a good set of starting phases for reconstructing the missing phase information for the remaining reflections. We have implemented a GA search routine to search for the ideal set of initial phases which utilizes an artificial natural selection for optimization based upon a Figure of Merit (FOM) calculated for each initial set of phases.

Since surface diffraction data is inherently noisy and at times incomplete, it is not possible to formulate a single FOM and expect the solution with the lowest FOM to be correct. Instead of desiring a single global optimization routine, the GA has been tailored to seek all possible minima that occupy the solution space. The subset of possible solutions is then discriminated based upon chemistry, structure and χ^2 analysis. A small FOM should be a necessary condition for the correct solution, but may not be a sufficient one.

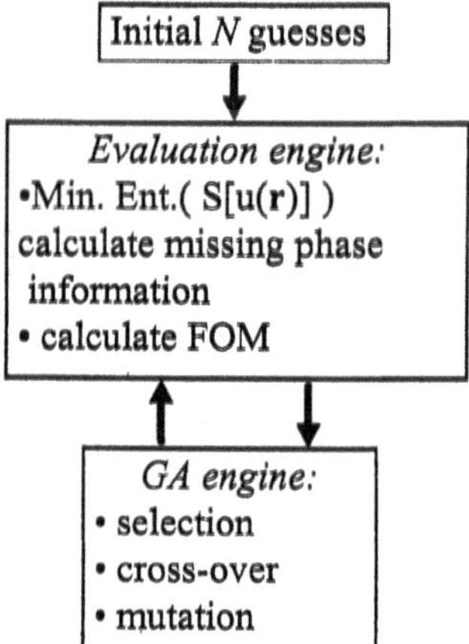

Figure 1. Abbreviated flow diagram illustrating how the Minimum Entropy and GA search solution space of a given starting set of *N* possible starting solutions

The GA is initialized by first guessing the phase for a subset of the total number of measured reflection consisting of the strongest reflections (usually 5-15% of the total number of reflections). This is repeated *N* times, resulting in a collection of *N* possible starting solutions; referred to as a population. For each possible solutions the phase of the remaining reflections are generated using a Minimum Entropy approach, and a corresponding FOM is calculated.

Using the value of the FOM as a likelihood of survival, the GA selects two favorable "parent" solutions, switches a random number of phases between them (cross-over), generating two new "children" solutions and introduces a degree of mutation is which randomly changes the value of some of the phases. By this process a new population of plausible solutions are generated. This population is then evaluated using the Minimum Entropy algorithm, and the entire process is iterated for a given number of cycles.

3. Minimum Entropy

Using windowed unitary structure factors to account for a non-infinite number of reflections, U(**h**), the actual process of replacing the missing phase information is treated as a "Sharpening Function." The measured intensities and h,k's (positions) are Fourier transformed into real space with the available phase information, u(**r**).

$$S[u(\mathbf{r})]=u(\mathbf{r})\ln\{\langle u(\mathbf{r})\rangle\} \tag{1}$$

Application of the function S[u(**r**)] modifies the real space electron potential/density map and generates the missing phase information. By analogy the Sayre equation [14], which uses a convolution in reciprocal space to replace the missing phase information, is identical to squaring the electron potential/density maps in real space. After the sharpening function, the FOM is calculated and the process is iterated so long as the FOM continues to decrease.

In essence, the application of the Minimum Entropy is actively minimizing the noise in the real space maps, favoring "peaks" which would correspond to potential atom position.

3.1 CALCULATED FIGURE OF MERIT

The FOM is calculated using a modified form of the Karle FOM (equation 2) [13] $U_{m-1}(\mathbf{h})$ is the (known) amplitude and (variable) phase of the structure factor at the beginning of each cycle; $U_m(\mathbf{h})$ is the amplitude and new (variable) phase calculated at the end of the cycle; α is a scalar chosen to minimize R; summed over all reflections excluding the 0,0 reflection.

$$\mathrm{FOM} = R^n = \frac{\Sigma' \left| U_{m-1}(\mathbf{h}) - \alpha U_m(\mathbf{h}) \right|^n}{\Sigma' \left| U_{m-1}(\mathbf{h}) \right|^n} \tag{2}$$

Typically errors were not used, which improved the method in all cases. In addition, n=1 proved to be a more robust fitting parameter able to account for outlying data points.

4. Discussion

In general GAs implicitly search for relationships between individual variables (schemata) , which in the ideal case for DM which seeks to find relationships between the individual phases. It has also proven itself at successfully producing a starting set of atom positions from which χ^2 analysis can be performed for numerous surface structures.

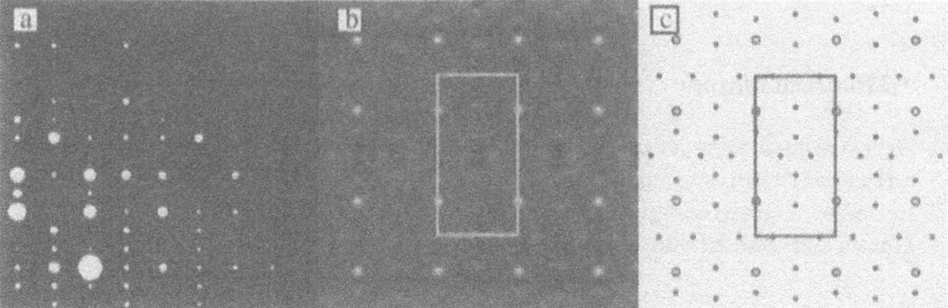

Figure 2. (a) Schematic of surface diffraction data for an In and Si 2x4 surface reconstruction with intensities that overlap bulk reflections or surface 1x1 reflections absent. Radius of the discs are proportional to the structure factor; area is proportional to the intensity (b) The generated direct phasing map and (c) the true solution a with 2x4 unit cell showing the indium (gray) and silicon (black) atom positions.

5. References

1. Marks, L.D. Plass, R. and Dorset, D.L. (1996) Imaging surface structures by direct phasing, *Surface Review and Letters*, **4**, 1-8.
2. Gilmore, C.J. (1996) Maximum entropy and bayesian statistics in crystallography: a review of practical applications, *Acta. Crystallographica* A **52** 561-589.
3. Marks, L.D and Landree, E. (1997) A minimum entropy algorithm for surface phasing problems, *Acta. Crystallographica* A Submitted.
4. Gutowski, M.W. (1994) Smooth genetic algorithm, *J. Phys. A: Math. Gen.* **27** 7893-7903.
5. Goldberg, D. E. (1989). *Genetic Algorithms in Search, Optimization, and Machine Language*, Addison-Wesley Publishing Company, Inc.
6. Landree, E and Marks, L.D. (1997) A multi-solution genetic algorithm approach to surface structure determination using direct methods, *Acta. Crystallographica* B. Submitted.
7. Marks, L.D. and Plass, R. (1995) Atomic structure of the Si(111)-(5x2)-Au from high resolution electron microscopy and heavy-atom holography, *Physical Review Letters* **75** 2172-2175.
8. Gilmore, C.J., Marks, L.D., Grozea, D., Collazo, C., Landree, E. and Twesten, R.D. (1997) Direct solution of the Si(111) 7x7 structure, *Surface Science* In Press.
9. Collazo-Davila, C., Marks, L.D., Nishii, K. and Tanishiro, Y. (1997) Atomic structure of the In on Si(111) (4x1) surface, *Surface Review and Letters* **4** 65-70.
10. Marks, L.D., Grozea, D., Feidenhans'l, R., Nielsen, M. and Johnson, R.L. (1997) Au 6x6 on Si(111): evidence for a 2D pseudoglass, *Physical Review and Letters* Submitted.
11. Landree, E., Marks, L.D., Zschack, P. and Gilmore, C.J. (1997) Structure of the TiO_2(100)-1x3 surface by direct methods, *Surface Science* Submitted.
12. Tanishiro, Y. and Takayanagi, K. (1989). Validity of the kinematical approximation in transmission electron diffraction for the analysis of surface structures, *Ultramicroscopy* **27** 1-8.
13. Woolfson, M. M. (1987) Direct methods - from birth to maturity, *Acta Crystallographica* A **43**, 593-612.
14. Woolfson M. M. & Hai-fu, F. (1995). Physical and non-physical methods of Solving Crystal Structures. Cambridge University Press.

MODULATED STRUCTURE DETERMINATION OF Pb-1212 AND Pb-1223 BY ELECTRON CRYSTALLOGRAPHIC IMAGE PROCESSING

J. LIU[1], X. J. WU[2], F. H. LI[1*], Z. H. WAN[1], H. F. FAN[1], T. TAMURA[2], K. TANABE[2]

[1] Institute of Physics & Center for Condensed Matter Physics, Chinese Academy of Sciences, Beijing 100080, P. R. China

[2] Superconductivity Research Laboratory, International Superconductivity Technology Center, Tokyo 135, Japan

(Pb, Sr, Cu)Sr_2(Ca, Sr)Cu_2O_y (Pb-1212) and (Pb, Sr, Cu)Sr_2(Ca, Sr)$_2Cu_3O_y$ (Pb-1223) are high-T_c superconducting compounds having commensurate and incommensurate modulated structure, respectively[1]. Modulated structures are characterized by the periodic deviation of the atomic position and/or the occupation from the average value. The ratio of modulation wavelength to the periodicity of basic structure is an integer for commensurate modulated structures, i.e. superstructures, while for incommensurate modulated structures is an irrational number. Incommensurate modulated structure can be described as a three-dimensional section of a high-dimensional periodic structure[2]. In the present paper, it is reported that both commensurate modulated structure of Pb-1212 and incommensurate modulated structure of Pb-1223 were determined by using the two-stage image processing technique [3] which consists of image deconvolution and phase extension.

1. Experimental

Specimens for TEM observation were prepared by crushing. Two series of [010] electron diffraction patterns (EDPs) were taken from areas as thin as possible of the two compounds, respectively, with an H-9000NA electron microscope at the accelerating voltage 300 kV. The corresponding high-resolution electron microscope image of the two phases were obtained by using a JEM-4000EX electron microscope operated under 400 kV.

The negatives of the EDPs were scanned by using a Perkin-Elmer PDS microdensitometer with a raster size of 30 μm × 30 μm. The program PLANE [4] was

* The corresponding author.

[1] Supported by the National Nature Science Foundation of China.

[2] Supported by NEDO.

D. L. Dorset et al. (eds.), Electron Crystallography, 393–396.

used for calculating the integral diffraction intensities. The high-resolution electron micrographs were digitized by using a CCD camera.

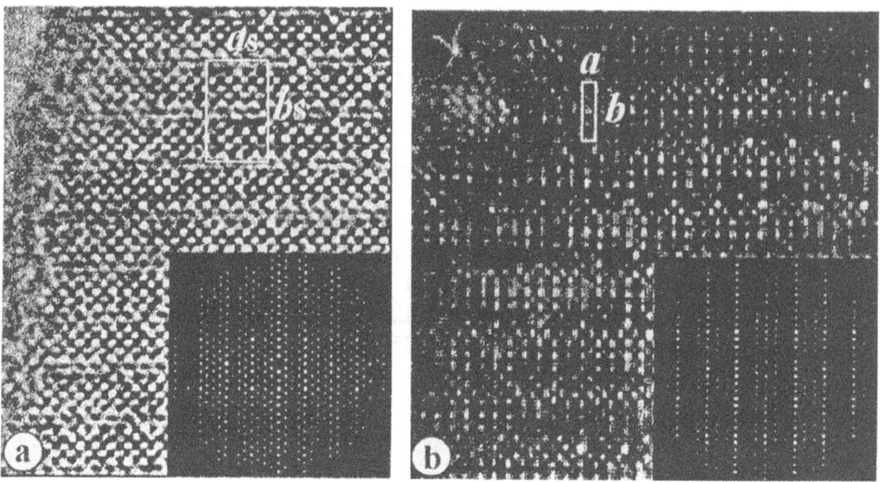

Fig. 1 Digitized experimental [010] high-resolution electron microscope images of (a) Pb-1212 and (b) Pb-1223. Insets are corresponding EDPs.

2. Commensurate modulated structure of Pb-1212

The basic structure of Pb-1212 belongs to the orthorhombic system with the lattice parameters a = 3.81, b = 3.83 and c = 12.13 Å [1]. The lattice parameters of superstructure are $a_s = 4a$, $b_s = a$, $c_s = 2a$ and the space group is Bmmm.

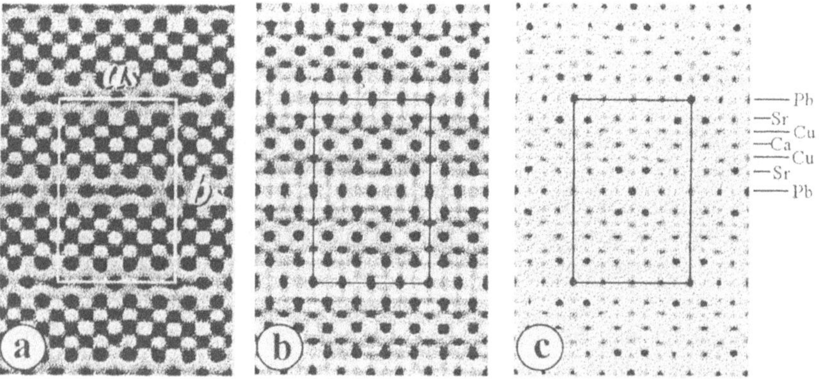

Fig. 2 (a) Average image obtained from Fig. 1a, (b) deconvoluted image and (c) final PPM of Pb-1212

Fig. 1a shows the digitized high-resolution electron microscope image of Pb-1212 projected along the [010] direction. The inset is the corresponding EDP. After noise filtering and symmetry averaging an average image as shown in Fig. 2a was obtained. Then the image deconvolution was carried out by using the program ECC [5]. The determined defocus value is -400 Å and the obtained low resolution structure image is shown in Fig. 2b. In such image all metallic atoms can be distinguished clearly. Before going to the phase extension for enhancing the image resolution an empirical method of electron diffraction intensity correction [6] was employed to reduce the diffraction intensity distortion due to various reasons. Fig. 2c is the final projected potential map (PPM) obtained after the phase extension and Fourier synthesis. All atoms including oxygen are clearly resolved. The modulation of Pb-1212 seems mainly caused by the partial substitution of Sr and Cu atoms for Pb atoms.

3. Incommensurate modulated structure of Pb-1223

The structure of Pb-1223 is one-dimensional incommensurate modulated. The lattice parameters of the orthorhombic basic structure are a = 3.82, b = 3.82 and c = 15.3 Å. The modulation wave vector q = 0.322a* + 0.5 c* [1].

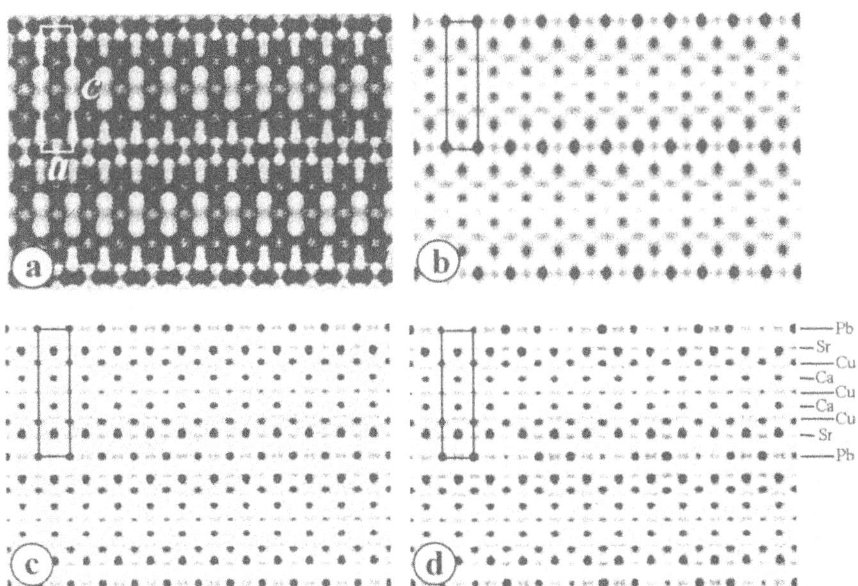

Fig. 3 (a) Average image obtained from Fig. 1b, (b)low-resolution average structure image, (c) high-resolution average structure image and (d) final PPM of the incommensurate modulated structure.

Fig. 1b shows the [010] digitized image together with the inset corresponding EDP. Fig. 3a shows the average image obtained after noise filtering and symmetry averaging according to the basic structure. It was transformed into the low resolution average structure image (Fig. 3b) by the image deconvolution, where the projection of average structure is seen with a resolution about 1.7 Å. All metallic atoms are resolved clearly. The inverse Fourier transform of the average structure image yields the initial phases of the main reflections. After correcting the diffraction data as mention above, the program DIMS [7] was used for deriving phases of all main reflections and the high resolution average structure image (PPM in Fig. 3c) was obtained. Phases of all satellite reflections were derived by using the program DIMS again. The Fourier synthesis was used to improve the PPM. All atoms including oxygen are resolved in the final PPM of the incommensurate modulated structure shown in Fig. 3d. The modulation of individual atom can be seen clearly. Similar to the case of Pb-1212, the modulation is mainly caused by the partial substitution of Sr and Cu atoms for Pb atoms.

4. Conclusion

The commensurate and incommensurate modulated structure of Pb-1212 and Pb-1223 were determined, respectively, from a single high-resolution electron microscope image and the corresponding EDP. All the unoverlapped atoms are resolved from the final PPMs. The structure modulation is seen directly.

This indicates that the two-step image processing technique combining with the empirical method of electron diffraction intensity correction is an efficient approach to *ab initio* modulated structure determination up to the resolution about 1 Å even for crystals containing very heavy atoms, such as Pb and Sr.

5. Reference

1. Wu, X. J., Tamura, T., Adachi, S., Jin, C. Q., Tatsuki, T. and Yamauchi, H. (1995) Structure study on high-pressure synthesized Pb-Sr-Ca-Cu-O system, *Phisica* C **247**, 96-104.
2. De Wolff, P. M. (1974) The pseudo-symmetry of modulated crystal structure, *Acta Cryst.* **A30**, 777-785.
3. Li, F. H. (1994) Two-stage image processing in HREM, in: *Proc. 13th Int. Congr. on Electron Microscopy*, Paris, vol. 1, 481-484.
4. Cheng, T. Z., Wang, J. F. and Wan, Z. H. *et al.* (1994) A program for electron diffraction intensity measurement from photographic data, *J. Appl. Cryst.* **27**, 430-432.
5. Wan, Z. H., Li, F. H. and Fan, H. F. Electron Crystallographic Computing (ECC) program, unpublished.
6. Huang, D. X., Liu, W., Gu, Y. X., Xiong, J. W., Fan, H. F. and Li, F. H. (1996) A method of electron diffraction intensity correction in combination with high resolution electron microscopy, *Acta Cryst.* **A52**, 152-157.
7. Fu, Z. Q. and Fan, H. F. (1994) DIMS-a direct method program for incommensurate modulated structures, *J. Appl. Cryst.* **27**, 124-127.

TEM STUDIES OF THE EARLY STAGES OF PRECIPITATION IN Al-Mg-Si ALLOYS IN COMPARISON WITH ELECTRON RADIATION DAMAGE EFFECTS

B. MINGLER and H. P. KARNTHALER
Institut für Materialphysik, University of Vienna
Boltzmanngassse 5, A-1090 Wien, Austria

1. Introduction

Al alloys containing Mg and Si are used in an increasing number of technical applications. These alloys have a high specific strength (yield strength / density) that is caused by the formation of small precipitates (Guinier-Preston zones GPZ). There is some confusion in the literature about the first stages of the precipitation process, since it was not possible to image the very small precipitates. It is the aim of the present study to investigate the early stages of precipitation by high resolution transmission electron microscopy (HRTEM). The GPZ have to be distinguished from defects induced by the electron radiation during the investigation in the electron microscope.

2. Experimental Procedure

A ternary Al alloy with 1wt.%Si, 0.9 wt.%Mg (containing no Cu) was solution heat treated for 1 h at 810 K, ice water quenched and investigated after different ageing processes. Pure Al (99.998%) was used for comparison to exclude confusion of the radiation induced defects with precipitates and radiation induced precipitates that can occur in the alloy. The TEM study was carried out either in a Philips CM200 or in a Philips CM30ST; the latter was equipped with an ion trap (using tilted dynodes) to exclude ion damage of the specimen which might cause disturbing effects.

3. Experimental Results and Discussion

3.1. PRECIPITATION IN Al-Mg-Si ALLOYS

Fig. 1 shows a HRTEM image of the alloy after 8 days of ageing at RT. Since the beam direction (**BD**) is [001] the (200) and (020) planes of the fcc Al matrix form a cross grating with a spacing of 0.2 nm. In the middle of the figure two or three neighbouring planes show a bright contrast that is about 4 nm long; the dark contrast in the neighbourhood of these planes seems to be caused by a distortion leading to a local

D. L. Dorset et al. (eds.), Electron Crystallography, 397–400.

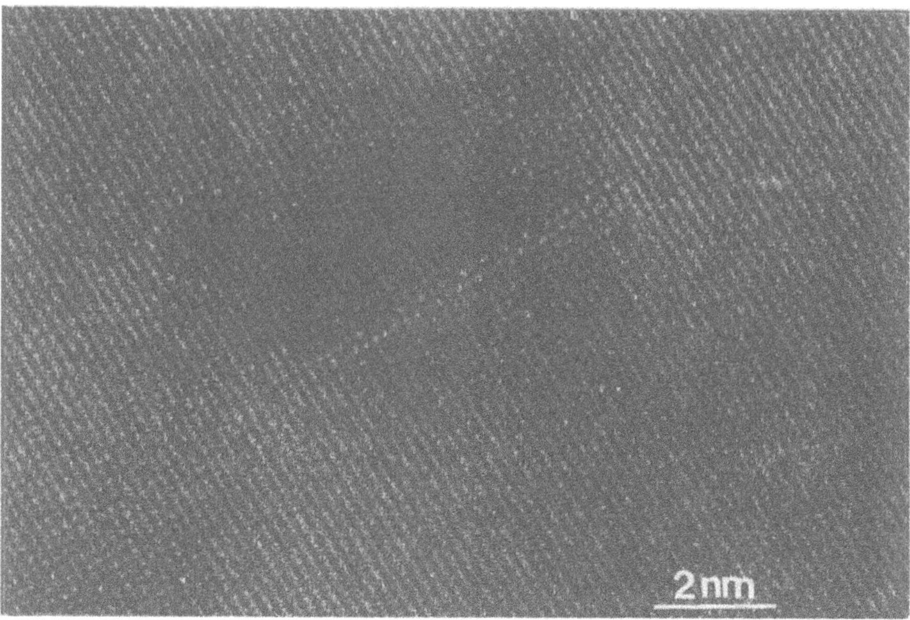

Fig. 1 HRTEM image of the AlMgSi alloy after 8 days of RT ageing; the contrast in the centre of the image is interpreted as a distortion of the matrix caused by a disc-shaped precipitate lying on {100}; **BD** = <100>.

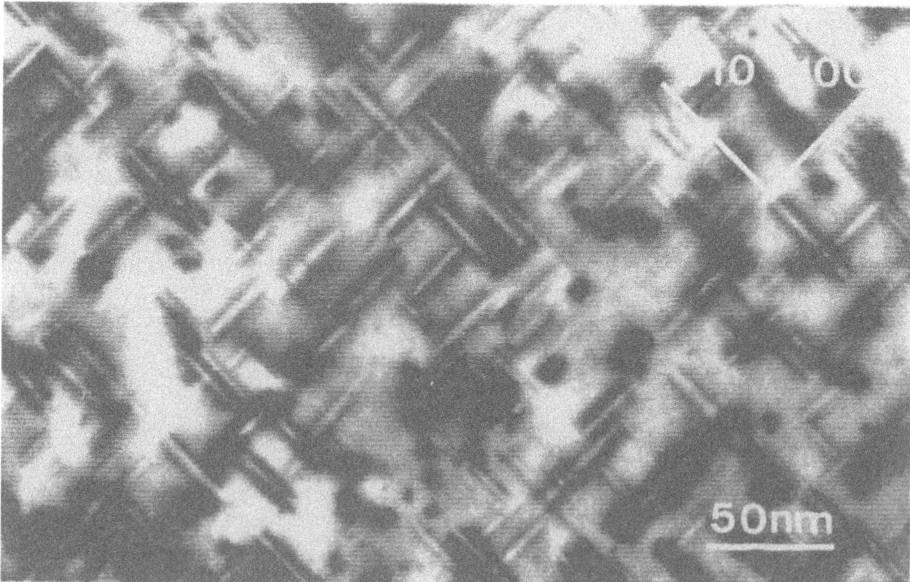

Fig. 2 Multi-beam bright field image of the AlMgSi alloy after 20 h ageing at 160°C. The two families of the needle-shaped precipitates lying perpendicular to the electron beam show the typical double contrast; **BD** = [100].

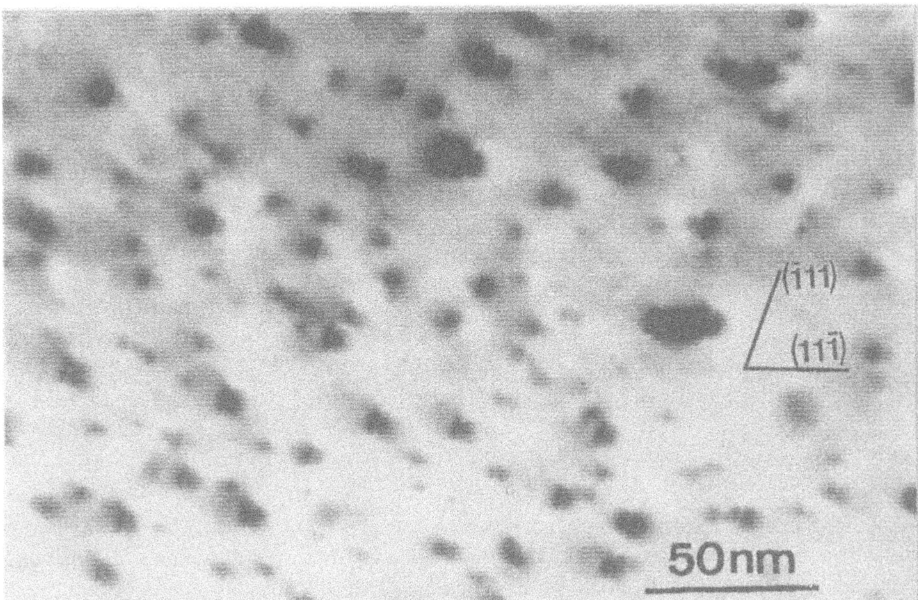

Fig. 3 Bright field multi-beam image (**BD** ~ [101]) of radiation induced defects in pure Al taken after an irradiation time of 30 ks at a voltage of 110 kV.

Fig. 4 HRTEM image of an extrinsic Frank loop (as determined by a Burgers circuit) in pure Al; **BD** = <110>.

bending of the thin foil. Since a ternary alloy was investigated that contains no Cu it is safe to assume that the precipitate imaged in Fig.1 is not a copper GPZ. It is proposed that the observed contrast variations are caused by a very small, fully coherent precipitate of Mg and Si that has the shape of a thin disc. This result is at variance with assumptions made in the literature postulating that the first precipitates of Mg and Si should be spherical to minimize their surface energy. The observation of thin disc shaped precipitates seems to agree with that of copper GPZ in Al which also lie on <100> planes. Large Mg_2Si precipitates have the form of needles or rods and a crystal structure (hcp) different from that of the matrix (fcc) [1]. Fig. 2 shows a multi-beam bright field image of the alloy after 20 h ageing at 160° C. The needle shaped precipitates that are aligned along [010] and [100] are both perpendicular to the electron beam (**BD** = [001]) and show the characteristic double contrast [2]. The needles which are parallel to the electron beam are imaged as dark dots.

3.2. RADIATION DAMAGE IN PURE AL AND IN THE ALLOY

HRTEM images require a high acceleration voltage (200 to 300 kV) and a focused electron beam. Therefore in Al radiation-induced atomic displacements can occur at considerable rates [3]. Both in pure Al and in AlMgSi alloys, the defects induced by electron radiation were studied. It is interesting to note that radiation damage was observed in pure Al at an acceleration voltage as low as 110 kV when the beam direction is <101>. Fig. 3 shows a multi-beam bright field image (**BD** = [110]) of pure Al after an irradiation time of 30 ks at an acceleration voltage of 110 kV. Many defects lying on {111} planes are visible in the area irradiated by the electron beam. During the investigation the nucleation, growth and dissolution of the radiation-induced defects was observed. Fig. 4 shows a HRTEM image (taken at 300 kV) of one of these defects in pure Al (**BD** = [101]). The extrinsic character of this Frank loop (containing an inserted close packed layer) was checked by drawing a Burgers circuit.

3.3. COMPARISON OF THE PRECIPITATES AND THE RADIATION DAMAGE

Both the first stages of the precipitation process and the radiation induced defects have the morphology of thin discs (only one or two atom spacings thick). Since the GPZ and the radiation-induced defects lie on different planes ({100} and {111} respectively) they can be distinguished unambiguously.

4. References

1. Sagalowicz, L., Lapasset, G. and Hug, G. (1996) Transmission electron microscopy study of a precipitate which forms in the Al-Mg-Si system, *Phil. Mag. Lett.* **74**, 57-66.
2. Mingler, M. and Karnthaler, H. P. (1993) TEM Investigations of Al 6061 and of Composites Reinforced with Al_2O_3 Particles, *Z. Metallkd.* **84**, 313-319.
3. Mingler, B. and Karnthaler, H. P. (1994) Electron Radiation Damage in Al-Mg-Si Alloys Caused by TEM Studies, in B. Jouffrey and C. Colliex (eds.), *Proc.13th Intern. Congr. on Electron Micr.*, Les Editions de Physique, Les Ulis, Paris, **2A**, 103-104.

STRUCTURAL MODELS OF τ^2-INFLATED MONOCLINIC AND ORTHORHOMBIC Al-Co PHASES

Z. M. Mo, X. L. Ma, H. X. Sui AND K. H. Kuo
Beijing Laboratory of Electron Microscopy
P.O.Box 2724, Beijing 100080, P. R. China
Department of Materials Engineering, Dalian University of Technology
Dalian, Liaoning 116024, P. R. China

1. Introduction

It is generally believed that quasicrystals and their approximants have almost the same local structures, arranged aperiodically in quasicrystals and periodically in approximants. In the Al-Co binary system, besides monoclinic $Al_{13}Co_4$[1] and orthorhombic $Al_{13}Co_4$[2] (also called orthorhombic Al_3Co[3]), phases determined by X-ray single crystal diffraction earlier, a series of τ-inflated monoclinic and orthorhombic Al-Co decagonal approximants, such as τ^2-, τ^3 -, τ^4-$Al_{13}Co_4$ and τ^2-, τ^3 -, τ^4-Al_3Co, were found recently[4]. These phases all have a composition around Al_3Co, but the lattice parameters a and c enlarged by about τ^2, τ^3 and τ^4 times than those of monoclinic $Al_{13}Co_4$ and orthorhombic Al_3Co, respectively. Their electron diffraction patterns (EDP), at least the strong spots, resemble the corresponding EDPs of the Al-Co decagonal quasicrystal (DQC). In theory, if such a τ-inflation continues infinitely, the unit cell will become infinitely large. A DQC with a layer structure of quasiperiodically arranged pentagons will thus form. In other words, the τ-inflated Al-Co approximants phases show more resemblance to a DQC.

It is important for the understanding of the structure of DQC to investigate the structure of these τ-inflated approximants. Unfortunately, due to the difficulty to obtain proper single crystals and rather large lattice parameters, their structure remains unsolved. Since the tessellation of subunits can be observed directly by high resolution electron microscope (HREM), attempts have been made to derive the unknown structure of an approximant from the known structure of a structurally close-related approximant. Here we first derive a structural model of the monoclinic τ^2-$Al_{13}Co_4$ based on HREM and the understanding of the relationship between $Al_{13}Co_4$ and its τ^2-inflated phase, then propose a structural model of orthorhombic τ^2-Al_3Co on the basis of the model of τ^2-$Al_{13}Co_4$.

D. L. Dorset et al. (eds.), Electron Crystallography, 401–404.

2. Structure of the monoclinic $Al_{13}Co_4$ phase

The structure of τ^2-$Al_{13}Co_4$ is closely related to $Al_{13}Co_4$. It is worthwhile to describe briefly the structure of $Al_{13}Co_4$ before constructing the structural model of τ^2-$Al_{13}Co_4$.

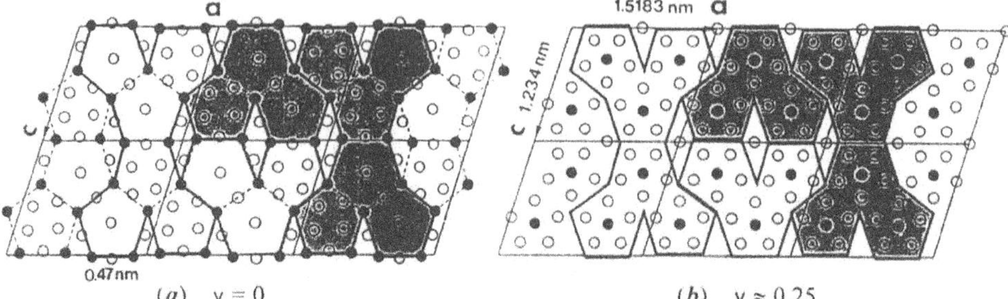

(*a*) y = 0 (*b*) y ≈ 0.25

Figure 1. (a) Flat layer at y=0 and (b) puckered layer at y≈0.25 of the monoclinic $Al_{13}Co_4$ phase

The structure of the monoclinic $Al_{13}Co_4$ phase is similar to that of θ-$Al_{13}Fe_4$, but has the space group ***Cm*** instead of ***C2/m***. The lattice parameters are a=1.5183 nm, b=0.8122 nm, c=1.2340 nm and β=107.9 °. The whole structure can be described by stacking the flat (F) layer at y=0 and the puckered (P) layer at about y=0.25 (varying from 0.210 to 0.293) along the b axis in a sequence of FPF'P^m... , in which F' is related to F by a translation of $a/2+b/2$ and P^m is obtained from P by a mirror at the flat layer. The atomic arrangement (Al atom: open circle; Co atom: solid circle) of the flat and puckered layers are shown in figures 1a and 1b , respectively. It can be seen that the Co atoms in the flat layer form a network of pentagons with an edge length of 0.47nm (shown with broken lines in figure 1a), while those in the puckered layers are located above the centers of these pentagons. Four such pentagons are connected in each unit cell. Two kinds of dumb-bell formed subunits composed of six pentagons (shown with thick solid lines in figure 1a and 1b, shadowed or non-shadowed) can be recognized.

3. Structural model of the monoclinic τ^2-$Al_{13}Co_4$ phase

The *C*-centered τ^2-$Al_{13}Co_4$ has a monoclinic structure, based on X-ray powder diffraction, with a=3.984 nm, b=0.8148 nm, c=3.223 nm and β=107.94° and belongs to the space group ***Cm***. Like monoclinic $Al_{13}Co_4$, τ^2-$Al_{13}Co_4$ also possesses a similar type of layer structure, i.e., four layers along the b axis, two flat (F,F') layers and two puckered (P, P^m) layers in a sequence of FPF' P^m.... Figure 2a is a HREM image of τ^2-$Al_{13}Co_4$ taken along the [010] direction. It is obvious that the bright image points form pentagons of two different sizes. The large pentagons are marked with solid circles and some of the small ones with dots. It should be pointed out that the small centered pentagons with an edge length of 0.47 nm are the same as those observed in $Al_{13}Co_4$. In each large pentagon with an edge length of 1.23 nm, there are six small pentagons, one inverted at the center and five of the same orientation as the large one surrounding the central one. Each unit cell, indicated by octagonal stars, consists of four such large pentagons.

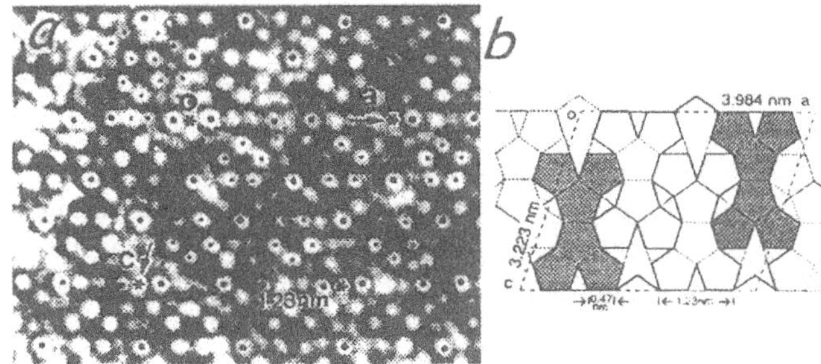

Figure 2. (a) [010] HREM image and (b) a schematic illustration of the Co-atom network in the flat layer of the monoclinic τ^2-$Al_{13}Co_4$ phase

From the HREM study of τ^2-$Al_{13}Co_4$[4], it is known that the bright points in the [010] HREM image correspond to the projected position of Co atoms. Thus the schematic illustration of the Co-atom network in the flat layer of τ^2-$Al_{13}Co_4$ (shown in figure 2b) can be extracted from its [010] HREM image. The small pentagons are shown with thin broken lines and the large ones with thin solid lines. Similar to that of $Al_{13}Co_4$, the structure of τ^2-$Al_{13}Co_4$ could be constructed mainly by dumb-bell formed subunits, though the tessellation is different.

Since $Al_{13}Co_4$ and τ^2-$Al_{13}Co_4$ have a similar composition and both are approximants of Al-Co DQC, it might be reasonable to assume that their subunits have the same Al atomic decoration. Therefore, we can use the subunits existing in figure 1 to construct the main part of the flat and puckered layers of τ^2-$Al_{13}Co_4$. The atomic arrangement of the structural model is shown in figure 3.

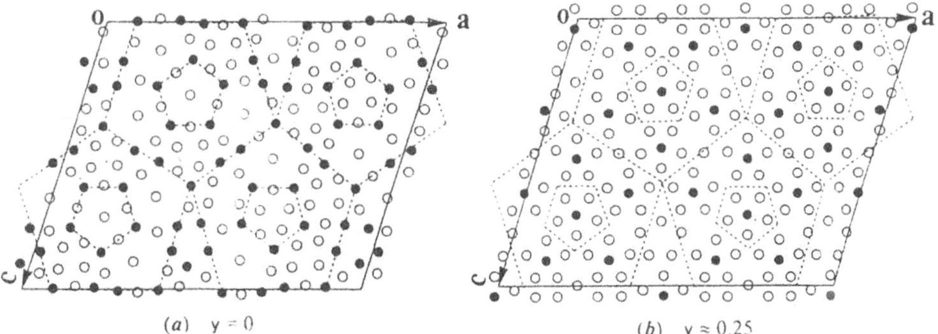

Figure 3. (a) Flat layer at y=0 and (b) puckered layer at y≈0.25 of the structural model of τ^2-$Al_{13}Co_4$

There are 520 Al and 168 Co positions in each unit cell, satisfying the symmetry of the space group ***Cm***. The agreement between simulated and experimental results, including EDPs and HREM images, is satisfactory.

4. Structural model of the orthorhombic τ^2-Al_3Co phase

The orthorhombic lattice of τ^2-Al_3Co and the monoclinic lattice of τ^2-$Al_{13}Co_4$ are closely related:

$$a_O \approx a_M \cos(\beta_M - \pi/2) = 3.79\text{nm};\ b_O \approx b_M = 0.814\text{nm};\ c_O \approx c_M = 3.23\text{nm};$$

where a_M , b_M , c_M and β_M are the parameters of τ^2-$Al_{13}Co_4$ and a_O, b_O and c_O are of τ^2-Al_3Co. The same relationship between $Al_{13}Co_4$ and Al_3Co has been reported and a structural model of Al_3Co, which was confirmed by recent X-ray single crystal diffraction, was suggested based on the shear mechanism[3]. Here, by introducing repeated (100) faults, which have the shift value of about $0.38c_M$ (=1.23nm), in the monoclinic lattice of τ^2-$Al_{13}Co_4$, we obtained the orthorhombic lattice of τ^2-Al_3Co. Figures 4a and 4b show the atomic arrangement of the flat layer (F) at y=0 and the puckered layer (P) at y≈0.25 (Al atom: open circle; Co atom: solid circle), respectively. The stacking sequence is also FPF'P^m..., where the flat layer F' relates to F by a twofold screw axis and the puckered layer P^m to P by a mirror on the flat layers . The space group of this model is ***P2₁mn***, which is the same as that of Al_3Co. The simulated EDPs and HREM images agree well with the experimental ones.

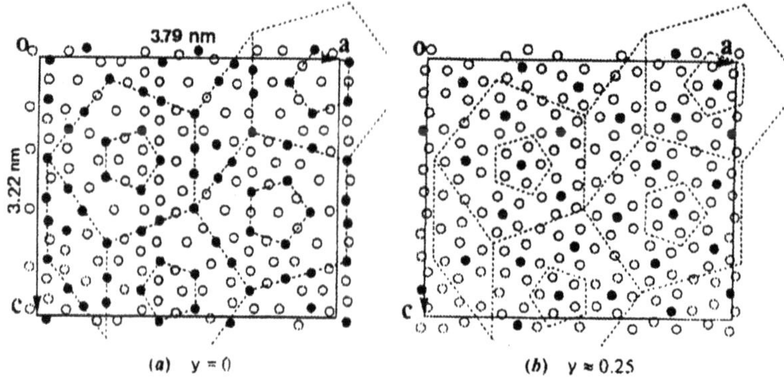

Figure 4. (a) Flat layer at y=0 and (b) puckered layer at y≈0.25 of the structural model of τ^2-Al_3Co

Acknowledgment

The authors acknowledge the financial support of the National Natural Science Foundation of China.

References

1. Hudd, R. C. and Taylor, W. H. (1962) The structure of Co_4Al_{13}, *Acta Cryst.* **15**, 441-442.
2. Grin, J., Burkhardt, U., Ellner, M. and Peters, K. (1994) Crystal structure of orthorhombic Co_4Al_{13}, *J. Alloys Comp.* **206**, 243-247.
3. Li, X. Z., Ma, X. L. and Kuo, K. H. (1994) A structural model of the orthorhombic $Al_{13}Co_4$ derived from the monoclinic $Al_{13}Co_4$ by high resolution electron microscopy, *Phil. Mag. Lett.* **70**, 221-229.
4. Ma, X. L., Li, X. Z. and Kuo, K. H. (1995) A family of τ-inflated monoclinic $Al_{13}Co_4$ phases, *Acta Cryst.* **B51**, 36-43.

ELECTRON MICROSCOPY OF THIN PROTEIN CRYSTALS FROM VAPOUR DIFFUSION 'HANGING DROPS' PROVIDES STRUCTURAL INFORMATION AT INTERMEDIATE RESOLUTION

R.H.NEWMAN and P.S.FREEMONT
Protein Structure Laboratory,
Imperial Cancer Research Fund,
PO Box 123, London, WC2A 3PX, England.

I. Introduction

Electron crystallography has emerged as a powerful method for the determination of molecular structures at intermediate to high spatial resolutions (3 - 20 A). Together with direct imaging by high resolution electron microscopy this technique provide an excellent tool for protein crystal analysis. However, application of these methods depends on the availability of relatively large (micrometre size), well-ordered and thin crystalline arrays of the protein molecules. One source of such crystals is that from vapour diffusion 'hanging drops' which have been set up in order to grow crystals suitable for X-ray diffraction. Crystals grown under these conditions are sometimes not suitable for X-ray analysis but when harvested onto an electron microscope grid prove to be sufficiently thin for analysis by electron microscopy.

2. Technique for handling Micro-crystals

Micro-crystals, barely visible under the light microscope, are taken from a drop of mother liquor containing a suspension of the crystals and deposited on carbon/plastic coated em grids. This is accomplished by inverting the cover slip on which the 'hanging drop' crystals are suspended, whilst being careful not to crack it when removing it from the crystallisation plate, and by touching the surface with a freshly prepared em grid. This grid should have a hydrophilic surface which can be acheived by glow discharge just before use. After blotting most of the water/salt from the surface of the grid a drop of 1 or 2% uranyl acetate solution is added to the micro-crystal resdue on the grid surface and blotted off after 20 to 30 seconds. The grid is then ready for observation in the electron microscope. Using an accelerating voltage of at least 100kV the grid can be examined at low magnification in

D. L. Dorset et al. (eds.), Electron Crystallography, 405–406.

order to locate suitable micro-crystals which are apparent by their electron translucency and visible lattice. Once a suitable crystal has been located it can be examined at a higher resolution, normally 30,000 to 50,000 times, in order to visualise the protein lattice. Micrographs of the micro-crystals can then be taken before image processing. Negatives can be examined using an optical bench in order to find well ordered areas of the microcrystals for subsequent image enhancement using Fourier analysis methods.

3. Some examples of Micro-crystals obtained from 'hanging drops'

We have used the above simple methodology to obtain structural information about various protein 'micro-crystals'. These include the proteins: annexin VI [1], protein kinase C [2], HAP1 endonuclease V, phospholipase C [3] and estrogen receptor hormone binding domain-glutathione-s-transferase fusion protein [4]. These protein microcrystals have revealed, by the simple application of negative stain, their lattices and molecular shapes. In some cases amplitude and phase information from the Fourier analysis of these negatives has been used to produce a more refined structure, albeit at intermediate resolution. Electron microscopy can be used to obtain initial structural information from protein crystals grown in hanging drops as well as encouraging X-ray crystallographers to persevere with these, or similar conditions in order to grow bigger crystals more suitable for their requirements.

4. References

1. R. Newman, A. Tucker, C. Ferguson, D. Tsernoglou, K.Leonard, M. Crumpton (1989) Crystallisation of p68 on lipid monolayers and as three-dimensional single crystals. J.Mol.Biol. 206, 213-219.

2. R.H.Newman, E.Carpenter, P.S.Freemont, R.L.Blundell, P.J.Parker (1994) Microcrystals of the beta1 isoenzyme of protein kinase C: an elcetron microscope study, Biochemical Journal, 298,391-393.

3. R. Newman (1995) (unpublished data)

4. J.M.Lally,R.H.Newman,P.R.Knowles,S.Islam, A.I.Coffer, M. Parker, P.S.Freemont (1997) Crystallisation of an intact GST-Estrogen receptor hormone binding domain fusion protein (in press)

STRUCTURAL MODULATIONS IN THE Sr-Ca-Cu-O SYSTEM CHARACTERIZED BY HRTEM.

J. RAMIREZ-CASTELLANOS[1], J. M. GONZALEZ-CALBET[1], M. VALLET-REGI[2] AND Y. MATSUI[3].

[1] *Dpto. Q. Inorgánica, Fac. Químicas (UCM), 28040-Madrid (Spain).*

[2] *Dpto. Q. Inorgánica y Bioinorgánica, Fac. Farmacia (UCM), 28040-Madrid (Spain).*

[3] *National Institute for Research in Inorganic Materials (NIRIM), Tsukuba, 305-Ibaraki (Japan).*

1. Introduction

$SrCuO_2$ prepared at room pressure (RP) shows an orthorhombic structure [1], which can be described as formed by Cu atoms in planar square coordination, which are connected via edge-sharing, giving rise to the formation of $[Cu_2O_2]_\infty$ double chains. By applying high pressure (HP), an infinite layer (IL) $SrCuO_2$ structure is obtained which has a tetragonal symmetry, with corner-sharing CuO_2 planes, and a density about 8% higher than the RP phase.

Superconducting materials structurally related to the IL-$SrCuO_2$ structure can be stabilized by high pressure conditions during the sample preparation but also by partial cationic substitution of the alkaline earth in Sr sites [2].

Recently, new phases in the Sr-M-Cu-O (M=Ca, Sm, ...) system have been described [3], consisting of an ordered distribution of oxygen-deficient perovskite structural blocks. Novel structures have been proposed based on the high resolution electron microscopy observations, whose framework is built up from the intergrowths of diferent kinds of Cu-O chains.

In the present work, two samples in the Sr-Ca-Cu-O system were prepared with the same cationic ratio but under different pressure synthesis conditions. The superstructures found by means of selected area electron diffraction (SAED) and high resolution electron microscopy (HRTEM) are discussed.

2. Experimental

The RP sample with nominal composition $Sr_{0.5}Ca_{0.5}CuO_2$ was prepared by the metal nitrates co-decomposition route, at 950 °C for 96 hours in air [4]. The average cation ratio in the obtained sample as analyzed by inductive coupling plasma (ICP) is almost identical to that of the nominal composition, as confirmed by EDX analysis, which showed a homogeneous Ca distribution in all crystals. Volumetric analysis indicate that only Cu(II) is present in the sample, leading to a chemical composition $Sr_{0.5}Ca_{0.5}CuO_2$. Most of the X-ray diffraction (XRD) reflections can be indexed on the basis of orthorhombic symmetry and space group Amam characteristic of

D. L. Dorset et al. (eds.), Electron Crystallography, 407–410.

the orthorhombic $SrCuO_2$ structure [1]. Moreover, some superstructure reflections cannot be indexed on the basis of such a unit cell. SAED and HRTEM studies were carried out on a JEOL 4000-EX transmission electron microscope, operated at 400 kV.

The HP sample was prepared at 5.5 GPa and fired at 1250 °C for 1 hour, and then quenched to room temperature, as previously reported [5]. XRD measurements indicated that the obtained high-pressure sample was a multiphase mixture, including $Sr_2Ca_3Cu_4O_{9+\delta}$ and an unknown phase, which was by far the most abundant. In order to identify the unknown phase, SAED and HRTEM studies were performed using a Hitachi H-1500 high voltage electron microscope operated at 800 kV.

3. Results and discussion

3.1. MICROSTRUCTURAL CHARACTERIZATION OF THE RP SAMPLE.

Figure 1a shows the SAED pattern along $[100]_o$ (o subindex is referred to the orthorhombic $SrCuO_2$ starting subcell) for x=0.5 sample. It can be observed that c parameter of the basic cell is doubled. Figure 1b shows the SAED pattern along the $[101]_o$ obtained by tilting around b* axis. $(1/2\ 1\ 1/2)_o$ and equivalent reflections show that both a and c parameters are double. Since chemical analysis suggests neither the absence of anionic vacancies nor the presence of intersticial oxygen, it can be concluded that the anionic sublattice is not responsible for such a new cell. It seems, then, that some kind of ordering between Ca and Sr atoms in $Sr_{0.5}Ca_{0.5}CuO_2$ must exist in order to get the new cell with $a=2a_o=7.71$ Å, $b=b_o=16.15$ Å and $c=2c_o=6.92$ Å parameters.
Figure 2a shows the HRTEM image of a well-ordered crystal along $[100]_o$ (corresponding SAED is shown in the inset). However, no different constrast can be observed in the Ca and Sr positions, as a result of the ordering of both cations. To distinguish the contrast between both cations in the HRTEM image along this projection, an image calculation (figure 2b) of this model was performed using the multislice method [6], under the following conditions: sample thickness t=40 nm and focus $\Delta f=-80$ nm. On the other hand, another image calculation was also performed considering the Sr-substitution by Ca without any kind of ordering between both cations, under the same conditons. In both cases, identical calculated images were obtained and it was not possible to detect any variation in the contrast, in agreement with the experimental HRTEM image along $[100]_o$ projection. This may be attributed to the very close scattering factors of both Sr and Ca cations. By this reason, a study by means of neutron diffraction seems to be necessary in order to confirm this structural model.

3.2. MICROSTRUCTURAL CHARACTERIZATION OF THE HP SAMPLE.

Figure 3a shows the SAED pattern of HP sample along $[100]_o$. Strongest spots suggest a tetragonal structure and weaker extra spots along both b* and c* reciprocal directions indicate that a six-fold superstructure along both b and c directions of the basic cell exists in this material. Moreover, a different intensity in the extra spots can be observed, suggesting that the symmetry is, in a first approximation, pseudotetragonal.
Figure 3b shows a complex contrast distribution along $[100]_o$, indicating that this compound has a novel wavy modulated structure. It can be very clearly seen that the

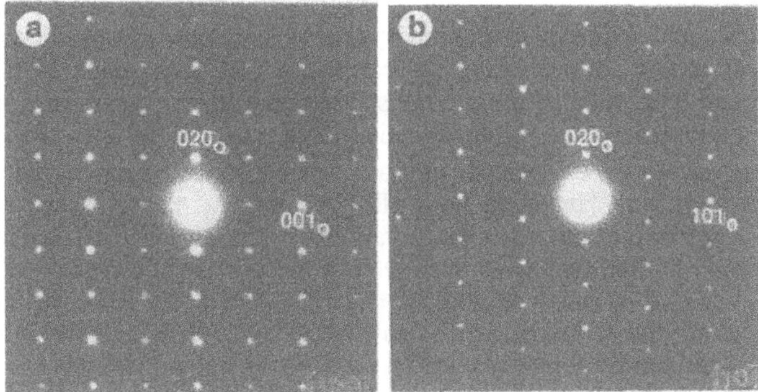

Figure 1. SAED patterns corresponding to the RP phase $Sr_{0.5}Ca_{0.5}CuO_2$ along the (a) $[100]_o$ and (b) $[10\bar{1}]_o$ directions.

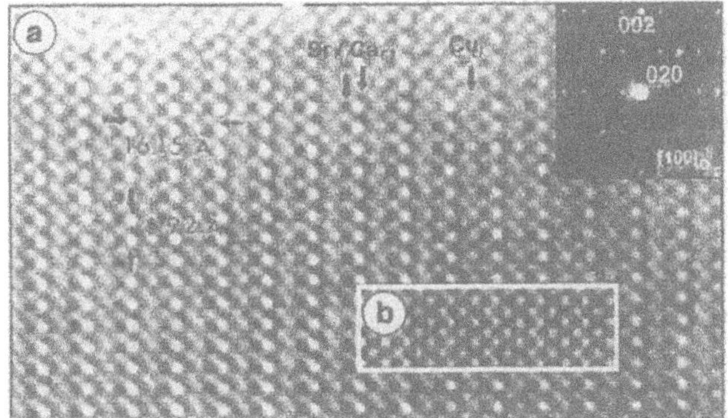

Figure 2. (a) HRTEM micrograph along the $[100]_o$ direction and (b) calculated image.

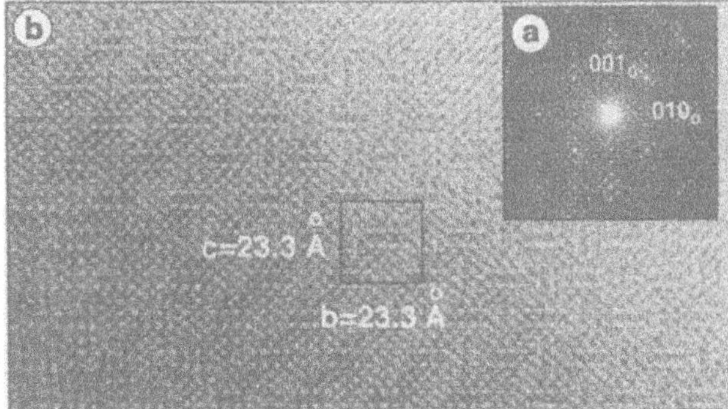

Figure 3. (a) SAED pattern along the $[100]_o$ direction for the HP phase and (b) corresponding HRTEM micrograph.

areas of three strong dots alternate along both b and c directions with the areas of three weak dots. Assuming that these dots are correlated to the atomic positions of strontium and calcium atoms, respectively, we concluded that this material has an ordered distribution of both alkaline-earth cations, according to the sequence [Sr-Sr-Sr-Ca-Ca-Ca]∞ along both $[010]_o$ and $[001]_o$ directions. We also assume from the image contrast that some replacements of Sr atoms by Ca can occur, inducing lattice distortions.

Hiroi *et al.* [2] reported from HRTEM images, the existence of both planar defects in a fairly random way, and pyramidally coordinated Cu ions in those defect layers arising from the alkaline-earth deficiency irrespective of oxygen content. However, the possibility that a small amount of extra oxygen atoms were incorporated into a couple Sr atom sheets might not be ruled out. According to this information, the oxygen content per unit formula in $Sr_3Ca_3Cu_6O_{12+\delta}$ is close to 15, since 1 extra oxygen atom is introduced per CuO_2 unit in the Sr-rich block, leading to a pyramidal coordination for Cu atoms in this block, leading to an average composition close to $Sr_3Ca_3Cu_6O_{15+\delta}$.

We can conclude that new copper compounds, $Sr_{0.5}Ca_{0.5}CuO_2$ (RP sample) and $Sr_3Ca_3Cu_6O_{15+\delta}$ (HP sample), were found under different pressure. The application of high pressure introduces some extra oxygen, giving rise to a different Sr/Ca ordering in both phases, as observed by SAED and HRTEM studies. Both materials exhibit new superstructures, which can be described by an ordered distribution of both Sr and Ca atoms, where Cu atoms are found in square planar coordination for RP sample and both pyramidal and square-planar coordinations for HP sample.

4. References

1. Teske, C.L. and Muller-Buschbaum, H. (1970) Uber Erdalkalimetall-Oxocuprate. V, *Z. Anorg. Allg. Chem.* **379**, 234-241.
2. Azuma, A., Hiroi, Z., Takano, M., Bando, Y. and Takeda, Y. (1992) Superconductivity at 110K in the infinite-layer compound $(Sr_{1-x}Ca_x)_{1-y}CuO_2$, *Nature* **356**, 775-776.
3. Mercey, B., Gupta, A., Hervieu, M. and Raveau, B. (1995) A new ordered oxygen-deficient perovskite $Sm_2Sr_6Cu_8O_{17+\delta}$: HREM study of PLD thin films, *J. Solid State Chem.* **116**, 300-306.
4. Ramírez-Castellanos, J., Matsui, Y., Vallet-Regí, M. and González-Calbet, J.M. (in press) Room and high pressure synthesis in the Sr-Ca-Cu-O system, *Solid State Ionics*.
5. Ramírez-Castellanos, J., Matsui, Y., Kawashima, T., Takayama-Muromachi, E. and Kirkland, A.I. (1996) Structural study of $Sr_3Ca_3Cu_6O_{15+\delta}$, *Physica C* **262**, 285-291.
6. NCEMMS program (1991) National Center for Electron Microscopy, Materials and Chemical Science Division, Lawrence Berkeley, CA (USA).

INVESTIGATION OF DEFECTS IN PLASTICALLY DEFORMED Ni_3Al BY TEM TILTING EXPERIMENTS

C. RENTENBERGER and H. P. KARNTHALER
Institut für Materialphysik, University of Vienna
Boltzmanngasse 5, A-1090 Wien, Austria.

1. Introduction

The intermetallic alloy Ni_3Al has the ordered $L1_2$ structure, which is based on the fcc lattice (Fig. 1). Ni_3Al shows an anomalous behaviour of both critical resolved shear stress (CRSS) and work-hardening rate with temperature. Since the magnitude of the anomaly depends on the orientation of the compression axis single crystalline specimens with different orientations were investigated. In Fig. 2 the results for the [001] compression axis are shown.

To understand the mechanical behaviour of ordered alloys it is necessary to analyse the fine structure of the defects of plastically deformed crystals. For the anomalous behaviour of Ni_3Al the non-planar structures of dislocations are decisive, since they can act as obstacles during the deformation process. The investigation of these defects was done by transmission electron microscopy (TEM) methods using weak-beam diffraction contrast images.

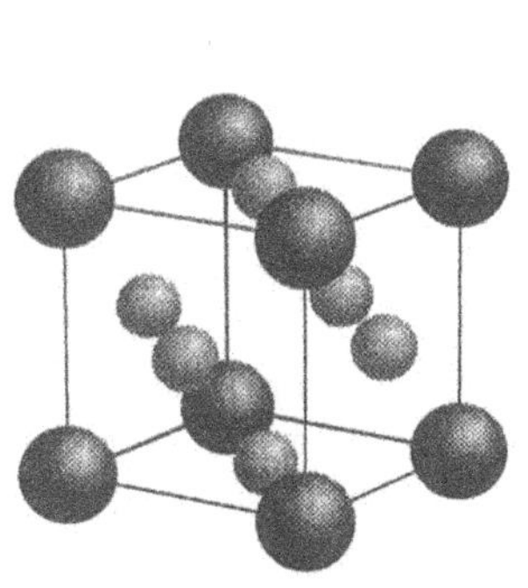

Figure 1: $L1_2$ ordered structure of Ni_3Al; Ni atoms sit at the corners, Al atoms at the face centred positions.

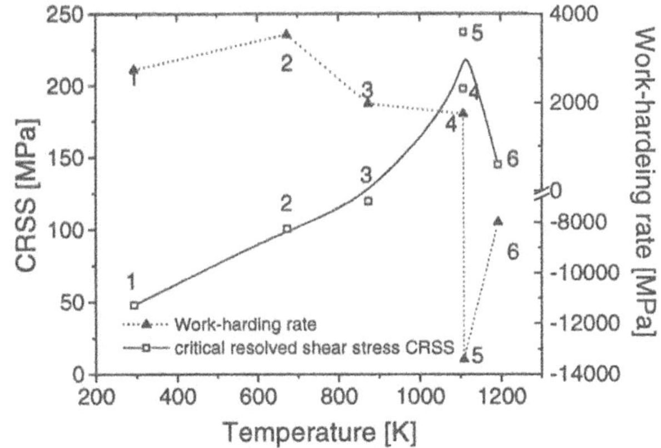

Figure 2: Anomalous behaviour of CRSS and work-hardening rate with temperature in binary Ni_3Al with [001] orientation.

D. L. Dorset et al. (eds.), Electron Crystallography, 411–414.

2. Experimental Procedure

Binary Ni_3Al single crystals, oriented for single and multiple slip (compression axis parallel $[\bar{1}23]$ and [001], respectively) were deformed at different temperatures T_{def} within the anomalous temperature regime. For the TEM investigation thin foils parallel to the (111) glide plane and to the (010) cube cross-slip plane were cut by spark erosion and prepared by jet electropolishing methods. The investigation was carried out using a Philips CM30ST electron microscope operating at 150kV in order to achieve systematic weak-beam diffraction conditions. To analyse the different non-planar dissociation modes of the superlattice dislocations different beam directions (BD) had to be used. To achieve the necessary resolution of the dislocations high deviation parameters hat to be applied using **g**(4**g**), **g**(5**g**) and 2**g**(5**g**) diffraction conditions.

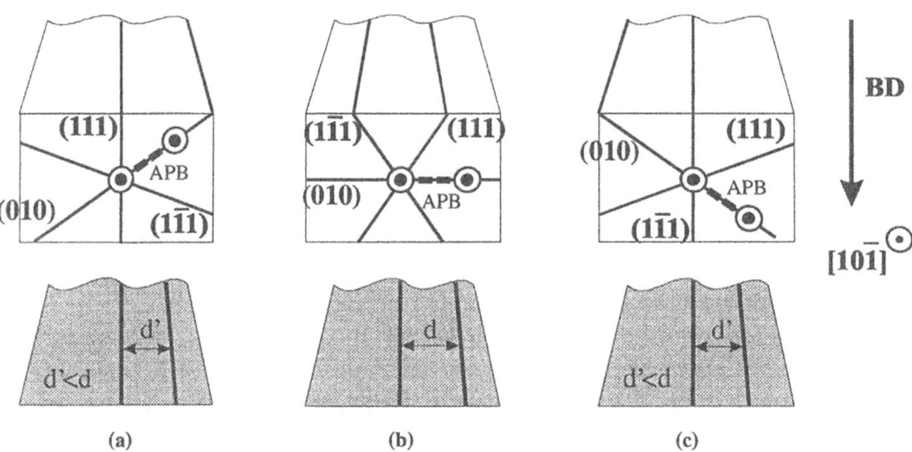

Figure 3: Schematic drawing of the tilting experiment to determine the plane of antiphase boundary fault (APB) dissociation by tilting around the $[10\bar{1}]$ screw direction: **(a)** BD~$[\bar{1}2\bar{1}]$, (111) plane end on; **(b)** BD~[010]; **(c)** BD~[121], $(1\bar{1}1)$ plane end on. When the plane of dissociation lies inclined to the incident beam BD (a and c), the distance d' between the contrast lines representing the superpartial dislocations is smaller than the dissociation width d.

3. Experimental Results and Discussion

Ni_3Al deforms by the glide of superlattice dislocations (**b**=±a<110>{111}) within the anomalous regime. In the specimens with $[\bar{1}23]$ compression axis only one slip system is activated, whereas [001] compressed samples show eight different glide systems. The glide dislocations are dissociated into two superpartial dislocations $\mathbf{b_p}$=±a/2<110> bounding an antiphase boundary fault (APB). Since the dissociation

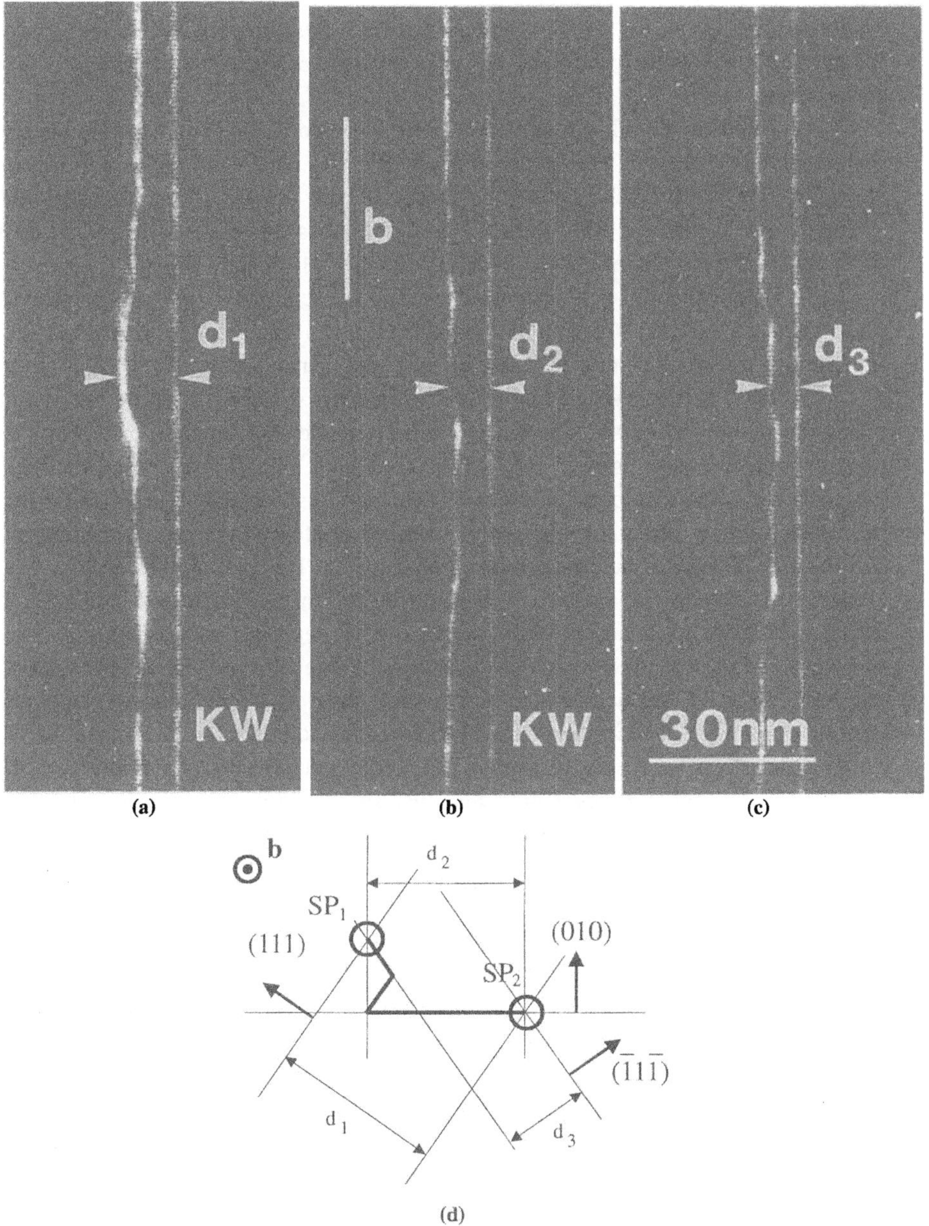

Figure 4: Ni_3Al, [001] compression axis, T_{def}=600°C, ε=2.6%, foil normal FN~[010] (150kV, **g**∥**b**): Sequence of **g**(4.2**g**) weak-beam images of a dissociated superlattice screw dislocation under different tilting conditions. **(a)** BD~[$\bar{1}2\bar{1}$], (111) end on; **(b)** BD~[010]; **(c)** BD~[121], (1$\bar{1}$1) end on. **(d)** Schematic drawing of a cross-section of the configuration. The bold line indicates the APB fault; d_i are the projected distances of the superpartial SP_1 and SP_2 using different beam directions.

width (depending on dislocation character and APB fault plane) is in the range of 5 to 20 nm [1], weak-beam conditions are necessary to resolve the fine structure of the dislocations.

In single and multiple slip specimens a predominance of superlattice screw dislocations is observed. Tilting experiments (tilt axis parallel to the screw direction) show that the screws have undergone cross-slip and are in a Kear-Wilsdorf (KW) configuration having the APB fault on the cube cross-slip plane (010), whereas the dislocations of other characters are dissociated in their (111) glide plane.

In $[\bar{1}23]$ oriented specimens the cross-slipped screws can bow out on the (010) plane since the external stress on that plane is high [1]. The amount of deformation on this plane is still small compared to that on the (111) plane.

In [001] oriented specimens where no external stresses are acting on any cube planes the screws remain straight. Since in this orientation the stresses on all four {111} planes are high, a stress assisted cross-slip process back onto different octahedral planes can occur. This leads to configurations of the screw dislocations showing segments of their APB fault on different planes: on the octahedral plane (111), the octahedral cross-slip plane $(1\bar{1}1)$ and on the cube cross-slip plane (010). The micrographs of Fig. 4 (a to c) show the result of an analysis carried out by the method indicated in Fig. 3. The sketch of Fig. 4d shows that the APB fault consists of segments on 3 different planes. In addition a different configuration, the so called SISF converted KW lock, is observed; in this case the APB segment on the octahedral plane transforms into a superlattice intrinsic stacking fault SISF. Since the SISF energy is considerably lower than the APB energies the stress necessary to generate an additional segment containing a SISF is rather low. At higher temperatures the analysis of tilting experiments shows the occurrence of extended screw dislocations which are formed by multiple cross-slip on different octahedral planes [2].

All these observed non-planar configurations contribute to the anomalous mechanical behaviour since they are locked and can act as obstacles for the glide dislocations. This investigation shows that tilt experiments under weak-beam conditions are the appropriate method to analyse the three-dimensional fine structure of defects.

4. References

1. Karnthaler, H. P., Mühlbacher, E. Th. and Rentenberger, C. (1996) The influence of the fault energies on the anomalous mechanical behaviour of Ni_3Al alloys, *Acta mater.* **44**, No. 2, 547-560.
2. Rentenberger, C. and Karnthaler, H. P. (1995) What determines the Peak Temperature in binary Ni_3Al, *Mat. Res. Soc. Symp. Proc.* Vol. 364, 689-694.

OXIDATION IN-SITU OF $Nb_{12}O_{29}$ INTO A HIGH RESOLUTION MICROSCOPE

M.J. Sayagués and J.L. Hutchison
Department of Materials University of Oxford, Parks Road,
Oxford OX1 3PH, U.K.

A non-stoichiometric $Nb_{12}O_{29}$ ($NbO_{2.417}$) block structure oxide (1-4) with monoclinic symmetry has been oxidised (with O_2) into a controlled environmental high resolution electron microscope. Such microscope is based in a JEOL 4000 EX HRTEM equipped with a gas reaction cell microscope (5) and it offers the possibility to analyse the changes in a solid state reaction at the unit cell level, directly while a reaction is being carried out at the microscope stage.

Suitable crystals were located with the cell under vacuum and orientated in the [010] zone axis, later, the oxygen gas was introduced into the chamber and the electron beam was used (by changing the condenser aperture) to heat the crystal. The images were recorder at magnification 300,000x and 400,000x using objective lens defocus values close to Scherzer-focus.

The $Nb_{12}O_{29}$ monoclinic structure can be seen in the micrograph shown in fig. 1a and the corresponding SAED (selected area electron diffraction) pattern in fig.1b. Observing carefully the white dots contrast in the micrograph (which correspond to the holes in the structure), it is possible to appreciate some staking faults (marked with arrows). As well, it can be explain as a **n** monoclinic blocks joined each other in an orthorhombic way in the perpendicular direction to **c** axis. The staking fault is depicted on the bottom of the micrograph, the monoclinic and orthorhombic blocks are marked.

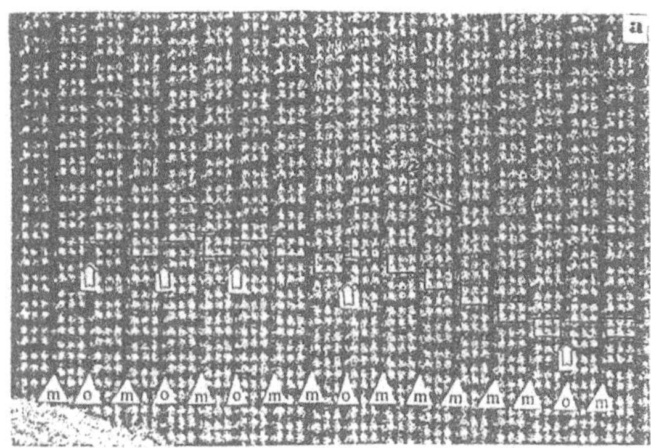

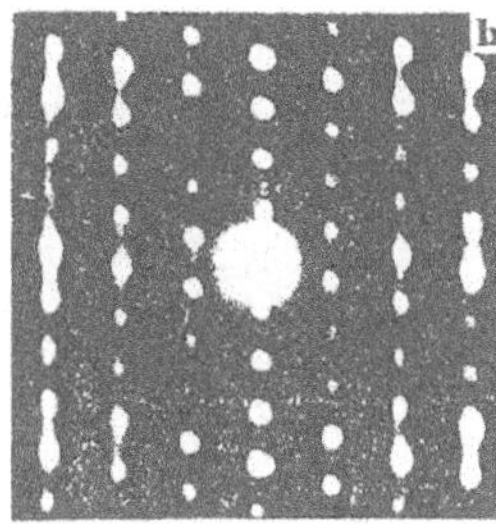

Figure 1. (a) Micrograph of a $Nb_{12}O_{29}$ crystal under vacuum along [010] and (b) corresponding SAED pattern.

D. L. Dorset et al. (eds.), Electron Crystallography, 415–418.

Oxygen (15 mb) was introduced in the cell during 10 minutes and then, the same area of the crystal was imaged. The micrograph and corresponding SAED pattern can be seen in the figs 2a and 2b. Some changed can be appreciate, which seem to start in paralell lines to **c** axis exactly where the monoclinic blocks were joined in orhorhombic manner. The marked area is mapped in fig. 2c where can be seen clearer the block movements. Each line of propagation of reaction apperars to start with the formation of a (5X3) block and the desplacement of block interface can be propagate in either directions from this, leading to ripples, which can be seen en the micrograph, there must necessarily be some defect structure with (3X3) blocks at the end of the *lamella* defect, which can be observed in the micrograph.

The first movement of atoms gives place to *lamella* defects $Nb_{11}O_{27}$ ($NbO_{2.450}$) in diferent areas, producing elongated white contrast in the micrograph due to the formation of rectangular tunnels produced when two blocks are linked in the [100] direction. This *lamella* defect is very simillar to that found in the $NaNb_{13}O_{33}$ ($NbO_{2.538}$) structure (6), altough in such structure the rectangular tunnels are ocuppied by Na cations and the blocks are (5x3) octahedra.

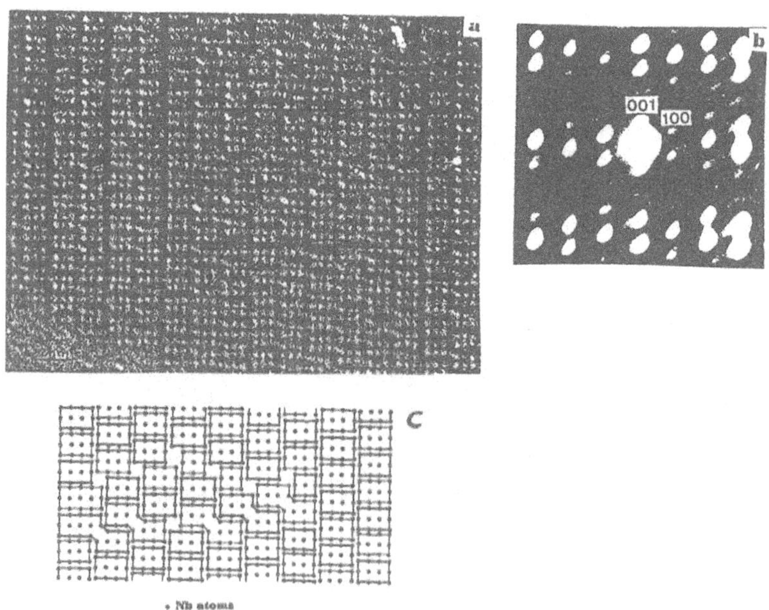

Figure 2. (a) Micrograph from the above crystal (same area) after oxidation. (b) Corresponding SAED. (c) Schematic representation of block structure from the marked area.

A further oxidation and heating were carried out in the same area, different changes could be observed, such a small domains of (3x3) blocks, which indicated that the formation of $Nb_{22}O_{54}$ ($NbO2._{454}$) structure has already begun.

To confirm the formation of the $Nb_{22}O_{54}$ structure, the corresponding SAED pattern is shown in fig. 3b, which is compared with the SAED patterns of the same $Nb_{12}O_{29}$ crystal in vacuum in the same area (fig. 3a) and a $Nb_{22}O_{54}$ crystal in vacuum (fig. 4c). The streaking appearing in fig. 3b seems to be in the same direction that [001] $Nb_{22}O_{54}$ structure (fig. 3c). $Nb_{22}O_{54}$ structure (fig. 3d) can be consider as a mixter of $Nb_{12}O_{29}$ ($NbO_{2.417}$) and $Nb_{10}O_{25}$.($NbO_{2.500}$). Also can be described as formed by (4x3) blocks alternating with (3x3) blocks of infinite length and the tetrahedral sites are ocuppied by Nb^{5+}.

Figure 4 shows a different area of the same crystal after second oxidation. A row of (3x3) blocks and black contrast dots corresponding to the tetrahedral sites (ocuppied by Nb^{5+}) can be very clearly observed (marked between two lines). Therefore, these changes can be interpretated as an indication of $Nb_{22}O_{54}$ structure formation in some areas of the crystal.

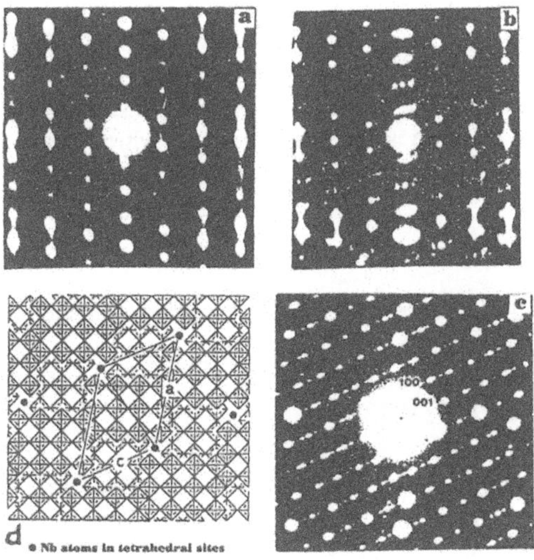

Figure 3. SAED patterns of (a) the above crystal in vacuum (b) the same crystal, same area after farther oxidation, (c) a $Nb_{22}O_{54}$ crystal in vacuum and (d) Schematic representation of $Nb_{22}O_{54}$ crystal structure (7) (a=2.12 nm; c=1.06 nm).

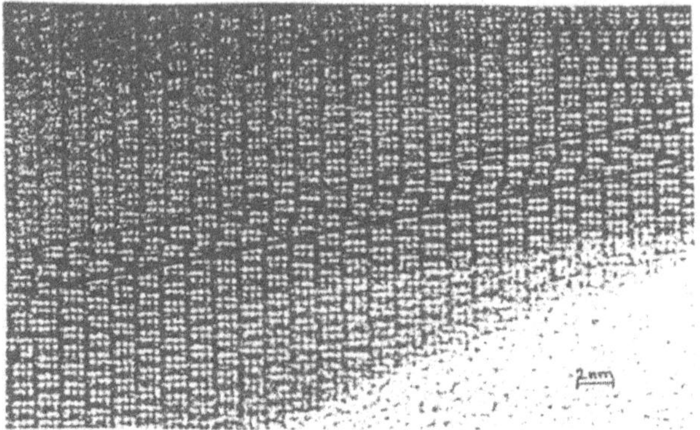

Figure 4. Micrograph showing the same $Nb_{12}O_{29}$ crystal in another area after second oxidation. The $Nb_{22}O_{54}$ structure is marked between two lines, the black contrast corresponds to Nb^{5+} in tetrahedral sites.

We can conclude that the oxidation process seems to start in the region where some stacking fault perpendiculars to the **c** axis appear (in which the structure has orthorhombic symmetry). The first movement of atoms involving a formation of *lamella* defect could be elucidated with great certainty. Such defect provides empty rectangular tunnels, similar to those found in $NaNb_{13}O_{33}$ structure.

The oxidation reaction implies a minimal rearrangement, usually involving a simple unit jump (the primitive diffusion step) by atoms in certain sites. Then, the co-operative unit jumps along rows of atoms propagate the transformation through the crystal. A further oxidation gives rise to $Nb_{22}O_{54}$ ($NbO_{2.455}$) structure in some areas of the crystal, which could be appreciated in the HREM image as well as in the electron diffraction patterns.

REFERENCES.

(1) J.G. Alpress, J.V. Sanders and A.D. Wadsley, Physics Status Solid., 25, 541 (1968).

(2) J.M. Browne, J.L. Hutchison and J.S. Anderson, Proc. 7th Symp. on Reactivity of Solids, 116 (1972).

(3) J.S. Anderson, Chemica Scripta, 14, 129 (1979).

(4) M.J. Sayagués and J.L. Hutchison, J. Solid State Chem., 124, 116 (1996).

(5) R.C. Doole, G.M. Parkinson, J.L. Hutchison, M.J. Goringe and P.J.F. Harris, JEOL News 30E, 30 (1992); R.C.Doole, G.M.Parkinson & J.M.Stead, Inst.Phys.Conf.Series No.118 157 (1991) Institute of Physics (Bristol) (1991).

(6) J. O. Bovin, D.X. Lee, L. Stemberg and H. Annehed, Z. Kristallgr. 168, 98 (1984).

(7) S. Horiochi and S. Kimura, Japanese, J Apl. Phys. 21, L97 (1982).

ELECTRON DIFFRACTION PATTERNS OF NATURAL ANTIGORITES: A STILL UNKNOWN MODULATED CRYSTAL STRUCTURE.

CECILIA VITI
Dipartimento di Scienze della Terra di Siena
Via delle Cerchia, 3 53100 Siena, Italia.

1. Introduction

Serpentine minerals (lizardite, chrysotile and antigorite) are 1:1 layer silicates, with a basic structure consisting of tetrahedral and octahedral sheets, succeeding along [001]. 1:1 layers are flat in lizardite, rolled in chrysotile (fibrous crystals) and corrugated in antigorite. The serpentine end member composition is $Mg_3Si_2O_5(OH)_4$, possibly with Al and Fe substituting for Si and Mg.

Serpentine minerals are typically affected by several problems, arising both from their low crystallization temperature and from the weakness of interlayer connections. Common drawbacks are the mutual fine intergrowth of different serpentines, the small crystal size, the crystal disorder. Crystal structure refinements from single crystal X-ray data have been obtained only for lizardite, whereas no three-dimensional crystal-structure refinement has been at now obtained for antigorite and chrysotile.

Owing to their defective nature, serpentine minerals are often investigated by TEM, even though they suffer by quick damage under the electron beam: as a consequence, the careful regulation of defocus, astigmatism, crystal orientation cannot be often achieved. In that sense, low-dose mode for high resolution imaging would be essential.

2. Antigorite structure

The corrugation of the antigorite layer gives rise to a superstructure: depending on the corrugation wavelength, antigorite superperiodicity may range from 30 to 60 Å, (Uehara and Shirozu, 1985; Mellini et al., 1987; Uehara and Kamata, 1994). Together with structural modulation, antigorite is also characterized by chemical modulation (polysomatism): the actual antigorite composition is $Mg_{3m-3}Si_{2m}O_{5m}(OH)_{4m-6}$ with m = number of tetrahedra along a wavelength (typically 17), thus significatively different from the end-member serpentine composition $Mg_3Si_2O_5(OH)_4$.

To account for the antigorite superperiodicity three plausible models were proposed: the alternating, the rectified and the zig-zag models. These were tested by Zussman (1954), by means of optical diffraction: he compared the X-ray diffraction pattern of

D. L. Dorset et al. (eds.), Electron Crystallography, 419–422.

natural antigorite to the optical diffraction patterns of the models. A good fit was obtained for both alternating and rectified waves, not for the zig-zag model.
Later, Kunze (1956, 1958, 1961) supported the alternating-wave model, on the basis of a two-dimensional Fourier synthesis of the X-ray amplitudes: his model is still largely adopted. Antigorite consists of a sinusoidal octahedral sheet and of a tetrahedral sheet which reverses its polarity every half-wavelenght. Two different tetrahedral reversals alternate in antigorite structure: the six-membered reversal (rings of 6 tetrahedra) and the eight-membered reversal (rings of 8 tetrahedra).

3. Microtextures and electron diffraction patterns of natural antigorites

Unfortunately, details on antigorite structure (for instance, the curvature and the shape of the wave, the bond geometry at tetrahedral reversals and so on) are still matter of debate. This is mainly due to the lack of suitable antigorite crystals, to be used for single crystal refinements.
Antigorite is always affected by defects and structural troubles, such as (001) polysynthetic twinning, polysomatic disorder (that is, intracrystalline variation of superperiodicity; *Figure 1*), polytypic disorder (that is, disorder in the stacking sequence along [001]) and other defects (for instance, (100) offsets or modulation dislocations).
At difference from the weaker X-ray diffraction, electron diffraction patterns of antigorites always show several intense superstructure effects, clustered around the subcell effects along the a* direction. Furthermore, the diffracting volume may be selected so as to avoid defective areas (for instance, untwinned areas, with no polysomatic disorder).
Electron diffraction patterns of natural antigorites are shown in *Figure 2*.

4. Investigation of antigorite structure by image processing and electron crystallography

More recently, antigorite structure has been investigated by image processing of high resolution TEM images (Yada, 1979; Yada et al., 1980; Otten, 1993). For instance, experimental images have been compared to simulated images (on the basis of the Kunze alternating model), at variable conditions of defocus and thickness.
At present, no attempts have been undertaken following the electron crystallography approach.
By contrary, electron crystallography could represent a basic tool for the study of antigorite structure, as well as of other natural poorly-crystalline minerals, whose structure cannot be determined by using single crystal X-ray refinement.
In particular electron crystallography may shed light on the actual geometry of antigorite structural modulation, at least by discriminating between the previously discussed models of rectified wave and alternating wave and, possibly, on the features of tetrahedral reversal.

Figure 1. Polysomatic disorder in antigorite; superlattice ranges from 30 Å to 60 Å.

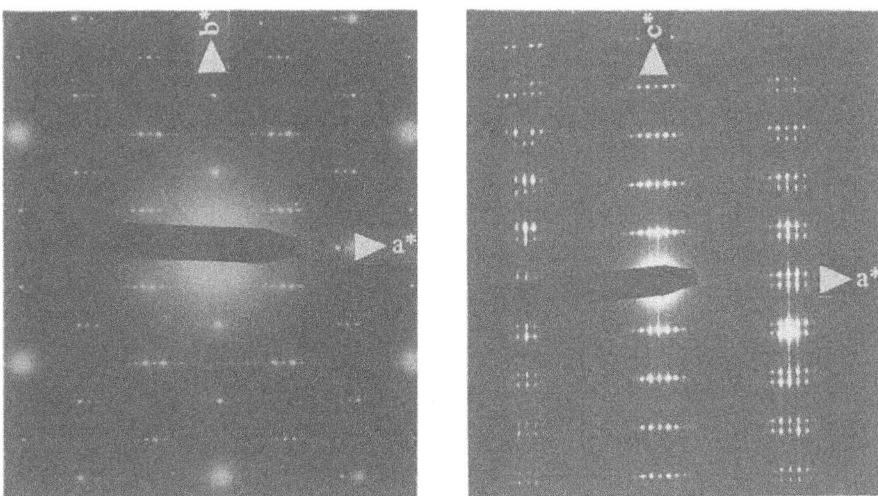

Figure 2. Electron diffraction patterns a*b* (left) and a*c* (right; twinned crystal) of natural antigorites.

References

1. Kunze, G. (1961): Antigorit. Strukturtheoretische Grundlagen und ihre praktische Bedeutung fur die weitere Serpentin-Forschung. *Fortschr. Mineral.*, **39**, 206-324.

2. Kunze, G (1956, 1958): Die Gewelte Struktur des Antigorits I-II. *Zeitschrift fur Kristallographie.*

3. Mellini, M., Trommsdorff, V. and Compagnoni, R. (1987): Antigorite polysomatism: behaviour during progressive metamorphism. *Contrib. Mineral. Petrol.*, **97**, 147-155.

4. Otten, M.T. (1994): High resolution transmission electron microscopy of polysomatism and stacking defects in antigorite. *Amer. Mineral.*, **78**, 75-84.

5. Uehara, S. and Shirozu, H. (1985): Variations in chemical composition and structural properties of antigorites. *Mineral. Mag.*, **12**, 299-318.

6. Uehara , S. and Kamata, K. (1994): Antigorite with a large supercell from Saganoseki, Oita Prefecture, Japan. *Can. Mineral.*, **32**, 93-103.

7. Yada, K. (1979): Microstructures of chrysotile and antigorite by high resolution electron microscopy. *Can. Mineral.*, **17**, 679-691.

8. Yada, K., Tanji, T and Nissen, H.U. (1980): Direct observation of antigorite at atomic resolution. *Fourth International Conference on Asbestos*, Torino.

9. Zussman, J. (1954): Investigation of the crystal structure of antigorite. *Mineral. Mag.*, **30**, 498-512.

EXACT ATOM POSITIONS BY ELECTRON MICROSCOPY?

- a quantitative comparison to X-ray crystallography

TH. E. WEIRICH, R. RAMLAU, A. SIMON
Max-Planck-Institut für Festkörperforschung
D-70506 Stuttgart, Germany

AND

S. HOVMÖLLER
Structural Chemistry, Stockholm University
S-106 91 Stockholm, Sweden

Direct imaging of structures is possible by high-resolution electron microscopy (HREM). This analytical method is therefore of special attraction in the field of structural chemistry. X-ray crystallography has been extremely successful for the determination of crystal structures, suggesting that there was no necessity to use high-resolution electron microscopy for that purpose. However, crystals that are thousands times smaller in size than that needed for single crystal X-ray measurements can be studied by EM since electrons interact much stronger with matter than X-rays. On the other hand, the strong scattering of electrons causes severe problems to obtain quasi-kinematical data. Therefore it is always necessary to take great care on the proper experimental conditions in quantitative EM. It was shown more than one decade ago that precise heavy atom positions for metal oxides could be determined if the technique of crystallographic image processing (CIP) is combined with EM [1]. Since then, the CIP technique for structure determination has been applied successfully to other metal oxides.

Several condensed-cluster compounds were identified in the course of an EM study on selected metal-rich systems of early transition elements with S, Se, Te and P [2]. Among those, the new structure $Ti_{11}Se_4$ was solved from HREM images and refined against SAED data [3]. Here we show the results from an EM investigation of several such structures that had been solved also by X-ray crystallography. The purpose of this study was to check the accuracy of the EM-CIP method for that class of compounds. HREM images and electron-diffraction patterns were recorded on film, digitized and later processed with the CRISP program package [4,5]. The results from CIP were compared with those from X-ray investigations on single crystals.

D. L. Dorset et al. (eds.), Electron Crystallography, 423–426.

The atomic coordinates differ on average around 0.2 Å (max. deviation around 0.4 Å, Table 1).

TABLE 1. Deviation Δ (in Å units) between atomic coordinates determined from single crystal X-ray diffraction and from EM.

		Space group	N_{asym}*	Δ_{min}	Δ_{max}	$\Delta_{average}$
Ti_9Se_2	[6]	*Pbam* (55)	5	0.02	0.23	0.11
Ti_2S	[7]	*Pnnm* (58)	9	0.05	0.43	0.27
Ti_2Se	[8]	*Pnnm* (58)	9	0.05	0.42	0.21
Zr_2Se	[9]	*Pnnm* (58)	9	0.04	0.23	0.13
Ta_2P	[10]	*Pnnm* (58)	9	0.02	0.32	0.15
Ti_8S_3	[11]	*C*2/*m* (12)	22	0.05	0.33	0.17
Ti_8Se_3	[12]	*C*2/*m* (12)	22	0.03	0.27	0.14

* N_{asym} = number of atoms in the asymmetric unit

Crystallographic image processing

In the following we sketch the principles of how atomic coordinates can be derived from HREM images. The compounds were investigated in a Philips CM30/ST microscope operating at 300 kV; HREM images and electron diffraction patterns along the [001] zone axis were recorded on film. The negatives were digitized and then processed using the CRISP program. A square area of about 100 to 200 unit cells (1024 × 1024 pixels) was chosen from the HREM image and its Fourier transform was calculated. Amplitudes and phases were extracted from the refined lattice positions in the Fourier transform (resolution ca. 1.9 Å). The lattice averaged images were calculated from the extracted amplitudes and phases (*Figure 1a.*). The first crossover of the contrast transfer function (CTF) was estimated from the amorphous region at the crystal edge and used to determine the defocus value. The phase shift by the CTF was corrected and a new potential map with *p1* symmetry was obtained (*Figure 1b.*). Atoms appear now as black features in the map. For each compound the 17 plane groups were tested and the one which gave the lowest phase residual indicated the correct symmetry.

Figure 1. (next page) Image processing of Ti_2Se. **(a)** original HRTEM image, **(b)** CTF corrected, **(c)** *p2gg* symmetry imposed. The structure of Ti_2Se projected along the short *c* axis as obtained from electron microscopy **(d)** and by X-ray crystallography **(e)** are virtually identical (● = Ti, ○ = Se). The Unit cell dimensions are a = 11.737 Å, b = 14.550 Å, c = 3.451 Å, space group *Pnnm* (58).

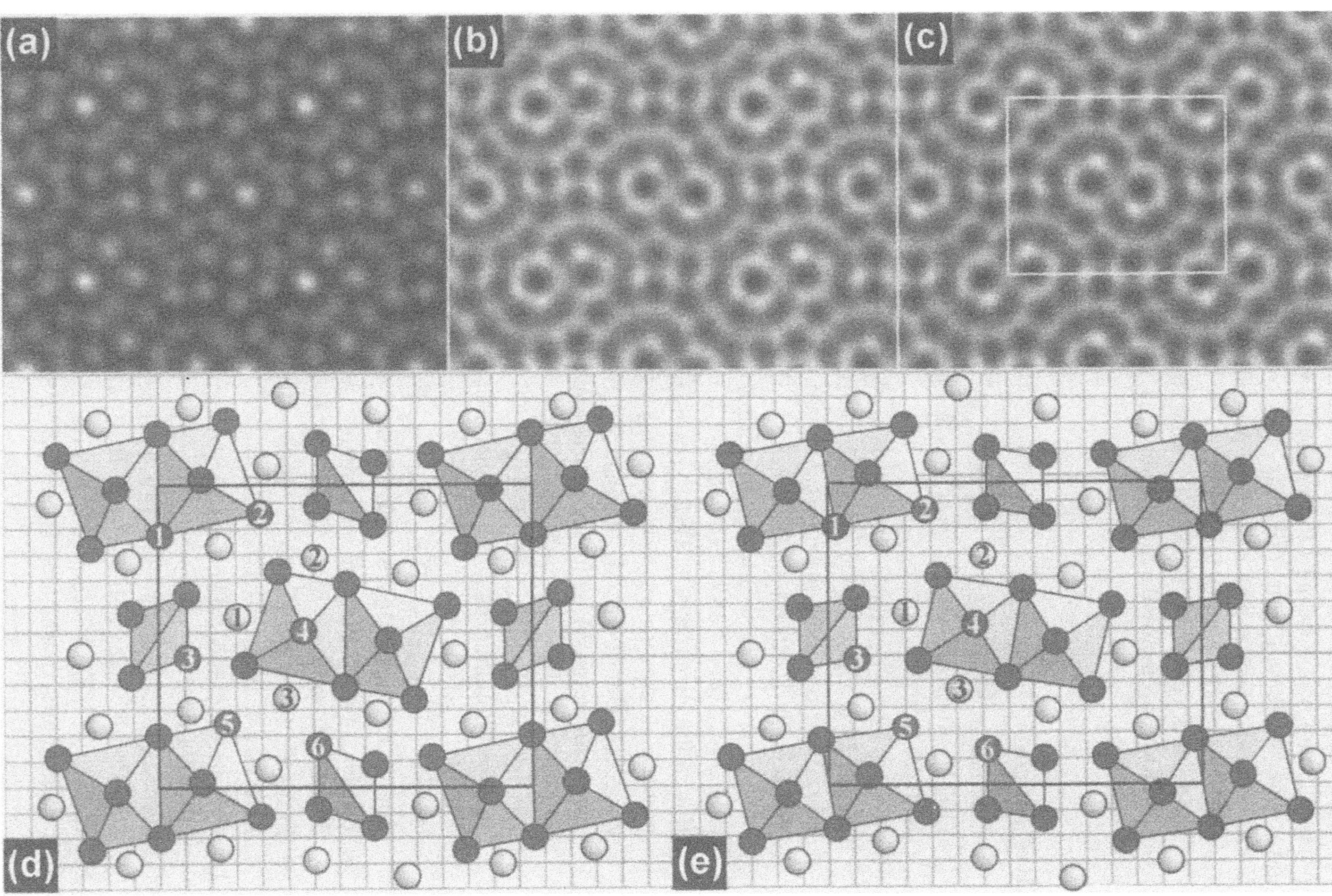
(a)
(b)
(c)
(d)
(e)

The plane symmetry was imposed on the amplitudes and phases and a new potential map was received (*Figure 1c.*). Two dimensional atomic coordinates were extracted from this map and compared to those obtained by X-ray crystallography. The described procedure can be followed in Figure 1., which shows the results for Ti_2Se.

The results compiled in Table 1 show that electron microscopy can be used to obtain accurate atomic coordinates, even for crystals containing no light elements.

References

1. Hovmöller, S., Sjögren, A., Farrants, G., Sundberg, M., Marinder, B.O. (1984) 'Accurate atomic positions from electron microscopy', *Nature* **311**, 238-241.
2. Weirich, Th.E. (1996), *'Metallreiche Systeme mit kondensierten Clustern'*, PhD thesis, Universität Osnabrück / Max-Planck-Institut für Festkörperforschung Stuttgart, Germany.
3. Weirich, Th.E., Ramlau, R., Simon, A., Hovmöller, S., Zou, XD. (1996) 'A crystal structure determined with 0.02 Å accuracy by electron microscopy', *Nature* **382**, 144-146.
4. Hovmöller, S. (1992) 'CRISP: crystallographic image processing on a personal computer', *Ultramicroscopy* **41**, 121-135.
5. Zou, XD., Sundberg, M., Larine, M., Hovmöller, S. (1996) 'Structure projection retrievel by image processing of HREM images taken under non-optimal defocus conditions', *Ultramicroscopy* **62**, 103-121.
6. Weirich, Th.E., Simon, A., Pöttgen, R. (1996) 'Ti_9Se_2 - Eine Verbindung mit kolumnaren ${}^1_\infty[Ti_9]$-Baueinheiten', *Z. anorg. allg. Chem.* **622**, 630-634.
7. Owens, J.P., Conard, B.R., Franzen, H.F. (1967) 'The Crystal Structure of Ti_2S', *Acta Cryst.* **23**, 77-82.
8. Weirich, Th.E., Pöttgen, R., Simon, A. (1996) 'Crystal structure of dititanium monoselenide, Ti_2Se', *Z. Krist.* **212**, 928.
9. Franzen, H.F., Norrby, L.J. (1968) 'The crystal structure of Zr_2Se', *Acta Cryst.* **B24**, 601-603.
10. Nylund, A. (1966), 'The Crystal Structure of Ta_2P', *Acta Chem. Scand.* **20**, 2393-2401.
11. Owens, J.P., Franzen, H.F. (1974) 'Preparation and Structure Determination of Ti_8S_3', *Acta Cryst.* **B30**, 427-430.
12. Weirich, Th.E., Pöttgen, R., Simon, A. (1996) 'Crystal structure of octatitanium triselenide, Ti_8Se_3', *Z. Krist.* **212**, 929-930.

WINREKS - A COMPUTER PROGRAM FOR THE RECIPROCAL LATTICE RECONSTRUCTION FROM A SET OF ELECTRON DIFFRACTOGRAMS

MAREK WOŁCYRZ and MAŁGORZATA ANDRUSZKIEWICZ
Institute of Low Temperature and Structure Research,
Polish Academy of Sciences,
ul. Okólna 2, 50-950 Wrocław, Poland

1. Introduction

It is relatively easy to obtain good quality electron diffraction patterns using electron microscope. It is usually the first step before obtaining HRTEM images. The diffractograms carry information about crystal symmetry, lattice parameters and ordering effects - superstructure or modulation. However, for the reliable conclusions concerning all these features, a set of several or even more then ten coherent (*i.e.* coming from the same crystallite) difractograms is necessary. Reflections recorded should span a satisfactory part of the reciprocal lattice. In such a case an interpretation of the series of diffractograms becomes difficult and demands a computer aid.

2. Program description

WINREKS is a computer program designed for Windows 3.1 and 95 which enables to reconstruct and evaluate reciprocal lattice (RL) from a set of electron diffractograms (EDs).

The necessary step before using the program is the digitization (when EDs are obtained in traditional, analogue form) and vectorization of EDs (calculation of reflection positions and intensities). *WINREKS* itself does not contain any vectorization procedure but other image processing packages can be used in this purpose, *e.g.* CRISP [1] with its *ED Processing* routine. In the *WINREKS* input file each ED is represented by coordinates and intensities of reflections as well as by two angles (φ and χ) specifying the position of the actual plane (section) in RL. At present, WINREKS is designed for the microscopes equipped with a rotation goniometer holder, but double tilt geometry will be also supported.

A coherent set of EDs can be bound together and placed properly in the reciprocal space forming RL (Fig. 1). Position of each individual ED can be corrected in order to avoid errors due to unprecisely determined φ and χ angles.

The bundle as a whole can be rotated in 3-D space in order to explore RL and to find RL unit cell. This last aim is the most important application of *WINREKS*. Now, it must be done by hand only but during a further program development, there will be made an effort in order to automatize this step and to implement numerical routine searching for

D. L. Dorset et al. (eds.), Electron Crystallography, 427–430.

optimal RL unit cells. On each stage of the work a *Measurement* option allows to control *d*-values for individual reflections.

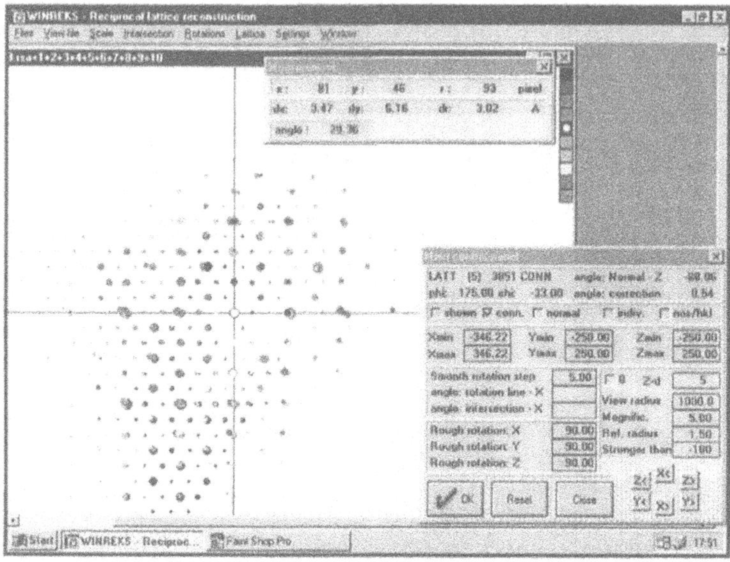

Figure 1. Screenshot from *WINREKS* showing 10 coupled diffractograms (indicated by different shades of gray) forming RL. *Main control panel* contains all parameters controlling RL rotation and display. *Measurement* window supplying *d*-values and angles between reflections is also shown.

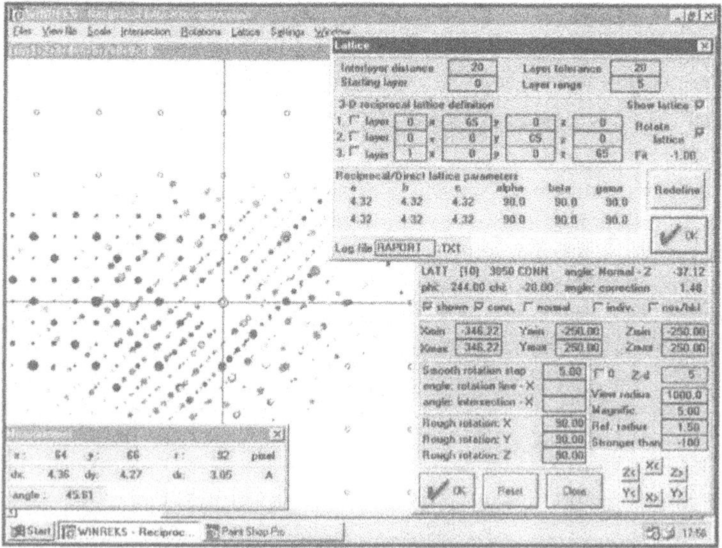

Figure 2. RL input. *Lattice* window is used for RL unit cell input and contains calculated reciprocal and direct lattice parameters. RL knots (in this case forming basic lattice) are denoted by small open circles.

When the optimal RL unit cell is chosen, a 3-D RL can be spanned and appropriate direct unit cell can be determined (Fig. 2).

The projections of the RL along the direct lattice direction [*uvw*] or on the (*hkl*) plane can be made (Fig. 3). For each RL orientation zero- as well higher-order layers can be easily observed and analyzed. Each ED can be observed individually or it can be excluded from the picture. Its color can be changed and the whole picture can be scaled.

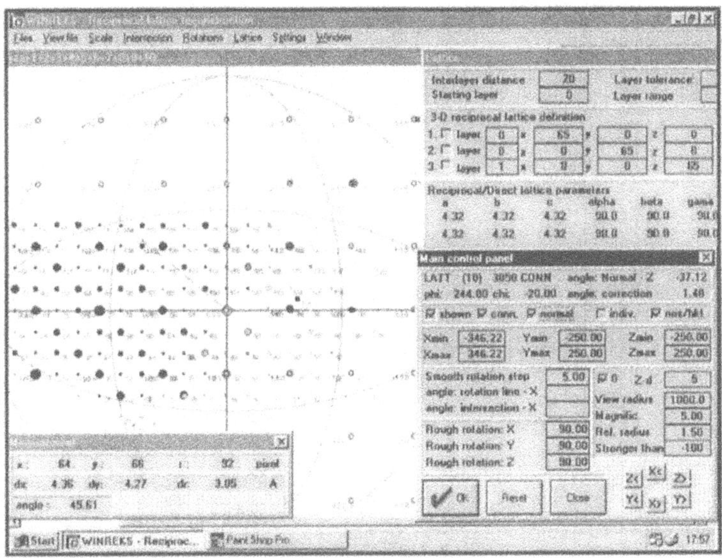

Figure 3. RL projection along chosen [*uvw*] direction. Three large circles show position of the chosen section in 3-D space in respect to the projection plane.

WINREKS is useful especially in the analysis of complex crystals when RL exploration and interpretation is difficult according to the large number of Bragg or superstructure/modulation reflections.

A good example of WINREKS application is a picture (Fig. 4) showing two perpendicular projections of the RL of $BiLa_xSr_{3-x}O_{5.5-y}$ [2]. It can be clearly seen that proper choice of RL unit cell of the superstructure and its relation to the basic lattice would be extremely difficult without computer help.

3. Program availability

Program *WINREKS* is free of charge and can be obtained from the Authors on request.

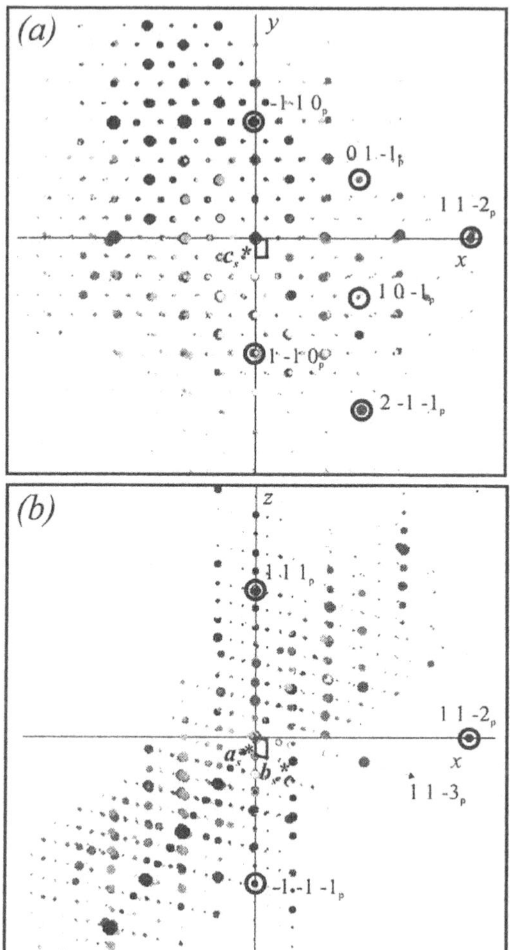

Figure 4. Projection of the RL of $BiLa_xSr_{3-x}O_{5.5-y}$ along two perpendicular directions: *z* (*a*) and *y* (*b*). Chosen RL unit cell is shown. Open circles and Miller indices denote reflections coming from the basic lattice.

4. Acknowledgments

This work was supported by State Committee for Scientific Research - grant no. 3 T09A 122 08.

References

1. Hovmöller, S. (1992) CRISP: crystallographic image processing on a personal computer, *Ultramicroscopy* **41**, 121-135.
2. Wołcyrz, M. and Kępiński, L. (1997) Electron microscopy and X-ray powder diffraction studies of the long-range ordering in α, β and γ phases of $BiLa_xSr_{3-x}O_{5.5-y}$, *J. Solid State Chemistry*, submitted.

CRYSTALLOGRAPHIC PHASE TRANSITIONS IN LASER IRRADIATED CERIUM DIOXIDE

F. VASILIU and C. SARBU
National Institute for Material Physics
P. O. Box MG-7, R-76900, Bucharest-Magurele (Romania)

1. Introduction

Two polymorphic Ce_2O_3 compounds are reported: A-Ce_2O_3 (hexagonal, space group P $\bar{3}$m1) and C-Ce_2O_3 (cubic, space group T_h^7 or Ia3) [1] . If A-Ce_2O_3 was obtained by difficult carbothermic reduction reactions, the C-Ce_2O_3 occurrence is yet controversial [2]. Structure and sintaxy of R_2O_3 (R = rare earth element) oxides and their relationship to the fluorite structure of RO_2 have been investigated in great detail. The intermediate oxides associated with various rare-earth elements (including cerium) can be described by the equivalent basic formulae: RO_x ($1.5 \leq x \leq 2$) and R_nO_{2n-2} ($4 \leq n < \infty$) where n=4 leads to R_2O_3 and n=∞ to RO_2. The homologous series R_nO_{2n-2} contains a great number of structures (especially hexagonal, monoclinic and triclinic) built on the fluorite-like lattice of RO_2 compound, which are due to various oxygen-vacancy rearrangement mechanisms. Some papers were devoted to the preparation of intermediate oxides [3] and the direct reduction of higher oxides during heating by means of an electron beam [4]. EM experiments demonstrated that electron beam irradiation induced unexpected phase transitions in thin films of rare earth oxides [5].

The aim of this work is to describe the various phase transformations initiated by laser irradiation in pure stoichiometric polycrystalline CeO_2 powders. A similar laser-induced reduction of V_2O_5, consisting essentially in the appearance of V^{4+} and V^{3+} ions has been investigated [6] .

2. Experimental

The specimen irradiation was made by a CO_2 laser operating in continous wave at power densities between 0.4-1.3 kW/cm^2 (irradiation times in the range 7-50 s). Conventionally, deposited energy densities (doses) were classified as: low (< 7 kJ/cm^2), mean (7-14 kJ/cm^2) and high (20 or 40 kJ/cm^2) doses.

Specimen characterization was made by CTEM and SAED in a TEMSCAN-100 CX (Jeol) electron microscope, operating at 200 kV.

D. L. Dorset et al. (eds.), Electron Crystallography, 431–434.

3. Experimental results and discussion

Laser irradiation induced a variable non-stoichiometry in pure f.c.c. CeO_2 depending on irradiation parameters. Generally, all the specimens contained as major phase a f.c.c. CeO_{2-x} (x very small) having an increased lattice parameter (a_0 = 5.53 - 5.55 A) (Fig. 1). A saturable effect of particle size growth in irradiated polycrystalline CeO_2 by the laser power and/or energy density increasing has been observed . For low and mean doses (< 9 kJ/cm^2) only CeO_{2-x} polycrystalline diffraction rings were registered but , for mean and high doses, many thin single crystal CeO_{2-x} platelets occurred. Two or three single crystal particles were joined after a prolonged irradiation.

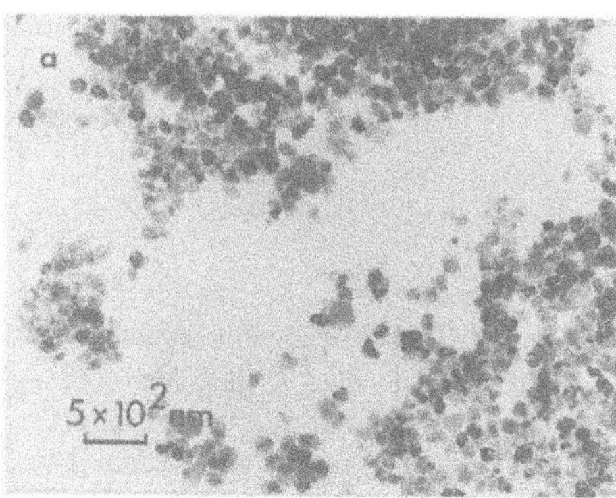

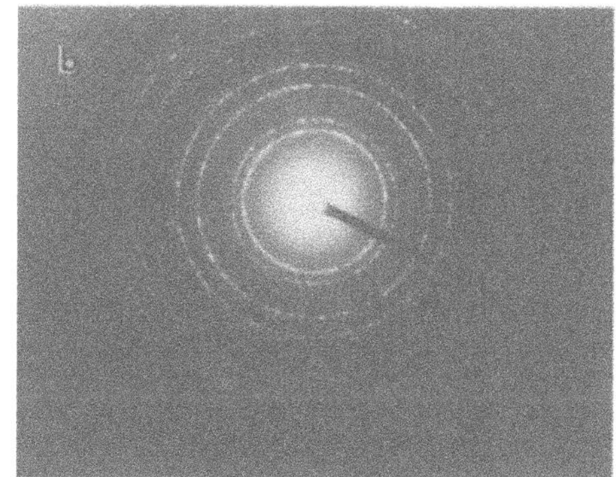

Figure 1. Small fcc CeO_{2-x} particles obtained by laser irradiation at mean doses (9 kJ/cm^2): (a) TEM image, (b) SAED pattern

The possibility of CeO_2 reduction to A-Ce_2O_3 by laser irradiation at low doses was demonstrated by SAED since < $\bar{1}20$ > and < $\bar{4}$ 2 3 > zone axis single crystal Ce_2O_3 particles were formed (Fig. 2). A simple mechanism, based on the clustering of the simplest defect, consisting from an oxygen vacancy and two trivalent cations ,which is virtually a Ce_2O_3 nucleus, could explain the A-Ce_2O_3 appearance in a CeO_{2-x} lattice containing very high anion vacancy concentrations. Complete conversion of CeO_2 to A-Ce_2O_3 was not possible at least in the range of the used irradiation parameters.

The formation of stable Ce_7O_{12} phase (Fig. 3) was detected only for low power densities (0.4 kW) but for mean doses and high doses.

For temperatures below 875^0 C, the vacancy ordering leaded to the formation of the homologous series Ce_nO_{2n-2} but the well-identified structures appeared only for n= 7, 11 and 12. In theCe_7O_{12} ordered phase, all Ce atoms had coordination numbers (c.n.) of 6 or 7 and the oxygen vacancies were localized on the <111> directions from the CeO_2 unit cell. The Ce atoms (c.n. = 6) were placed in the {135} planes, each of them having

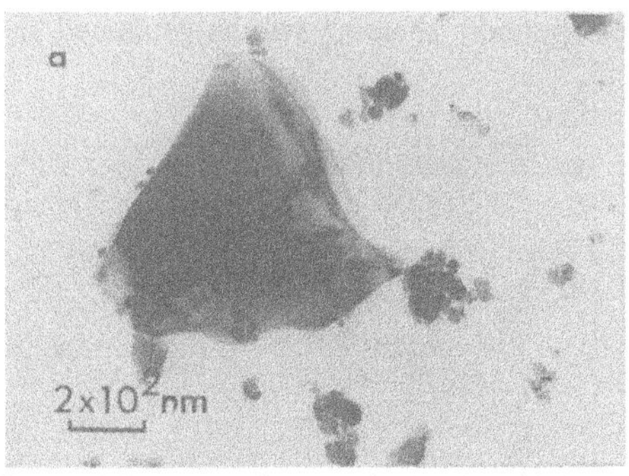

Figure 2. Large single-crystal A-Ce_2O_3 platelet appearing in laser-irradiated CeO_{2-x} at mean doses: (a) TEM image, b) $\langle \bar{4}\ 2\ 3 \rangle$ zone axis pattern

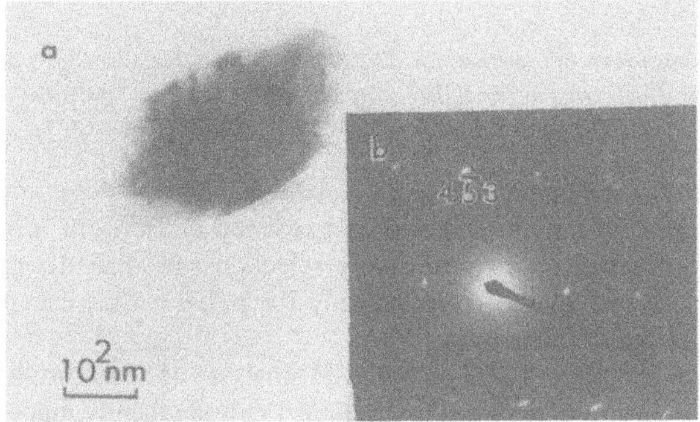

Figure 3. Ce_7O_{12} particle observed in a specimen irradiated at a high dose (40 kJ/cm²): a)TEM image, (b) $\langle \bar{8}\ 3\ 14 \rangle$ zone axis pattern.

as neighbours two oxygen vacancies, whereas other six metallic atoms (c.n. = 7) were localized along the [111] F direction. The six metallic atoms, three up and three down, together with two oxygen vacancies, formed a so-called "Bevan cluster" [7].

Compounds of strange non-stoichiometry, similar to those formed in oxidized rare-earth thin films irradiated by energetic electron beams [5] were observed for low power densities and short irradiation times. An epitaxial relationship between triclinic $Ce_{11}O_{20}$ and cubic $Ce_{12}O_{22}$ phases was found (Fig.4). The controversial C-Ce_2O_3 (Ia3) phase, having a_c = 11.12 A, was identified as a thick quasiepitaxial layer grown at the limits of a prismatic particle with an unusual b.c.c. structure ($a_0 \approx a_c/2$).

The phase transitions induced by laser irradiation, leading to A-Ce_2O_3 or C-Ce_2O_3, could be correlated to the obvious structural relationships between the fluorite structure of CeO_2 and the A and C types of the Ce_2O_3 sesquioxide, derived by the introduction of oxygen vacancies in some ordered arrangements. If a quarter from the oxygen atoms were removed along of four non-intersecting <111> directions and each Ce atom had a c.n. equal to 6, a C-Ce_2O_3 lattice will be obtained. On other part, if the oxygen substructure in CeO_2 was preserved but metallic atoms were displaced in interstitial sites, the result will be a lattice of A-Ce_2O_3 where periodic shearing will be produced and Ce ions had a c.n. equal to 7.

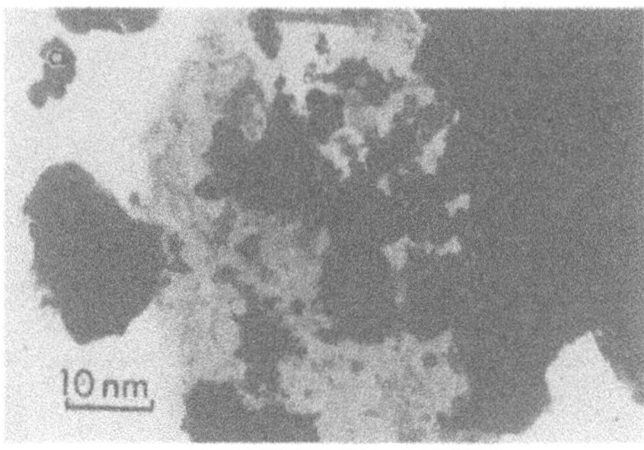

Figure 4 The occurrence of triclinic $Ce_{11}O_{20}$ phase from cubic $Ce_{12}O_{22}$ particles after laser-beam irradiation : (a) TEM image, (b) SAED pattern exhibiting the orientation relationships: <110> $Ce_{11}O_{20}$ // <110> $Ce_{12}O_{22}$ and [1 $\bar{1}$ 4] $Ce_{11}O_{20}$ // [$\bar{1}$ 1 1] $Ce_{12}O_{22}$.

An another unknown phase of needle-like morphology, strongly related to C-Ce_2O_3, was registered only in a specimen irradiated at 1kW/cm^2 (7s). The real structure seemed to be governed by a more severe selection rule than the space group T_h^7- (la3) because the observed lines correspond only for h+k+l = 4n , excepting the very intense (222) line of the fluorite structure.

A more complete TEM and SAED analysis of laser irradiation effects in pure CeO_2 [8] and binary rare-earth oxides systems [9] was recently made.

4. References

1. Gasgnier, M. (1989) Rare earth compounds (oxides, sulfides, silicides, boron,...) as thin films and thin crystals, *Phys. Stat. Sol. (a)* **114** , 11-71.
2. Gasgnier, M. (1980) Thin films of rare earth metals, hydrides and oxides, *Phys. Stat. Sol. (a)* , **57**, 11-49
3. Knappe, P., and Eyring, L. (1985) New metastable phases belonging to the Ce_nO_{2n-2} homologous series, *J. Solid State Chem.* **58**, 312-322.
4. Schweda,E., Kang, Z. C., and Eyring, L. (1987) HVEM study of ROx (R-rare earth) thin crystals, *J. Microscopie* **145**, 45-59.
5. Gasgnier, M., Schiffmacher, G., Svoronos, M., and Caro, P. (1988) Unusual phases in rare-earth oxide thin films:evidence for acoustic wave generation during electron microscopy observations , *J. Microsc. Spectrosc. Electron.* **13**, 13- 30.
6. Khawaja, E. E., Khan, M. A., Al-Adel, F. F., and Hussain, Z. (1990) Laser- produced reduction of pentavalent vanadium in aqueous solutions and V_2O_5 powder, *J. Appl. Phys.*, **68**, 1205-1211.
7. Sorensen, O. T. (1981) *Non-stoichiometric oxides*, Academic Press, New York.
8. Vasiliu, F., Parvulescu, V., and Sarbu, C. (1994) Trivalent Ce_2O_3 and CeO_{2-x} intermediate oxides induced by laser irradiation of CeO_2 powders, *J. Mat. Science* **29**, 2095-2101
9. Vasiliu, F., Sarbu, C, and Parvulescu, V. I. (1997) Investigation of some phase transformations induced by CO_2-laser irradiation in some rare-earth oxides, *Solid State Ionics* **95**, 107-112.

INDEX

GPSR Compliance
The European Union's (EU) General Product Safety Regulation (GPSR) is a set of rules that requires consumer products to be safe and our obligations to ensure this.

If you have any concerns about our products, you can contact us on

ProductSafety@springernature.com

In case Publisher is established outside the EU, the EU authorized representative is:

Springer Nature Customer Service Center GmbH
Europaplatz 3
69115 Heidelberg, Germany

www.ingramcontent.com/pod-product-compliance
Ingram Content Group UK Ltd.
Pitfield, Milton Keynes, MK11 3LW, UK
UKHW021859190726
13853UKWH00003B/1343

* 9 7 8 9 4 0 1 5 8 9 7 2 7 *